GEOGRAPHY
a reference handbook

GEOGRAPHY
a reference handbook

SECOND EDITION REVISED AND ENLARGED

C B MURIEL LOCK
BA PhD FRGS

LINNET BOOKS & CLIVE BINGLEY

016.91
L789g2

FIRST PUBLISHED 1968
REPRINTED WITH MINOR CORRECTIONS 1969

THIS SECOND EDITION REVISED AND ENLARGED
FIRST PUBLISHED 1972 BY CLIVE BINGLEY LTD
AND SIMULTANEOUSLY PUBLISHED IN THE USA
BY LINNET BOOKS, AN IMPRINT OF
SHOE STRING PRESS INC,
995 SHERMAN AVENUE, HAMDEN, CONNECTICUT 06514
PRINTED IN GREAT BRITAIN
COPYRIGHT © C B MURIEL LOCK 1972
ALL RIGHTS RESERVED
0 – 208 – 01173 – 0

FOREWORD

This handbook had its origin in the belief that a quick reference book covering some of the main focal points of geographical study would be useful to both geographers and librarians. It has not been the intention to maintain a balanced coverage either regionally or thematically (this the author hopes to attempt in a future publication), but to draw attention to the outstanding scholars whose achievements have helped to shape the modern concept of geographical studies and to some of the organisations and sources of the greatest continuing significance within the framework of world geography.

Selection of entries is international in scope, with a British bias. Terms are not included, as there are now several excellent dictionaries of geographical terms, but some notes are offered on abstracts, cartobibliography, cataloguing, classification, globes and guides. Biographical entries, which do not include living geographers, are concerned with academic and bibliographical achievements rather than with personal details. Titles of works in foreign languages have been given in the original when these are considered readily recognisable by English-speaking readers, otherwise in the most familiar form.

CBML

FOREWORD TO THE SECOND EDITION

The success of the first edition of this handbook has made a new edition desirable, at the same time enabling a wider concept to be carried out. While keeping the basic entries of the first edition (revised as necessary), representing a body of material which should be known to all practising geographers and relevant specialist librarians, the scope now covers a number of works of essential value to scholars concentrating on the geography of particular regions or systematic studies. Among the new entries is an extended one on national bibliographies of interest to geographers. New publications emphasise the establishment of current trends in geographical thought—the qualitative and quantitative approaches, expansion of facsimile publication, renewed emphasis on field work, environmental studies and conservation, also a revived preoccupation with the value and potentialities of maps, the care of maps and map collections and the use of automation in their preparation.

Several outstanding geographers have died since the first edition appeared; I would again stress the policy behind the inclusion of biographical entries, as stated in the foreword to the first edition; one or two reviewers failed to take note and regretted the omission of some names whose owners were still very much alive!

Inclusion of all the items herein can be justified; it is in the area of omissions that differences of opinion are bound to arise. Suggestions for future editions will be welcomed. Two fundamental works only have been included from the fast-growing literature on the Moon. Factual data to be found in *Orbis Geographicus* has not been duplicated—for instance, 'Chairs and institutions of university geography' or 'Hydrographic departments'. Selection of key documents is now being made increasingly acute by the greater numbers of reports issued in mimeographed form, having limited circulation, and by the publication of geographic research in non-geographic journals.

Comments in the entries are not to be considered as comparable with reviews or full bibliographical descriptions; the intention is to point out why the works are important, naming the significant features; these may be the presentation of new ideas or factual knowledge, the summary of otherwise scattered material, notes on the development of some aspect of geography, or a work in some way unique.

CBML

1 'About Sweden 1700-1963: *a bibliographical outline'*, compiled by Bure Holmbäck and others, was published by The Swedish Institute, Stockholm. About five thousand titles have been systematically divided among a number of subject fields, including special sections on Geography and Maps; the former *Books in English on Sweden,* by Nils Afzelins, 1951, has been incorporated. Arrangement within the sections is alphabetical by author or title.

2 'Abr-Nahrain: *an annual published by the Department of Middle Eastern Studies'*, University of Melbourne, 1959-60-, usually contains in each issue at least one article of interest to geographers. The chief editor is J Bowman. In volume I, for example, appeared 'Aspects of Phoenician settlement in the West Mediterranean', by W Culican; and in volume X, 1970-1971, issued on the occasion of the 28th International Congress of Orientalists held in Canberra in January 1971, is 'Asian studies in Australia, 1970', by R R C de Crespigny, and 'The traders of the pearl . . .', by B E Collers. There are plates and figures in the text.

3 Abstracts: the central abstracting service is *Geographical abstracts (qv).* Two main groups of other abstracting services should also be scanned for relevant articles: (a) those covering a wider field, such as *Geophysical abstracts; GeoScience abstracts; Historical abstracts; Economic abstracts; Tropical abstracts;* and (b) those dealing with specific fields of interest to individual geographers, of which examples are *African abstracts; Copper abstracts* (*see* Copper Development Association); *Forestry abstracts; Meteorological abstracts and bibliography; Rubber abstracts; Zinc abstracts* (*see* Zinc Development Association); *Aluminium abstracts* (*see* Aluminium Federation); *Industrial diamond abstracts* (*see* Industrial Diamond Information Bureau). Some of the leading geographical periodicals including abstracts are *Annales de géographie (qv); Erde (qv); Geographical journal of the Hungarian Academy of Sciences; Referativnyi zhurnal geografiya (qv); Soviet geography (qv); Revue Canadienne de géographie (qv); Revue de géographie de Lyon; Journal of the Società Geografica Italiana; Cartography (qv); Geografisch tijdschrift; Geographical bulletin,* Department of Mines and Technical Surveys (*see* Canada, Department of Mines and Technical Surveys); *L'information géographique (qv); Nigerian geographical journal; Polar record (qv).* Other relevant abstracting services may be identified in the *Index bibliographicus,* in *Abstract-*

ing services in science, technology, medicine, agriculture, social sciences and humanities (FID) and in *Ulrich's periodicals directory*.

4 Academy of Sciences, USSR, the highest Soviet institute of learning, counting among its members the most outstanding Soviet scientists and scholars, was founded in St Petersburg in 1724; it has been known as the Russian Academy of Sciences, the St Petersburg Academy of Sciences and, until 1917, the Imperial Academy of Sciences. In 1917, the original name was resumed, until, in 1925, it was finally renamed as above. The headquarters has been in Moscow since 1934. The academy co-ordinates the work of scientific organisations and institutes of higher learning and establishes and maintains contacts with scientific institutions abroad; it is a member of more than forty international bodies. High level conferences, meetings, discussions and research projects are sponsored. Each subject department has its own branches, in addition to the Scientific Research Institute on Sakhalin Island and the Institute of Physics in Krasnoyarsk. The branches study the local natural resources and economy. Publications include seventy learned journals and numerous scientific papers. In addition to the All Union Academy of Sciences, the constituent republics have their own academies, also numerous academies in specialised fields, such as agriculture.

5 'Acta cartographica' (Theatrum Orbis Terrarum) is the title given to a series of unabridged reprints of monographs and studies of a cartographical-historical nature, drawn from some fifty of the foremost European and American historical and geographical journals since 1800. Three volumes have been published each year from 1967, each having its own table of contents; an annual index is published. Later, the annual indexes are to be cumulated in a separate general index of authors and subjects. The project has an international advisory board. Leaflets giving the contents of each volume are available from the publishers.

6 'Acta geographica', the title of a publication of the Department of Geography, University of Stellenbosch, 1967-, is mostly, though not entirely, concerned with Africa. All articles are in Afrikaans, with no summaries in other languages; edited by Professor A Nel and produced irregularly.

7 '**Acta historiae Neerlandica**', published annually since 1966 by Brill of Leiden, is a periodical intended to make more widely known the progress of historical work in the Netherlands. From the articles selected, reprinted in English, French or German, one or two in each volume are of interest to specialist geographers, for example, 'The balance of trade of the Netherlands in the middle of the sixteenth century', 'Village and hamlet in a sandy region of the Netherlands in the middle of the eighteenth century'. Tables, graphs and sketchmaps are included.

8 Admiralty: *Geographical handbooks* series (*qv*).

9 '**The Admiralty chart:** *British naval hydrography in the nineteenth century*', by G S Ritchie, Rear Admiral, Hydrographer of the Navy (Hollis and Carter, 1967), has become one of the central works on its subject, illustrated with a few photographs and maps. The statements in the introduction—'I hope I have presented a broad picture of the progress of British Hydrography throughout the nineteenth century, accompanied by detailed sketches here and there to show something of the men, the ships in which they sailed, and the conditions under which they worked in the ever changing theatres of hydrographic interest dictated by politics, war, and, above all, trade . . .', and 'I trust that I have shown that throughout the nineteenth century the naval surveyors were alert to every aspect of the physical properties of the sea and the floor beneath, and that these properties were observed, measured and recorded to limits imposed only by the instruments and equipment available to them . . .'—summarise the theme of the book. The charting of waters in all parts of the world is chronicled, under the successive Hydrographers of the Navy, Alexander Dalrymple, Thomas Hurd, Sir William E Parry, Sir Francis Beaufort, John Washington, Sir George H Richards, Sir Frederick J O Evans and Sir William J L Wharton.

Refer also H C Anderson: 'User requirements for modern nautical charts' *in Surveying and Mapping,* June 1970.

Vice-Admiral Sir Archibald Day: *The Admiralty Hydrographic Service 1795-1919* (HMSO, 1968).

S Fawcett: 'Problems on the maintenance of Admiralty charts', ICA Technical Symposium, Edinburgh, 1964.

G A Magee: 'The Admiralty chart: trends in content and design' *in The cartographic journal,* June 1968.

C H Martin: 'Present and future trends in the maintenance of nautical charts' *in Canadian cartography*, 1964.

L N Pascoe: 'Some problems on charts' *in The cartographic journal*, June 1968.

G S Ritchie: 'Developments in British hydrography since the days of Captain Cook' *in Journal of the Royal Society of Arts*, April 1970.

'Surveyors of the oceans' *in The geographical magazine*, October 1969.

10 'Advanced practical geography', by Arthur Guest (Heinemann Educational Books, 1968), is based on the study of eight maps— 'The Sussex Weald and South Downs', 'Dovedale', 'Crackington Haven', 'The Cairngorms', 'The upper Neath and Taff Basins', 'Alluvial valleys', 'Settlement and population' and 'Climatic data'. At each stage, there are worked examples and a wide variety of exercises based on the maps and photographs. Special sections on the diagrammatic representation of statistical material and on climate and weather maps are also included. The geomorphological features shown on the regional maps are analysed and explained, correlating these features with the methods of portraying them cartographically and by diagrams.

11 'Aerial surveys and integrated studies', the proceedings of the Toulouse Conference, September 1964, under the joint sponsorship of Unesco, the National Centre for Scientific Research and the University of Toulouse (Unesco, in English and French, 1968), contains reviews of research concerning the applications of aerial photograph interpretation to the investigation of natural resources, major surveys and sample surveys, other surveys of an integrated nature and, finally, a discussion on survey principles.

12 'Afghanistan', a quarterly, has been published since 1946 at Kabul, in separate English and French editions, later, in a single edition, each issue containing some articles in French, some in English. Contributions are from local scholars or public officials.

Refer also Afghanistan: development in brief (London Information Bureau, Royal Afghan Embassy, 1958); a well illustrated review of recent developments, followed by brief sections on geography, history, culture and people.

Afghanistan at a glance (Kabul: Government Printing House, 1957) presents a general description of the country.

Afghanistan present and past (Kabul: Government Printing House, 1958), a general survey.

13 'Africa: maps and statistics', complete in twelve volumes by the Africa Institute, Pretoria, 1962-, contains in each part a statistical text, in English and Afrikaans, and a bibliography relating to the whole of Africa, grouped under the following subject headings: Population, Vital and medical aspects, Cultural and educational aspects, Transport and communication, Energy resources, production and consumption, Agriculture and forestry, Pastoral and marine products, Mining, industry and labour, Manufacturing industries, Commerce and finance, Economy, Technical and scientific development in Africa. The text is illustrated with maps, tables and figures.

14 'Africa on maps dating from the 12th to the 18th century' was edited by Egon Klemp for the Deutsche Staatsbibliothek, Berlin (Edition Leipzig, 1969; available in the United Kingdom from Sweet and Maxwell). The maps are intended to present 'a selection of the more important maps of the continent' in reproduction. Some medieval world maps, such as the Hereford and Catalan, are also included. A set of notes and brief bibliographies accompany the maps.

15 'Africa research bulletin', a cumulative monthly booklist having detachable pages, contains summaries of news from African countries; a systematic reference work, issued by Africa Research Limited, London and Exeter, England.

16 'Africa south of the Sahara', by A T Grove (OUP, 1967), covers physical environment, vegetation, soils, fauna, pests and disease, the people, traditional trading, farming and fishing, followed by sections on African prehistory and the development of each country. The Indian Ocean islands are included, the Malagasy Republic, Mauritius, Réunion, the Comores and the Seychelles. The text is illustrated and there is a bibliography.

There are many guides to the vast literature on this area. *Refer*, for example, *Africa south of the Sahara: a select and annotated bibliography* 1964-1968, *comp* K M Glazier (Stanford, Hoover Institution Press, 1969).

17 'African bibliography' of current publications, which had been published quarterly in *Africa,* journal of the International African

Institute, has from 1971 appeared as a quarterly separate publication entitled *International African bibliography*. Books, articles, conference and other papers are included.

Refer also Sanford H Bederman: *A bibliographic aid to the study of the geography of Africa*.

18 'African heritage: *studies in African culture, history and anthropology'* is the general title of a series of foundation documents edited by Weston La Barre for reproduction by the Johnson Reprint Corporation, which together present a picture of the development of African life against the varied landscapes of this vast country. *The present state of the Empire of Morocco*, for example, written by Louis Sauer de Chavanne and published in English in London in two volumes, 1788, describes the climate, soil, cities, ports, animals and products.

19 'African notes', in printed form from 1970, beginning then with volume six, number one, is a semi-annual issued by the Institute of African Studies, University of Ibadan in January and July. It aims to serve primarily as a forum for contributions for people who are professionally involved with the study of traditional culture.

20 African studies centres are located, among others, at the School of Oriental and African Studies, University of London, which issues a *Bulletin;* Institute of Commonwealth Studies, University of Oxford; the Centre of African Studies, University of Edinburgh; African Studies Unit, University of Leeds, which publishes the *Leeds African studies bulletin;* Centre of West African Studies, University of Birmingham; African Studies Centre, University of Cambridge; School of African and Asian Studies, University of Sussex; African Studies Centre, Boston University; African Studies Centre, University of Chicago; African Studies Centre, University of California; Centre of African Studies, University of Warsaw, its chief publication being the *African bulletin*. Special collections, in addition to those in Africa itself, include the Africa Collection, Hoover Institution, Stanford University; the Africa Department, Northwestern University, where a joint acquisition list of Africana is issued, the contribution of nineteen libraries in Evanston, Illinois, and others in the United States, Canada and the United Kingdom; the African Department, University of Yale; the Scandinavian Institute of African Studies, University of Uppsala and the African Institute at the

University of Geneva, with its publication *Genève-Afrique*. Important work is carried out by the Committee for the Comparative Study of New Nations, University of Chicago; the Programme of East African Studies, Syracuse University; the Committee on African Studies in Canada, University of Alberta.

Refer also Directory of Afro-American resources (Bowker, 1971), which lists 2,108 organisations and institutions with 5,365 collections of primary and secondary source materials.

21 'An African survey, revised 1956: *a study of problems arising in Africa South of the Sahara'*, by Lord Hailey, was originally published in November 1938, reprinted in 1939 as essentially a new and more comprehensive work and issued again in 1957 by the Oxford University Press for the Royal Institute of International Affairs. Besides ' The physical background ', ' Meteorology and climatology ', ' Survey and mapping ' and ' Geographical research and study ', other sections of particular interest to geographers include ' The African peoples ', ' Population records ', ' The land ', 'Agriculture and animal husbandry ', ' Forests ', 'Water supply and irrigation ', ' Soil conservation ', ' Economic development in Africa ', ' Projects of economic development ', ' Minerals and mines ', ' Transport and communications ' and ' The organization of research '. Folded maps in the text illustrate leading topics.

22 'Africana catalogues ', issued by the Africana Center, International University Bookseller Inc, New York, from January 1967-, are designed not only as sales catalogues but as bibliographic guides to publications on Africa. Information is given on books and pamphlets and on serial publications produced in the United States and elsewhere by research institutions and associations concerned.

Refer also The *Africana catalog* of the Ibadan University Library, reproduced in two volumes by G K Hall, 1972.

23 'Afrika Kartenwerk ' is an ambitious West German project, of which the four representative 1 : 1M sheets of Tunisia, Nigeria, East Africa and Mozambique-Transvaal were selected for display at the 21st Geographical Congress of the International Cartographic Association. For each individual area, eighteen thematic maps have been prepared to show a wide range of environmental phenomena as well as various aspects of human activity. In addition, larger scale maps, at 1:50,000 or 1:10,000, have been produced of selected sub-areas, each set accompanied by a descriptive memoir.

24 'Afrikaforum' (or *Internationales Afrikaforum*), the most comprehensive German-language periodical concerned with African affairs, is published ten times a year, the issues for July/August and September/October being double numbers, by the Weltform Verlag Gmbh, Munich. Articles, commentary, news items, short reports and a 'Market studies' feature giving the latest information on African industry, are prepared in close collaboration with the *Africa research bulletin* editors, of London and Exeter (*qv*), the *Bulletin de l'Afrique Noire,* Paris and *Problèmes Africains,* Brussels. From time to time, information on an individual industry is presented in the form of a detailed portrait or an intensive analysis of some particular aspect. Also published by the Weltforum Verlag Gmbh is *Africa studies,* monographs in English and German from the Ifo Institute for Economic Research, Munich.

Refer also *Africana notes and queries.*

25 Agostini, Giovanni de (1863-1941), founder of the Istituto Geografico De Agostini (*qv*), was an outstanding cartographer, who exerted great influence on Italian cartography. His most widely known single achievement is probably the *Grande atlante geografico* (*qv*).

26 'An agricultural atlas of Scotland' was begun in 1966 with the help of grants from the Carnegie Trust, the Leverhulme Trust and the University of Edinburgh; the information portrayed therein was intended for use by both geographers and all those concerned with agricultural administration and advisory work. The maps are monochrome at scales appropriate to a quarto page; most of them were based on the June 1965 parish summaries, which offered the latest overall data available. Seven main categories of data are considered; physical features, economic characteristics, land uses, crops, livestock, types of farming and crofting. In the preparation of distribution maps, an atlas autocode programme, later named Camap, was devised for use with a KDF9 computer and the maps were drawn by a line printer.

> Refer also 'An agricultural atlas of Scotland', by J T Coppock, *in The cartographic journal,* June 1969, which includes sample sheets.

27 'Agricultural development in Nigeria 1965-1980', a report prepared at the Nigerian government's request by FAO, has resulted in an important analytical study considering the food requirements of the rapidly growing population, the raw materials necessary for the expansion of the country's interests, the exports needed to pay for imports of capital goods and the means of livelihood for the additional working population that must find employment in agriculture by 1980. The text, divided into four parts, deals with ' The policy framework ', ' Policies and programmes for development of agricultural production ', ' Organizational and institutional aspects ' and 'Appendices and annex tables ', accompanied by the essential maps and charts; an important document, not only for students of Nigerian geography, but also for those concerned with similar problems in other parts of the world.

28 Agricultural Economics Research Institute, University of Oxford, founded in 1913, provides facilities for research and for various forms of undergraduate and post-graduate teaching. Occasional talks are given to farmers. The field of interest covers the economics of agricultural production, financial accounts and enterprise costing, land economics, marketing studies, agricultural income, statistical studies, rural social organisation, international trade, the world markets and the demand for food, agricultural legislation, administration and economic intelligence. *The farm economist,* published three times a year, contains articles on current research work by members of the Institute and some outside contributors; the quarterly *Digest of agricultural economics* includes summaries of all important publications on agricultural economics which have appeared in Britain during the previous quarter, together with selected studies from overseas having application to Great Britain. *A record of agricultural policy,* by agreement with the Farm Economics Branch, Cambridge University, who formerly published it is issued by the Institute biennially. Other publications include: Colin Clark: *The economics of irrigation in dry climates;* K E Hunt and K R Clark: *The state of British agriculture, 1959-1960: a comprehensive collection of statistics;* and, by the same two authors, *Poultry and eggs in Britain, 1961-62.*

29 'An agricultural geography of Great Britain', by J T Coppock (Bell, 1971), in the *Advanced economic geographies* series, traces the variations in agricultural activity under the following headings:

'The changing context of British farming', 'Land and weather', 'Farm and fields', 'Men and machines', 'Markets and marketing', 'Land and livestock', 'Dairying', 'Beef cattle', 'Sheep and lambs', 'Pigs and poultry', 'Crops', 'Horticulture', 'The pattern of farming', with a number of small distribution maps in the text and end of chapter references.

30 'Agricultural geography symposium: *a report on the proceedings'* of the symposium held at Liverpool and Nottingham in 1964 as part of the programme of the 20th International Geographical Congress, was edited by E S Simpson, Department of Geography, University of Liverpool (Research paper no 3).

31 'Agricultural meteorology' is the title given to the Proceedings of the World Meteorological Organization Seminar held at the Melbourne Bureau of Meteorology in 1966 (WMO, 1968, in two volumes). Mainly of interest to countries in South East Asia and the South West Pacific, this was the third WMO seminar to deal with agricultural meteorology, the two previous meetings having taken place in Venezuela, 1960, and in the United Arab Republic, 1964.

32 'Agricultural research in tropical Africa', by St G C Cooper, FAO Regional Adviser, and others, is a searching analysis of contemporary African research into agricultural problems (East African Literature Bureau, 1970). Problems investigated include ' The diversity of Africa ', ' Research organization, aspects and defects ', ' The regionalization of research ', ' The rice dilemma ', ' Wheat in the African economy ', ' Research and food production ', ' The protein issue ', ' The fertilizer perspective ', ' The financing of agricultural research ', and ' The manpower resources, its amount and distribution ', followed by an evaluation of the special cases of individual countries.

33 'Agricultural typology and land use mapping' was the subject of a symposium organised by the Commission for Agricultural Typology of the International Geographical Union held in New Delhi, 1968; the following papers and discussions, together with the proceedings of the symposium, were published in a special issue of *Geographia Polonica*—' Land use and types of farming ', ' The land of Hungary and types of its utilization ', ' Farming systems of the world ', 'A conceptual model of four types of world agriculture ', 'A new approach to the study of changes in cropland use: a case study of Uttar Pradesh, India ', ' Land use studies as a basis for agricultural typology of East-Central Europe '.

34 'Agroclimatological methods', Proceedings of the symposium held at Reading in 1966, was published in English and French (Unesco, 1968). The symposium reviewed the problems of agroclimatology on macro- or micro-scales, the published text including graphs and illustrations, and was complementary to the Unesco symposium held in Copenhagen, 1965, which dealt with ecosystems, involving research on a micro scale.

35 'Aguilar nuevo atlas de España' (Madrid, 1961) consists of four main sections, all subdivided as necessary: ' General geography ', ' Historical geography ', ' Regional geography ' and ' Provincial geography '. The work is a fine production and an excellent introduction to the geography of Spain, with text in Spanish, illustrated with black and white photographs, statistical diagrams, graphs and charts and including a bibliography; the cartography shows a bold use of colour, especially in the relief representation by contours and layer colouring.

36 'Aids to geographical research: *bibliographies, periodicals, atlases, gazetteers and other reference books'*, compiled by John Kirtland Wright and Elizabeth T Platt (American Geographical Society, first edition, with sub-title ' Bibliographies and periodicals ', by J K Wright, 1923; second edition with E T Platt, completely revised, Columbia University Press for the American Geographical Society, 1947; Research series, no 22), remained for many years the only comprehensive English language attempt to cover the subject objectively, that is, from other than the teaching aspect. The detailed annotations are particularly helpful. ' This book is intended to be of help to anyone who has occasion to make a serious study of geography ' begins the Introduction. Grouped in three main parts— ' General aids ', ' Topical aids ', ' Regional aids and general geographical periodicals '—each is subdivided as necessary. An appendix gives as an explanatory note a ' Classified index of American professional geographers, libraries of geographical utility and institutions engaged in geographical research ' and there is a substantial index.

37 Akademiya Nauk SSSR, the Institute of Geography of the Academy of Sciences, is the major research body for geographical studies in the USSR. Work is carried out on a vast scale and with practical application. More than three hundred geographers work in ten divisions: physical geography; geomorphology; climatology and hydrology; biogeography; glaciology; cartography; economic geography of the USSR; geography of people's democracies; geography of capitalist countries; history of geography. One of the most important functions is the investigation of research problems posed by the government. The organisation of field expeditions is also important, often in conjunction with the universities and with other institutes of the academy. A vast publishing programme, including the results of the field research, comprises much of the most valuable achievements of Soviet geography. *Izvestiya, seriya geograficheskaya,* 1951—bi-monthly, has the contents page in English and is illustrated with fine maps, photographs and charts. The book reviewing section is detailed and authoritative. The second great geographical journal of the institute is the *Referativnyi zhurnal geografiya (qv).*

38 'The Aldine university atlas' (Aldine Publishing Company, 1969) is a version of Philip's *Library atlas* prepared for the American market; detailed maps of the British Isles are replaced by thematic maps of the States and Southern Canada and statistical information on the nations of the world takes the place of some of the meteorological graphs.

39 Al-Idrisi (*c* 1099-1154), one of the best known of the great Arab school of geographers, travelled in north Africa and in Asia Minor, before settling in Sicily, where, on the order of King Roger II, he compiled a description of the world, based on a synthesis of contemporary knowledge, and a world map. He divided the world into seven latitudinal climates or zones between the equator and the north pole, each zone being sub-divided into ten parts by lines at right-angles to those of the latitudes. In the best known of the two editions of his work, there are seventy maps, each representing one of the parts. The Mediterranean coastline is shown in particular detail. The earliest known translation was published in Rome, 1619, based on an incomplete abridgement of the work. A complete edition of al-Idrisi's work has long been overdue. G Levi della Vida and Francesco Gabrieli presented a general outline of such a project at the Convegno internazionale di studi Ruggeriani held at Palermo in 1954, but, unfortunately, it proved impossible to find any way of financing the execution of the work. Subsequently, the Italian Institute for the Middle and Far East, the University Oriental Institute at Naples and the Institute of Oriental Studies of the University of Palermo formed an editorial committee and in 1970, published by Brill of Leiden, the first fascicule appeared; further fascicules will follow, in Arabic, with translation, commentary, a glossary and indices, accompanied by a separate section of maps.

Refer also al-Idrisi: *Description de l'Afrique et de l'Espagne: Texte arabe*, published for the first time, from *ms* at Paris and at Oxford, by R Dozy and M J de Goeje, with a translation, notes, a glossary, from the 1866 edition (1968).

40 All-Union Geographical Society of the USSR, the most important organisation for co-ordinating the country's research in geographical subjects, was founded in Leningrad in 1845. The society now comprises twenty-four main branches, eighty-four minor branches and sixteen sub-branches; membership numbers some ten thousand, composed entirely of professional geographers. The Leningrad branch remains the headquarters, with a membership of over a thousand; here is located a fine library holding more than three hundred thousand volumes, a great map collection and archives, including the private collections and manuscripts of many eminent Russian geographers of the past. Other important branches are the Moscow, Ukrainian and Georgian. A number of sections are devoted to individual aspects of geography. A vast number of scientific conferences and expeditions is organised; in addition, the quinquennial congresses of the society, begun in Leningrad in 1947, have been of the utmost significance. The main part of the publishing programme comes within the responsibility of the Academy of Sciences and the State Publishing House of Geographical Literature; this includes journals, reports of research and scholarly monographs on all aspects of geography, especially physical, regional and economic. *Izvestiya vsesoyuznogo geograficheskogo obshchestva* is the journal of the main society; *Voprosy geografii* that of the Moscow branch, both of which are included in the translation policy of *Soviet geography* . . . (*qv*).

41 Almagià, Roberto (1884-1962), was a scholar of international repute, who exerted a unique influence on geographical studies through his academic work and publications. His interests, reflected in the geography departments of the universities of Padua and Rome, ranged from geology and oceanography to the history of geographical science, especially cartography, and the whole field of human geography. He became president of the Società di Studi Geografici, Florence, president of the Italian National Committee for Geography and a vice-president of the International Geographical Union. Among the best known of his publications are *L'Italia*, in two volumes, 1959; *Il mondo attuale*, in two volumes, 1953; *I primi esploratori dell'America*, 1937. His *Monumenta cartographica vaticana*, in four volumes, 1944-55, was prepared while he was working in the Apostolic Library during the second world war. As co-editor of the *Rivista geografica Italiana*, he helped to make this one of the great geographical periodicals.

42 Aluminium Federation, London, formerly the Aluminium Development Association, is one of the many examples of special organisations producing source material of central interest to economic geographers. The publication programme includes information bulletins, brochures, technical memoranda and other papers, and the *Research reports* series; regular contributions are made to the quarterly *The aluminium courier* and to *Aluminium abstracts*, produced fortnightly by the Centre International de Développement de l'Aluminium. The library contains upwards of two thousand books, besides a vast collection of pamphlets, patents and periodicals. An information department and technical advisory service are maintained.

43 'Amazonia', a new periodical devoted to the limnological landscape ecology of Amazonia, established close scientific collaboration between the Instituto Nacional de Pesquisas de Amazonía in Manáus, Brazil, and the Hydro-biologische Anstalt der Max-Planck-Gesellschaft zur Fönderung der Wissenschaften in Plön (Federal Republic of Germany), published by the Kommissions-Verlag Walter G Mühlau, Kiel, from 1965.

44 'The American city: an urban geography', by Raymond E Murphy (McGraw-Hill, 1966), presents a systematic analysis of many aspects of urban geography. The text outlines classifications and comparisons of cities as unit areas and analyses the patterns of the interiors of cities. Source material is mentioned for further study of past, present and future urban geographical research.

45 American Geographical Society of New York, founded in 1852 by a group of businessmen to advance the science of geography by discussion, publication and lectures, to establish in the chief city of the United States a place where accurate information on every part of the globe could be obtained and to encourage exploration and research, has developed into one of the greatest influences on world geography. Its history is best traced in the file of *The geographical review* (*qv*) and in the anniversary volume by J K Wright, *Geography in the making: the American Geographical Society, 1851-1951* (The Society, 1952). A *Newsletter,* issued four times a year, includes notes on projects in progress; *Focus* (*qv*) and *Soviet geography . . .* (*qv*) are two more of the society's periodical publications and the total vast publishing programme includes research monographs, maps and atlases. For many years, the society has promoted studies relating to questions of wide public interest; for example, pioneer settlement, polar exploration and conservation of resources. Current research includes studies on perception of the environment, on glacier fluctuations and on the automation of cartographic generalisation. The *Serial atlas of the marine environment* (*qv*) and *Antarctic map folio* series (*qv*) present in continuing map form results of interdisciplinary research in the physical sciences. Books and monographs appear at intervals in the *Research* series, *Special publications, Studies in urban geography* and other series. The various departments are always engaged in research, especially the Cartographic Department, which makes maps for all kinds of purposes. The continuing grant from the World Data Survey A, Glaciology, was withdrawn at the end of 1969 and arrangements were made shortly after to transfer the collection of maps, photographs and reports to the Water Resources Division of the US Geological Survey at Tacoma, Washington. G K Hall published *Index to maps in books and periodicals,* prepared by the Department, in ten volumes, 1967.

The society's library is one of the largest geographical collections in the world, including monographs, pamphlets, periodicals, photographs, atlases and maps. Since 1923, a research catalogue has been maintained, arranged by a scheme of classification specially devised to serve geographical research; the catalogue contains references to monographs, periodical articles, government documents and maps, both separately published and those included in books and periodicals. In 1962, G K Hall reproduced the catalogue in fifteen volumes, which are available as a complete set or in eight individual sections. From the research catalogue has developed the bibliographical tool, *Current geographical publications* (*qv*).

46 ' The American Neptune ', quarterly journal of maritime history, is published by the Peabody Museum of Salem, Massachusetts. Articles are frequently of special interest, such as ' Early steam navigation in China ', a main theme in 1966. Book reviews are included and there is a pictorial supplement, an inset section of photographs and other illustrations, in addition to those in the text.

47 Amtmann, Bernard, Inc, Montreal, antiquarian booksellers, specialise in very useful catalogues, including out of print and rare books, scientific books and periodicals, arranged alphabetically by author or by title in the absence of a known author, each entry bearing a running number for ordering purposes. Each catalogue deals with one region, with emphasis on Canada, the Arctic and the Antarctic, but also with other areas, such as ' Pacific Northwest ', which was in three parts, or ' Voyages and travels '. Many of the catalogues have interesting reprints of title pages on the front cover.

48 ' The analysis of geographical data ', by W H Theakstone and C Harrison (Heinemann Educational Books, 1970), provides an interesting text on the statistical analysis of the many kinds of data handled by geographers, broadly concerned with spatial distribution. Figures and diagrams are included, also a glossary of symbols and a glossary of terms.

49 ' The ancient explorers ', by M Cary and E H Warmington (Methuen, 1929; revised edition, Pelican Books, 1963), a classic saga of exploration, begins with a discussion of the objects of the ancient explorers and the materials and equipment available to them. The following chapters deal with the opening up of the first known parts of the world, with a final chapter summing up the results of ancient explorations. There are fifteen maps in the text and extensive notes.

50 'Annales de géographie', founded by Paul Vidal de la Blache and Marcel Dubois in 1891, soon became and has remained one of the foremost of the scholarly geographical journals devoted to the development and principles of the subject, its aim being ' to reason, to link, to interpret '. In 1941, it joined with the *Bulletin* of the Société de Géographie de Paris under the title *Annales de géographie et bulletin de la Société de Géographie*. In the six issues each year, the articles maintain the highest standard of original scholarship or synthesis; the summaries of contemporary knowledge of the continents in the early numbers are of special interest. The notes and reports of projects are invaluable, as are the numerous maps; ' Books received ' has been a feature since 1933, statistics sections since 1937; the importance given to bibliographical work was early shown in the inclusion of the annual bibliographies, which subsequently developed into the independent *Bibliographie géographique internationale* (*qv*). Annual and ten-yearly indexes are prepared.

51 'Annotated bibliography of Afghanistan ', compiled by Donald N Wilber (New Haven: Human Relations Area Files Press, third edition, 1968), contains general sources of information and reference books, in broad subject groupings, including geography and history, each section having a brief introduction; the entries, frequently annotated, are numbered and there is a name index.

52 'Annotated bibliography of Burma ', prepared by the Burma Research Project at New York University, was edited by Frank N Trager and others (New Haven: Human Relations Area Files Press, 1956). Included are bibliographies, books, pamphlets and other separates (these merely listed); periodicals; government publications; general works, geographical and historical monographs, etc. The entries are numbered for reference and many are briefly annotated.

53 'An annotated world list of selected current geographical serials in English: *with an appendix of major periodicals in various languages regularly providing English summaries of articles, or periodicals partly in English and partly in other languages'* is a useful pamphlet, compiled by C D Harris (Department of Geography, University of Chicago, 1960). The third edition concentrated on periodicals in English, French and German.

See also the author's *International list of geographical serials*.

54 'Annual summary of information on natural disasters', 1970-, continues the work of a series of publications by the Union Internationale de Secours which came to an end in 1966. Of the four sections: *Earthquakes* is compiled by Professor J P Rothé; *Tsunanis,* compiled from information supplied to Unesco by the Tsunanis Warning Centre in Honolulu; *Storm surges,* from information supplied by the Tidal Institute, observatory of the University of Liverpool; and *Volcanic eruptions,* from information given by the Vulcanological Society of Japan and by Dr Latter of the Institute of Geological Sciences, Edinburgh.

55 'Antarctic', the quarterly news bulletin of the New Zealand Antarctic Society, 1956-; book reviews are included.

There is also the *Antarctic record* issued by the Ministry of Education, Tokyo; and the *Information bulletin* of the Soviet Antarctic Expedition, in addition to the following major reference works on the Antarctic.

56 'Antarctic bibliography', prepared at the Library of Congress, under the editorship of George A Doumani, sponsored by the Office of Antarctic Programs, National Science Foundation, Washington DC (Washington DC, Superintendent of Documents, 1965-), originated in the bibliographic card service of Antarctic abstracts compiled and disseminated by the Cold Regions Bibliography section of the Library of Congress, Science and Technology Division, 1963-. All topics relevant to the region south of latitude 60 degrees south and the sub-Antarctic islands were included, the first volume covering mainly 1962-1964. Since 1963, polar libraries and interested scientists, both within and outside the United States, have availed themselves of the service, entries being printed on four by five inch cards. The first two thousand entries have been published by the Office of Antarctic Programs. Entries represent the publications of some thirty countries, in thirteen languages, with emphasis on the USA, arrangement being by subject groupings, with author, subject and grantee indexes.

There is also the bibliography bearing the same title, compiled by John Riscoe, sponsored by the US Navy Department, Bureau of Aeronautics (Washington, DC; Government Printing Office, 1951), which includes maps.

57 'Antarctic ecology', in two volumes, edited by M W Holdgate, of the Nature Conservancy, London, is a major reference work (Academic Press, 1970), dealing comprehensively with the Antarctic and offering comparisons with the Arctic regions. The text represents the papers from the second symposium of the Biological Working Group of the Scientific Committee on Antarctic Research, the first volume dealing with past environments and biotas, marine plankton and its pelagic consumers, marine benthos, seals and seabirds; the second with freshwater systems in the Antarctic and the Arctic, terrestrial soils, vegetation and fauna, concluding with a review of conservation.

58 'Antarctic journal of the United States', 1966-, is issued bi-monthly by the Office of Antarctic Programs of the National Science Foundation and the United States Naval Support Force, Antarctica, of the Department of Defense, jointly, to provide a forum for the exchange of information on the Antarctic.

59 The 'Antarctic map folio' series, begun by the American Geographical Society in 1964, sponsored by the National Science Foundation, is planned to comprise about twenty folios, each containing several sheets of maps and text on a specialised subject within the whole, under the general editorship of V C Bushnell. The aim is to summarise existing knowledge of the Antarctic continent and adjacent waters. Two of the latest folios to appear were no 12 and no 13, published in August 1970: The first, 'Geologic maps of Antarctica', is the work of Campbell Craddock and twenty-four other contributors, comprising two groups of maps, regional maps of bedrock geology at 1:1M scale, and continental maps of fossil sites, tectonics and additional features at a smaller scale. The second, 'Circumpolar characteristics of Antarctic waters . . .', by A L Gordon and R D Goldberg, covers all the waters surrounding the continent as far north as 40 degrees south.

60 'Antarctic research: a review of British scientific achievement in Antarctica', edited by Sir Raymond Priestley and others, covers the period 1944 to the end of 1963 (Butterworth, 1964), with illustrations, diagrams, bibliographies and maps in a separate folder. Leading authorities have contributed twenty-one informative articles on various aspects, presented in non-technical language.

61 'Antarctica', edited by Trevor Hatherton (Methuen, 1965) and published with the co-operation of the New Zealand Antarctic Society, surveys the present state of knowledge of the continent. The twenty-one scholars involved have together made a comprehensive coverage, arranged under four main headings: 'The nations in Antarctica', which includes a summary of mapping; 'The southern ocean'; 'The Antarctic continent' and 'The south Polar atmosphere'. The work has been accepted as a standard treatise, which has been translated into Russian and Spanish. There are numerous magnificent photographs throughout the text, also sketchmaps and diagrams where useful, and a folded 'Map of the Antarctic region' fits into a back cover pocket. Two appendices set out 'The Antarctic Treaty' and 'National stations in Antarctica since 1957'. There is also a bibliography.

Refer also, from the vast literature existing: ANARE, 'Australian National Antarctic Research Expeditions'.

Frank Debenham: *Antarctica: the story of a continent* (Jenkins, 1959).

Sir Vivian Fuchs and Sir E Hillary: *The crossing of Antarctica, the Commonwealth Trans-Antarctic Expedition, 1955-1958* (Cassell, 1958).

D M Goodall: *The seventh continent* (Royston: Priory Press, 1969).

H G R King: *The Antarctic* (Blandford Press, 1969).

62 'Antique maps and their cartographers', by Raymond Lister (Bell, 1970), begins with a study of map-making among primitive peoples, tracing its development to the eighteenth century; following chapters are devoted to the achievements of individual countries —the Low Countries, France, Great Britain, Scandinavia and Russia, Africa, Asia, America and Australasia, accompanied by numerous black and white reproductions, end of chapter bibliographies and a 'General bibliography'. Maps are described in the manner expected of a life-long lover and collector of early maps; seldom is any linkage attempted with contemporary techniques or with geographical developments. The work, eminently readable, is general in approach; nor does it take into account some more recent specific publications, such as the *Atlantes Neerlandici* . . . (*qv*).

63 'Arcano del mare', the first marine atlas, produced by Sir Robert Dudley in three volumes, 1646-47. The charts, drawn on the Mercator projection and finely engraved by Antonio Francesco Lucini in Florence, marked a great advance on those in previous works. A second, corrected edition, was issued in 1661.

64 'Arctic', quarterly journal of the Arctic Institute of North America, 1948-, contains papers summarising original research, notes and reviews, with at least one erudite article of central interest.

65 'Arctic and Alpine research', issued quarterly, was initiated in 1969 by the Institute of Arctic and Alpine Research, University of Colorado, ' to provide a vehicle for scientists, at the international level, with a special interest in Arctic and Alpine environments, to give emphasis to INSTAAR development and to assist in the experiment of environmental teaching being undertaken within the Institute '. The high latitude and mountainous areas of the world and related topics of the Pleistocene era form the central theme.

66 'The Arctic basin', in a revised edition by J E Sater (Arctic Institute of North America, 1969), originated with a report from the Institute prepared in 1960 for the US Army to provide an account of the environmental factors relevant to operations in the Arctic. Much material was added and the revised edition comprises a comprehensive general account of the physical features, the history of exploration and the nature of human settlement in the area.

67 'The Arctic bibliography', edited by Marie Tremaine for the Arctic Institute of North America, 1953- (McGill-Queen's University Press), contains abstracts and index entries in Russian, English, Scandinavian, German and other languages, as appropriate. Coverage of geophysical studies has increased through the years, as more detailed research has been made possible, with the expansion of transportation. The publication is recognised as the standard, unified research tool for the area; entries are numbered and the location of each document is noted with full bibliographical description, annotated as necessary. Preliminary work had been begun under an Office of Naval Research contract with the Arctic Institute of

North America, with the support of the Department of the Army, the Department of the Navy and the Department of the Air Force, under the supervision of a directing committee. The developed project, from 1953, was prepared in co-operation with the Department of Defense under the direction of the Arctic Institute.

68 Arctic Institute of North America was established after the second world war to encourage scientific research into Arctic conditions and to act as a repository for information, with emphasis on the North American Arctic; activities have since been extended to include the Antarctic. The main centre is in Montreal, with a large section also in Washington. Publications include reports and special studies; *Arctic* (*qv*) and *The Arctic bibliography* (*qv*) are of international importance. A library was established at McGill University in 1949, moving subsequently to new premises. Holdings include some six thousand books, more than twenty thousand pamphlets, several hundred reports on research projects, periodicals, photographs and more than five thousand maps. A unique reference service, operated since 1959, enables the library to handle research and information projects on a cost plus basis; indexes, abstracts and bibliographies are compiled, also translations by request, and accessions lists are issued.

69 'Area', the house journal of the Institute of British Geographers, 1969-, when a new periodical was considered desirable in addition to the institute's *Newsletter,* aims to encourage free and impartial discussion of ideas and techniques, ' to probe, to report and to examine the implications of work being done in these expanding fields of study '. *Area* appears four times a year; the first issue, of fifty-six pages, showed the variety of content, being full of information and comment on ' Computer graphics ', a ' General report of the Edinburgh British Honduras-Yucatan Expedition ', ' Social science research in geography ', ' Centre for Environmental Studies ' (*qv*), ' Research at the Countryside Commission ', ' Metrication in the *Transactions*', 'The Institute of Agricultural History at the University of Reading' and several reports of conferences and symposia. A few sketchmaps usually illustrate the text.

70 'La Argentina: suma de geografia', under the direction of Francisco de Aparicio, Horacio A Difrieri and others, is a superb publication in nine volumes on all aspects of geography and cartography

(Buenos Aires: Ediciones Peuser, 1958-1963). Maps were contributed by Hildebrando O Boccio and the aerial photography was the work of the Instituto Fototopográfico Argentino, the Instituto Geográfico Militar and the Ministerio de Marina. The text is in Spanish, but the folding maps, photographic reproductions, diagrams and the superb aerial photographs facilitate understanding. Included is a detailed description of the *Atlas aerofotográfico*.

71 'Arid lands: a geographical appraisal', edited for Unesco/ Methuen by E S Hills, 1966, was planned to present a broadly based conspectus of the vast area of dry country in the world. The book deals with all aspects of the problem of aridity, ranging from water supply, use and management, through industrialisation and social life in the arid zones.

Refer also H E Dregne: *Arid lands in transition* (American Association for the Advancement of Science, 1970).

Patricia Paylore, *comp: Seventy-five years of arid lands research at the University of Arizona* (Office of Arid Lands Research, University of Arizona, 1966): *a selective bibliography of arid lands research 1891-1965.*

72 'Arid lands in perspective: *including AAAS papers on water importation into arid lands,* edited by William G McGinnies and Bram J Goldman (American Association for the Advancement of Science and the University of Arizona Press, 1969), is one of the contributions to the Unesco project, Arid Zone Research, illustrated and including maps, diagrams and bibliographies. The first section contains nineteen papers on a wide range of topics; a second group, on physical geography, comprises studies of quantitative analysis of desert topography and the classification of arid land soils. There are a number of regional studies. There are two bibliographical papers: one by Andrew Warren on desert dunes, the other by Patricia Paylore on the sources for arid lands research. A final section comprises twelve papers and comments deriving from a symposium on water importation into arid lands, sponsored by the AAAS, held in Dallas, Texas, in December 1968.

73 'Arid land research institutions: *a world directory'*, by Patricia Paylore (University of Arizona Press, 1967), formed part of an inventory of geographical research on desert environments being conducted by the Institute of Arid Lands Research of the University of Arizona. The project as a whole seeks to determine in detail what topics are being or have been investigated relating to the world's deserts and what remains to be done. Subsequently, a compendium will be issued based on a comprehensive critical review of the published literature, augmented by consultations with leading scientific specialists. Topics to be covered are physical features, flora and fauna, weather and climate, coastal deserts and regional types. The work above-mentioned is the first directory of arid zone institutions to appear since the Unesco *Directory of institutions engaged in arid zone research,* 1953. Since that time, the rate of research in this field has increased considerably and the number of entries in the present publication is more than double those listed by Unesco— more than two hundred organisations in thirty-nine countries. Arrangement is by countries and each entry is annotated with notes on the scope and type of work in the organisation concerned. A first supplement appeared in 1968, prepared also by Patricia Paylore (Natick: Earth Sciences Laboratory, United States Army Natick Laboratories, 1968), which included yet more institutions and updated the existing entries as necessary.

74 Arnold, Edward, (Publishers) Limited, London, have a special interest in books on geography, mainly devoted to the more scientific aspects of the subject and having a bias towards the educational. A catalogue, *Geography,* published at intervals, lists all titles designed for sixth form level and above. Recent titles of particular interest include David Harvey: *Explanation in geography (qv)*; Peter Haggett and Richard J Chorley: *Network analysis in geography (qv)*; a new edition of F J Monkhouse: *A dictionary of geography (qv)*; and the series *Progress in geography (qv)*.

75 'The art of navigation in England in Elizabethan and early Stuart times', by David W Waters (Hollis and Carter, 1958), a thorough and scholarly work, considers 'The development of the art of navigation in Europe in the fifteenth and early sixteenth centuries'; and 'The English contribution'; followed by thirty-three appendices on specific topics integral to the subject. The text is illustrated by many plates, reproductions and diagrams and there is a bibliography.

Compare similar works by E G R Taylor *(qv)*.

76 'The Asia bulletin: *a monthly review from Asia Publishing House'* frequently includes books, documents and reports of interest to the geographer—for example, P L Mehra: *The Younghusband Expedition,* 1968—or on economic conditions, the methodology of economic research and on trade and industry in India and other Asia countries.

Refer also A T Embree *et al, comp: Asia: a guide to basic books* (New York: Asia Society, 1966) and a supplement, a guide to paperbacks on Asia, compiled by J H Bailey and others.

Peter L A Gosling: *Maps, atlases and gazetteers for Asian studies: a critical guide* (University of the State of New York, 1965).

C A Fisher: *Modern Asian studies* (CUP for the School of Oriental and African Studies, University of London, 1967).

77 'Asian survey: *a monthly review of contemporary Asian affairs'* has been published from 1960 by the Institute of International Studies, University of California at Berkeley. The Institute, established in 1955, carries on organised research in comparative and international affairs, provides facilities for research for individual scholars and serves as an administrative agency which assists in promoting new research interests among the faculty. The journal reflects these aims in major articles, shorter notes and comments. The first three volumes of the journal have been reprinted by Kraus.

78 Asia's lands and peoples: *a geography of one-third the earth and two-thirds its people',* by George B Cressey (New York: McGraw-Hill, 1944; third edition, revised 1963), follows a regional treatment after general opening comments—' China ', ' Japan and Korea ', ' Soviet Union ', ' Southwestern Asia ', ' India and Pakistan ', ' Southeastern Asia ', subdivided as necessary. Photographs, maps and diagrams abound throughout the text and there is a substantial section of ' Suggested readings ' limited to the more readily accessible literature.

Refer also Crossroads: land and life in Southwest Asia, by the same author (Syracuse University; Lippincott, 1960; Feffer and Simons International University Editions).

79 Associated Publishers, Amsterdam, an independent publishing company, was founded in the Spring of 1967 by a group of Dutch reprint publishers—A Asher and Company; John Benjamins N V Scientific Periodical Trade; E J Bonset; S Emmering; De Graaf, publisher, Nieuwkoop; Gerard Th. Van Heusden, publisher and bookseller; B M Israël N V; Fritz Knuf, publisher, Hilversum; Meridian Publishing Company; Editions Rodopi N V; P Schippers N V (B R Grüner); and Theatrum Orbis Terrarum Limited (*qv*), all, with the two noted exceptions, at Amsterdam, having the aim of working in closer mutual co-operation. Each specialises in one or more fields. Reference works, maps, atlases and periodicals of permanent value have been reproduced and catalogues have been issued representing the output of an individual publisher or of the new group.

80 Association of American Geographers, the organisation for professional geographers, was founded in 1904. The annual conferences, the opportunities for publication and the awards offered have proved a great stimulus to academic work. A quarterly journal, *Annals,* has been issued since 1911, carrying abstracts, bibliographies, statistics and reviews. A recent innovation has been the issue of special volumes of the *Annals* in which symposia beyond the normal scope of the journal may appear. After amalgamation with the Society of Professional Geographers, a bi-monthly journal, *Professional geographer,* was issued, 1949-   and in 1961 a new journal, the *Southeastern geographer,* was launched by the southeast division of the Association.

81 Association of Geography Teachers of Ireland was founded in 1962 with the aim of giving every Irish child a sound knowledge and understanding of geography. Constant efforts have been made to improve syllabuses, teaching methods and textbooks. The association's journal, *Geographical viewpoint,* has been issued since 1964, annually, five issues constituting a volume; lectures, discussions and field meetings are arranged and travelling exhibitions prepared.

82 'The astronomical and mathematical foundations of geography', by Charles H Cotter (Hollis and Carter, 1966), a small introductory work setting out the fundametals of this basic aspect of geography, reviews the development of such studies, proceeding to clarify 'Ideas on the size and shape of the earth ', ' The sphere and the ellipsoid ', ' The triangle ', ' Problems of sailing on the earth ', ' The earth's orbit ', ' The earth's rotation and problems of finding latitude and longitude ', ' The elements of surveying ' and ' Map projections '. Line drawings accompany the text and there is a selected list for further reading.

83 Athens Center of Ekistics, Athens Technological Institute, Graduate School of Ekistics, has organised an annual international seminar on 'Ekistics and the future of human settlements' from 1965. The aim is to provide an opportunity for interdisciplinary exchange and discussions on growth and change within the subject. The Centre consists of five divisions: Research, Education, International programmes, a documentation unit and Administration. Library facilities are available. The pioneer of the study, ekistics, was Dr C A Doxiadis, expert on development planning, whose book: *Ekistics: an introduction to the science of human settlements,* gave impetus and precision to this emerging aspect of population study. *Ekistics,* the monthly periodical issued by the Centre, includes original articles as well as abstracts from relevant articles in other journals, each issue being devoted to a particular aspect; also *Ekistics index,* monthly, consists of a list of cross-referenced articles selected from some six hundred periodicals from fifty countries; annually, a selection of *Scientific reports* is produced and there is the monthly ATO-ACE Newsletter, which is circulated gratis.

84 'Atlante internazionale della Consociazione Turistica', a comprehensive reference atlas, was originally conceived by Luigi V Bertarelli, carried out with the collaboration of Olinto Marinelli and Pietro Corbellini and presented to the tenth Italian Geographical Congress in 1927. Between 1927 and 1938, five editions were published, the 1938 edition being particularly carefully revised. The edition of 1955-1956 was issued in serial instalments, directed by Manlio Castiglioni, to mark the sixtieth aniversary of the Club's foundation and the eighth edition was edited by Sandro Toniolo, 1964-1968. The work, in two volumes, is an atlas totalling more than 170 plates and a gazetteer-index of about 250,000 entries. The plates, with many large-scale insets, were each drawn and engraved by hand and printed by the most modern lithographic techniques. Relief is by hachuring and hill shading, with frequent spot heights. A brief summary of the source material is shown on the back of each plate, and useful notes, glossaries and diagrams are included. The atlas has been used as the base for other world atlases, notably for the *Gyldendals verdens-atlas,* Copenhagen, 1951.

Refer Sandro Toniolo: ' The new edition of the "Atlante Internazionale" issued by the Touring Club Italiano' *in International Yearbook of Cartography,* 1970.

85 'Atlantes Neerlandici: *bibliography of terrestrial, maritime and celestial atlases and pilot books, published in the Netherlands up to 1880'*, compiled and edited by Dr Ir C Koeman, in three volumes (Amsterdam: Theatrum Orbis Terrarum Limited, 1967, 1969), under the authority of the Committee for the Bibliography of early Dutch atlases, is to be in five volumes when completed. ' Maps, like paintings and prints, belong to the characteristic manifestations of seventeenth century Dutch culture ' states the key theme. Each entry contains a complete bibliographical description, including any distinguishing characteristic, of editions and issues of atlases published in the Netherlands and by Dutch cartographers elsewhere, arranged in alphabetical order of cartographers. Volume I comprises ' Van der Aa to Blaeu; II, Blussé-Mercator '; III Mercator-; these being land atlases and town books. The fourth volume will deal with sea atlases, pilot guides and celestial atlases. Biographies of the outstanding publishers and map-makers, chronological lists, personal, publishers', cartographers' and geographical names indexes and, in the fifth volume, a cumulative index, will complete the work, which is generously illustrated with reproductions. A useful point is the indication of location of copies.

Refer also G R Crone: ' Seventeenth century Dutch charts of the East Indies' *in The geographical journal,* November-December, 1943.

86 'Atlas ', the journal of Dublin University Geographical Society, is not confined in scope to articles on Ireland. Features include ' Society news ' and a modest review section. There are usually sketchmaps in the text.

87 'Atlas aérien ' is an exciting and valuable experiment by Gallimard, in five volumes: 1, Alps, Rhône, Provence, Corse, 1955; 2, Bretagne, Loire, Sologne et Berry, Entre Loire et Gironde, 1956; 3, Pyrénées, Languedoc, Aquitaine, Massif Central, 1958; 4, Paris et la vallée de la Seine, Ile-de-France, Beauce et Brie, Normandie, de la Picardie à la Flandre, 1962; 5, Alsace, Lorraine, Morvan et Bourgogne, Jura, 1964.

88 'Atlas botanic ', compiled by Professor Lucia Popovici and Professor Dr Constanta Moruzi (Bucharest: Didactic and Pedagogic Publishing House), is in four main sections: Plant evolution, Plant morphology and physiology, Plant systematics and Geobotanics,

Nature protection and curiosities from the plant world. 181 coloured tables accompany an explanatory text and all the plates are printed in six to eight colour offset. A classification of plants is included, also explanations of some botanic terms, an alphabetical index and bibliographical references.

89 'Atlas Českoslovovenské Socialistiché Republiky', edited by Antonin Gotz (Prague: Československa Akademie Vedústrědni Sprava Geodézie a Kartografie, 1966), consisting of fifty-eight map sheets and a gazetteer, is a beautiful example of a national atlas. The length of Czechoslovakia east-west suggested two-page spreads, on which the whole country could be shown at 1 : 1M. Much detail is included and the judicious selection of symbols and colouring makes the detail effective. When desirable, for comparison, as with crop production, sixteen maps at 1 : 4M are shown on the double page. All the usual aspects are covered, with, in addition, a survey of housing conditions, equipment, floor space, electricity consumption, etc.

Refer also J Demek *et al*: *Geography of Czechoslovakia* (Prague: Academia, Publishing House of the Czechoslovak Academy of Sciences, 1971).

90 'Atlas de Colombia', edited by Eduardo Acevedo Latorre, of the Sociedad Geográfica de Colombia (República de Colombia, Instituto Geográfico Agustín Codassi, Bogotá, Academia Colombiana de Historia, 1967), is a glorious production, having the aim of ' compiling more or less all cartographic works about geographical aspects . . . to summarize the present status of geographical investigations in the country '. As indicated, therefore, this is not an original atlas; most of the plates have appeared elsewhere, a large number being from the four-volume *Atlas de Economía Colombiana* (Banco de la Repúblico, Bogotá, 1959-1964), also edited by Acevedo Latorre. Much of the work is intended for the general public rather than the research-minded geographer; but for geographers particularly there are three special sections: on historical cartography, including some fine reproductions; a section on natural environments and one on economic and demographic aspects. 1964 census material was not included and this is a pity, for there were great changes in population after 1951. The colour printing is of a very high quality and the production has throughout a multitude of illustrations, extra maps and diagrams. There are an excellent section of drawings of products shown alongside statistics relating to them, a dictionary of astronomical, geodetic and physical terms relating to the world as a whole, an introduction and a bibliography. The final section contains hypsometric maps of the individual states and plans of their capitals, with a special section on Bogotá. Based on this atlas is the *Atlas escolar de Colombia,* of which a first edition was for limited circulation only, a revised definitive edition being planned for distribution to secondary schools.

91 'Atlas de France', one of the most interesting of the national atlases, was first published 1934-1945, with a second edition, 1953-1959, in loose-leaf form, by the Comité National de Géographie, printed by the Service Géographique de l'Armée. Some plates in the second edition were reprinted from the first, but by more modern methods; the new plates show types of agriculture, livestock distribution, electrical energy and density of railway traffic. Population data are based on the 1946 figures. Field work was of very high quality and the whole production demonstrates the best of French cartography. Interesting new techniques include the use of proportional spheres to denote urban centres; and useful plans of medieval towns are included. Scales range between 1:1,250,000 and 1:8M.

92 'Atlas de la France de l'est ', an excellent example of a regional atlas, was prepared by l'Association pour l'Atlas de la France de l'Est and edited by La Librairie Istra, Strasbourg, 1959 (Les Editions Berger-Levrault, Nancy, 1960). Seventy loose-leaf maps, accompanied by explanatory notes in a binder, give a more detailed treatment of the region than was possible in the national atlas. In the introduction by Professor Henri Baulig, the unity of the region from the population aspect is brought out; its frontier position and its independence of spirit are stressed.

Refer also Atlas de la France rurale, edited by Jean Duplex (Colin, 1968).

93 'Atlas de Paris et de la région Parisienne ', edited by Jacqueline Beaujeu-Garnier and Jean Bastié (L'Association Universitaire de Recherches Géographiques et Cartographiques—Édition Berger-Levrault, 1967), is one of the most interesting in a long line of regional atlases of France. Governmental ministries have contributed to its preparation, also organisations such as the Centre

National de la Recherche Scientifique, the Institut National de la Statistique et d'Etudes Economiques and the Institut de Géographie of the Université de Paris. The governments of the District and the City of Paris and of the Department of the Seine subsidised the atlas financially and also provided substantial research assistance. The completed work comprises some 350 major maps, 961 pages of explanatory text, which complements and elaborates on the maps, making the whole a geographical study of the first order. Innumerable smaller maps and charts and statistical information are inserted throughout and there are bibliographical references at the end of each section. Particularly interesting is the examination of population patterns and the relationship of Paris to the rest of France, with an emphasis on economic factors.

94 'Atlas de Venezuela', published by the Republica de Venezuela Ministerio de Obras Publicas, Direccion de Cartografia Nacional, 1969, is a beautiful production, finely planned on the basis of historical cartography, current cartography, Caracas and Federal maps, thematic maps dealing with physical features, population, economy, communications, town plans; minor maps, sketchmaps, diagrams, photographs and statistics are placed throughout the work, with explanatory notes. Aerial photographs highlight the variety of relief. Included is a summary of the progress of mapping in Venezuela, accompanied by delicately reproduced early maps, for example, the Blaeu map of 1635. The use of colour is bold and imaginative, particularly in the town plans and every kind of skill has been brought to the whole production. There is a bibliography.

95 'Atlas de Schweiz', the national atlas of Switzerland, is in continuous publication at Berne, by Eidgenössische Landestopographie, 1965- . The atlas, edited by Professor Imhof, with the co-operation of numerous authorities, is being produced in the highest standard of Swiss cartography, using the large-scale maps of Switzerland recently completed; explanatory notes and statistical tables accompany each map.

96 'Atlas du Maroc', prepared by the Comité National de Géographie du Maroc (Rabat: Institut Scientifique Chérifien, 1968-), is still in progress. When complete, the atlas is planned to consist of eleven sections and fifty-four sub-sections, covering all aspects of the physical and human geography of the country. The maps so far produced are detailed and attractive; they, together with the accompanying text, have already added greatly to knowledge of the Moroccan economy.

97 'Atlas for anthropology', compiled by Robert F Spencer and Eldon Johnson (Dubuque, Iowa: William Brown, 1960), provides a useful collection of folded maps in black and white, held within a spiral binding; they deal with cultural and tribal groups and language families of the world, with sections devoted to the Old World and to the New World and the racial distribution of mankind.

98 'Atlas général Larousse' was published in 1959 and distributed in Great Britain by Harrap. Seventy-two full or double plate maps are printed in six colours, in addition to many in two colours. City plans are included, also statistical tables, commentaries and a name gazetteer of some fifty-five thousand entries. Historical maps comprise fifty-five pages and thirty articles by eminent historians are a valuable feature.

99 'Atlas général Vidal-Lablache', first produced in 1894, with further issues in 1909, 1918, 1922, 1933, 1938 and 1951, is a fine example of French geography and cartography, under the direction of Paul Vidal de la Blache and Emmanuel de Martonne. The 1951 edition (Paris: Armand Colin) carries an increased number of maps in two parts, 'cartes historiques' and 'cartes géographiques', with an index gazetteer of thirty-one pages.

100 'Atlas historique (Atlas Belfram): *Provence, Comtat, Orange, Nice, Monaco'*, compiled by E Baratier and others (Paris: Colin, 1969), numbers more than three hundred maps, plans of towns and buildings and textual commentary. Patron of the new atlas was President Pompidou and contributions were selected from among archaeologists, prehistorians, economists, sociologists, geographers and philologians, working under the direction of R H Bautier, of the Ecole des Chartes. Introductory maps present an outline of the geological, physiographic and archaeological background and a number of specialist studies enhance the reference value of the work. Also included are a biographical repertory and topographical dictionary.

101 'Atlas internationale Larousse', a useful reference atlas, in a second edition, 1957, contains nearly seventy folding maps, with accompanying text in French, English and Spanish.

102 'Atlas mira': an outstanding event in the development of Soviet cartography was the publication of the *Great Soviet world atlas,* 1937-40, on which a specially organised research committee had worked. The atlas was integrated in character, containing political, historical, topographical and special-purpose maps. On the outbreak of war in 1939, the two volumes so far available were withdrawn. A new atlas, edited by A N Baranov and others, was issued in its place, published in 1954 by the Chief Directorate of Geodesy and Cartography, Moscow. The USSR is covered comprehensively and other countries are well represented. The emphasis is on relief maps, plus general political and communications maps. Economic maps are not included; these may be found in the world atlas designed for school use, the *Geograficheskii atlas,* 1954. Maps of areas outside Russia are arranged by continents; each section includes general maps of politico-administrative divisions, communications and physical features, followed by detailed regional maps. Nearly all major towns and areas of particular interest have been given inset maps. The index comprises more than two hundred thousand entries. In 1962 and 1967 the State Publishing House for Geodesy and Cartography, in connection with the Ministry for Geology and Natural Resources of the USSR, brought out new editions of the great atlas, with additional economic maps, those showing natural resources and power being especially valuable. The section covering the USSR forms, in effect, a national atlas. A medium-sized *Atlas mira,* edited by S I Shurova and others, was published in Moscow, 1959; and a small *Atlas mira,* in a third edition, 1958, is a handy atlas of political maps on smaller scales, edited by I M Itenberg and others. A pocket edition of the great atlas is available also, and still another *Atlas mira* is a military atlas, published in Moscow, 1958.

103 'Atlas nacional de España', prepared at the Instituto Geográfico y Catastral, Madrid, 1965, is in loose-leaf form, provided with a box container. Topographic and thematic maps are large and very fine, the subject matter being detailed, yet clearly visible, due particularly to the judicious choice of several type faces, of symbols and colouring. Accompanying the atlas are two bound volumes, an index of placenames and an excellent review of the geography of Spain, entitled 'Reseña geográfica; the latter following a general introduction, is divided into main sections: Orography, Coastal

Regions, Frontiers, Hydrography, Natural regions, and Spanish Territories in North Africa. Each of these sections includes many subdivisions, the subject content being linked with the relevant map sheets.

104 'Atlas narodov mira' (Atlas of the peoples of the world), edited by S I Bruk and V S Apenchénko (Moscow: Glavnoye Upravleniye Geodezii i Kartografii Gosudarstvennogo Geologicheskogo Komiteta sssr and Institut Etnografii im N N Miklukho-Maklaya Akademii Nauk sssr, 1964), shows the distribution of ethnic groups throughout the world magnificently presented on 106 coloured sheets. A series of world maps portrays the distribution of states, density of population, the ethnic groups, languages and races of mankind, followed by ninety-seven regional maps showing in detail the contrasts between the density and distribution of population between countries or parts of countries. The Soviet Union is given particular attention, but there is an adequate treatment of the rest of the world. An English translation of the table of contents and the legends on the plates has been issued in a separate volume by the Telberg Book Corporation of New York. The colours and symbols on the maps were carefully considered to suggest relationships; description of the methods and much of the information had been previously published by the Institute of Ethnology. Statistical tables show details of 910 ethnic groups, as of 1961, arranged by continents and countries, also data on ethnic structure in each country of the world. An alphabetical index of 1,600 peoples gives the English equivalents of Russian names and the documentation of the whole work is excellent. Reference should be made also to [Numbers and distribution of the peoples of the world], also edited by S I Bruk, 1962, including maps, bibliographies, more detailed statistical tables and textual material, arranged as in the atlas; 1959, or the nearest year to it, has been taken as the base—for example, for India, the censuses of 1931 and 1951 were used.

105 'Atlas of Alberta', compiled by the Department of Geography, University of Alberta, was edited by T A Drinkwater and others (University of Alberta Press and University of Toronto Press, 1969), one of several atlases of the Canadian provinces recently published or in preparation. In many ways, the overall plan follows the conventional regional form and it is a true atlas also in the sense that it includes no text apart from the preface. Relief and geology, climate, water, vegetation, soil and wild life are followed by history and population, land use, agriculture, forestry, minerals, power, manufacturing and service industries, including transport and communications. There is an obvious concentration on the analysis of settlement patterns, emphasising the fact that in Alberta large areas have a small and scattered population. Double page maps are at the 1:2M scale, those on single pages are as 1:3,300,000 and there are many smaller scale maps. Production of the atlas is very fine and the symbols and colour ranges have been carefully selected.

106 'Atlas Antarktiki', a loose-leaf atlas with a cover-to-cover English translation, edited by Dr U G Bakaev and others (Glavnoe Upravlenie Geodezii i Kartografii, 1966), consists of 225 plates carrying more than three hundred maps, charts, diagrams and comprehensive legend. The outline runs as follows: Historical matter; General aspects; Aeronomy and Physics of the Earth; Geology and Topography; Climate; Waters of the Pacific Ocean. Seven regions are distinguished, of which the maps are in a separate portfolio. Layer colouring is used to portray relief and the colouring systems throughout are very fine.

107 'Atlas of the Arab world and the Middle East' was published by Macmillan in 1960. The maps, prepared by Djambata, Amsterdam, cover physical features, political data, climate and natural regions, town plans and settlement types. North Africa is included, Libya, Egypt, Sahara region of the UAR, the Nile region, the Sudan, Near East, Syria, Northern region of the UAR and Lebanon, the Jordan region, with a separate map of Jerusalem at 1:22,500, the coastal zone of Lebanon and the region of Syria, Iraq, the Arabian Peninsula, Iran and Turkey. Endpapers depict ' The spread of Islam in the early twentieth century: refugee camps in the Near East ' and ' Middle East oil concessions and the world of Islam in the Middle Ages '. There is a useful introduction and text, with photographs, throughout the work. The maps are not too cluttered with detail; the river systems and irrigation projects being especially well executed.

108 'Atlas of Australian resources', the first comprehensive atlas of Australia, was prepared by the Department of Natural Development, Canberra, Commonwealth Government of Australia, and edited by Dr Konrad Frenzel (Angus and Robertson, first series 1952-1960). Thirty plates, with accompanying text, were issued as loose sheets, mounted or unmounted, and with a binder or box container, as required. The base map was on a polyconic projection, scale 1:6M. The sheets cover structure, geology, physical features, drainage and climate; land use and agricultural production; transport, all aspects of population, industries and manufactures; a special feature being a chart of 'Major developmental projects'. Commentary by a specialist accompanies each large map, illustrated by informative diagrams and special maps. In 1961, work began on a second series, planned for continuous revision, replacing the original thirty map sheets. Projections were changed as necessary and drawing and reproduction techniques improved where possible. The reverse of some sheets was used for additional maps and lists of references; a special booklet accompanies each map sheet.

109 'An atlas of Australian soils' is being published by the Melbourne University Press in association with the Soils Division of the Commonwealth Scientific and Industrial Organization. Planned in ten sheets, the first sheets to appear were those for the Port Augusta-Adelaide-Hamilton area, 1960; the Melbourne-Tasmania area, 1962; Canberra-Sydney-Bourke-Armidale, 1966. Handbooks compiled by K H Northcote accompany each sheet; the maps are based upon his scheme of classification of Australian soils, using a system of primary profile forms, subdivisions, subsections and classes. The complete atlas was exhibited at the Ninth International Congress of Soil Science at Adelaide, 1968.

Refer K H Northcote: 'A factual key for the recognition of Australian soils' (CSIRO Divisional Report, 1960, second edition, 1965).

C G Stephen: 'The soil landscapes of Australia' (CSIRO, Soil publication, no 18, 1961).

110 'Atlas of birds in Britain' is being prepared by the British Trust for Ornithology. The fieldwork for the atlas is being carried out by voluntary observers who, since 1968, have been plotting on a ten km National Grid square basis, as was done in the case of the *Atlas of the British flora (qv)*. The atlas should be completed by 1972.

111 'The atlas of Britain and Northern Ireland' (Oxford: Clarendon Press, 1963) was a publication of the greatest cartographical interest. In the words of the publishers, the atlas is ' a statement, on a general basis, of modern Britain's resources, physical, economic and industrial—a complete and ordered portrait of this country from the rocks beneath to the industry above '. Historical geography is not included. Data are taken from the 1951 census, from 1955 agricultural and fisheries figures, with 1948-57 averages, and from 1954-56 figures for the majority of the industrial maps; the latest information possible has been incorporated where this is particularly vital, as with atomic energy and air traffic. A fold-out section at the back gives details of authorities and sources. While preserving the traditional form, an original approach is revealed, both in the four-fold overall design and in the treatment and colouring of individual maps. The standard scale adopted is 1:2M, which allows the whole area to be shown on one page; this is also the scale of the very useful ' transparent reference overlay '. The 1:1M scale and the 1:500,000 are used for successively greater detail. The larger-scale maps and the overlay are marked with the national grid, which is used also in the gazetteer. Imagination is one keynote of the atlas, revealed in the choice of subjects—for example, the map of coastal relief—and in the variety of cartographic techniques and use of colour. Another key-note is the practical approach throughout, the constant invitation to purposive use, shown, for example, in the placing of pairs of maps for comparative study on facing pages, the additional information on the map sheets both in word and diagrammatic form, the notes at the end and the evocative use of symbols.

112 'Atlas of the British flora', edited by F H Perring and M Walters (Nelson for the Botanical Society of the British Isles, 1962), consists of black and white distribution maps portraying 1,700 flowering plants and ferns; included also are twelve transparent overlays concerned with climate, topography and geology. The text begins with an introductory history of the mapping of plant distribution, followed by an outline of the method used in preparing the maps for this work and the interpretation of the maps.

113 'Atlas of Canada', in its current form, undertaken by the Geographical Branch of the Department of Mines and Technical Surveys and published in 1959 (dated 1957), is a superb achievement covering all aspects, with special emphasis on economic factors— ' an outline of the physical background and the economic development of the nation at mid-century' (Foreword). Particularly useful is the series of urban land use maps, linked with the World Land Use Survey of the International Geographical Union. The previous official atlas of Canada was published in 1906 by the Department of the Interior of the Canadian Government, revised and enlarged in 1915.

114 'Atlas of Denmark' is a particularly excellent example of co-operation in the production of a national atlas, by geographers associated with the University Geographical Laboratory, a number of firms and individuals, the Royal Danish Geographical Society, the Carlsberg Foundation and the Danish Government. Edited by Niels Nielsen, in five volumes, 1949- , the atlas includes information on historical evolution and occupational changes, together with detailed treatment of the ten sections into which the country has been divided and accompanying text in Danish and English. Air photography has been used to great effect.

Refer W M Gertsen: 'Danish topographic mapping', *in The cartographic journal,* December 1970, which is illustrated by reproductions of map excerpts.

115 'Atlas of diseases', produced in sheets, 1950- , by the American Geographical Society, under the direction of Dr Jacques M May, was issued with the quarterly *Geographical review,* in which there appeared descriptions supplementary to the annotations on the map sheets themselves. The maps include world distribution of major diseases such as cholera, leprosy, and two studies of human starvation: 'Sources of selected foods' and 'Diets and deficiency diseases'. With each map are bibliographies and references; in addition, Dr May's three-volume work, *Studies in medical geography,* 1958, 1961, should be used in conjunction with the maps. The maps portraying studies of poliomyelitis and cholera data and of the distribution of malaria and yellow fever all demonstrate different fields of research in which progress has been encouraged by the application of techniques demonstrated here. The First Report of the Commission on Medical Geography (Ecology of Health and Disease) of the International Geographical Union was presented to the seventeenth International Geographical Congress, 1952, by Jacques M May; in it, he considers the study of diseases and the progress of their correlation with the physical environment. He discusses the presentation of such data on maps, mentions the Society's *Atlas of diseases* and future research in the field.

Refer also L D Stamp: *The geography of life and death,* 1964. *Studies in medical geography,* which have been published irregularly from 1958 by the American Geographical Society.

116 'Atlas of Edinburgh', published by the Edinburgh Branch of the Geographical Association, 1965, comprises thirty-nine pages of maps, text and illustrations in a paper cover, with a spiral binding; the material was provided by members of the Branch, mainly by teachers of geography in Edinburgh schools, organised and introduced by Professor J Wreford Watson, including a brief outline of ' The rise and growth of Edinburgh '. All aspects of physical features and climate, land use, social and statistical data are incorporated and the maps are interspersed with annotated photographs, graphs, diagrams and drawings.

117 'Atlas of European birds', compiled by K H Voous (Nelson, 1960), originally published earlier in the year as *Atlas van de Europese vogels* by Elsevier of Amsterdam, was translated by Professor Voous into English. For each species of bird native to Europe there is a map showing its breeding range, not only within the continent but extra-limitally also. The fine collection of 355 photographs covers the essential features of the bird and each bird is also described, giving factual information of habitat, characteristics, nesting and migration, all from a geographical point of view.

118 'Atlas of European history', edited by E W Fox and H S Deighton (OUP, 1969), contains two types of maps. The first are concerned with topography, expressed by layer colouring and shading, as it has formed the basis for the location of historical settlements and other developments. These maps have purposely been kept as clear of detail as possible, so that only the really important names can be quickly assimilated and remain in the mind's eye. Secondly, there are the historical thematic maps, either inset with the physical maps or as separate sheets. Good colour arrangements in the land areas set off the grey sea and allow details to stand out well, such as the changing political condition of Europe, shown in bold black. The editor states—' The first requirement for this Atlas, then, was that it should provide information ... as simply and clearly as possible '. There are several interesting innovations; for example, the double page of ' The great explorations ' and ' Renaissance Italy '.

119 'Atlas of evolution', compiled by Sir Gavin de Beer (Nelson, 1964), consists of 202 pages of concise text including many line diagrams, photographs and drawings, in which the bold use of colour makes an immediate visual impact. The quality of the map content could be improved, but the printing, done in Holland, is very fine; there is a bibliography and an adequate index.

120 'Atlas of Finland', first published by the Geographical Society of Finland in 1899, was issued ' to assist the people of Finland to know themselves and their country '. A second edition appeared in 1911 and the third, 1925-28, comprised two volumes, thirty-eight plates of maps and a volume of text. The fourth edition, compiled by the Geographical Society and the Institute of Geography of the University of Helsinki under the direction of Leo Aario, was completed in 1960, the cartographic work and reproduction of the plates being done by the General Survey Office. The atlas contains 445 maps and diagrams, on scales ranging from 1:1M to 1:9M, mostly on double-page sheets; comparison of the maps is facilitated, as groups of related maps are shown together. The atlas is a splendid example of a national atlas, covering all aspects of the nature and economy of the country and, from the first, has been of special interest as a survey of a northern country in detail and at such uniformly high standard. The meteorological section, for example, includes frost data and the section on forests and water power is of great practical value. Population density and analysis are particularly well treated, as are agriculture and industry. Throughout, the atlas gives an indefinable impression of enthusiasm and pride of country. Experimental symbols have been used with effect. There is a foldout ' Communes ' map, with inset 'Administrative provinces, 1960 ', which can be kept open while consulting other sheets. The notes in the 1960 edition are briefer than those for former editions; in this edition, all references to matters not presented in the atlas itself have been omitted, since the publication of *Suomen Maantieteen Käsikirj*—Handbook of Finnish Geography, 1936 and 1951, there has been no need for such copious texts. Also, *Explanatory notes,* a separate volume published in 1962, is in English. Legends and marginal information on the map sheets are in Finnish, Swedish and English and there are detailed contents pages. The Preface was translated into English by Richard Ojakangas and others.

121 The 'Atlas of Florida', by Erwin Raisz and associates with text by John R Dunkle (Gainsville, 1964), was prepared in the Department of Geography of the University of Florida. It contains 57 pages of colourful information and three supplementary pages of statistics. An enterprising range of methods is used to present this State, 1845-, which has had the fastest growing population of any in the Union; time charts, bar graphs, divided circles and flow diagrams complement quantitative and non-quantitative maps, accompanied by notes of past events and present scenes; a popular atlas, significant of the current trend to reproduce local characteristics geographically.

122 'Atlas of Germany', see *Die Bundesrepublik Deutschland in Karten* and *Deutscher Planungsatlas*.

123 'Atlas of glaciers in South Norway', prepared by G Ostrem and T Ziegler (Oslo: Hydrologisk Avdeling, 1969), is available from the Hydrological Division, Norwegian Water Resources and Electricity Board, Oslo. The work, compiled as a result of a glacier inventory carried out in accordance with a resolution of the International Hydrological Decade, contains maps of each drainage basin, showing the glacier outline and existing water gauges. A description is given of the various areas, together with photographs of typical glacier types. Text is in Norwegian and English.

124 'Atlas of the Great Barrier Reef', compiled by W G H Maxwell (Elsevier, 1969), is mainly a geological appreciation of the Australian Great Barrier Reef province, illustrated with maps, photographs and figures—a book, rather than an atlas proper. Following an introductory chapter, chapters II and III provide short and conventional reviews of the morphology and structure of the South West Pacific, which are amply supported by maps and cross-sections, together with a summary of the tectonic framework and evolution of eastern Queensland and its continental shelf. Chapter IV presents a series of detailed maps and cross-sections showing the tectonic evolution of the continental shelf and continental islands. There follows a long chapter VI on the morphology and distribution of reefs, with maps, cross-sections, aerial and underwater photographs and the final chapter summarises the results of extensive fieldwork and data analysis, accompanied by a group of maps covering the one thousand miles of Queensland's broad continental shelf.

125 'Atlas of the history of geographical discoveries and explorations', edited by K B Martova and others (Moscow, 1959), prepared under the direction of an editorial committee, of which the Chairman was K A Salishchev. Ninety-two pages of maps are especially valuable for coverage of Russian exploration.

126 'Atlas of Islamic history', compiled by Harry W Hazard, with maps executed by H Lester Cooke, jr, and J McA Smiley (Princeton University Press, 1954, third edition, in the *Oriental studies* series, volume 12), is intended, from the first appearance of the work in 1951, for students, businessmen, government officials and all concerned with Near and Middle East affairs. It assembles material otherwise scattered and the historical analyses in the third edition are brought up to date to 1953. The Islamic world from Morocco to Iran is the area considered in the main part of the work, from the seventh century—the first Islamic century—to the twentieth century, the thirteenth Islamic century, with notes on Moslem sea power, the Moslem calendar, the Crusades and the Ottoman Empire. A final section deals with Islam in the Middle and Far East, with India, Central Asia, South-eastern Asia and Indonesia. Each map is faced by explanatory text; there are conversion tables and an index.

127 'Atlas of Israel', the national atlas, published in a second (English language) edition, 1970, was designed to combine and present in concise form a vast amount of information about every aspect of the country. The contents span the whole range of the history of the Holy Land from pre-historic times to the present. An official work of the State of Israel, prepared by the Survey and Elsevier Publishing Company, Amsterdam, it is the most reliable, accurate and comprehensive work on Israel yet produced, the result of co-operation by a large team of eminent scientists, economists and historians from the Hebrew University at Jerusalem, working in concert with the staff of the Department of Surveys and other Israeli government institutions. The first (Hebrew) edition was published sheet by sheet between 1956 and 1964 by the Department of Surveys, Ministry of Labour (Survey of Israel) and the Bialik Institute of the Jewish Agency. Sub-titles were in English. In compiling the maps for the English edition, coverage was extended in many cases to include the entire area under Israeli administration since 1967; it includes Judea and Samaria, as well as the Gaza Plain and the Golan Heights. Production was given the most intensive consideration. The 120 gram paper was manufactured in the Hadera Paper Mill, the inks were specially made in Haifa and the colours mixed at the Reproduction Department of the Survey of Israel. The double-page sheets are grouped as follows: Cartography, Geomorphology, Geology, Climate, Hydrology, Botany, Zoology, Land Utilization, History, Population, Settlements, Agriculture, Industry and trade, Communications, Services, such as cultural features and tourism. Each section is sub-divided and each introduced by concise, informative text. A number of the historical maps have never appeared before. One of these is a half-page map of Palestine as it was under the Arabs, 640-1099, another that of Palestine under the Mamluks, 1291-1516 and a third of the period 1516-1900. The spread of Christianity is clearly recorded and there are maps showing archaeological excavation sites, by period. The Preface outlines the approach to the atlas, transliteration of place-names, a short pronunciation guide and a description of the early maps, especially of *The Medeba map,* with reproductions, including 'Map of the Palestine Exploration Fund' and 'British maps of the early twentieth century', each with a bibliography.

128 'Atlas of London and the London region' was prepared by Emrys Jones and D J Sinclair at the London School of Economics and Political Science (Pergamon Press, 1968). The total of seventy maps fit into a container. Work on the atlas had been in progress by a team of experts during the previous seven years, beginning with a programme of research into socio-economic topics, supported by the Anthropological and Geographical Research Division of the School. In the earliest stages, further financial assistance was given by the Frederick Soddy Trust. In 1965, a grant from the Department of Economic Affairs made actual production practical and this was later backed by an additional grant from the Greater London Council. The maps in the atlas are at various decimal scales, showing a series of areas from Central London to the whole of south-east England. The statistical units at which the demographic and economic data are mapped vary according to the scales. The choice of topics was governed in part by the funds available. All the maps are on DD standard sheets, so that they can be stored flat, once folded. Printing is on one side of the sheet only and each map is accompanied by a brief text panel and explanatory key. Two sections, one on physical background and the other showing historical background, preface the main body of the atlas, which deals with aspects of population distribution, density, housing standards, land utilisation, industry, employment and transport. Such sources as the 1961 Census, the 1966 Sample Census and the London Traffic Survey were used, in addition to newly gathered data. A base map is printed in pale grey; in all the other maps, colouring is bold and, probably, controversial.

129 'Atlas of meteorology: *a series of over four hundred maps . . . ,*' compiled by J G Bartholomew and A J Herbertson and edited by Alexander Buchan, was issued under the patronage of the Royal Geographical Society (Bartholomew, 1899) as volume III of Bartholomew's *Physical atlas*. Still basically useful and evocative of ideas, its information must now be supplemented by reference to, for example, the us Navy *Marine climatic atlas of the world,* in four volumes, 1955-1958; the *Climatological atlas of the world,* by H Walter and H Lieth, published in a binder by G Fischer, Jena, as the first part of a three-volume work; and the *World map of climatology,* by H E Landsberg and others, edited by E Rodenwaldt and H Jusatz (Springer-Verlag, 1963; second edition 1965).

130 'Atlas of New Zealand geography', compiled by G J R Linge and R M Frazer (A and H Reed, Wellington, 1966), contains sixty-one black and white maps and thirty pages of text. The arrangement of topics treated runs from geology and surface relief, climate and soils, agriculture and land use to population and industry; a series of topographic maps illustrates particular regions and there are special land use and industrial maps of Auckland. There are many insets and diagrams. Each map is explained by a page of text and bibliographies, including articles from journals, are included for each topic.

131 'An atlas of North American affairs', compiled by D K Adams and H B Rodgers (Methuen paperback, 1969), is a handy reference volume featuring topics of current interest, such as the problem of ' Megalopolis ', as well as traditional subjects—climate, population, etc. All the maps are monochrome, varying in scale with the subject, some having insets. The Canadian part of the continent is not comprehensively covered and Alaska is confined to an inset. A recurrent theme is the transformation of the economies of both Canada and the United States by large-scale industrialisation.

132 'Atlas of physical, economic and social resources of the Lower Mekong Basin' (New York: United Nations, 1968) was prepared under the direction of the United States Agency for International Development, Bureau for East Asia, by the Engineer Agency for Resources Inventories and the Tennessee Valley Authority for the Committee for Co-ordination of Investigations of the Lower Mekong Basin, the United Nations Economic Commission for Asia and the Far East. Areas covered are Cambodia, Laos, the Republic of Viet-Nam and Thailand. This unique collection of maps provides an inventory, in cartographic and narrative form, of the natural and man-made resources within the drainage basin of the Mekong. The finished product represents years of cumulative effort by many organisations and individuals in the fields of mapping, aerial photography and geography. Individual maps will be periodically reviewed and a companion bibliography is in process of preparation.

133 'An atlas of population change 1951-66—*the Yorkshire and Humberside Planning Region'*, compiled by D G Symes and E G Thomas, with R R Dean (Hull University, Department of Geography, Miscellaneous series, no 8, 1968), presents a successful example of contemporary work in population mapping. Here are seventeen folded maps portraying the demographic analysis of the 1961 Census statistics compared with population changes during 1951-1961. The mapping symbols of three maps are the now familiar proportional spheres; the rest of the maps are compiled on an experimental system of choropleth mapping. There is little explanatory text. Local authority boundary lines are printed on a transparent overlay.

134 'Atlas of Saskatchewan' edited by J Howard Richards (University of Saskatchewan, 1969), provides a wealth of information put together by a number of specialists, with K I Fung as cartographic editor. The statistical maps are interesting for the mathematical techniques used in their compilation. Text is included on each aspect treated and the whole atlas is most attractive and carefully documented.

135 '**Atlas of social and economic regions of Europe**' is in process of preparation by the Soziographisches Institut an der Universität Frankfurt am Main, under the direction of Ludwig Neundörfer (Verlag August Lutzeyer, Baden-Baden, 1964-), under the aegis of the Council of Europe and with the assistance of the government of the Federal Republic of Germany and the Deutsche Forschungsgemeinschaft. Intended as a basis for the social and economic planning of Europe in the second half of the twentieth century, the work is being issued in instalments, in separate folders, each containing a hundred maps, accompanied by texts and keys in German, English and French, detailed gazetteers and lists of statistical sources.

136 '**Atlas of South-East Asia**', with introduction by D G E Hall (Macmillan: St Martin's Press, 1964) and maps by Djambatan, Amsterdam, consists of sixty-four pages of maps portraying the physical characteristics and the political and land-use conditions applying in the Philippines, Indonesia, Singapore, Malaya, Thailand, Indochina and Burma. Scales range from 1:25,000 for the town plans to 1:40M for general and distribution maps. Historical maps have been used as endpapers. Photographs accompany the explanatory text.

137 '**Atlas of Tanzania**', compiled, published and printed by the Surveys and Mapping Division, Ministry of Lands, Settlement and Water Development, Dar-es-Salaam, 1967, replaces the *Atlas of Tanzania* of 1957. The aim of the later work is stated in the Foreword—' not as a luxury or an academic publication . . .' but to present ' Tanzania's human and natural resources for use in development planning'. The maps are printed in many colours on high quality paper; there are some omissions compared with the previous atlas and some new material has been added, such as the maps of potential land use and fisheries. Explanatory notes face the maps and it would be helpful to refer to some of the *Research papers* and *Notes* issued by the Bureau of Resource Assessment and Land Use Planning at the University College, Dar-es-Salaam, also to the reports of the second Five Year Development Plan, 1969-74.

138 '**Atlas of the Union of South Africa**', prepared by A M and W J Talbot in collaboration with the Trigonometry Survey Office and under the aegis of the National Committee for Social Research

(Government Printer, 1960) is in English and Afrikaans. This is a comprehensive atlas, covering numerous aspects of South Africa; maps show relief, geology, mining, soils, vegetation, climate, water resources, agriculture, industries, population, occupations, transportation and trade. The introductory texts are valuable, illustrated by sketchmaps, tables and diagrams.

It may be helpful to refer also to the smaller *Road atlas and touring guide of Southern Africa,* published by the Automobile Association of South Africa, in a second edition, 1963. In addition to the detailed maps, diagrammatic maps are included of the entrances and exits of more than two hundred towns.

139 'Atlas of the United States': *see 'National atlas of the United States'.*

140 'Atlas of the universe', compiled by Patrick Moore (Mitchell Beazley with George Philip, 1970), includes such information on the history of scientific thought and discovery, on physics and astrophysics, celestial mechanics, chemistry, geology and technology as will help in the understanding of the universe in which the earth is but one planet. Maps are accompanied by photographs, graphs and diagrams, grouped in five sections: ' Observation and exploration of space '; 'An atlas of the earth from space '; ' The Moon '; ' The solar system '; and ' The stars '. Each section includes explanatory text, illustrations and captions. A foreword was contributed by Sir Bernard Lovell.

141 'Atlas of Western Europe', compiled by Jean Dollfus and an advisory committee drawn from several European universities, has been issued in English by Murray and in French, German, Italian and Dutch by the Istituto Geografico de Agostini, Novara. The atlas has been constructed with the Common Market in mind. Paul-Henri Spaak has contributed a preface. Twenty-four colour maps and three in monochrome, accompanied by illustrations, depict the relief and climate, geological and mineral deposits, aspects of demography, land use, communications and industries, followed by an index of principal place-names and their equivalents. The monochrome maps are of 'Administrative divisions ', ' European Community-Council of Europe-NATO ' and ' Europe and Africa '.

142 'Atlas of the world commodities, production, trade and consumption: *an economic-geographical survey',* by Olof Jonasson and

Bo Carlsund (Göteborg: Akademiforlaget/Gumperts ...; Scandinavian University Books, 1961, for the Göteburg Graduate School of Economics, where Olof Jonasson is Professor), contains maps of the staple goods in world trade, with major emphasis on animal and vegetable raw materials, also minerals, coal and petroleum. The section on international sea-borne shipping is most useful. The maps are diagrammatic, in black and white, accompanied by brief text and statistics.

143 'An atlas of world history', compiled by S de Vries and others (Nelson, 1965), comprises sixty-four maps, each designed to clarify one topic, in black and white and with textual commentary. There are useful plans and a section 'Battlefields of antiquity'.

144 'Atlas of the world's resources', issued by the Department of Geography, University of Maryland, in three volumes (Prentice-Hall, 1952-1954), with notes and accompanying texts, under the chief editorship of William van Royen, consists of: I The agricultural resources of the world; II The mineral resources of the world; III Forest and fishery resources of the world.

145 'Atlas Östliches Mitteleuropa' was prepared by T Kraus, E Meynen, H Mortensen and H Schlenger (Velhagen and Klasing, 1959). It is a superb, loose-leaf production, very well documented, containing clear maps, the product of fine cartography and a most judicious use of colour. The regional maps, at the scale 1:300,000, are in a pocket at the end and a set of aerial photographs complements the maps.

146 'Atlas över Sverige' a magnificent atlas giving a clear picture of the country's natural and cultural geography, population and economy, trade and port traffic, was first prepared in 1900, with a new edition, published in sheets, 1953- by Svenska Sällskapet för Antropologi och Geografi, Stockholm (Kartografiska Institutet, General-Stabens Litografiska Anstallt, Stockholm). Text is in Swedish, with English summaries, map titles and legends.

147 'Atlas Porrua de la Republica Mexicana' was prepared in collaboration with Professor Jorge Hernandes Millares and Professor Alejandro Carillo Escribano (Editorial Porrua, S A, Mexico, 1966). Included are physical maps, general and historical monographs concerning the federal states and black and white plans of the main settlements. Vital statistics are usefully gathered together and there are some fine photographs and an explanatory text.

148 'Atlas van Nederland' is a fine example of the work of the Topographic Service, Delft, which is responsible for the production ('s-Gravenhage: Government Printing and Publishing Office, 1963-   ). Work began in 1963, following planning by the Ministry of Education, the Royal Netherlands Geographical Society and the Society for Economic and Social Geography, and the complete atlas is to contain 109 sheets, loose-leaf in a sturdy binder. Most sheets carry Dutch and English explanatory texts, sometimes illustrated with diagrams. The sheets so far issued represent all sections of the atlas, beginning with a sheet demonstrating the history of cartographic techniques in the Netherlands by sixteen fragments of printed maps from the sixteenth century to the present, compiled by Dr C Koeman. Contemporary social and economic conditions are meticulously mapped, with special inset maps and diagrams as required. The treatment of soils is particularly outstanding. The clear colouring throughout is striking and aesthetically pleasing, both in all the thematic maps and in the relief map, ' The Netherlands and surrounding countries' on 1:500,000, in which eight layer colours are used, with contouring, to show land relief, and five colours to show sea depths and tidal flats. Great skill has been shown in the selection and placing of names to avoid over-crowding.

149 'Atmosphere, weather and climate', by R G Barry and R J Chorley (Methuen, 1968), aims to bring together the latest ideas on synoptic and dynamic climatology in a concise text, generously illustrated with sketchmaps and diagrams of all kinds. The first three chapters deal with the nature of the atmosphere, continuing with a discussion on the air masses and the processes which lead to the development of frontal and other depressions and concluding with a brief consideration of the modifications of climate produced by urban and forest environments and of the inherent variability of climate with time. Appendix I presents a summary of the major schemes of climatic classification for reference purposes and Appendix II sets out ' Nomograms of height, length and temperature'. There is a short bibliography.

150 'Atmospheric tides, thermal and gravitational', by Sydney Chapman and Richard S Lindzen (Dordrecht-Holland: D Reidel

Publicity Company, 1970), summarises the latest known facts and theorises on the travelling waves and daily oscillations caused by the sun's heat and the moon's attraction acting on the rotating earth's atmosphere, beginning with a historical introduction. Data, methods of data analysis and the mathematical methods for solving the theoretical equations are all discussed.

151 'Australia: a study of warm environments and their effect on British settlement', by Griffith Taylor, was published by Methuen in a seventh edition, revised 1958, reprinted 1961, in the *Advanced geographies* series. Sections have been added on central Australia, on Australian Antarctica and recent industrial growth.

Refer also Erwin and Gerda Feeken: *The discovery and exploration of Australia* (Nelson, 1971).

152 'Australia, New Zealand and the South Pacific: *a handbook'* (Anthony Blond, 1960), was prepared by fifty distinguished contributors under the editorship of Charles Osborne. Nine articles deal with the physical conditions and characteristics of the States and Territories of Australia; eleven deal with the economy. Social patterns form a major topic. Three sections are given to New Zealand—'Political background', 'The economy' and 'Social patterns'. Much informed comment is included by C H Newbury on the Pacific Islands, viewed in their relation to Australia and New Zealand.

Note: Further volumes in this series, *Handbooks to the modern world*, are entitled *Africa, Western Europe, Latin America, The Soviet Union, Eastern Europe, The United States and Canada* and *The Middle East*.

153 'Australian bibliography: *a guide to printed sources of information'*, compiled by D H Borchardt (Melbourne: F W Cheshire, 1963, second revised edition, 1966), covers library catalogues and general retrospective bibliographies, current national newspapers, bibliographies of subject areas, regional bibliography, government publications and the state of bibliography in Australia. A list of works referred to is included.

154 'An Australian bibliography of agricultural economics 1788-1960', compiled by J L Dillon and G C McFarlane (Government Printer's Office, New South Wales, 1967), includes some ten thousand items, arranged within three sections: a listing classified by subject matter, an author listing and a subject listing of major statistical sources.

155 'The Australian encyclopedia', in ten volumes, edited by A H Chisholm and others, was published by the Michigan University Press (Sydney: Angus and Robertson, 1958). Particularly detailed treatment is given to the Australian aborigines and those of Papua, New Guinea and New Zealand.

156 'The Australian environment' prepared by CSIRO (Melbourne University Press, 1960; third edition, reprinted, 1966), has become a standard reference work, owing its origin to an agricultural conference, 1949, on plant and animal nutrition. In the revised, illustrated editions, Australian plant and animal life is examined against the background of the environment.
> *Refer also* Allen Keast: *Australia and the Pacific Islands: a natural history* (Hamish Hamilton, 1966).

157 'Australian geographer', journal of the Geographical Society of New South Wales, has appeared twice a year since 1928; it contains original articles and book reviews. Annual and approximately three-yearly indexes are prepared.

158 The Australian Geography Teachers Association, founded in 1969, is a national body aiming to promote the teaching and study of geography and to provide a forum for Australian geography teachers; its journal, *Geographical education,* also began, on an annual basis, in 1969.

159 Australian Institute of Cartographers has the main headquarters at Melbourne, where also is the headquarters of the Victoria division. The New South Wales division is based on Sydney, the South Australian division on Adelaide and a Tasmanian division is centred at Hobart. The Institute's journal is *Cartography (qv).*

160 'Australian maps' has been published since January 1968 by the National Library of Australia. Issued quarterly, with an annual cumulation, the publication lists sheet maps and atlases prepared in or which represent areas within the Commonwealth of Australia and its external territories, or which are associated with persons or corporate bodies throughout this area.
> *Refer also* R V Tooley: *One hundred foreign maps of Australia 1773-1887* (The Map Collectors' Circle, 1964).

161 'Background notes on the countries of the world' (Bureau of Public Affairs, Department of State, Washington, DC) are excellent factual pamphlets, each in a binder, about countries and territories; each *Note* contains information on land, people, history, government and political conditions, economy and foreign relations. There are maps in the text and, usually, a bibliography. The *Notes* are available on subscription or any special part may be had separately.

162 'Background to geography', by G R Crone (Museum Press, 1964; second edition, 1968; paperback edition, 1969), 'is an attempt to interest the general reader in the study of geography and to explain the value of a geographical outlook on the world today' (Preface).

163 'Background to political geography', by G R Crone (Museum Press, 1967) aims 'to discuss a representative selection of contemporary political problems', including such pressing topics as 'The world's food supplies', 'Frontier zones and boundaries', developing countries and the role of such diverse countries as China, Latin America, Australia, Western Europe and the United States, concluding with an assessment of the United Nations. A section, 'Books for further reading', is appended and there are a number of photographs and sketchmaps.

Among a growing body of studies on this aspect of geography, a useful complementary work is

J R V Prestcott: *The geography of state policies* (Hutchinson University Library, 1968).

164 Bagrow, Leo (1881-1957) was one of the great scholars whose work so stimulated the present interest in historical cartography; during a life-long study of early cartography, in Russia until 1918, in Berlin, 1918-1945, and in Stockholm, 1945-1957, he published more than seventy studies, making important contributions to cartographic history and encouraging international activity and co-operation. He wrote numerous papers on the early maps of the Black Sea, Siberia and Asiatic Russia, monographs on the history of ancient geography and on the literature of the history of cartography, and bio-bibliographical studies on the sixteenth century cartographers who contributed to the *Theatrum* of Abraham Ortelius, which were published as supplements to *Petermann's Mitteilungen,* 1928 and 1930. Probably his best-known work in the west is *Geschichte der Kartographie,* 1951 (*qv*) and the *Imago mundi: a review of early cartography* (*qv*), which he founded. Bagrow himself made a fine collection of maps, especially of Russian maps, which were acquired by the Houghton Library of Harvard College in 1956.

165 Bartholomew, John and Son Limited was founded in 1797 by George Bartholomew, map engraver, and is now in the fifth generation of an unbroken family tradition of high quality cartography. The first John Bartholomew (1831-1893), son of George, founded the Geographical Institute in Edinburgh, which has remained the headquarters of the House of Bartholomew; John George Bartholomew (1860-1920) is noted for his development of layer colouring as a method of relief representation; the late John Bartholomew (1890-1962), was himself a geographer, being honorary secretary of the Royal Scottish Geographical Society and president 1950 to 1954, and was influential in the establishment of the Chair of Geography at Edinburgh University. Outstanding among the firm's contributions to mapmaking have been the editions of *The Times atlas of the world* (*qv*); the ' half-inch ' map of Britain, the first topographic series to make use of layer colouring, 1878-; *The Edinburgh world atlas* (*qv*); numerous school and university atlases, such as the *Comparative atlas* (*qv*); the *Atlas of Scotland,* published in 1895 for the Royal Scottish Geographical Society; the *Physical atlas,* planned in five volumes, of which the *Atlas of meteorology* (*qv*), 1899, and the *Physical atlas of zoogeography* (*qv*), 1911, were particularly important; and the *World reference map* series of topographic maps in their frequent revisions, now in a new format under the collective title *World travel* series. The Bartholomew *Road atlas of Great Britain* is now revised annually. Specially designed for motorists and tourists on a scale of a fifth of an inch to the mile, a double-page is given to each map section, so that a large area can be studied at one time. Full contour colouring shows relief, with spot heights. Six road classifications are indicated and Ministry of Transport road numbers are used throughout. An index of some four thousand names quotes map references. Useful also is Bartholomew's *Roadmaster motoring atlas of Great Britain,* which is regularly updated. The *Reference atlas—Greater London* has become a standard work; and the *Gazetteer of the British Isles,* with its supplement, contains details of more than ninety thousand cities, towns, villages, hamlets, etc of the British Isles and the Channel Islands, arranged in alphabetical order. A sixteen-page summary includes the 1961 population census figures. *The road atlas of Europe* is a later atlas specially designed for tourists, including all kinds of information likely to be useful. New maps for motorists and tourists are frequently prepared; *The tourmaster maps of Britain,* for example, and the GT series.

From 1969, the firm has undertaken a ' Bartholomew early maps ' series; in co-operation with the Royal Geographical Society, London, maps from the Blaeu *Atlas novus* have been republished in colour, on specially selected antique cartridge paper. The maps are issued in a carrying tube. Bartholomew has been the United Kingdom wholesale distributor for the Cappelen *Road and tourist map of Norway* and has more recently been appointed distributor of the Mair/Shell maps of Europe. A valuable collection of atlases dating from the sixteenth century has been built up by the firm.

The catalogue of publications, revised annually, is clear and includes index maps; the latest editions contain a ' Table of map scale equivalents ' in fractions, mileage and kilometres.

Refer ' Bartholomews—on the map ' *in British printer,* September, 1963.

166 Behaim, Martin (*c* 1459-1506/7) was a famous globe-maker. One of his globes, made in Nuremburg about 1492, still survives.

Refer E G Ravenstein: *Martin Behaim: his life and his globe,* 1908.

See also Globes.

167 ' Bell's advanced economic geographies ', edited by R O Buchanan, are organised in two series, systematic and regional. Both series have the same aims—' to demonstrate the kind of thinking, critical and constructive, that informs current research in economic geography and to make the results of such research as widely available as possible '. Each volume is prepared by a recognised authority in the field. Systematic studies include Michael Chisholm: *Geography and economics;* P P Courtenay: *Plantation agriculture;* Peter R Odell: *An economic geography of oil;* L J Symons: *Agricultural geography;* among the regional studies are R C Estall: *New England: a study in industrial adjustment;* F E Ian Hamilton: *Yugoslavia: patterns of economic activity;* J E Martin: *Greater London: an industrial geography;* Wilfred Smith:

An historical introduction to the economic geography of Great Britain; H P White and M B Gleave: *An economic geography of West Africa.*

168 'Berge der Welt', edited by H R Muller, was published in Munich by the Nymphenburger Verlagshandlung, 1969, for the Swiss Foundation for Mountain Research; the whole set of volumes, 1947- is extremely fine, the first eight volumes having been edited by Marcel Kurz. At the end of the set is a consolidated index of authors, biographies, placenames and subjects.

169 Berghaus, Heinrich Karl (1797-1884), one of the great German cartographers of the nineteenth century, worked with Alexander von Humboldt and Carl Ritter and at the Cartographic Institute Justus Perthes at Gotha. His outstanding achievement was the *Physikalischer Atlas . . . (qv).*

170 Bertelsmann Cartographic Institute at Gütersloh, West Germany, is one of the most influential of the European cartographic agencies. Four great atlases are especially outstanding: the family reference atlas, *Bertelsmann Hausatlas,* 1960, 1962; the large and small world atlases, *Der Grosser Bertelsmann Weltatlas (qv)* and the *Kleiner Bertelsmann Weltatlas;* and the *Bertelsmann Atlas International,* 1963.

Refer Werner Bormann: 'Aus der Arbeit des Kartographischen Institutes Bertelsmann', *International yearbook of cartography,* 1964.

171 'Bibliographia geodaetica' has been produced monthly from 1963 by the National Committee for Geodesy and Geophysics of the German Democratic Republic, in co-operation with the International Documentation Centre for Geodesy at Dresden Technical University, the International Association of Geodesy and the International Federation of Surveyors (Berlin: Akademie-Verlag); it is referred to also as the International Geodetic Documentation. The publication continues the *Bibliographie Géodésique Internationale (qv)* (1928-1960) and the *Monatwissenschaftliche Literatur-berichte,* which covers 1962, and includes about ninety abstracts in each issue, each repeated in English, French, German and Russian.

172 'A bibliographic aid to the study of the geography of Africa', compiled by Sanford H Bederman for the Bureau of Business and

Economic Research at Georgia State University, comprises a selected list of recent literature published in English. The work contains some 1,900 items listed under countries, thence alphabetically by authors, using different coloured papers for the sections.

173 '**A bibliographic guide to population geography**', by Wilbur Zelinsky (University of Chicago Department of Geography, Research Paper, no 80, 1962), includes 2,588 items from the latter half of the nineteenth century to mid-1961; the work is intended as a finding list, no attempt being made at critical evaluation.

Refer also Hans Dörries: ' Siedlungs-und Bevölkerungsgeographie (1908-1938) ' *in Geographisches jahrbuch,* 1940.

H T Eldridge: *The materials of demography: a selected and annotated bibliography* (Columbia University Press for the International Union for the Scientific Study of Population and the Population Association of America, 1959).

174 '**A bibliographical and historical essay on the Dutch books and pamphlets** *relating to New-Netherland and to the Dutch West India Company, and to its possessions in Brazil, Angola, etc, as also on the maps, charts, etc of New-Netherlands*' compiled by G M Asher, was reprinted by N Israel, Meridian and Theatrum Orbis Terrarum in 1966 from the Amsterdam edition of 1854-1867. It is thought to be the only special bibliography of the Dutch expansion in the western hemisphere and in West Africa in the seventeenth century. There is a folding map.

175 '**A bibliographical guide and index to the principal collections of town plans and views published in the sixteenth and seventeenth centuries in Die Alten Städtebilder:** *ein Verzeichnis der graphischen Ortansichten von Schedel bis Merian*', by Von Friedrich Bachmann, provides an invaluable first study on this topic, first published in 1939 in an edition of five hundred and re-issued in 1965 in Stuttgart by Anton Hiersemann. The bibliographical history of each collection is traced and the contents are analysed town by town, with an emphasis on German collections.

176 '**Bibliographie cartographique internationale**' was first proposed in 1938, but due to the war, publication did not begin until 1949, by Armand Colin, under the auspices of the Comité National Français de Géographie and the International Geographical Union, with the support of Unesco and the Centre National de la Recherche

Scientifique. It remains the only official international record of maps and atlases published throughout the world, including also monographs of cartographical interest. The first volume listed maps published in 1946-47 in eight countries; the geographical societies of more than twenty countries now collaborate in listing local and regional maps, and editing is done at the Bibliothèque National. Each entry is accompanied by a brief annotation and the whole work is extensively indexed.

177 ' Bibliographie de cartographie ecclésiastique ' (Leiden: Brill, 1968-) began with the publication of the *Bibliographiekirchengeschichtlicher Karten Deutschlands,* prepared at the Institut für Landeskunde, Bad Godesberg, by Karl-Georg Faber, Bertram Hartling and Hans-Peter Kosach, together with the *Bibliographie kirchengeschichtlicher karten Österreichs,* under the direction of Ernest Bernleithner and Rudolf Kienauer. In brief, the background to this monumental work was the creation of the Commission Internationale d'Histoire Ecclésiastique Comparée at the International Congress of historical sciences at Rome in 1955; a subcommission, ' Cartographie ecclésiastique ' was at the same time formed to prepare an atlas of the historical geography of religions. At the 1960 Congress at Stockholm, the Commission Internationale d'Histoire Ecclésiastique Comparée accepted the offer of Professor Hermann Heimpel, Director of the Max Planck Institute at Göttingen to set in train a programme of bibliographical work to parallel the cartographic.

178 ' Bibliographie de l'oeuvre de Lucas Jansz Waghenaer ', an outstanding bibliographie compiled by Thomas James I Arnold, was reprinted from the *Bibliotheca Belgica,* first series, Ghent, 1880-1890 by N Israel, Meridian and Theatrum Orbis Terrarum in 1961. All the editions of the *De Spieghel der Zeevaerdt* are described page by page, noting all known variants of the charts as well as of the text.

See also under Waghenaer.

179 ' Bibliographie des ouvres relatifs à l'Afrique et à l'Arabie ' was compiled by J Gay in 1875. The work covers about 3,700 titles, with bibliographical references, notes, cross-references, an index of names and one of towns. It has been reprinted for the second time in 1970 by Theatrum Orbis Terrarum.

180 ' **Bibliographie de Tahiti et de la Polynésie française** ', compiled by Patrick O'Reilly and Edouard Reitman, was published in 1967 by the Musée de l'Homme, Paris, being number 14 of the *Publications de la Société des Océanistes*. ' This weighty volume completes the series of bibliographies dealing with New Caledonia, the New Hebrides, Wallis and Fortuna '. In the standard format adopted by the editors, works of reference are included, also voyages, monographs on geology, botany, zoology, geography, ethnology (including social anthropology and linguistics), history (including administration and missions) and economic life. More than ten thousand items, with numerous analytical sub-headings, are accompanied by evaluative notes and further references.

181 ' **Bibliographie de Népal** ', published by the Centre National de la Recherche Scientifique, Paris, forms the first volume in the series *Cahiers Népalais,* compiled by L Boulnois and H Millot. References are to western literature from the eighteenth century to 1968.

182 ' **Bibliographie géodésique internationale** ' in its first series was published by the International Association of Geodesy, as follows: 1, 1928-30, 1935; 2, 1931-34, 1938; 3, 1935-37, 1939, all edited by G Perrier and P Tardi; 4, 1938-40, 1947, edited by G Perrier, P Tardi and G Laclavère; 5, 1941-45, 1952; 6, 1946-48, 1954; 7, 1949-51, 1956, all edited by P Tardi and G Laclavère; 8, 1952-55, 1962, edited by the Central Bureau of the International Association of Geodesy (under the chief direction of J Villecrosse), as were volumes 9 and 10, 1956-57, and 1958-60, published respectively in 1963 and 1965. Beginning with the volume for 1961, the bibliography was published in a new format, as a monthly magazine, *Bibliographia geodaetica* (qv) and on cards, International Geodetic Documentation, edited by the Geodätisches Institut of Dresden. The bibliography aims to be a truly international publication; under the aegis of the International Association of Geodesy, to date the following regional institutes collaborate in the work: the Finnish Geodetic Institute (Sweden, Norway, Denmark, Finland); the Central Bureau of the IAG (Spain, Portugal, Italy, Belgium, Greece, France, French-speaking Africa); Institut für Angewandte Geodäsie, Frankfurt a/Main (Netherlands, Austria, Luxembourg, Switzerland, West Germany); Technische

Universität Dresden Geodätisches Institut (Poland, Czechoslovakia, Bulgaria, Romania, Yugoslavia, Hungary, East Germany); Military Survey—Geodetic Office of Great Britain (Great Britain, Ireland, Commonwealth, South Africa); the Soviet Committee for Geophysics (USSR); the US Coast and Geodetic Survey (USA, Japan) and the Inter-American Geodetic Survey—Canal Zone (South America). *Bibliographia geodaetica* is published in French, English (Butterworths Scientific Publications), Russian and German, with abstracts in the four languages. Entries, arranged by the UDC, are each accompanied by a brief annotation. The cards are prepared and sold in two different forms: 'simple series', one card per title; 'complete series', two cards per title, in either English, French, Russian or German, grouped according to three main subjects, higher geodesy, surveying, and photogrammetry (in German only). Only group I is published by the International Association of Geodesy; the second and third groups are published with the cooperation of the International Federation of Surveyors and other relevant bodies.

183 'Bibliographie géographique internationale' is the central current geographical bibliography. The first issues, 1891-1913-14 were published as separate parts of the *Annales de géographie* (*qv*) under the direction of L Raveneau. After the first world war, the bibliography was published under the present title by the Association de Géographie Français (Armand Colin) and, since 1931, it has been a publication of the International Geographical Union in international co-operation, with the assistance of Unesco and the Centre National de la Recherche Scientifique, the latter assuming increased publishing responsibility with the 1954-55 volume, 1958. All aspects of geography are covered, except purely technical or local topics; useful reference works, such as glossaries and yearbooks, are included. Arrangement is primarily by country, with subject divisions, under three main headings: historical; general geography; regional geography. Most of the entries carry brief annotations. Author indexes are included and the later issues have also detailed subject indexes.

184 Bibliographies: national control, from the point of view of geographical studies

Arrangement by continent, then alphabetically by country

EUROPE

Albania The National Library publishes a quarterly current bibliography of books and pamphlets, official publications, maps and atlases, also bibliographies of bibliographies. University theses and dissertations are listed by the State University of Tirana and published in special booklets or in its bulletin. Periodicals are indexed by the National Library and published in the *Bibliographie trimestrielle du périodique. The Albanian book* has been published by the National Library from 1958, first annually, quarterly from 1960; and *Articles from Albanian periodicals* is also published by the National Library, quarterly 1961-1964, bi-monthly 1965 and monthly from 1966.

Austria The national current bibliography, *Österreichische bibliographie,* has been published fortnightly since 1946; official documents, new journals, theses and maps are included. *Refer also* to the *Österreichische bibliographie,* compiled by C Junker and L Jellinek, 1899-1901, which includes publications in German only, classified systematically by subject, issued first fortnightly, then monthly.

Belgium The *Bibliographie de Belgique* is prepared by the Royal Library. Official publications and government documents are to be found in two retrospective bibliographies—Paul de Visscher and Jaques Putzeys: *Répertoire bibliographique du Conseil d'Etat,* with a supplement covering 1960-62, compiled by Jacques Putzeys (Institut Belge des Sciences Administrative, 1963); and Denise de Weerdt: *Bibliographie retrospective des publications officielles de la Belgique, 1794-1914* (Louvain: Editions Navwelaerts, 1963). Since 1963, the Centre International de Documentation Economique et Sociale Africaine has regularly published a *Bulletin of information on theses and studies in progress or proposed. Bibliographie géographique de la Belgique,* compiled by M E Dumont and L De Smet, with a first supplement in 1960, is issued by the Commission belge de bibliographie; the *Bibliographie de Belgique: liste mensuelle des publications belges ou relatives à la Belgique* has been published since 1875, a new series being started from 1967; and *Bibliotheca Belgica: bibliographie générale des Pays-Bas,* edited by Marie-

Thérèse Lenger was published in two volumes, 1966. *Abstracts of Belgian geology and physical geography,* 1967-, edited by L Walschot, began publication in 1969 by the Ghent Geological Institute of the University of Ghent. The volumes are to appear annually, each containing some 250 abstracts, a subject index and a geographical index.

Bulgaria The national current bibliography, *Balgarski knigopis,* published monthly since 1897, includes maps, new periodicals and any changes in the publication policy of periodicals. Since 1962, a quarterly supplement has listed official publications and dissertations. The Centre for Scientific Information and Documentation also publishes a series of abstracting journals devoted to specific subjects; geography is not named separately. The Centre for Agricultural Scientific Information and Documentation of the Academy of Agricultural Science issues a series ' Rural economy and forestry ' within the *Referativnyi bjuleten. Refer also* to the publication of the Library of Congress, Slavic and Central European Division, Reference Department, *Bulgaria: a bibliographic guide,* 1965, compiled by Marin V Pundeff, especially the sections ' General reference works ', ' Land and people ' and ' Economy and social conditions '.

Czechoslovakia There are two current national bibliographies, one weekly, ' Czech works ', published by the National Library of Prague, the other, monthly, ' Slovak works ', published by the Matica Slovenská. Some official documents are included. Bibliography of theses began with a special number each year of *České knihy,* 1965- (1964). Since 1965, maps and atlases have appeared regularly in *Česka grafika a mapy,* an annual index which lists the productions of the previous year. A similar list is in progress in Slovakia.

Denmark Biblioteca Danica, the national retrospective bibliography, published by the Royal Library 1872-1931, covering the period 1482-1830, was followed by a new and enlarged edition 1960-63. Annual catalogues for the years 1959-63, as well as a cumulative volume of the Danish current bibliography, *Dansk bogfortegnelse,* 1955-59, were prepared by the Bibliotekscentralen and published by G E C Gad; the volumes covering the periods 1841-58 and 1859-68 were republished by anastatic printing in 1960 and 1961 respectively. An index of official publications and government

184 Bibliographies: national control, from the point of view of geographical studies

documents is prepared by the Danish Institute of International Trade and published by the Bibliotekscentralen. *Danish theses . . . a bibliography* was made available in 1929 and 1961, the two volumes covering the period 1836-1958. Bibliographies concerning geography and topography are so far devoted to specific themes, such as *Bibliography of Old Norse-Icelandic studies,* prepared by Hans Bekker-Nielsen and Thorkil Damsgaard Olsen, in collaboration with the Royal Library, 1963(1964)-. *Denmark: literature, language, society, education, arts: a select bibliography* was one of the publications of the Royal Library, 1966.

Finland The Finnish national bibliography, prepared by Helsinki University Library, is a current bibliography which has, from 1932, included all literature published in Finland, regardless of language. The first volume of a retrospective bibliography, 1878, covers literature in Finnish for the years 1544-1877; each of the succeeding volumes lists works published during the previous three or eight years. This bibliography includes maps. From 1961, official publications have been listed in various forms by the Library of Parliament. The University of Helsinki publishes an annual list of theses and academic publications of all Finnish universities, 1966-; there is a special systematic section for maps. Catalogues of maps of the General Cartographic Service of Finland have appeared irregularly.

France The *Bibliographie de la France,* the national current bibliography, celebrated its 150th anniversary in 1961, when the Cercle de la Librairie issued a special number devoted to the history of this work, which appeared originally under the title *Journal général de l'imprimerie et de la librairie.* The technical development of the bibliography was described in the *Bulletin des bibliothèques de France,* November 1961. Since 1960-61 the annual volumes have included an index of current serials; and since 1963 the official part of the publication has been classified according to the UDC rules.

See also under Bibliothèque Nationale.

Germany Deutsche *Bibliographie: Wöchentliches Verzeichnis,* 1947-, a weekly systematic national bibliography, began in 1965 to separate into three series: books on sale, books not on sale and maps. There are monthly and quarterly indexes; five-yearly indexes record not only books and maps published in the two parts of Germany, but all published in German and issued abroad. Official publications and government documents are listed as a separate part of the *Deutsche Bibliographie* . . ., issued every other year. Information on maps and atlases is issued also in the *Berichte zur Deutschen Landeskunde,* Stuttgart.

Greece has no current national bibliography, only individual catalogues, such as the *Bulletin analytique de bibliographie hellénique,* published by the French Institute in Athens.

Hungary The national current bibliography, 1946-, *Magyar nemzeti bibliográfia,* was published by the National Széchényi Library monthly to 1961, then fortnightly; it records, by UDC arrangement, books, pamphlets, maps and theses. A cumulative volume appears every two or three years. A cumulative retrospective bibliography of works published after the second world war appears under the title *Magyar könyvészet 1945-1960* (1964); volume IV contains the entries for art, literature, geography and history. A catalogue of Hungarian periodicals is published separately; and articles published in Hungarian periodicals are listed monthly in *Magyar folyóiratok repertóriuma.* There are various specialised abstracts reviews.

Ireland Irish publications are included in the *British national bibliography.* In addition, the Stationery Office, Dublin, issues an annual list of official publications; the National Library of Ireland, Dublin, prepares a national retrospective bibliography, which includes printed works, manuscripts, periodical articles and maps. *Irish historical studies* contains an annual bibliography, *Writings in Irish history.*

Italy The *National bibliography* has been published monthly between 1886 and 1957, arranged by subject, with annual indexes of authors and subjects; in 1958, a new series began, also monthly, entitled *Bibliografia nazionale Italiana.* Periodicals and official publications are noted in separate supplements. Maps published in Italy are appended as a special supplement to the monthly issues.

Luxembourg The annual *Bibliographie Luxembourgeoise* lists publications received by the National Library, also works written by Luxembourg nationals or relating to the Duchy. Since 1958, it has included the contents of periodicals and from 1962 official publications have also been included, distinguished in the text by a special sign. There is no current bibliography of special subjects.

Netherlands The Royal Library issues *Bibliografie van in Nederland verschenen officiële en semi-officiële uitgaven.* Netherlands periodicals are indexed monthly, with annual cumulations, in

184 Bibliographies: national control, from the point of view of geographical studies

Nijhoff's index op de Nederlandse periodieken van algemene inhoud. Brinkman's catalogus van boeken, a private undertaking, began in 1846 with monthly issues with annual and quinquennial cumulative lists; books and periodicals published in the Netherlands have been included, also those in Dutch published in Belgium. Since 1945, university theses have been listed. There is also a select classified bibliography, *Nieuwe uitgaven in Nederland.* Another useful guide is *Sijthoff's adresboek voor de voor de Nederlandse boekhandel,* with its separate lists of periodicals and official publications.

Norway The *Norsk bokfortegnelse* includes books, pamphlets, periodicals, university theses, official publications and maps. Official publications are also listed in a special current bibliography, *Bibliografi over Norges offentlige publikasjoner,* published annually since 1956. In the annual cumulation of the national bibliography, maps and atlases are treated separately. For retrospective bibliography, the *Bibliotheca Norvegica* was compiled by Hjalmar Pettersen in four volumes (Christiania, 1899-1924) listing works printed in Norway before 1814, works of Norwegian authors and translations of works in Norwegian published abroad, as well as foreign works dealing with Norway. Vilhelm Haffner compiled retrospective lists of works from 1814 to 1924 (1925) and from 1925 to 1935 (1936).

Poland The *Przewodnik bibliograficzny,* which includes maps and atlases, is published weekly, with a monthly list of authors and an annual cumulation, containing a list of subjects. On the basis of this publication, the Bibliographical Institute prepares the official statistical yearbook of Polish publications, *Ruch wydawniczy wliczbach,* having text in Polish and English. An annual list of periodicals is published by the Bibliographical Institute, also a monthly bibliography of periodical articles. University theses are listed by the Ministry of Higher Education, beginning with a first volume in 1962, which covered the years 1959-1961, thence annually. Special bibliographies are being steadily compiled: an annual *Bibliografia geografi polskiej* has already been initiated by the Bibliographical Institute and similar works have been started for economics, geology and forestry. The Bibliographical Institute is also preparing a retrospective bibliography of Polish works and works on Poland.

Portugal The national bibliography, *Boletim de bibliografia portuguesa,* has been compiled by the National Library since 1935, except for the war years, annually until 1954, then monthly; it includes books, pamphlets, maps and official publications. A number of special bibliographies have also been published, on geology, seismology and meteorology, among others. The Portuguese National Library Commission, working in the Lisbon Academy of Science, is compiling a retrospective ' Portuguese general bibliography '. Lisbon National Library publishes also the monthly ' Portuguese bibliography bulletin ', which includes the most important articles from major periodicals.

Romania The Central State Library issues the *Bibliografia Republicii Populare,* fortnightly; it includes books, pamphlets and maps. A separate fortnightly publication lists periodicals. A Centre of Information and Documentation in Social and Political Sciences began operations in July 1970, as an institution of the Academy of Social and Political Sciences. The Centre took over the whole activity carried out by the Scientific Documentation Centre of the Academy of the Socialist Republic of Romania in these fields. The ' Social sciences ' section of *Romanian scientific abstracts* would be of interest to geographers.

Spain The monthly *Boletín del depósito legal* and the *Bibliografía española,* annually, both began in 1958 and include maps and atlases; the latter also lists periodicals and official publications. The Ministry of Information and Tourism (Press Department) publishes an *Anuario de la prensa española,* of which Part I is concerned with periodicals. From the numbers of special bibliographies, two are of direct interest: *Guía bibliográfica para una geografía agraria de España,* compiled by José Muñoz Pérez and Juan Benito, 1961, and *Bibliografía de geografía económica de Galicia,* by F J Rio Baija, 1960.

Switzerland Das Schweizer buch, bibliographical bulletin of the National Library, 1901-, cumulates into the *Répertoire du livre suisse,* in two parts, authors and subjects. Maps and atlases are included. *Bibliographie des publications officielles suisses* has also been prepared by the Bibliothèque Nationale since 1946, and the *Catalogue des écrits académiques suisses,* from 1897. The Société Suisse des Libraires et Editeurs has issued irregularly the *Répertoire des périodiques suisses.*

USSR Two current bibliographies, both long-established, give systematic coverage of all books published in all languages in the USSR: *Knižnaja letopis,* weekly with a monthly supplement and *Ežegodnik knigi SSSR,* in two volumes annually. Current Union records of periodicals, theses and official publications appear regularly

184 Bibliographies: national control, from the point of view of geographical studies

and geographical maps and plans are included in the *Kartografičeskaja letopis,* published annually by the All-Union Book Chamber. In addition, all the Republics keep their own records. Since 1925, books, pamphlets, official publications, maps and atlases have been listed in the *Annals of publications of the Byelorussian SSR,* issued in monthly parts. Publications dealing with the Byelorussian SSR, but appearing in other republics of the Soviet Union, have since 1946 been listed in a special section of the *Annals.* Every year, cumulative indexes annexed to the *Annals* contain lists of periodicals published in the Byelorussian SSR, including new publications and those that have changed their titles. The Ukrainian SSR national bibliography began in 1924, based on the Ukrainian SSR Book Chamber at Kharnov, set up in 1922. An index of periodicals has been issued for the years 1918 to 1950 and the Chamber is responsible also for the national retrospective bibliography.

Note The Bibliographical Institute of the National Library of Poland, Warsaw, published in 1969 a work by Pelagia Girwic on the organisation and network of current Soviet bibliography.

United Kingdom The *British national bibliography,* 1950-, based on legal deposits, does not include maps. A *Guide to British periodicals,* by Mary Toase, was published by The Library Association, 1962. Official publications are made known by daily, weekly, monthly and annual cumulations published by HMSO; *Sectional lists,* dealing with individual department publications, are revised as necessary. Aslib issues an annual *Index to theses accepted for higher degrees in the universities of Great Britain and Ireland* and periodical articles from major periodicals are listed in the *British humanities index,* the *British technology index* and the *Index to articles in British technical journals,* all begun in 1962 by The Library Association. The *Bibliotheca celtica* lists annually all publications concerning Wales, Celtic works and the Celtic language, prepared on the basis of the acquisitions of the National Library of Wales, arranged by subjects, with a list of authors. *A bibliography of works relating to Scotland,* compiled by P D Hancock in two volumes, was published by the Edinburgh University Press, 1960. There is a rapidly increasing number of local bibliographies, such as the *East Anglian bibliography* . . . (The Library Association, 1960-). Individual institutions issue their lists of maps and atlases; there is no centralised publication.

Refer also Statistical news: developments in British official statistics, published quarterly by the Government Statistical Service 'to provide a comprehensive account of current developments in British official statistics and to help all those who use or would like to use official statistics'. Every issue contains up to three articles dealing with special subjects in depth. Shorter notes give news of the latest developments in many fields, including international statistics. Frequent reference is made to other works which, though not issued by government organisations, are closely related. Each issue is indexed and a feature is the reference to 'Articles in recent issues of *Statistical news* . . .'

Yugoslavia Since 1950, the Jugoslovenski Bibliografiski Institut has published a current national classified bibliography, the *Bibliografija Jugoslavige* . . ., fortnightly from 1954, which lists books, pamphlets and maps, succeeding the annual bibliography *Jugoslovenska bibliografia,* 1945-1950. Yugoslav periodicals have been listed, since 1956, in the quarterly *Bibliografija jugoslovenske periodike,* as well as in the national bibliography.

ASIA: MIDDLE EAST: FAR EAST

Afghanistan has remained isolated and until recent years did not attract scholarly research. Now the Afghans themselves are attempting research projects and documentation. The Russians made a beginning in bibliography for the general area in the nineteenth century, but up to 1968 only one important work was published, the *Bibliographie analytique de l'Afghanistan,* by Mohammed Akram, 1947-; only the first volume was issued then, dealing with works published outside Afghanistan and it is believed that no other volumes were completed. Afghanistan is still inadequately mapped and there is no official list of maps as yet.

Burma The *Burmese national bibliography* has been undertaken by the Daw Nyunt Myint National Library, Rangoon. The plan is for cumulative volumes to be published every two or three years and further details should be available soon.

Ceylon The national bibliography has been published quarterly from 1962 by the National Bibliography Division of the Government Archives Department in Gangodawila; it includes books, pamphlets, published theses, dissertations and new periodicals. Three languages are used, Sinhala script, Tamil and English.

184 Bibliographies: national control, from the point of view of geographical studies

China A Bibliographic Centre was set up in the National Central Library in Taipei in 1960; the Centre publishes a *Monthly list of Chinese books,* in Chinese and English, and an annual *National bibliography of the Republic of China,* prepared by the National Central Library which has also published a national retrospective bibliography in two volumes listing more than fifteen thousand documents published in Taiwan between 1949 and 1963. An *Index to the Chinese periodical literature* was published by the Library of the National University of Taiwan.

India The *Indian national bibliography* is based on a very complex system, involving numerous publications including works in the main languages; the Ministry of Education prepared a list in 1958, subsequently handing over the work to the Central Reference Library, Calcutta, from which annual volumes appeared between 1960 and 1962, quarterly issues in 1963 and monthly issues from September 1964, cumulated annually. A second part deals with government publications. Many publications include lists of the subjects of theses and there is also the *Bibliography of current reports,* published by the Atomic Energy Establishment, Information Division. A *Guide to Indian periodical literature* (*Social sciences and humanities*), 1964-, has been prepared by the Prabhu Book Service, in quarterly issues, with annual cumulations.

Refer also N N Gidwani and K Navalani: *Indian periodicals: an annotated guide,* 1969.

Indonesia The current bibliography, *Berita bulanan,* is prepared by the Bureau of National Bibliography; an index of Indonesian scientific periodicals was initiated in 1960 and a national commercial bibliography, *Berita bibliografi,* has been published since 1955 by a wholesale bookseller, Gunung Agung.

Iran Index Iranicus, compiled by Iradj Afshar, is among the series of works published by the University of Tehran; it comprises a systematic list of Iranian articles on Iranological studies published in journals and other periodical publications, the first volume, covering 1910 to 1958, having been completed in 1961. The University of Tehran has also published a *Bibliography of Iranian bibliographies,* also compiled by Iradj Afshar. The same editor prepares the annual *Bibliography of Persia,* published by the Persian Book Society, 1955-, a systematic current bibliography of all books and periodicals published in Iran; a retrospective bibliography is in progress. A *Geographical bibliography,* under the direction of M Gandjei, lists geographical articles throughout the Iranian press.

Israel Kirjath sepher is the main current bibliography. The Publishers' Association of Israel has inaugurated the ' General catalogue of books ', of which a new edition was published in 1969. The Centre of Scientific and Technological Information has issued a *Directory of current research in Israel: physical and life sciences;* the second survey, listing some four thousand projects, was published in Tel Aviv in 1969. A *Directory of serials in pure and applied sciences and economics published in Israel* was published in Tel Aviv in 1963, in Hebrew and English; and an *Abstracts of theses . . .* is prepared two-yearly by the Hebrew University, Jerusalem. The Government Press Office issues *Newspapers and periodicals appearing in Israel* and there is also the *Index of articles on Jewish studies,* by Issachar Joel, 1966—published by the Magnes Press, Jerusalem, 1969-. A project aiming at a complete Hebrew retrospective bibliography is being prepared at the Jewish University, with the cooperation of the National Library and under the auspices of the Hebrew Institute of Bibliography.

Japan The National Diet Library is responsible for the current national bibliography, *Zen Nihon Shuppan-butsu Sô-mokuroku,* which lists all works published in Japan, in two parts, official and unofficial publications; it contains several indexes and a directory of publishers. The Library issues also *Nohon Shuho,* a weekly list of works deposited with the Library and, jointly with the Ministry of Education, the *Directory of Japanese periodicals: natural sciences, medical sciences and industry,* first issued in 1962. From 1963, the Shuppan News Company has issued a general catalogue of Japanese periodicals and from 1957 the Board responsible for issuing official publications has prepared a monthly list. Theses are listed by the Research Institute of Educational Administration, which puts out an annual publication; maps and atlases are mentioned in the national bibliography. There are many specialised catalogues and lists compiled by national bodies, for example, the ' Wheat information service '.

Korea The Korean national bibliography, published monthly by the National Central Library, gives detailed bibliographical information on all categories of books, pamphlets and periodicals published during the preceding month. In November 1964, the National Central Library issued a contribution to the national retrospective

184 Bibliographies: national control, from the point of view of geographical studies

bibliography, covering 1945 to 1962. A list of periodicals and one of official publications are issued by the Ministry of Information. Lists of university theses are prepared by the Korean Research Centre. The Centre has also published a *Selected list of Koreana collections. Bibliography in business and economics (1945-1960)* was published in 1961 by the Industry Management Research Centre, which is sponsored by Yunsei University.

Laos The only bibliography at present available is the *Bibliographie du Laos,* compiled by Thao Kéne, a listing of books and review articles dealing with Laotian civilisation.

Malaysia has no complete current national bibliography. 'Current Malayan serials', compiled by Lim Wong Pui Huen, was included in the *Singapore library journal,* October 1963, and a 'Guide to Singapore government departments and serials', by E Sormitasagama, in the issue for April 1964. Useful is the *Catalogue of government publications* issued at intervals by the Government Printing Office. The Singapore National Library has issued many specific bibliographies, including 'Books about Malaysia', 1963, revised in 1964 and 1967, and 'Urban community development', 1962. *Books about Singapore and Malaysia, a selective bibliography,* prepared by the Reference Section of the National Library, was completed in 1965; it includes references to works on economics, commerce and industry, physical features and natural history, travel and serials. There is also *Malaya, a background bibliography,* compiled by Beda Lim and published in 1962 by the Malaya Branch of the Royal Asiatic Society.

Mauritius Bibliography of Mauritius (1502-1954) covering the printed records, manuscripts, archivalia and cartographic material (Port Louis, Government Printer, 1956) has been continued since 1955 in the form of annual supplements containing a list of all works and printed matter printed in Mauritius during each year. These supplements appear in the *Annual Report of the Archives Department* and contain a list of acquisitions concerning Mauritius, recording also scientific publications relating to it, theses, etc. The *Annual Reports of the Mauritius Institute* for 1960 and 1961 contain information on the publications concerning Mauritius, with special emphasis on zoology.

Nepal The Nepalese Study Centre at the Centre National de la Recherche Scientifique, Paris, is engaged on a comprehensive bibliography of Nepal, arranged by subjects. Volume I, 1969, comprised

'Sciences humaines—Références en langues Européenes', compiled by L Boulnois et H Millot.

Pakistan There is a current bibliography in preparation, but no information concerning it has come to hand. Otherwise, apart from PANSDOC, the only contribution to a national bibliographic coverage is the annual catalogue of official publications issued by the Government Printer.

Philippines A current bibliography is being prepared and the retrospective bibliography has been compiled to 1957. The University of the Philippines, Institute of Public Administration, Inter-Departmental Reference Service prepares the *Index to Philippine periodicals* and the Research Co-ordination section of the University edits *Thesis abstracts,* of which the first issue appeared in 1962, covering 1947 to 1954. A *Checklist of Philippine government publications* is prepared by the National Library.

Singapore There is a separate annual national bibliography, issued by the National Library, which includes government publications and maps.

Thailand There is no systematic current bibliography, but some materials are covered. *Bibliography of Thai government publications* has been issued by the Thai Library Association since 1962; an *Index to Thai periodical literature 1957-62* was published in 1962 by the Ministry of Education, Department of Teacher Training, Bangkok; *Bibliography of material about Thailand in western languages* was published in 1960 by the Chulalongkorn University, Faculty of Arts; and a *Statistical bibliography: an annotated bibliography of Thai government statistical publications* was prepared in the Office of the Prime Minister (National Office of Statistics, 1964).

Turkey The current national bibliography, *Turkiye biblioyografyasi,* was begun in 1928 by the Directorate of Legal Deposit at the Ministry of Education; in 1953, the work was undertaken by the National Library. It includes periodicals, official publications, theses, maps and atlases. A bibliography of articles in Turkish periodicals began quarterly production in 1952.

Viet-Nam Current national bibliography is undertaken by various organisations and comes in various forms. The *National bibliography of Viet-Nam,* issued quarterly by the Directorate of Archives and National Libraries, contains lists of non-periodical publications deposited with the Copyright Department; since 1961, these lists have been included in the *Journal officiel* of the Republic; also

184 Bibliographies: national control, from the point of view of geographical studies

Sách Mó, 1962-, a monthly bulletin designed to keep the public informed of new acquisitions of the National Library and the General Library attached to the Directorate. The Ministry of Information publishes a monthly bulletin giving a list of publications submitted to the Ministry for purposes of censorship. The Directorate of Archives and National Libraries is preparing a retrospective bibliography, incorporating some individual separate existing studies. The *Catalogue des cartes et plans* published by the National Geographic Department of Viet-Nam is in the form of general tables of sheets on scales of 1:25,000, 1:50,000, 1:100,000, 1:250,000, 1:500,000, with current price lists.

AFRICA

Naturally, no bibliography covers the whole of the continent, but the following references help in general studies. *Sub-Saharan Africa: a guide to serials* was published in 1970 by the Library of Congress, Washington, DC; compiled by the African section of the Library's General Reference and Bibliography Division, the guide contains 4,670 entries recording a selection of serials published before 1969 in western languages and in African languages using the Roman alphabet. It includes many of the titles appearing in *Serials for African studies,* issued by the Library in 1961, except that publications specifically on North Africa have been excluded. The journal *Afrique équatoriale,* 1955-, includes mention of works of all kinds, including those published abroad dealing with central Africa. There are also H A Rydings: *The bibliography of West Africa* (Ibadan University Press for the West African Library Association); Mary Jane Gibson: *Portuguese Africa: a guide to official publications* (Library of Congress, General Reference and Bibliography Division, Reference Department, 1967); Julian W Witherall: *French-speaking West Africa: a guide to official publications* (Library of Congress, General Reference and Bibliography Division, Reference Department, 1967).

Algeria The Algerian National Library publishes the *Bibliographie de l'Algérie.*

Egypt Since the National Library in Cairo was made the centre for legal deposit in 1954, it has been responsible for compiling the current national bibliography, *Egyptian publications bulletin,* 1956-. In 1962, cumulative lists were issued: 1955-60; 1961-62; 1963-December 1964. There is also available a *Subject catalogue of Egypt,* an index of all foreign publications on Egypt in the National Library, in three volumes. A collection entitled 'Series of bibliographical lists of books and references about the Arab world', 1960-, began with *Algeria, Palestine, Syria, Lebanon, Iraq, Sudan, Al-Maghreb, Tunisia, Libya* and *Arabian peninsula*.

Ethiopia The bibliographies produced by the Ethiopian Research Institute at present do duty for a national bibliography.

Ghana The Ghana national bibliography is issued annually by the Ghana Library Board, by whom a retrospective bibliography is in progress; the Board issued also in 1960 *Books about Ghana*. The Director of Surveys, Accra, issues a catalogue of maps, with supplements. Government publications are listed annually in *Government publications: price list* and the University of Ghana Library prepares a *List of theses*.

Refer also A F Johnson: *A bibliography of Ghana 1930-1961* (Longmans, 1964).

G M Pitcher: *Bibliography of Ghana 1957-1960* (Kumasi: Kwame Nkrumah University, 1960).

The Padmore Research Library issues a *Special subject bibliography* series, which includes *A select annotated bibliography of Ghana*.

The Research Library on Africa Affairs, University of Ghana Library, Institute of African Studies, has prepared a *Union list of Africana periodicals*, which is revised at intervals.

The Ministry of Agriculture Reference Library, Accra, issued *Ghana agriculture, 1890-1962: bibliography of crop and stock, co-operation and forestry, food and fisheries,* 1962.

A W Cardinall: *A bibliography of the Gold Coast* (Accra: Government Printer, 1931).

Guinea The national bibliography is prepared by the National Library and published as a supplement to the review *Recherches Africaines* (Institut National de Recherches et de Documentation). A first five-year survey, from September 1958 to December 1963, was attached to no 1, 1964, followed by annual cumulations.

Madagascar The University Library and the National Library began an annual current bibliography in 1965 (1964-), under the supervision of M S de Nuce and J Ratsimandrava. A retrospective bibliography, published by Grandidier, covers 1800-1904; 1903-1933 and 1934-1955; and another section, covering 1956-1963, was prepared with the help of Unesco.

184 Bibliographies: national control, from the point of view of geographical studies

Morocco An annual list, *Informations bibliographiques Marocains,* is published, consisting of the publications received by the Bibliothèque Générale et Archives du Maroc, based on fortnightly lists prepared by the librarians. Contributions toward a retrospective bibliography appear in the review *Hespéris.*

Nigeria The National Library produces *Nigerian books in print* and *Serials in print in Nigeria.* Lists are also issued of theses and dissertations accepted for higher degrees in Nigerian universities. The University Library, Ibadan, also functions as a national bibliographic centre; a current bibliography is prepared which includes books and pamphlets, also official publications. The University Institute of Librarianship published *Nigerian library resources in science and technology,* compiled by F Adetowun Ogunsheye, 1970.

> *Refer also* Margaret Amosu, *compiler: Nigerian theses* (Ibadan University Press, 1965).
> Helen F Conover: *Nigerian official publications, 1869-1965: a guide* (Library of Congress, 1959).
> L C Gwam, *compiler: A handlist of Nigerian official publications in the National Archives Headquarters, Ibadan,* 1964.
> W John Harris, *compiler: Books about Nigeria: a select reading list* (Ibadan University Press, fourth edition, 1963, fifth edition, 1969).

Senegal Official publications are listed by the Library of the Archives Service, Dakar; university theses by the University Library, Dakar. The National Geographical Institute, Hann, compiles information concerning maps and atlases. The Centre de Documentation Economique et Sociale Africaine (CIDESA) acts as a bibliographic centre.

Sierra Leone *Sierra Leone publications,* published annually by the Sierra Leone Library Board, does not include maps or audiovisual materials. *Official publications of Sierra Leone and Gambia,* by A Walker, was published by the Library of Congress, 1963; and *A bibliography of non-periodical literature on Sierra Leone 1925-1966,* compiled by H M Zell, excluding government publications, was published by the Fourah Bay College Bookshop Limited, Freetown, 1966.

Uganda Publications received in the University College of Makerere are recorded in the library bulletin; in fact, all official publications of East African countries are included.

Note It may be noted here that the second *Tanzania national bibliography* began in 1970—no details available as yet.

The Rhodesias and Nyasaland: a guide to official publications, compiled by A A Walker (Library of Congress, General Reference and Bibliography Division, Reference Department, 1965).

NORTH AMERICA

Canada Since 1950, the National Library has been responsible for the preparation of the national current bibliography, *Canadiana: publications of Canadian interest noted by the National Library,* which is published monthly, with annual cumulations; publications issued by commercial organisations are listed, also those published by the Federal Government and the ten Provincial Governments. The first issue of every periodical published in Canada is mentioned. In 1962, the Canadian Library Association published a cumulative list covering a period of twelve years, *Canadian index to periodicals and documentary films: an author and subject index, 1948-1959.* Official publications and government documents are listed by the Department of Public Printing and Stationery, as follows: *Canadian government publications monthly catalogue,* 1953-, of which an annual cumulation is published; and *Daily checklist of government publications,* 1952-. An annual list of theses is compiled by the National Library and published by the Queen's Printer under the title *Canadian theses,* arranged by broad subject headings. The National Film Board publishes every two years *Canadian filmsstrips catalogue* and, since 1948, the Canadian Library Association has issued monthly *Canadian index to periodicals and documentary films,* with annual cumulations.

Refer also Raymond Tanghe: *Bibliographie des bibliographies,* under the auspices of the Bibliographic Society of Canada (University of Toronto Press, 1960-), with annual supplements.

United States There is no current national bibliography in a formal sense, but a number of publications serve the purpose. The Library of Congress has been given public recognition as a national bibliographic centre, many other centres concentrating on individual subjects. The *National union catalog: a cumulative author list representing Library of Congress printed cards and titles reported by other American libraries* is compiled by the Library of Congress with the assistance of the American Library Association Committee

184 Bibliographies: national control, from the point of view of geographical studies

on Resources of American Libraries. A quarterly and annual companion volume—*Library of Congress catalog—book: subjects . . . a cumulative list of works represented by Library of Congress printed cards*—is prepared by the Catalog Maintenance Division of the Library of Congress. A *National union catalog* covering the years 1958-1962 was published in fifty-four volumes by Rowman and Littlefield, Inc, New York; this work includes the titles indexed by the Library of Congress during these years, about 830,000 titles, plus some eight hundred monographs published between 1956 and 1958 recorded by other American libraries. *United States government organizations manual* contains a special twenty-five page section devoted to the main publications of the Federal Government departments and services; *United States research reports* are published fortnightly by the Office of Technical Services, and the Library of Congress, Exchange and Gift Division issues a *Monthly checklist of State publications*. In addition, the United States Superintendent of Documents prepares a *Monthly catalog of United States government publications*. *Dissertation abstracts* are published monthly, in subject sections, with an annual cumulation; *Maps and atlases* forms part six of the *Catalog of copyright entries* and acquisitions of atlases are listed also in the *National Union Catalog* and in the *Library of Congress catalog—books: subjects.* The General Land Office issues a *List of cartographic records.*

LATIN AND CENTRAL AMERICA; WEST INDIES

El Salvador An official current national bibliography is in the early stages of establishment, prepared by the National Library. A national retrospective bibliography, covering the years 1830 to 1954, has been completed under the title *Bibliografía salvadoreña 1830-1954.*

Guatamala Printed matter subject to legal deposit has been listed since 1932 in the *Boletín de la Biblioteca Nacional*. The National Library publishes *Indice bibliográfico guatemalteco,* which has listed publications of all kinds, including periodical articles, since 1951; the Library also contributes to the *Bibliografía de Centro América y del Caribe,* with assistance from Unesco. Government publications are recorded in legislation compendia; university theses in *Publicaciones de la Escuela de Bibliotecología de la Facultad de Humanidades de la Universidad de San Carlos de Guatemala.* For more than fifty years, periodicals have been indexed by the National

Library of Guatemala and, since 1960, a special section, ' Hemeroteca nacional ', has been created.

Guyana A list of maps and plans is issued by the Cartographic Section, Lands Division.

Honduras Books, pamphlets, official publications, periodicals, maps and atlases are listed in the *Anuario bibliográfico,* 1961-.

Jamaica Plans are in train for a systematic current bibliography. In addition, *Jamaica: a select bibliography 1900-1963* was prepared by the Jamaican Library Service, the Institute of Jamaica and the University of the West Indies (Jamaica Independence Festival Committee, 1963); *A selective bibliography of Jamaica,* revised to December 1967, compiled by the Institute of Jamaica, was included in the *Handbook of Jamaica* (Government Printing Office, 1967) and is to be revised annually. *The printed maps of Jamaica to 1825,* the work of K S Kapp, was published by the Bolivar Press, Kingston, 1968.

Mexico The *Bibliografía mexicana* has been produced fortnightly since 1967, also a *Bibliografía histórica Mexicana* from the same date, by the Mexico University. The latter includes thesis material prepared in Mexican universities. Theses from American and Mexican universities are included in the *Latin American research review* supplement, compiled by Allen D Bushong. Retrospective bibliography has been compiled for certain periods, the most important being that for 1900-1958.

Peru The National Library of Peru publishes the *Anuario bibliográfico peruano,* which includes all Peruvian books and pamphlets, books and periodical articles about Peru, official publications, maps and atlases and printed theses. The Library issues also a *Boletín,* the review *Fénix* and reprints under the title, *Memoria.* Special bibliographies are published in the *Boletín bibliográfico* of the Universidad Nacional Mayor de San Marcos and in the Library's *Boletín.*

Trinidad and Tobago The *Current Caribbean bibliography* is published at Porto Rico by the Caribbean Organization. Apart from this work, retrospective bibliographies include *A guide for the study of the British Caribbean, 1763-1834,* compiled by Lowell J Ragatz (United States Government Printing Office, 1932); *Bibliography of the West Indies (excluding Jamaica),* by Frank Cundall (Institute of Jamaica, 1909); and the *Catalogue of the library of the West India Committee,* 1941. *Tropical agriculture,* quarterly review of the

184 Bibliographies: national control, from the point of view of geographical studies

Faculty of Agriculture in the University of the West Indies, lists research reports and theses.

Uruguay The National Library has published *Anuario bibliográfico uruguayo* since 1949, and the Biblioteca del Poder Legislativo began publication in 1962 of the *Bibliografía uruguaya* in three issues a year, of which the last is cumulative.

AUSTRALIA

The Australian Advisory Council on Bibliographic Services was established in 1956; its secretariat, the Australian Bibliographical Centre, forms part of the Commonwealth National Library. Since 1961, the Library has published the *Australian national bibliography*, which lists not only publications of Australian origin, but also publications relating to Australia. *Australian government publications,* monthly 1952-1960, was cumulated in the *Annual catalogue of Australian publications* in 1961 and has appeared annually since that date. In addition, the National Mapping Council of Australia publishes annually its *Pictorial index of activities* and the Hydrographic Service of the Royal Australian Navy has issued a *Catalogue and index of Australian charts* since 1962. In 1961, the Department of National Development published an *Index to Australian resources maps of 1940-59*. The Commonwealth National Library prepares, in separate parts, a *Guide to collections of manuscripts relating to Australia,* which mentions the place of deposit and gives a description of private documents as well as of documents from government archives. *Marshall's union list of higher degree theses in Australian University libraries* was inaugurated in 1959 and there have been three supplements since; and *Current Australian serials: a select list* has been issued by the National Library since 1963. The standard retrospective bibliography is Sir John Ferguson's *Bibliography of Australia . . . (qv).*

New Zealand The *Bulletin of the General Assembly's copyright publications* was published annually between 1934 and 1949, when it became a monthly, with annual cumulations; it includes books, pamphlets, official publications, maps and new serials. The *New Zealand national bibliography,* complete from 1890, has been published by the Government Printer and the National Library of New Zealand. The *Copyright publications bulletin* records all New Zealand official publications and the New Zealand Printing and Stationery Department also publishes monthly a special selective

Government publications. The *Union list of theses of the universities of New Zealand* has been undertaken since 1963 by the New Zealand Library Association.

185 'Bibliography and index of geology exclusive of North America', 1933- has been published by the Geological Society of America, 1934-. It is comprehensive; arrangement is alphabetical by authors, with a detailed subject and area index. Professor K M Clayton has prepared abstracts of current British geological literature, sponsored by OSTI, as a basis for the United Kingdom contribution to the bibliography.
Note The complementary bibliography, *Bibliography of North American geology,* 1919-, was published by the Society, 1931-.

186 'Bibliography of Africa', proceedings and papers of the International Conference on African Bibliography, Nairobi, December 1967, was edited by J D Pearson and Ruth Jones (Cass, 1970).

187 'Bibliography of African hydrology', compiled by J Rodier (Unesco, 1963), deals with precipitation, general climatological data and the effect of climatological factors, run-off, rivers, streams and lakes. The text is in English and French.

188 'Bibliography of agriculture', the monthly publication of the United States Department of Agriculture, ceased with the December 1969 issue. The work was taken over by the CCM Information Corporation, a subsidiary of Crowell Collier and Macmillan Inc. Data input will continue to be generated by the National Agriculture Library, the indexing and printing being handled by the CCM Corporation. The information contained remains substantially the same, but the subject index is to be enlarged by additional thesaurus terms. Production is by photo-composition.
Refer also World bibliography of agricultural bibliographies, which covers libraries, societies, dictionaries of terminology etc, kept up to date by regular supplements in the *Quarterly bulletin of the International Association of Agricultural Librarians and Documentalists.*
Agrarmeteorologische bibliographie, prepared annually by Maximilian Schneider (Offenbach am Main, Deutscher Wetterdienst—Bibliographien des Deutschen Wetterdienstes).

189 'A bibliography of arid-lands bibliographies', prepared by P Paylore (Natick, Earth Sciences Laboratory, US Army Natick Laboratories, 1967), brings together arid-lands bibliographies from citations in subject chapters comprising the compendium *An inventory of geographical research on world desert environments,* adds appropriate bibliographies from other sources and indexes the whole by geographic area and subject.

190 'Bibliography of Australia', compiled in four volumes by Sir John Alexander Ferguson (Angus and Robertson, 1941-1955), runs as follows: 1, 1784-1830; 2, 1831-1838; 3, 1839-1845; 4, 1846-1850. A fifth volume, covering 1851-1901, was published in three volumes 1963-, including printed books, pamphlets and broadsides, with certain exceptions.

191 'Bibliography of the Australian aborigines and the native peoples of Torres Strait to 1959', was compiled by John Greenaway (Angus and Robertson, 1963). The numbered entries are arranged alphabetically by author and a map is included—'Location of Australian tribes and key to location of tribes'.

192 'Bibliography of bibliographies on the Arctic', compiled by Artheme A Dutilly (The Catholic University of America, Washington, DC, 1945; 1946), is concerned with the region circumscribed by the 10 degrees C (50° F) isotherms for July. An appendix lists prominent names in Arctic exploration and their ships.

193 'Bibliography of books on Alaska published before 1868', the work of Valerian Lada-Mocarski, was published by the Yale University Press; it includes primarily first editions, in chronological order, illustrated by annotated facsimiles and giving a critical history in all major languages.

194 'A bibliography of British Columbia: *laying the foundations 1849-1899'* was prepared by Barbara J Lowther and Muriel Laing under the auspices of the Social Sciences Research Centre, University of Victoria, 1968. Arrangement is chronological by imprint and the data included brings in works published up to the end of 1964 which are concerned with the period of coverage; these may be significant general books, books with relevant chapters and serials.

Items are numbered and annotated except when the titles are sufficiently descriptive and a useful feature is the inclusion of location.

195 'A bibliography of British railway history' compiled by George Ottley (Allen and Unwin, 1965), is a classified arrangement of books, parts of books and pamphlets on the history and description of rail transport in the British Isles from the earliest known records to the present. Annotations are appended as necessary. The classification used is printed on the endpapers and there is an extensive introduction on the scope and use of the bibliography. Appendices set out The Railways Act 1921, First Schedule and ' Railway undertakings nationalised under the Transport Act 1947 '. Folded charts show the hierarchies of the London Passenger Transport Board, the various Midlands, Manchester and Great Western Railways and there is a very adequate index.

196 'A bibliography of Canadiana: *being items in the Public Library of Toronto, Canada, relating to the early history and development of Canada'*, edited by Frances M Staton and Marie Tremaine, was published by the Library in 1934. Books were selected from the Reference collection of the Library to form a chronological record of the history of Canada from its discovery to the Confederation of the Province in 1867 (1534-1867). Included are the journals of explorers, traders and missionaries from the Atlantic to the Pacific and from the Great Lakes to the Arctic Circle; the records of soldiers and administrators and the narratives of travellers. There is a special guide to government documents. The entries are numbered and the index refers to author or authority, title or catchword and subject. Following the same plan, a supplement edited by Gertrude M Boyle and Marjorie Colbeck was issued in 1959; this text includes Newfoundland and some errata.

197 'A bibliography of Fiji, Tonga and Rotorua', edited by Philip A Snow, was published by the Australian National Press in 1969; some ten thousand entries provide sources for study of the central Pacific.

198 'A bibliography of Ghana 1930-1961', compiled by A F Johnson, was published for the Ghana Library Board by Longmans in 1964. It is, in some respects, a continuation of *A bibliography of the Gold Coast,* prepared by A W Cardinall as a supplement to the Census Report of 1931. In the later work, an attempt is made to list all publications on the Gold Coast and Ghana, including selected periodical articles and translations. The work is arranged in sections according to category of material—bibliographies, periodicals, maps, government publications—also by some subject headings, such as 'Geography', 'Cocoa', 'Forestry', 'Fisheries', 'Agriculture' and 'Travel'; altogether an invaluable compilation.

199 'Bibliography of Kerala State, India' was compiled by William A Noble for the Geography Department, University of Missouri, 1970, and is an example of the local studies which are gradually leading to greater knowledge of the sub-continent. 1878 items, in European languages, are arranged within fourteen subject groups, with a special section for periodicals and an index to maps.

200 'Bibliography of the literature relating to New Zealand', compiled by T M Hocken (Government Printer, 1909) has been regarded as a standard work. Arrangement is chronological, with an Addenda, a Maori bibliography, 'Varia' and an index to authors and subjects, not titles. Full bibliographical details are given, also annotations when necessary. A supplement was issued by Whitcombe and Tombs in 1927.

201 'Bibliography of Malaya: *being a classified list of books wholly or partly in English relating to the Federation of Malaya and Singapore'* was edited by H R Cheeseman and published by Longmans Green, 1959, for the British Association of Malaya, 1959.

202 'Bibliography of maps and charts published in America before 1800', a valuable production with scope wider than indicated in the title, was compiled by James Clements Wheat and Christian Brun (Yale University Press, 1969). It is in fact an exhaustive, annotated list covering the territory which is now the USA, also parts of Latin America, Europe, Africa, the West Indies and Asia. The date of publication is included in the entries, description and size and scale and place of publication. Eighteen rare maps are reproduced.

203 'The bibliography of Mauritius (1502-1954)' was compiled by Dr A Toussaint and H Adolphe, superseding the *Select bibliography of Mauritius* by Dr Toussaint, which appeared as publication no 4 of the Société de l'Histoire de Maurice. This later bibliography includes printed records, manuscripts, archives and maps and is kept up to date by annual supplements.

204 '**A bibliography of New Zealand bibliographies**' was published in 1967 by the New Zealand Library Association. The association has also sponsored a *Guide to New Zealand reference material and other sources of information,* compiled by William John Harris and published in a second edition in 1950; supplements, by A G Bagnall, were issued in 1951 and 1957.

205 '**Bibliography of place-name literature:** *United States and Canada*', edited by R B Sealock and P A Seely (American Library Association, 1948; revised edition, 1967), contains references to 3,600 books and articles. Arrangement is by state and province; there is a complete author index and a sensibly compiled subject index.

206 '**Bibliography of printed Maori to 1900**', a valuable work edited by H W Williams (Government Printer, 1924, with a supplement, 1928), includes all books printed wholly in Maori or in Maori with a translation, but excluding works of wider scope. Full bibliographical details are given, frequently with annotations; arrangement is chronological, with indexes to authors and titles, but not to subjects.

207 '**Bibliography of reference materials for Russian area studies:** *a preliminary checklist to which is appended a selective list of periodicals and newspapers published in the USSR and by emigrés abroad*', compiled by Peter Gay (The City College Library, University of New York, 1962), was prepared in the course of developing the college collection in the area of Russian studies. Annotations are added as desirable and the work, divided into two main parts, relative to material in the college library and other material not at the time available, includes guides, bibliographies of periodicals and newspapers, general bibliographies, biographies, encyclopedias and research lists, bibliographies of bibliographies, subject bibliographies, with a separate section for 'Geography', publishers, library resources and a list of periodicals in English or Russian.

208 '**Bibliography of seismology**' has been compiled by the Canada Department of Mines and Technical Surveys, half-yearly from

1929 (Dominion Observatory, Ottawa). More than sixty-two journals have been scanned and the items are arranged alphabetically by author. A cumulative index covers 1947-1956.

209 ' Bibliography of soil science, fertilizers and general agronomy ', 1931/1934-, prepared by the Imperial Bureau of Soil Science (subsequently the Commonwealth Bureau of Soils), Harpenden, Hertfordshire, Commonwealth Agricultural Bureaux, has been issued triennially from 1935. The bibliography in each volume concentrates on a specific topic, arranged according to an adaptation of the decimal classification, plus a geographical sequence for regional items, with author and subject indexes.

210 ' Biblio-mer: *bibliographie de la mer, des marines, des eaux de mer et intérieurs: sciences, technique, enseignement, litérature'*, *'Livres et périodiques 1967-1968'* (Ostende, Editions Biblio-mer, 1969), was edited by R Roze and M Lelarge. The work is intended as an annual, classified bibliography, with analytical and critical notes of all French-language books, periodicals and other publications, wherever issued, on the subject of the sea and inland waters in the broadest sense; the sea, war and merchant fleets, instruction, history, science, techniques, organisation, economy and law. Full bibliographical descriptions are given, brief reviews and condensed analyses of most of the items. To allow mounting on standard cards, the type area of the pages does not exceed 103 mm.

211 ' Bibliotheca Americana: *a dictionary of books relating to America from its discovery to the present time'*, compiled by Joseph Sabin and others, was reprinted by N Israel of Amsterdam in twenty-nine volumes in 1961-62 from the New York edition of 1868-1936. This is an indispensable tool, being the total printed material published in England, France, Holland, Germany, Italy, Spain, Portugal, Scandinavia and the American countries relating to the western hemisphere including the Pacific Ocean. Joseph Sabin worked on the bibliography for twenty years prior to the beginning of publication in 1867. Thirteen volumes had been brought out at the time of his death in 1881; Wilberforce Eames continued the work and published six more volumes between 1884 and 1892. Then followed a break until 1927, when, with the assistance of the Bibliographical Society of America, the work was continued under the editorship of Wilberforce Eames and R W G Vail; it was completed in 172 parts in 1936. The twenty-nine volumes have been reprinted in fifteen volumes, using fine quality thin paper.

212 '**Bibliotheca Australiana**' is the collective title given to four series of facsimile editions which chronicle four hundred years of Pacific exploration from Magellan's voyage at the beginning of the fifteenth century to the highly organised scientific expeditions at the end of the nineteenth century. Significant journals and histories written during these four centuries by the mariners themselves and their crews are included, providing a chronological record of the development of knowledge concerning the Pacific Ocean. The series is published jointly by N Israel, and the Da Capo Press of New York; related volumes are issued in groups of ten to twelve, two to four groups appearing annually. For each work, the most important edition is reproduced, giving preference to English language editions, but in its own language if a work has not been translated into English. Whenever possible, works are published in English and in the original language. A feature is the reproduction of many maps and charts.

213 '**Bibliotheca cartographica**', an international bibliography covering all new publications relating to all aspects of the theory and practice of cartography, has been published twice a year from 1958-. The work was planned by a working group, the Arbeitskreis Bibliographie des Kartographischen Schrifttums (Commission on the bibliography of cartographic literature), set up in 1956 by the German Society for Cartography. More than five hundred publications are examined for each issue, by more than fifty experts and international organisations throughout the world; editing is done by the Institut für Landeskunde within the Bundesanstalt für Landeskunde und Raumforschung, with the Deutsche Gesellschaft für Kartographie, and publication is based on the Institute for Cartography and Topography, University of Bonn. Publication languages are German, English, French and Russian. A *General catalogue of cartographic literature* is also in preparation.

214 '**Bibliotheca geographorum Arabicorum**' is the title of a comprehensive selection of works originally published between 1870 and 1894 reprinted by Brill of Leiden, from 1967, edited by M J de Goeje. Eight volumes are planned, including descriptive and biographical works, systematic studies and a number of dictionaries and glossaries, such as the edition of T G J Juynboll of the great *Lexicon geographicum,* 1850-64, which together provide an invaluable reference source for the study of Arab geography and the contribution of Arab geographers.

215 ' Bibliothèque Nationale ', Département des Cartes et Plans, Paris, holds one of the world's great collections of maps, including a rich collection of ancient maps and portolan charts. Since 1830, the total French cartographical production has been kept, together with the most important of that of other countries. In 1963 was published an invaluable *Catalogue des cartes nautiques sur vélim conservées au Département des Cartes et Plans,* by Myriem Foncin and others; this includes the basic collection of the Bibliothèque Nationale, collections transferred to the library from the archives of the Service Hydrographique de la Marine and from the library of the Société de Géographie, with considerable descriptive annotation and bibliographies. The library sponsors and contributes to a vast amount of bibliographical work, both national and international.

216 ' Biogeography and ecology in South America ', edited by E J Fittkau and others (The Hague: W Junk NV, 1968 (1969)), in two volumes, is a collection of essays demonstrating the natural wealth of South America, in English, German, Spanish or Portuguese. Volume one considers the wider aspects of physical conditions, climate, agriculture, environmental change; the second treats of individual groups of flora, fauna, birds and fishes. The two volumes form Volume XIX of the *Monographiae Biologicae.*

Refer also Joy Tivy: *Biogeography: a study of plants in the ecosphere,* 1971.

David Watts: *Principles of biogeography: an introduction to the functional mechanics of ecosystems,* 1971.

217 ' Biological and agricultural index ', the successor to the H W Wilson Company *Agricultural index,* began monthly publication in 1964 (except September). The coverage of the 115 periodicals already indexed has been expanded to include the US Department of Agriculture publications, experiment station publications etc, making the present total of 146 periodicals scanned, from the United States, Canada and the British Commonwealth. The form of indexing is similar to that used in most of the Wilson indexes, with subject headings based on those used in the dictionary catalogue of the Library of Congress, and numerous subheadings and cross-references. There are annual cumulations.

218 Blaeu, Willem Janszoon (1571-1638) developed at Amsterdam a cartographical publishing house of the greatest importance. The House of Blaeu, carried on by his son, Jan Willenz (1596-1673), was responsible for some of the finest works of the seventeenth century, notably *Le grand atlas ou cosmographie Blaviane en laquelle est exactement descritte la terre, la mer, et le ciel*, originally published in Amsterdam, 1663, in twelve volumes, which were reprinted in facsimile by Theatrum Orbis Terrarum (*qv*), Amsterdam. This atlas is considered by scholars to be one of the most magnificent and important atlases known; Amsterdam being a centre of learning and geographical information, it was possible to create the atlas with the collaboration of artists, engravers and men of learning and experience from many countries. The text is in French, including geographical and historical descriptions of each of the 609 double-page maps. Editions of the atlas in French, Latin, Spanish and Dutch had been made available before a fire in the printing works terminated Blaeu's career and dealt an irreparable blow to Dutch cartography. The contents of the twelve volumes are as follows: ' World map, North Polar region, Denmark, Norway, Sleswick ' (61 maps); ' Sweden, Russia, Poland, Greece ' (41 maps); ' Germany ' (97 maps); ' Belgium and the Netherlands ' (63 maps); ' England ' (58 maps); ' Scotland and Ireland ' (55 maps); ' France, Switzerland ' (two volumes, 73 maps); ' Italy ' (60 maps); ' Spain, Portugal, Africa ' (41 maps); 'Asia ' (28 maps); 'America ' (23 maps). The reproduction is printed on specially designed paper, matching the paper used by Blaeu in texture and colour. The fine copies of *Le Grand atlas* in the Amsterdam and Leiden University libraries were used; all the beautiful frontispieces, some of Tycho Brahe's astronomical plates and thirty-two of the maps have been reproduced in full colour following the hand-coloured copy presented by Blaeu to Colbert, Minister to King Louis XIV. Dr C Koeman contributed an introductory study, in book form, to be used with the Atlas and Miss E L Yonge wrote about the colour plates in *The Library journal*, 1 June 1967. The Royal Geographical Society, London, and John Bartholomew and Son have co-operated to publish a series of reproductions of maps from the Blaeu Atlas of 1648, printed in full colours on antique cartridge paper.

Refer also Edward Luther Stevenson: *Willem Janszoon Blaeu 1571-1638: a sketch of his life and work with an especial reference to his Large World Map* . . . (The Hispanic Society of America, 1914).

C Koeman: *Joan Blaeu and his Grand Atlas* (Philip, 1970), in which chapters deal with ' Biography of Joan Blaeu ', ' Printing house and cartographical institutes', 'Origin and growth of Blaeu's atlas, ' The Atlas Maior ', ' The cartographical contents of the Grand Atlas ', ' The fire of 22nd February 1672 ' and ' Further history of the Atlas Maior '.

'A catalogue by Joan Blaeu: a facsimile with an accompanying text by Dr. Ir. C Koeman', published by N Israel, Theatrum Orbis Terrarum Limited and the Meridian Publishing Company for their friends at the occasion of the New Year, Amsterdam, December 1967 '. This is 'A catalogue of atlases, globes and maps published by Joan Blaeu (1670-71), preserved in the Plantin-Moretus Museum, Antwerpen ' with commentary, which includes further references.

S J Tyacke: ' Blaeu and Kepler ' *in The British Museum Society Bulletin,* October 1971.

219 ' Blumea ': ' Tijdschrift voor de systematick en de geographie der planten ', a research journal of plant taxonomy and plant geography, international in scope, was begun at the Leyden Rijksherbarium in 1954. Articles are mainly in English, illustrated and with frequent bibliographical references.

220 Board of Agriculture: *The review and abstract of the county reports to the Board of Agriculture,* by William Marshall, which were brought together in five volumes as a uniform series in 1818, have been reprinted by David and Charles of Newton Abbot. Together the volumes summarise the findings of the county by county surveys of the Board of Agriculture in the closing years of the eighteenth century and the opening of the nineteenth century. They run as follows: Northern Department; Western Department; Eastern Department; Midlands: Southern and Peninsula Departments. The reprinting of the actual *Reports* is in progress; each is being printed in its entirety, no new material being added, but, in due time, a book is planned which will discuss the project as a whole, together with critical comment. Already available are the volumes for Lincolnshire and Sussex, both by Arthur Young.

221 Board of Trade Library (London) of books, pamphlets and periodicals on all aspects of home and overseas trade and commerce is of the greatest value to geographers. When the library was re-organised in 1946, special divisions were created, including the statistics division library for the United Kingdom, Commonwealth and foreign countries, and a section devoted to the economic statistics of the United Nations; the whole collection of statistics is one of the best available. The board's publications are listed in the HMSO sectional list no 51, 1932- ; the weekly issues of the *Board of Trade journal*, 1886- and its indexes provide a central source of information. Since 1970, the Library has come within the newly formed Department of Trade and Industry.

222 Bodleian Library, University of Oxford; the Map Room in the New Bodleian, opened in 1947, has acquired during recent years more than eighteen thousand maps a year, making the total more than two million, invaluable for the study of the development of cartography. Holdings include many geographical treasures collected through several centuries; manuscripts, early local surveys of the British Isles and Europe and many famous collections of topographical books, such as the Gough Collection. The Gough Map (*qv*) is on view in the reading room. On open access are reference works and a selection of geographical texts, representative periodicals from many countries and some of the general and national atlases. The map catalogue, with its 'Area-scale-date' and 'Subject' headings, has been devised to give the reader the maximum help and information.

223 'Boletin de información científica Cubana' has been issued twice a year from December 1969; it contains abstracts of articles on all aspects of science and technology appearing in Cuban publications, arranged according to the UDC classification. The Institute can supply microfilms or photocopies of the original articles.

224 'The book of British topography', produced by John P Anderson in London, 1881, contains a classified catalogue of all topographical works in the library of the British Museum relating to Great Britain and Ireland. An unabridged reprint of the first edition has been published by N Israel, Meridian Press and Theatrum Orbis Terrarum.

225 Bowman, Isaiah (1878-1950) exerted a wide influence on world geography during a period of rapid change, both in the international scene and in the profession, especially as director of the American Geographical Society between 1915 and 1935 and as president of the International Geographical Union, 1931 to 1934. His best known published work, *The new world* (*qv*), reflects his theories and knowledge of international political geography. Under his guidance, the American Geographical Society became more influential, research and publications programmes were initiated, the monthly *Bulletin* became *The geographical review* (*qv*) and, in co-operation with other learned bodies, great impetus was given to American geographical work. The knowledge gained from wide travel and the far-reaching theories developed by his original mind are revealed also in his other published work, *Forest physiography: physiography of the United States and principles of soils in relation to forestry*, 1911, *The Andes of Southern Peru*, 1916, *Desert trails of the Atacama*, 1924, *Pioneer fringe,* an expansion of his ' pioneer fringe ' thesis, 1931 and *Limits of land settlement . . .,* 1937. Another influential publication was *Geography in relation to the social sciences,* 1934. The Isaiah Bowman School of Geography at Johns Hopkins University, Baltimore, was named in his honour.

226 ' Boundary layer meteorology ' began life in March 1970 (Reidel Publishing Company), edited by R E Munn and a panel of international experts. This quarterly journal, with others, such as *Agricultural meteorology* (*qv*), deals with those factors of meteorology which affect the environment; it is a handsome production, including diagrams and illustrations.

227 ' Bradshaw's canals and navigable rivers of England and Wales: *a handbook of inland navigation for manufacturers, merchants, traders and others',* compiled by H R de Salis, a director of a firm of canal carriers, 1904; the work has been reprinted by David and Charles of Newton Abbot, 1969. Detailed itineraries are given of 129 waterways, preceded by thirty pages of general information on various aspects of canals and their use. A glossary of canal terms is included.

228 Brill, E J, printers, publishers and booksellers of Leiden (Netherlands), are publishers of scholarly monographs and bibliographical works, with special interests in all aspects of Africa and the Near and Far East and in geographical subjects, biology, agriculture, economics and law. *Brill's weekly,* noting recent and forthcoming books, as well as new and second-hand books from stock, concentrates usually on a particular regional or thematic topic in each issue; *Brill's news,* each autumn, a most helpful publication arranged by broad subject headings, draws attention to works in preparation or recently published, especially series, frequently with full annotation and sometimes with accompanying plates. Through the year, catalogues are constantly issued on specific topics and a separate catalogue of periodical publications is available; the latest to be issued, no 440 (1971) is ' Beyond Calcutta—Far East, Indonesia, Pacific '. Individual leaflets are circulated, which give full particulars of every major book on publication. A recent work of the greatest significance to geographers in the study of geographical thought is the reprint *Bibliotheca geographorum Arabicorum* (*qv*). Special subject catalogues, published frequently, are particularly useful for regional studies, such as—' Peoples and travels '; ' Near East ': ' Books on South and Central Asia from E J Brill '; ' Eastern Europe '; 'Africa and ethnography—geography '; 'Africa-America —Europe '; and ' The Iberian Peninsula, the Maghrib and Africa (except Egypt)', the latter mainly from the collection of Professor R Bassett.

229 ' Britain and the British seas ', one of the most influential works of Halford J Mackinder (*qv*), was published in *The regions of the world* series (Heinemann, 1902), of which Mackinder was editor. In the preface, Mackinder states that Britain ' is known in such detail that it has been possible to attempt a complete geographical synthesis. The phenomena of topographical distribution relating to many classes of fact have been treated, but from a single standpoint and on a uniform method '. The book made a great impact, being the most detailed and successful study of this country to date, and it remains a pioneer work in regional treatment. The teacher is present throughout, in the clarity of the exposition and the inclusion of informative figures and sketch-maps in the text; some map plates were drawn specially by J G Bartholomew.

230 'Britain's green mantle, past, present and future', by A G Tansley (Allen and Unwin, 1949, 1968), traces the history of the natural vegetation of Britain from the end of the Ice Age, describing woodlands and grasslands, heaths and bogs, mountain vegetation, coastal dunes and salt marshes. An undoubted classic, written by the first chairman of the Nature Conservancy, it was completely revised for the second edition by M C F Proctor to adjust to interim changes. Though not written specifically with geographers in mind, the work, with its numerous fine photographs, makes essential background reading.

Refer also A G Tansley: *The British Islands and their vegetation,* 1939.

231 'Britannia', a topographical work of considerable interest to geographers, by William Camden, first published in Latin in 1586; two more London editions followed, then one published in Frankfurt, and an English edition appeared in 1610. The 1607 edition was the first to contain the county maps, engraved by William Hole and William Kip. The work has been reprinted in facsimile by David and Charles of Newton Abbot, 1971, with a new introduction by Stuart Piggott. The folio edition of 1695, prepared by Bishop Edmund Gibson, which is acknowledged to be the best, was used, retaining the splendid first edition English prose style, with careful editing; it includes a fine set of county maps engraved by Robert Morden. A bibliographical note by Gwyn Walters, Assistant Keeper of the National Library of Wales, is included.

232 'The Britannica world atlas' consists of 349 maps dealing with 'The world scene'—social and natural features, 'Travel and tourism' and 'World political geography', followed in greater detail by political and physical maps, geographical summaries and comparisons of most of the notable man-made features and an index gazetteer. There are numerous statistical tables and charts and explanatory text.

233 British Antarctic Survey, which originated in 1943, was known between 1945 and 1962 as the Falkland Islands Dependencies Survey. A *Bulletin,* published normally two or three times a year,

1963-, makes known the work of the survey; preliminary reports, notices of new projects, notes on papers and correspondence are also featured. The *Bulletin,* which is obtainable only from the survey, supplements the *Scientific reports,* the series in which are published results of the scientific work, except in hydrography, carried out in the British sector of Antarctica by the survey. The reports are published at irregular intervals, not necessarily in numerical order, and they are available from the survey or from HMSO. The *Bulletin* is illustrated by line blocks and half-tone plates, and valuable folded maps, plates and other illustrations accompany the reports.

234 British Association for the Advancement of Science: the monthly journal, *Advancement of science,* is necessary background reading for geographers, especially the annual address by the president of section E, ' Geography '. Valuable also are the scientific surveys prepared for the annual meetings held in different parts of the country. Traditionally, the work of editing and production has been entrusted to a geography department. The scope of the volumes has broadened during recent years and it has been interesting to trace progress both in method and presentation. Of many excellent regional studies, *Sheffield and its region: a scientific and historical survey,* edited by David L Linton, 1956, *Manchester and its region . . .* edited by Charles F Carter, 1962, *Nottingham and its region,* edited by K C Edwards, 1966 and *Durham County and City with Teesside,* edited by J C Dewdney, 1971 are perhaps particularly informative and well executed.

Refer O J R Howarth: ' The centenary of section E (Geography) ', *Advancement of science,* no 30, 1951.

235 ' The British bulletin of publications on Latin America, the West Indies, Portugal and Spain ' is published twice a year by the Hispanic and Luso-Brazilian Council, prepared in Canning House under the direction of the librarian, G H Green, with the collaboration of Dr A J Walford. Works are arranged under the headings of Latin America as a whole, then under an alphabetical listing of the individual countries and alphabetically by author. Bibliographical details are given for each entry, with a few lines of description.

236 British Cartographic Society was the first separate organisation in Britain concerned solely with cartography, founded at the end of

1963, following discussions at informal cartographic symposia at Edinburgh in September 1962, organised by the Departments of Geography and Edinburgh and Glasgow Universities, and a second at the University of Leicester in September 1963, ' to promote the development of cartography '. During the period 1965-66, a Map Curators' Group was formed and a Hydrographic Department was established. An annual general meeting is held in September; ordinary meetings, twice a year, and visits to map collections, also weekend symposia, have developed successfully. A *Newsletter* is being published as a valuable forum for the exchange of information. The progress of the society may be traced in the issues of *The cartographic journal* (*qv*). A Library has been established, housed at the Edinburgh Public Library; holdings of books and atlases are increasing rapidly, as well as current periodicals from all parts of the world, and accessions lists have been issued. In February 1971, Aslib, in conjunction with the Map Curators' Group, co-operated in arranging an advanced course on work with maps; it was intended for staff dealing with maps at professional level and included discussions on some of the more specialised problems of administration, recent advances in map reading, carto-bibliography and the study of maps.

237 ' British geological literature ', compiled by Edward L Martin and Anthony Harvey, has been issued quarterly from 1964 (Coridon Press, Bourne End, Bucks), with the aim of listing all books and articles on geology and allied subjects. It is available in two editions: A, a standard library edition, and B, a catalogue edition, supplied on loose sheets printed on one side of the paper only.

Refer also Kirtley F Mather and Shirley L Mason, *ed*: *A source book in geology 1400-1900* (Harvard UP/OUP, 1971).

238 British Geomorphological Research Group, now a Study Group of the Institute of British Geographers, was established by the Department of Geography, University College of Swansea, in 1961. Four working sub-committees investigate different aspects of geomorphology: a geomorphological map of Great Britain; slope studies; form of the sub-drift; surface and morphometric analysis. Publications include an annual *Register of current research in geomorphology,* 1963-, which lists specialists in Britain working on geomorphological topics throughout the world; and the *Bibliography of British geomorphology,* edited by K M Clayton (Philip, 1964), undertaken as a contribution to the work of the 20th International Geographical Congress, 1964. Relevant books and periodical articles published between 1945 and 1962 are included, with a selection of earlier works; the service is continued by *Geomorphological abstracts* (qv) and *Geographical abstracts* (qv). Occasional papers include *Geomorphology in a tropical environment,* papers given to the symposium of May 1968, edited by Anthony Harvey.

239 ' British interests in the Persian Gulf ', by Abdul Amir Amin (Leiden: Brill, 1967), describes British trade in the mid-eighteenth century by the British East India Company, the new developments in the Gulf and in India, the breakdown of Persian authority and the gradual removal of trade towards India. The work, which includes one folding map, provides a welcome piece of research into the economic history of the Anglo-Persian-Indian relationship.

240 British Museum: the original Cotton, Sloane and Harleian Collections all contained maps, the Royal Library was added in 1757 and the fine geographical and topographical collection assembled by George III was transferred to the museum in 1828. Maps, therefore, have formed an integral part of the collection, but the Map Room was not established as an independent department until 1868, and the manuscript maps remained in the Department of Manuscripts. In 1902, the Map Room became a division of the Department of Printed Books. The collection includes the whole of current British map production and a representative selection of topographical and special maps from other countries. The section of historical maps is especially strong, including a comprehensive series of maps and atlases from the fifteenth century, King George III's Topographical and Maritime Collections and the Crace Collection of London Plans, also a considerable number of manuscript maps, particularly early estate surveys. Some oriental maps are kept in the Department of Oriental Printed Books and Manuscripts. The current acquisition policy is to keep all maps on a scale of approximately 1:50,000 and all special maps of value, likely to be useful to students. Holdings of Ordnance Survey maps are complete; Admiralty charts are also deposited, but not the inter-edition corrections. Exchange of maps with the United States is comprehensive, but uneven for the rest of the world. A study room and open-access reference collection are available in the department and an extensive information service is maintained. On occasion, articles in the *British Museum quarterly* and *The British Museum Society bulletin* are of relevance, particularly on early topographical literature or maps; also the excellent catalogues of exhibitions, finely illustrated, such as *Prince Henry the Navigator and Portuguese marine enterprise,* 1960, which includes annotated entries.

See also Catalogue of the manuscript maps, charts and plans and of the topographical drawings held in the British Museum.

241 The British National Committee for Geography, set up by the Royal Society in 1920 to act as the British link with the International Geographical Union, has been re-constituted to make it more representative of geography in Britain today. The function of such committees, laid down in the statutes of the IGU (*qv*) is to nominate delegates to meetings of the Union and to promote the objects of the Union in the member countries. Professor J A Steers is the present (1971) Chairman and *ex officio* membership is extended to the Hydrographer of the Navy, the Director-General of the Ordnance Survey, the Director of Military Survey, the Director of Overseas Surveys, a Secretary of the Royal Society, the Secretary of the British Academy and the Chairman of the Cartography Sub-Committee, also elected members from many geographical organisations within the British Isles. A Study Committee has been set up to consider issues of topical importance concerning geography in the United Kingdom. The development of formal cartographic education in the United Kingdom was given a new impetus by a conference held at the Royal Society in September 1968, under the auspices of the Cartography Sub-Committee of the British National Committee for Geography. A full report was issued—' Education in cartography in the United Kingdom '—and the final resolution requiring the Department of Education and Science and the Scottish Education Department to set up a Joint Committee for the formulation of qualifications in the fields of cartography, photogrammetry and surveying at Ordinary and Higher National Certificate levels was duly carried into effect.

242 ' British weather in maps ' by James A Taylor and R A Yates (Macmillan, second edition, 1958; New York: St Martin's Press,

1967), presents a method of geographical analysis and interpretation of the 'primary documents' of British weather, especially the British Daily Weather Reports. The successive theories evolved to explain and classify the weather are described as a background to the appreciation of the format and function of the Weather Reports —'General considerations', 'Some fundamental properties of the atmosphere', 'The British Daily Weather Report', 'Air masses', 'Fronts', 'The anticyclone', 'Other synoptic types' and 'Classifications of British weather and climate'. Appendices present 'Dates and times of selected British Daily Weather Reports showing examples of standard front and air mass types'; 'A selection of symbols used on weather maps'; 'Procedure for the classroom analysis of weather maps'; 'Equivalents of temperature'. Throughout the work the weather codes used are those given in the publication *Instructions for the preparation of weather maps* (HMSO, 1954). There are figures and sketchmaps in the text; references are listed for chapter 7, 'Classifications of British weather and maps' and there is a short general bibliography, mainly of Meteorological Office publications.

243 Brunhes, Jean (1869-1930) studied with Paul Vidal de la Blache and in his turn contributed greatly to the reputation of the French school of geographical thought, especially in regional and human geography; a man of outstanding scholarship and a stylist, exerting very wide influence on academic work during his term as Professor of Human Geography at the Collège de France, 1912 to 1930. His chief published works include the pioneer study *La géographie humaine* (*qv*), *La géographie de l'histoire,* with C Vallaux, 1921 and *Géographie humaine de la France,* in two volumes, 1920 and 1926, each with a collaborator.

244 Brunswick International School Book Institute, created by the Council of Europe some years ago, includes a 'Europe centre'. The institute has so far collected about sixty thousand textbooks. Policy is for the institute to receive all new publications and new curricula and the German Research Community has sponsored the acquisition of books from North and Latin America. There are also plans for making even more international the *International yearbook for the teaching of history and geography,* which has been published by the institute for the past thirteen years (1971). The institute, headed by Professor Georg Eckert, has arranged more than a hundred bilateral or international conferences, many of them to discuss the standard and use of geography textbooks, atlases and maps. Five more European centres, based on the Brunswick model, are planned by the Council of Europe.

245 'Bulgaria: a bibliographic guide', compiled by Marin V Pundeff (Library of Congress, Slavic and Central European Division, Reference Department, 1965, 1968; New York: Arno Press), presents a bibliographic survey—a discussion of sources, grouped under seven subject headings: 'General reference works'; 'Land and people', etc, followed by a 'Bibliographic listing of publications discussed', each numbered and given a brief bibliographical description.

246 'Bulletin of material on the geography of the USSR' was published by the University of Nottingham Department of Geography between 1958 and 1961, while its usefulness seemed assured. Readers are reminded of *Soviet geography* . . . and the increased number of new texts in English on Russian geography and of English abstracts or translations in Russian works. The twelve issues of the Bulletin included 'Agricultural statistics' by C A Halstead, 1958; 'New economic regions in the USSR', by J P Cole, 1958; 'Industrial statistics', by J P Cole, 1958; 'Transport statistics and notes', by R E H Mellor, 1959; 'Population statistics' by J P Cole 1959, in two issues; 'Selected data from the *Soviet statistical yearbook* for 1958', by J P Cole, 1960; 'Nationalities of the USSR in 1959' by J P Cole, 1960; 'Agricultural statistics', by C A Halstead, 1961; 'Economic geography in the USSR, 1960' by F C German, 1961; 'The Soviet iron and steel industry', by R E H Mellor and J P Cole, 1962.

247 'Bulletin of quantitative data for geographers' edited by J P Cole and published by Nottingham University from 1965, aims to inform geographers on quantitative data not readily available due to language or other difficulties and to introduce and show the application of mathematical and statistical processes in geography.

248 'Die Bundesrepublik Deutschland in Karten' is the most comprehensive documentary cartographical work to date on all aspects of the land and communities of Germany, 1965- by the Institute for Space Research at Hanover, the Federal Office of Statistics and other authoritative bodies (W Kohlhammer, 1965-).

249 The Butterworth Group has published a number of books on geography, all well illustrated and produced: T R Tregear: *An economic geography of China,* 1970; J P Cole and F C German: *A geography of the USSR: the background to a planned economy,* second edition, 1970; J P Cole: *Latin America: an economic and social geography,* second edition, 1970; I B Thompson: *Modern France: a social and economic geography,* 1970; Sir Raymond Priestley and others: *Antarctic research: a review of British scientific achievement in Antarctica,* 1964; N C Pollock: *Studies in emerging Africa,* 1971; Chiao-Min Hsieh: *Taiwan-Ilha Formosa: a geography in perspective,* 1964.

250 ' Cahiers de géographie de Québec ', journal of the Institut de Géographie, Université Laval, Quebec, superseded the Institute's two series, *Cahiers de géographie* and *Notes de géographie,* in 1956. Two issues are published each year; articles are mainly in French, with English summaries, and informative book reviews are a feature. Indexes are prepared annually and five-yearly.

251 ' Cahiers d'outre-mer: *revue de géographie de Bordeaux'* is a most interesting and valuable journal, produced quarterly from 1948 by the Institut de Géographie de la Faculté des Lettres de Bordeaux. Annual indexes are published, with ten-yearly cumulations.

252 ' Cambrian bibliography: *containing an account of the books published in the Welsh language, or relating to Wales, from 1546 to the end of the eighteenth century, with bibliographical notes',* by W Rolands, edited and enlarged by D Silvan Evans, 1869, has been reprinted by the Meridian Publishing Company, Amsterdam, in collaboration with N Israel, 1970.

253 ' Cambridge expeditions journal ', annual publication of the Cambridge University Explorers' and Travellers' Club, 1969/70-, has taken in *Visa*. The journal exists ' to provide a definitive list and factual description of all expeditions planned and undertaken in the year: to provide a medium for expressing our appreciation of the services rendered to these expeditions by the firms of the

Cambridge Block Order Scheme, and of the financial and informational assistance given by countless individuals and organizations to each expedition'. The texts are accompanied by photographs, sketchmaps and figures.

254 Cambridge University Department of Geography was founded in 1888 with periods of re-organisation in 1919 and 1937. The library, classified by its own scheme, is primarily a working library for students and staff; in addition, some valuable collections are held, especially a collection of eighteenth and early nineteenth century books on discovery and travel. The Cambridge University Geographical Society has since 1920 made useful contributions to original work.

255 ' Canada: a geographical interpretation ', prepared under the auspices of the Canadian Association of Geographers, was edited by John Warkentin (Methuen, 1968) on the occasion of the centennial of Canadian Confederation and was subsidised by the Centennial Commission. The book is divided according to the following topics: ' The setting ', ' Lands and people ', ' People and places ', ' Relations and trends '. Each part is followed by a list of references and there is also a ' Note on sources '. Photographs and diagrams accompany the text and there are maps in a pocket.

256 Canada, Department of Mines and Technical Surveys, Geographical Branch, developed in 1951 from the Geographical Bureau of 1947, established ' to collect and organise data on the physical, economic and social geography of Canada '. Two divisions operate for the Canadian work, designated Regional Geography and Systematic Geography; the Foreign Geography Division works on studies of foreign areas of interest to Canada. A *Geographical bulletin* has been published irregularly from 1951, in English and French; the articles are authoritative and illustrated with photographs and maps in the text, also frequently with maps in a separate pocket. Abstracts are included. An index is cumulated for every four issues. *Memoirs* and *Papers* have also been published and a *Bibliographical series* began in 1950. The occasional monographs are valuable, as, for example, *An introduction to the geography of the Canadian Arctic,* 1951. Other divisions of interest to geographers include the Surveys and Mapping Branch, the Marine

Sciences Branch, the Dominion Observatories and the Mineral Resources Division.

See also Geological Survey of Canada.

257 'Canada in maps', with French and English text, contains documents drawn from an exhibition arranged by the National Map Collection, Public Archives of Canada, in October 1969-February 1970. The Map Division was created in 1908, but the collection began in the 1880s and it now provides a focal point and central management control for projects of national interest. Section I deals with ' The age of discovery ' and ' Canada emerges '; II, ' Exploration '; III, ' Political development '; IV, ' Settlement '; V, ' Urban development '; VI, 'Architectural, military, engineering and naval plans '; VII, ' Hydrography '; VIII, ' Canada's maps today ' and, finally, 'Atlases display ' from four cases. Entries include annotation and the location of each item. The Collection numbers more than five hundred thousand items, rapidly increasing, and the Division is responsible also for Canada's contribution to the *Bibliographie cartographique internationale* (*qv*).

258 Canadian Association of Geographers is the association of professional geographers in Canada, founded in 1950. The first conference was held in the following year, and the *Proceedings* of these conferences have been published in a semi-annual publication, the *Canadian geographer,* which is bi-lingual, with abstracts in either English or French. Book reviews are included.

259 ' Canadian cartography ', 1962-, is the title given to the Proceedings of the Symposium on Cartography, The Canadian Institute of Surveying, Ottawa, first held in Ottawa, 1962; the second volume reports on the Symposium of 1964, and so on every other year.

260 ' Canadiana ', monthly publication lists the works of Canadian interest received by the National Library (National Bibliography Division). A main development in *Canadiana* 1970 was the inclusion of sound recordings; Library of Congress classification numbers are also added to the entries in Part I, except those in the fields of Canadian history, literature and law. Queen's Printer catalogue numbers are appended. The first section includes the main body of material, catalogued and classified by DC, both adult and works for young people; the second section comprises pamphlet material and brochures; the third, microforms; 4, films; 5, publications issued by the Government of Canada and 6, publications of the Provincial Government of Canada.

261 'Canadiana before 1867', a series of reprints published to coincide with the centenary of Canadian Federation in 1967, issued by SR Publishers Limited, was prepared under the auspices of La Maison des Sciences de l'Homme, Paris, The Humanities Research Council of Canada, The Social Science Research Council, Ottawa, and the Toronto Public Library, on history and government. Also included are, for example, 'An account of the countries adjoining to Hudson's Bay', by Arthur Dobbs, 1744; 'General introduction to the statistical account of Upper Canada', by Robert Fleming Govilay, 1822; 'Statistical account of Upper Canada' by the same author, 1822; 'British Columbia and Vancouver Island', by W C Hazlitt, 1858 and 'An account of Prince Edward Island in the Gulf of St Lawrence, North America', by John Stewart, 1806, a survey of the geography, history, industries and natural resources of the Island.

262 'Cappelen road and tourist map of Norway', the first such series of an international standard to become available, covers the entire country. The map is based on the most recent material from the Norges Geografiche Oppmåling (The Norwegian Geographic Survey Department) and the tourist information represents the accumulated result of investigations carried out by several hundred contributors throughout the country. Six colours are used: road networks are classified according to the latest road regulations, highway and road numbers being in line with the new system brought into effect in June 1965. Thirty-five thousand placenames are inserted. The key, shown on the reverse of each map, includes some sixty symbols representing information likely to be of use to tourists, given in English, French and German. On the four most northerly maps, coastal steamer routes are indicated. The maps are folded into a stiff paper cover. In the United Kingdom, John Bartholomew and Son Limited are the distributors.

263 Caribbean Council developed from the former Caribbean Organisation, which itself succeeded the Caribbean Commission; all functions and the library have been transferred. A *Current Caribbean bibliography* was begun by the library in 1950 (1951-);

other bibliographies, catalogues, bulletins, pamphlets, reports and surveys are issued from time to time.

264 'Caribbean review' has been published quarterly in Puerto Rico since April 1969; articles, essays and bibliographies are included, in Spanish or English. Reviews are a main feature, with the intention of covering all the main books relating to Latin America and the Caribbean and each issue contains a list of recent books, classified by subject.

265 'Caribbean studies', a quarterly journal published by the University of Puerto Rico Institute of Caribbean Studies, includes current bibliographies of books, pamphlets and periodical articles.

266 'Caribbeana 1900-1965: *a topical bibliography*', compiled by Lambros Comitas (University of Washington Press for the Research Institute for the Study of Man, 1968), contains more than seven thousand references to books, monographs, reports, articles and miscellaneous publications, arranged on the following plan: a general introduction, ' The past ', ' The people ', ' Elements of culture ', ' Health, education and welfare ', ' Political issues ', ' The environment and human geography ', ' Socio-economic activities and institutions ', ' Soils, crops and livestock ', ' Economic and social projects '—each being sub-divided as necessary, with author and geographical indexes. Each entry bears a reference number and a geographical code.

267 'Cartactual', an international bi-monthly topical map service begun in 1965 by Cartographia of Budapest ('Kultura '), is produced in English, French and German, under the editorial direction of Professor Dr Sándor Radó. Bi-monthly booklets, each providing a set of maps on various topics, printed as loose-leaf double page maps, trace recent changes in geographical information throughout the world. The first issues were in monochrome, each map being carefully drawn in simple style, with adequate documentation and scale. In volume 2, 1966, red colour was introduced to show changes, black remaining for existing detail. Aspects especially noted are modification of administrative boundaries, expansion or reduction of transport systems, changes in the numbers of inhabitants of settlements and the foundation of new settlements, new geographical names and other such data.

268 'Carte de France' on a scale of 1:80,000 and, by photographic enlargement, at 1:50,000, was completed as early as 1878, the sheets being printed in monochrome. Relief was shown by hachures. It is instructive to compare this series with that of the first official British mapping programme as the earliest national surveys to be carried out on scientific principles. The *Nouvelle carte de France* at 1:20,000 now provides the basic map for all cartographic work.

269 Carto-bibliography as a term came into current use with the increasing variety of maps and the universal recognition of their uses in contemporary life, leading to a corresponding increase in the literature concerning them and in the guides to collections, both current and retrospective. The first General Assembly of the International Cartographic Association was held in 1961 and the *Bulletin,* containing reports of the cartographic activity of member nations, first appeared in 1962. Among the organisations making regular contributions to the subject are the International Association of Geodesy, the International Society for Photogrammetry, the Centre de Documentation Cartographique et Géographique, the Institute of Geodesy, Photogrammetry and Cartography of Ohio State University, the Ontario Institute of Chartered Cartographers and the British Cartographic Society (*qv*). Of central importance are *New geographical literature and maps* (*qv*), the 'Cartographic progress' section of *The geographical journal* (*qv*), the 'Geographical record' section of *The geographical review, Cartography* (*qv*) and the *Kartographische nachrichten* (*qv*) of the Deutsche Gesellschaft für Kartographie eV. *World cartography* (*qv*) is the first source of information for international work. Individual national coverage in carto-bibliography still varies considerably; maps are included in perhaps about half of the national bibliographies. Countries such as Portugal and the Netherlands have been particularly active in compiling retrospective bibliographies, with such major works as the *Atlantes Neerlandici* (*qv*). The reports of individual mapping agencies are contributions to national bibliographical completeness, such as the monthly *Publication reports* and the guides and other informative material issued by the Director General of the Ordnance Survey (*qv*). *Imago mundi* (*qv*) is the international journal covering historical cartography and the *International yearbook of cartography* (*qv*) is a unique publication including finely illustrated articles by leading world authorities. The

Bibliographie cartographique international (*qv*) provides an extensive coverage, as does the *Bibliotheca cartographica* (*qv*). A novel and highly successful venture is the *Zumstein katalog* . . . (*qv*). Retrospective bibliographies, each devoted to a particular period or type of material, include *Official map publications* . . . (*qv*), *The printed maps in the atlases of Great Britain and Ireland* . . . (*qv*) and *Map collections in the United States and Canada* . . . (*qv*). The catalogues and accessions lists of such great map collections as those of the British Museum, the Library of Congress, the Bibliothèque Nationale and the Bodleian Library are essential bibliographical sources, as are the catalogues of the great national cartographical publishing houses, such as Bartholomew, Esselte, Cartographia of Budapest, Rand McNally, Georg Westermann and the Stanford annual international catalogue. General bibliographical services usually include a section for maps and atlases, such as *Whitaker's cumulative book list* and section K of *British books in print*.

Refer also Denoyer-Geppert, Chicago: *Maps, globes, charts, atlases, models, transparencies*—annual.

Harold Fullard: 'Atlas production of the 1970's' *in The geographical magazine,* October 1969.

G R P Lawrence: *Cartographic methods* (Methuen, 1971; *The field of geography* series, which includes information on atlases and map series).

Muriel Lock: *Modern maps and atlases* . . . (Bingley, 1969).

Documentatio geographica, 1966-, bi-monthly with annual cumulations (Bad Godesberg; Institut für Landeskunde).

K A Salichtchev: ' Contribution of geographical congresses and the International Geographical Union to the development of cartography ' *in IGU Bulletin,* Volume XXII, no 2, 1971.

270 Cartographia of Budapest, one of the leading European cartographic houses, has made major contributions to modern mapping. Among their best known works are the London, Madrid and Rome sheets of the 1 : 2,500,000 world map, published by the East German Department of Geodesy and Cartography, 1962; and the ' Hydrographic map of Europe's surface waters ', 1 : 10M, 1965.

271 ' The cartographic journal ', issued twice a year from 1964 by the British Cartographic Society, is of the highest standard both in content and production. Technical articles, world-wide in scope, on all aspects of contemporary cartography are suitably illustrated; society news is included, also cartographic news and notices of meetings and symposia. Useful features are careful signed reviews, ' Recent literature ', including both books and periodical articles, and ' Recent maps ', which covers also atlases. An index for volumes 1 to 4 was distributed with the December 1968 issue; articles were subject classified, authors' names arranged alphabetically, while maps, diagrams and illustrations were indexed by topic and there is also a separate index of speakers whose papers have been briefly reported.

272 ' Cartography ', journal of the Australian Institute of Cartographers, Melbourne, has been issued irregularly from 1954, usually twice a year. Important features are the ' Cartographical abstracts ' and reviews of overseas publications in this field. An index is compiled every two years.

273 ' Cartography of the Northwest Coast of America to the year 1800 ', by Henry R Wagner, originally published in Berkeley, 1937, was reprinted by N Israel of Amsterdam in 1967. The two volumes, bound in one, contain thirteen folding maps. The first volume traces the evolution of the cartography of the region to the explorations of the Vancouver expeditions; the second comprises a critical and descriptive list of more than nine hundred maps, with an index, a list of place-names still in use, one of those now obsolete and a bibliography.

> Refer also H Harrisse: *The discovery of North America . . . with an essay on the early cartography of the New World,* 1892.
> E D Fite and A Freeman: *A book of old maps delineating American history,* 1926.

274 Cary, John (1754-1835), one of the most brilliant of English map-makers, made great advances in technique in his maps of the British Isles, plans of London, large and small county maps and a number of special-purpose maps depicting such features as canals and post-roads. He surveyed nine thousand miles of turnpike roads for the Postmaster-General, incorporating much of his observed detail in his atlases, especially in the *New and correct English atlas,* 1787 and many other editions to 1831, *Cary's travellers' companion,* 1790, *Cary's new itinerary,* 1798, and the *New British atlas,* in collaboration with John Stockdale, 1805. Cary's county maps of Cardiganshire, Middlesex, Oxfordshire, Glamorganshire and the plan of London and Westminster, published between 1786 and 1803, reveal new advances in geographical delineation. The firm of G and J Cary produced also guide books, road books, globes, celestial charts and instruments of many kinds, all set out in their *Catalogue of maps, atlases, globes and other works published by G and J Cary,* London, 1800.

Refer Sir H G Fordham: *John Cary, engraver, map, chart and print-seller and globe-maker, 1754-1835: a bibliography, with an introduction and biographical notes* (Cambridge, 1925).

275 Casella, C F, and Company, Limited, London, are not only one of the foremost firms making scientific instruments for meteorology, research and industry, but also issue useful catalogues on specific types of instruments, such as ' Meteorological instruments ', ' Humidity, dew point and frost point ', ' Barometers and manometers ', ' Photogrammetry '. Leaflets include 'A selection of meteorological instruments for schools ', ' Wind velocity and direction ', 'Atmospheric pressure ', ' Temperature ', ' Humidity ', ' Precipitation and evaporation ', ' Sunshine, solar radiation and cloud ' and 'Alnor dew point meter '. A range of colour films and slides has been produced with the aim of assisting in the selection of instruments for particular purposes, their setting up and use, including guidance on the choice of site, each with explanatory notes. Filmstrips include a ' Screen colour guide to clouds ', ' Unstable weather' and ' Stable weather '. The firm also produces the Air Photo Packs compiled by The Geographical Association.

276 Cassini, Jean Dominique (1625-1712) was the founder of the distinguished family of scientists. His son, Jacques (1677-1756), evolved the Cassini projection for the first scientific triangulation of France, and with his son, César François Cassini de Thury (1714-1784) initiated the Survey of France, the first precise national survey, which greatly influenced the progress of the Ordnance Survey in Britain and led in due time to the trigonometrical survey of France and Britain by General Roy in 1787.

277 ' Catalan atlas ' of 1375, traditionally attributed to Abraham or Yehuda Cresques and dedicated to King Charles V of France, is a masterpiece of the cartography of the period. The first two of the six vellum leaves comprise the text in Catalan; illustrations of men and animals are illuminated in gold, silver, red, orange, violet, blue and green. The remaining four leaves—the Nautical Atlas—contain four maps which, placed next to one another, give a picture of the world as then known, with many illustrations and explanatory texts. The maps of the areas from the British Isles to the Black Sea are of the portolan type and of a high standard of accuracy and cartography; the maps of the Near and Middle East and of north Africa show information gathered from Venetian and Genoese merchants, while towards central Asia and China the knowledge brought back by the Polos and other missionaries and travellers has been inserted.

>*Refer also* J A C Buchon and J Tastu: *Notice d'un atlas en Langue Catalane,* 1375 (Paris, 1839).
>Pinhas Yoeli: 'Abraham and Yehuda Cresques and the Catalan Atlas', *in The cartographic journal,* June 1970.

278 ' Catalog of African Government Documents and African Area Index' (Boston University Library, second revised edition, 1964, published by G K Hall) contains in one volume references to about four thousand monographs. The African Area Index is an alphabetical index to book materials by topics under the area to which they belong.

279 ' Catalogs of the Scripps Institution of Oceanography Library', University of California, San Diego, have been reprinted by G K Hall: the Author-title catalogue, in seven volumes, 1970; the Subject catalogue, in two volumes, 1970; and Shelf-list, in two volumes, 1970 and the Documents-Reprints-Translations Shelf-list, in one volume, 1970. The library holds outstanding collections in oceanography, marine biology and marine technology, with imprint dates ranging from 1633 to the present. In addition to a basic reference collection in mathematics, physics and chemistry, the collection is strong in atmospheric sciences, fisheries, geology, geophysics and zoology; it holds a major collection of oceanographic expedition literature, reports, special documents and translations.

280 ' Catalogue of books, maps, plates on America *and of a remarkable collection of early voyages . . . presenting an essay towards a Dutch-American bibliography',* by Frederik Muller, printed in Amsterdam 1872-1875, has been reprinted by N Israel, 1966, with a subject index and an index of personal names compiled by G J Brouwer.

281 'A catalogue of Latin American flat maps 1926-1964', compiled by Palmyra V M Monteiro (Institute of Latin American Studies, University of Texas at Austin, 1967), is arranged regionally; bibliographic details are given, but there are no annotations.

282 ' Catalogue of the Malaysia/Singapore Collection, University of Singapore Library ', was printed in one volume by G K Hall in 1969. The collection, of some 7,500 items, is particularly strong in source material tracing the various aspects of the development of the Malay States, Singapore and the Bornean regions, but excluding Brunei and Indonesian Borneo. Arranged by the Library of Congress classification, holdings include public records, government publications, long runs of serials, theses, manuscripts and translations. Titles are given mostly in English and some bear imprint dates as early as 1596.

283 ' Catalogue of the manuscript maps, charts, and plans and of the topographical drawings held in the British Museum ' was planned in three volumes. Volumes I and II were originally published in 1844, reprinted in 1962; volume III was printed in 1861, but never published, the impression being destroyed in a fire. The 1962 reprint was made from a copy comprising a set of surviving sheets and was in fact the first publication of volume III. The *Catalogue of the printed maps, plans and charts in the British Museum,* compiled by R K Douglas, published in two volumes, 1885, has been supplemented thereafter by annual volumes of the *Catalogue of printed maps in the British Museum: accessions,* 1884-. The working copies in the museum have been maintained in guard-books, now numbering eighty-four, incorporating the columns of the original catalogue and the accession entries added from year to year. A photolithographic edition of the *Catalogue of printed maps, charts and plans,* recording the holdings of the museum to the end of 1964, in fifteen volumes, 1967, is a valuable tool for carto-bibliographical work.

284 Cataloguing books on geographical subjects presents no particular problems, but the cataloguing of a map collection demands specialised knowledge. All the great map departments have formulated their own rules for cataloguing their collections; a comparison of their findings reveals interesting differences, not so much of opinion as of emphasis. The rules evolved by the British Museum, the Bibliothèque Nationale and the War Office are especially clear and detailed. In addition, the following works are informative: Samuel W Boggs and Dorothy Cornwell Lewis: *The classification and cataloging of maps and atlases,* Special Libraries Association, second edition, 1945; Mary Ellis Fink: ' A comparison of map cataloging systems ', *Bulletin,* Special Libraries Association, Geography and Map Division, December 1962; *The catalog library and the map librarian,* part IV of the *Cartographic research guide,* Special Libraries Association, Geography and Map Division, 1957; B M Woods: ' Map cataloging ', *Library resources and technical services,* Autumn, 1959; R J Lee: *English county maps: the identification, cataloguing and physical care of a collection,* Library Association, 1955, pamphlet no 13. Reports on American cataloguing practice regarding maps, atlases, globes and relief models were included in the *Bulletin* of the Geography and Map Division for December 1948, October 1953 and April 1956. *The Anglo-American Cataloging Rules,* 1967, for maps, relief models, globes and atlases are considered in Chapter 11; ' To avoid excessive handling of maps, the catalog entry should give as much aid as possible to the reader in the selection or rejection of a particular map '.

285 Cave Research Group of Great Britain was founded in 1946 for the encouragement of the scientific study of caves and of their past and present contents and inhabitants. Two general meetings are held each year for the reading of papers, which are printed in the *Transactions.* The *Occasional papers* have included such studies as E A Glennie and M Hazleton: *Cave fauna,* 1947; N Kirkham: *Derbyshire lead mining glossary,* 1949; and A L Butcher: *Cave survey,* 1950. The group possesses a Library, housed in Berkhamsted, Hertfordshire, which holds long runs of British and foreign speleological journals and a considerable collection of books on caves and limestone regions throughout the world.

Refer from the vast literature of journals and monographs: *Cave science,* quarterly, 1947-, journal of the British Speleological Association.

Annales de Spéléologie, quarterly, Centre National de la Recherche Scientifique.
Proceedings, annual, of the University of Bristol Speleological Society.

286 Central Office of Information *Reference pamphlets* series is among the most useful to geographers of all the publications of the Department, particularly *The making of a nation* booklets: Nigeria, 1960; Sierra Leone, 1960; Tanganyika, 1961; Jamaica, 1962; Trinidad and Tobago, 1962; Uganda, 1962; Kenya, 1963; Zanzibar, 1963; Malta, 1964; Malawi, 1964; The Gambia, 1964; Zambia, 1965; Guyana, 1966.

287 The Centre d'Analyse Documentaire pour l'Afrique Noire, Paris (CADAN), aims to provide documentary information of research workers in the social and human sciences on Africa South of the Sahara. The Centre systematically lists and analyses the relevant articles and scientific works, including some two hundred periodicals on African matters, plus as many of more general interest, together with national bibliographies, publishers' catalogues and the acquisition lists of the large libraries. This scanning yields some four thousand abstract cards a year, so far used in two different ways: either placed at the disposal of workers in standard card index files or processed by automatic data methods (SYNTOL), worked out in the CNRS automatic data section to facilitate automatic location of the information they contain.

288 The Centre for East Asian Cultural Studies, Tokyo, has published the following valuable aids: *A survey of Japanese bibliographies concerning Asian studies,* 1963; *Research institutes for Asian studies in Japan,* 1963; *Japanese researchers in Asian studies,* 1963; *Research institutes and researchers of Asian studies in Thailand,* 1964; *Bibliography of bibliographies of East Asian studies in Japan,* 1964; *A survey of bibliographies in Western languages concerning East and South East Asian studies,* 1966.

289 Centre for Environmental Studies, London, an independent body sponsored jointly by the Ford Foundation and the United Kingdom Government, was set up in 1967 to promote research and education in the planning and design of the physical environment and as a forum for discussion on an international basis. More than thirty researchers, full- or part-time, are working in eleven institutions on such problems, among others, as industrial location, population mobility, the housing market, activity networks, communication in re-housing and atmospheric pollution. The internal research programme of the centre itself began in 1968 with the appointment of a research team trained in town planning, sociology, transport engineering, mathematics, economics, physics, geography, systems analysis or market research. 'Patterns of urbanisation' is the subject of a current working party.

290 Centre National de la Recherche Scientifique publishes a number of series of interest in geographical studies. In addition to the *Bibliographie géographique internationale* (*qv*) and the *Mémoires et documents du Centre de Recherches et Documentation Cartographiques* (*qv*) are the *Annales de spéléologie, Annales de géophysique, Annuaire d'Afrique du nord* and individual maps and monographs. *Le bulletin signalétique* includes sections on earth sciences and agriculture.

291 'Ceres', the FAO review, published in six issues a year in three editions, English, French and Spanish, is concerned with the adaptation of peoples to their environment, urbanisation, re-housing, the role of technology in agriculture and the dramatic social and economic changes that precede and accompany development. 'State of the market' commodity reports are included, a 'World report', short topical articles, reports and news items, short notes in a section entitled 'Forum', reviews of appropriate books and a list of 'Books received', the whole being accompanied by illustrations and the occasional lively cartoon.

292 The Challenger expedition, 1872-1876, which sailed round the globe under the leadership of Charles Wyville Thomson (later Sir Wyville), heralded the large-scale, systematic science of the seas. Reports of the expedition, *The Challenger reports,* begun by Sir Wyville Thomson and completed by Sir John Murray, run to some fifty volumes. Sir John Murray compiled also *Report on the scientific results of the voyage of HMS Challenger during the years 1873-76* in two volumes.

293 The Challenger Society was founded in 1903 to promote the scientific study of oceanography; it is based now on the National

Institute of Oceanography, Wormley, Surrey (*qv*). The *Annual reports* and *Occasional papers* are documents of central importance. Scientific meetings are held quarterly in the rooms of the Linnean Society and joint meetings with representatives of marine laboratories are held three times a year at various laboratories and other institutions.

294 ' **Chambers's historical atlas of the world** ' is intended for the use of students as a companion to their studies, being of pocket-size format, containing 108 six-colour maps and a seven thousand entry index. Treating first the span of the earliest known civilisations from the third century BC, the maps trace the main streams of world history, influenced by geographical, economic and political factors, to the complex and numerous developments and movements of the twentieth century.

295 ' **Chambers's world gazetteer and geographical dictionary** ', 1954, was revised by T C Collocott and J O Thorne in 1957. Emphasis is on Great Britain. Information given includes location, description and population, with some longer articles on more important places and some short definitions of geographical terms. A supplementary index includes variant names and alternative spellings.

296 ' **The changing nature of geography** ', by Roger Minshul (Hutchinson University Library, 1970, in hardback and paperback editions), examines the nature of geography and of geographers, the content of geographical studies and the methodology which has evolved through the years to the new quantitative and qualitative techniques. There are references and notes at the end of each chapter and a separate bibliography, including articles as well as books.

297 ' **Character of a conurbation:** *a computer atlas of Birmingham and the Black Country* ', compiled by Kenneth E Rosing and Peter A Wood and published by the University of London Press, is a collection of some fifty maps derived from the sample survey of 1966. Explanatory text and diagrams interspersed with the maps show either the whole conurbation based on electoral wards or the City of Birmingham based on enumeration districts, coded by means of the SYMAP computer programme and reproduced directly from the output of the line-printer. The maps demonstrate the value of automation in making use of such data.

298 'A check list of Canadian literature and background materials 1628-1950', in two parts, presents a comprehensive list of books which constitute Canadian literature written in English, together with a selective list of other books of Canadian origin which reveal the backgrounds of that literature. The work was compiled by Reginald Eyre Watters for the Humanities Research Council of Canada (University of Toronto Press; OUP, 1959). Part II is particularly relevant here; it contains bibliographies with a Canadian focus, on local history and description, social history, travel and description. Entries are arranged alphabetically by author's name and there is an 'Index of anonymous titles' and an index of authors' names, initials and pseudonyms.

299 'Checklist of Southeast Asian serials', in the South East Asia Collection, Yale University Library, printed by G K Hall, 1969, includes all holdings catalogued before 1966; holdings catalogued after 1966 are given for those serials held in the main library. The collection is notable for government documents, society publications and limited editions of publications pre-1945 concerning Burma, Thailand, Laos, Vietnam, Cambodia, Malaysia, Singapore, Brunei, the Philippines, Indonesia, Portuguese Timor and New Guinea. Author and title references are given.

300 Chicago University, Department of Geography was established in 1904 and has taken its place among the most influential of academic centres of geographical studies. The research papers and irregular monographs issued since 1948 have proved invaluable; numerous works have been published by members of the staff, including the *International list of geographical serials* (*qv*) by C D Harris.

301 'China and Japan' is the title of a set of facsimile reproductions of nineteenth century British Parliamentary papers, involving fifty-two volumes, an invaluable guide to primary source materials concerning China and Japan and British - Far Eastern relations during the nineteenth century, edited by Professor P Ford, Mrs G Ford and W G Beasley, in the Irish University Press *Area Studies* series.

302 'China reconstructs' is published monthly in English, Spanish, Arabic and Russian, by the China Welfare Institute, Peking. The periodical, generously illustrated, covers all aspects of the nation's

life; of particular interest to geographers are the articles and photographs concerning agriculture, farming, the economy, industries and manufactures. An index is circulated with the final issue of the year.
Refer also China pictorial and the *Peking review*.

303 'Chisholm's handbook of commercial geography' has been a standard work since its first appearance in 1889. The work was entirely re-written by L Dudley Stamp and S Carter Gilmour in a fifteenth edition, 1956, and is now in the eighteenth edition, 1966.

304 'A chronological history of voyages into the Arctic Regions', by John Barrow, 1818, has been reprinted by David and Charles of Newton Abbot. John Barrow was responsible for the renewed search for the North West Passage—his name is commemorated in Point Barrow and the Barrow Strait—and his work is one of the best available histories of the quest. Beginning with the Norse discovery of Vinland, Barrow tells of the voyages of such explorers as Frobisher, Davis, Baffin and Hudson, bringing the story to the time of his own voyages.

305 Chubb, Thomas is best known for his monumental *The printed maps in the atlases of Great Britain and Ireland* . . . (*qv*). Other major works include *A descriptive list of the printed maps of Somersetshire, 1575-1914*, 1914, and *A descriptive list of the printed maps of Norfolk, 1574-1916*, 1928.

306 'Cities in evolution' was edited by Patrick Geddes for The Outlook Tower Association, Edinburgh, and The Association of Planning and Regional Reconstruction (Williams and Norgate, revised edition, 1949). Geddes, a pioneer in the developing concept of city planning with regard to environment, included illustrations from his second exhibition and text drawn from his Catalogue to the first exhibition. Appendices include 'The Geddes diagrams', 'Geddes final Dundee lecture' and a brief biography of him.

307 'Cities of destiny', edited by Arnold Toynbee (Thames and Hudson, 1967), well illustrated with reproductions, photographs and sketchmaps, considers selected examples of well defined types of cities, under the following headings: 'Cities in history'; 'The city-state'; 'Capital cities' and 'Megalopolis', each chapter being contributed by an expert. Notes are appended to the text, also a bibliography.

308 'Cities of the Soviet Union: *studies in their functions, size, density and growth',* by C D Harris (Rand McNally for the Association of American Geographers, 1970), assembles factual data collected over many years from Soviet census and other primary documents, dealing mainly with the 1,247 centres having more than 10,000 inhabitants for which data were published in the 1959 Census. A brief introduction describes the basic framework of the urban geography of the USSR and the work already completed on this topic by Soviet geographers; the growth of urban population as a whole is then considered, followed by comments on the development of individual towns from the early nineteenth century. A substantial bibliography contains references mainly to works in Russian, but many also in English, French or German.

309 'The citizen's atlas of the world', by John Bartholomew, is especially useful for library and educational use. Two hundred pages of coloured maps, the majority printed in band colouring for clarity, are designed primarily to show settlements and boundaries, both internal and international. Every second double page is presented as a single spread to give a larger unbroken map face. Introductory sections: 'Dates in history of exploration', 'World exploration chart', a glossary of geographical terms and other information are included for quick reference.

310 'City', 1967-, the bi-monthly journal of The National Urban Coalition, Washington, DC, contains illustrated articles, a 'Digest of recent urban commentary', a correspondence section and notes.

311 'The city in history', by Lewis Mumford, first published in America, 1961, and, in the same year, by Secker and Warburg, was reprinted by Pelican Books in 1966. The work expands the information and ideas in the historical sections of *The culture of cities* (*qv*); parts of the four original chapters have been incorporated in the eighteen chapters of the later work. Four groups of reproductions or photographs are included, also a bibliography.

312 'City, region and regionalism: *a geographical contribution to human ecology',* by R E Dickinson (Routledge and Kegan Paul,

1947, with a fourth edition, 1960), for the International Library of Sociology and Social Reconstruction, was superseded in 1964 by the title *City and region: a geographical interpretation,* in the same series.

313 'Civitates orbis terrarum', originally compiled between 1572 and 1618 by Georg Braun and Franz Hogenberg, was reproduced in black and white facsimile in the *Theatrum Orbis Terrarum* series, Amsterdam, 1965, with an introduction by R V Tooley. Six hundred towns in all parts of the world are included; the beautiful engravings are by Hogenberg and others, many of them representing the first known engraved views. Apart from its main topographical value, this magnificent work is of great interest as a record of the domestic life of the period.

314 The Clarendon Press, Oxford, Cartographic Department, became an individual concern in 1952, though still working in close collaboration with the Oxford University Press, and has since become noted throughout the world for consistent excellence of cartography and for experimental and pioneer work, both in conventional and automated cartography. The Department carries out all the processes of map and atlas preparation; the location, in Oxford, gives access to the advice of specialists in many disciplines and to the reference libraries of the University. Courses in the training of cartographers are offered. The Press also publishes monographs and series, such as the *Oxford studies in African affairs* and *A Pacific bibliography (qv).*

See also Atlas of Britain and Northern Ireland; The Oxford atlas; The Oxford economic atlas of the world; The Oxford junior atlas; Oxford plastic relief maps; The Oxford school atlas; Oxford system of automated cartography.

Refer David Bickmore and Experimental Cartography Unit, ed *Automatic cartography and planning* (The Architectural Press, 1971).

315 Classification: The variety of material in a geographical collection and the broad scope of the subject content have rendered inadequate any of the general bibliographical schemes of classification for the organisation of a specialised geography library. The Library of Congress classification, section G, and the Universal Decimal Classification 91 are used in some academic departments, usually with modifications; the latter has also provided the base for some special applications, such as that used by the Scott Polar Research Institute, in addition to the international classification below-mentioned. The British Museum Map Room, the Bodleian Map Room and the Bibliothèque Nationale have all evolved empirical schemes to suit their needs. The system used for the arrangement of entries in the *Bibliographie géographique internationale* dates, with modifications, from the system devised by Raveneau in 1891. The classification created by S W Boggs for the Council of the Association of American Geographers, published in 1937 under the title *Library classification and cataloging of geographical materials,* is used by the American Geographical Society of New York in its research catalogue and in the bibliographical service *Current geographical publications;* it has also provided the starting-point for the classifications devised for several other notable collections, both in America and Britain. The Oxford University School of Geography uses the American scheme for the classified index of periodical articles maintained, but has found the scheme as a whole too detailed for organising the collection itself, for which a straightforward scheme giving regional priority is used; a numeral notation denotes the regional scheme, letters the thematic aspects, with running numbers added to give precise location, so that each item in the collection bears a unique call number. G R Crone, former librarian and map curator of the Royal Geographical Society, worked out an interesting classification, which, unlike the majority of geographical schemes to date, has a philosophical basis; in Mr Crone's words, ' The ruling idea has been to move from the separate subjects in which the geographer may be interested, through their inter-relations and on to the final synthesis '.

The schemes evolved for the vast collections in the us General Staff and the British War Office again demonstrate the practical approach, using letters, numerals and symbols in combination, so that each item is identified individually both in location and in all indexes and references. The Geographical Association has given much thought to the whole question of arrangement of geographical material and for the annual conference of 1962 an exhibition was mounted to display the application of classification to the material of geographical teaching. The classification used by the Geographical Association itself is based on experience gained by studying the scheme used by the University of Southampton Department of Geography and that of the Association of American Geographers; again, a combination of numerals and letters is used, with detailed sub-division, and a regional preference when possible. A simpler scheme drawn up for school use is class P of the Cheltenham Classification, which is in three main parts, each subdivided into nine sections and using the decimal point as necessary for further sub-division. Among the more recently evolved special classifications has been the classification of world vegetation, prepared by the Unesco Standing Committee on classification and mapping of vegetation on a world basis, worked over especially at a meeting held during the eleventh International Botanical Congress, August 1969; it provides a comprehensive framework for the more important categories to be used on vegetation maps at scales of 1:1M or smaller. The categories in this classification are units of vegetation, including both zonal formations and the more extensive azonal and modified formations; the system may be applied to larger scales by expanding it through further sub-divisions. The framework is related to a scheme of colours, shades and symbols of world vegetation and an explanatory manual is under the editorship of Professor A Kücher. The Oxford forestry classification and the polar classification devised by the Arctic Institute of North America are two other widely accepted special classifications.

The Commission on Classification of Geographical Books and Maps in Libraries was constituted during the International Geographical Congress held in Washington in 1952. After much discussion at successive meetings, the section 'Geography' in the Universal Decimal Classification was adopted for expansion and modification. Professor E Meynen was foremost in the execution of this work, aiming at and, to a great extent, succeeding in integrating the different viewpoints in many countries into a synthesis. The resulting scheme, which has been recognised by the FID, was published in English, French and German, as the *Final report on the classification of geographical books and maps* for the XIth General Assembly and XXth International Geographical Congress, London 1964, preceded by a masterly article by Professor Meynen 'On the classification of geographical books and maps and the application of the Universal Decimal Classification (UDC) in the field of geography'.

Refer S W Boggs and D C Lewis: *The classification and cataloging of maps and atlases,* Special Libraries Association, second edition, 1945.

E J S Parsons: *Manual of map classification and cataloguing prepared for use in the Directorate of Military Survey,* War Office, 1946.

R T Porter: 'The library classification of geography', *Geographical journal,* March 1964.

316 'The classification and cataloguing of maps and atlases', by Samuel W Boggs and Dorothy Cornwell Lewis (Special Libraries Association, 1932; second edition, 1945), remains the most comprehensive single English-language work on this subject.

317 'The climate near the ground', by Rudolf Geiger (Harvard University Press, 1950, 1957, 1965, 1966; translated from the fourth edition of *Das Klima der bodennahen Luftschicht,* 1961), is a classic work, of which successive editions have taken account of new research. The following are the main topics: 'Heat budget of the earth's surface as the basis of microclimatology'; 'The air layer over level ground without vegetation'; 'Influence of the underlying surface on the adjacent air layer'; 'Quantitative determination of heat-balance factors'; 'The air layer near plant covered ground'; 'Problems in forest meteorology'; 'The influence of topography on microclimate'; 'Relation of man and animals to microclimate'. A section 'Hints on measurement techniques used in microclimatologic and micrometeorologic investigations' was contributed by Gustav Hofmann and there is a comprehensive bibliography of books and articles.

318 'The climates of the continents', by W G Kendrew, was a pioneer work in English when it was first published in 1922; now in a fifth edition (Oxford: Clarendon Press, 1960). Comparison may usefully be made with the *Handbuch der Klimatologie,* edited by W Köppen and R Geiger in five volumes, 1930-1938.

319 'Climatic atlas of Europe', published 1970- by Unesco, WMO and Cartographia of Budapest, is in looseleaf form, the corners of the sheets being reinforced. The volume contains two sets of thirteen maps each, showing the distribution of monthly and annual values of the mean atmospheric temperature and precipitation, together with a map representing annual temperature ranges. For its preparation, the meteorological services of the individual countries provided data obtained from several thousand stations; the information from which these maps were compiled covers the period 1931 to 1960, using a fuller body of data than has been available for any previous work of this kind and the maps have been prepared in accordance with the specifications laid down by WMO in the *Guide to climatological practices*. Colour scales have been applied consistently, place-names appear in the form used by the country concerned and headings and other explanatory matter have been given in the four working languages of WMO, English, French, Russian and Spanish. The map sheets are easily detachable from the metal spiral binding, 60 by 42 cm. This is the first volume of a world climatic atlas, under the direction of WMO, which will indicate the principal factors determining the climate of each continent.

320 'Climatic factors and agricultural productivity', a report on papers and discussion presented at a symposium held on March 13, 1963, edited by James A Taylor, was made available in duplicated form by the University College of Wales, Aberystwyth in 1964 (Memorandum no 6). Sketchmaps and diagrams are included; the main topics under discussion, following papers by a number of experts, were ' Economic and ecological productivity under British conditions: an introduction '; ' Light and temperature efficiency, with reference to forest grasses '; ' The effect of weather conditions on the growth of lucerne and clover '; Drought, soil water and grass growth '; ' Climatic factors in the development of local grass conservation '; ' Climatic factors affecting Danish and Irish agricultural development ' and ' Rainfall probability and agricultural productivity (overseas work with its possible implications in Britain) '. Edited extracts from the general discussion and conclusions are included.

Refer also Earlier memoranda: ' The growing season ', 1958; ' Shelter problems in relation to crop and animal husbandry, 1959; ' Hill climates and land usage, with special reference to the Highland Zone of Britain ', 1960; 'Aspects of soil climate ', 1961; Climatic factors and diseases in plants and animals ', 1962; ' Climatic factors and agricultural productivity ', 1963.

321 'Climatological atlas of Africa', compiled and edited in the

African Climatology Unit, University of Witwatersrand, under the direction of S P Jackson (Government Printer, 1961), contains fifty-five plates presenting data on monthly and annual rainfall, daily temperatures and humidity.

322 'Climatological atlas of the British Isles', published by the Meteorological Office, was begun before the second world war, in co-operation with the interested governmental authorities. A new edition, 1945- (1952), incorporated an additional feature; new information and maps were compiled on the recommendations of the National Agricultural Advisory Council, with the interests of agriculture in mind. 220 maps, arranged in ten sections, deal with all aspects of the climate and weather of the British Isles, each section being accompanied by an introduction and a bibliography, with tables and diagrams to supplement the maps. Most of the maps show average conditions 1901-1930, selected as being the standard period for climatological averages for all the meteorological services of the world. In some sections, data have been incorporated for later periods, covering as many years as possible; particularly interesting maps show snowfall, thunder frequency, monthly average means of vapour pressure, relative humidity and saturation deficit.

323 'Climatological atlas of Rumania', published by the Meteorological Institute, Bucharest, 1966, in English and French, presents a very detailed climatological picture, reflecting observations during the past sixty years of intensive research. Temperature, humidity, cloud cover, precipitation, winds and sunshine have been mapped at 1 : 1,500, with topographical and political supporting detail. More than twelve layer colours are used on some maps and the printing is excellent throughout.

324 'Climatological atlas of the world' *see* under *Atlas of Meteorology* . . .

325 ' Cloud studies in colour ', by Richard Scorer and Harry Wexler (Oxford: Pergamon Press, 1967), comprises a collection of fine coloured pictures of cloud types, with brief explanatory comments on the facing pages. Introductory text discusses ' Cloud names ' and ' Magnitudes '; appended are a ' Diagram of cloud mechanics ', a section ' Types of motion ' illustrated by figures and a combined index and glossary.

Refer also The colour encyclopaedia of clouds, by the same authors (Pergamon Press, 1968).

326 ' The coast of England and Wales in pictures ', with commentary by J A Steers (CUP, 1960), presents a collection of superb photographs, with geographical explanations.

Refer also J A Steers, *ed Applied coastal geomorphology* (Macmillan, 1971).

J A Steers, *ed Introduction to coastline development* (Macmillan, 1971).

327 ' Coelum Stellatum ' *see* under *Theatrum Orbis Terrarum.*

328 The Coffee Information Bureau, London, representing the Coffee Publicity Association Limited, provides a general information service on coffee in all its aspects. An educational visual aids section includes leaflets, cards on specific topics, colour films and demonstrations.

329 Colin, Armand, one of the leading French publishers, has a special interest in geographical subjects; for many years, individual geographical monographs have included a number based on doctorate theses. Outstanding works include L de Launay: *Géologie de la France,* 1921; Henri Baulig: *Le plateau central de la France et sa bordure méditerranéenne,* 1928; Philippe Pinchemel: *Géographie de la France,* 1964. Armand Colin is also associated with high-level bibliographical work, such as the current *Bibliographie géographique internationale (qv).*

330 Collet's Holdings Limited, Wellingborough, Northants, is an invaluable centre for the checking or purchase of Russian publications, such as the *Geologicheskaya Karten SSSR,* 1:2,500,000, published in sixteen sheets by the Ministry of Geology in 1965 (1968). Other notable publications introduced into this country by Collet's include the *Atlas of the lithological and palaeogeographical maps of the USSR,* in four volumes; *Deposits of useful minerals of the world,* four sheets at 1:20M; *Tectonic map of Eurasia,* twelve sheets at 1:5M, with legend in Russian and English; *Tectonic map of Europe with an explanatory book in Russian* (the French edition is out of print); *Tectonic map of the USSR,* sixteen sheets at 1:2·5M, with legend in English; and the *Atlas SSSR* in Russian.

331 Collins, Sons, and Company Limited, London, publishes one series of especial interest to geographers and cartographic works. The series is *The new naturalist: a survey of British natural history,* edited by a number of experts, which now numbers more than fifty titles, some having gone into new editions. As the series has progressed, the idea of conservation of wild life and natural conditions has been stressed; number 49 is Sir Dudley Stamp: *Nature conservation in Britain.* The photographic editor, Eric Hosking, has included magnificent studies, some in colour, and there are usually maps in the text, varying in number according to the topic; the texts are of high standard and eminently readable, the overall production also being pleasing, including some very useful bibliographies. A few titles of particular interest are J A Steers: *The sea coast,* 1953, fourth edition, 1969, dealing with the relation of the coast to the general structure of Britain, followed by examination of the factors which develop different kinds of coasts and studies of contrasting coastal forms in all parts of Britain; L A Harvey and D St Leger-Gordon: *Dartmoor,* 1953, second edition, 1962, 1963, 1970, covers all parts of the moor; S W Wooldridge and Frederick Goldring: *The Weald,* 1953, 1966, deals similarly with the origin and development of this unique region; Sir Dudley Stamp contributed also *Man and the land,* 1955, third edition, 1969, a follow-up to his *Britain's structure and scenery,* number 4 in the series, and, with W G Hoskins, *The common lands of England and Wales.* Other particularly relevant titles include H J Fleure: *A natural history of man in Britain;* Gordon Manley: *Climate and the British scene;* and K C Edwards: *The Peak District.* The second main value of Collins publications for geographers lies in atlas production. *Collins world atlas,* ' an atlas for the person who cares about the world we live in and who wishes to know more of the background to today's events ', contains general, regional and thematic maps. Collins-Longmans have published a graded series of educational atlases—*Atlas 1,* for the youngest scholars; *Atlas 2,* in a second edition, 1967, for upper primary or junior secondary school use; *Atlas 3, Atlas 4* and *Atlas advanced* for CSE, fifth form and o level, and for sixth form, college and university, respectively. The latter, 1968, contains seventy-two pages of thematic maps drawn from the researches of leading geographers, with a forty-two page section on the world's climate, vegetation and resources and country by country graphs of the world's produce. Educational atlases include also the *Study atlas* and the *Visible regions atlas* and special editions prepared for Malaysia (in Malay), Thailand, Kenya, Malawi, Sierra Leone and Australia.

332 The Colombo Plan, initiated by a number of Commonwealth foreign ministers, who met in 1950 with the aim of planning co-operative economic development of south and south east Asia, following the difficulties created by the second world war, established a Consultative Committee and a Colombo Plan Council, the former consisting initially of Ministers from Australia, Britain, Canada, Ceylon, India, New Zealand and Pakistan. Subsequently, with the increase of the area under discussion from Iran to South Korea and from the Maldive Islands to Indonesia, membership of the Council was also increased; it meets annually at Colombo. The Consultative Committee is the senior policy-making body, meeting annually to review economic, technical and cultural progress. The Colombo Plan Bureau consists of an international staff of a Director and three officers; their function is to service the Council for Technical Co-operation, to record all technical assistance given to areas under the Plan and generally to disseminate information, acting also as a centre through which countries can initiate special regional programmes. A journal is issued, *The Colombo Plan*.

Refer Frederic Benham: *The Colombo Plan and other essays* (OUP for the Royal Institute of International Affairs, 1956).

333 ' Columbia-Lippincott gazetteer of the world ', edited by Leon E Seltzer, was a joint project by the Geographical Research Staff of Columbia University Press and the American Geographical Society (Columbia UP; Oxford UP, 1952). The second printing, with supplement, brings the work up to date to the end of 1960. Articles are included on the nations, continents, regions and major cities of the world, with shorter notes on towns and villages; descriptions are given of geographic features, also the essential facts on such aspects as population and products. The first part of the supplement is a series of articles on the major political changes in the world since 1952; the second part lists figures for all towns in the

United States with a population of over a thousand. Frequent cross-references are made and the work is thumb-indexed.

334 ' Columbus ' (Allen and Unwin, 1967) traces ' the story of Don Cristóbal Colón, Admiral of the Ocean, and his four voyages westward to the Indies according to contemporary sources retold and illustrated by Björn Landström '. Details are taken from the journals, from the letters to Ferdinand and Isabella and from other primary sources, superbly complemented by full colour illustrations and maps.

The Columbus Collection in the Berio Library, Genoa, begun in 1892, is today the centre for research on the voyages and discoveries of Columbus. Its carefully edited catalogue is a valuable bibliographic tool also for the early exploration of America.

Refer also Christopher Columbus: documents and proofs of his Genoese origin (City of Genoa, 1932).

The journal of Christopher Columbus, translated by Cecil Jane (Anthony Blond, 1968).

Select documents illustrating the four voyages of Columbus, translated and edited by Cecil Jane (Hakluyt Society, two volumes, 1930).

E G R Taylor: ' The navigating manual of Columbus ', *in* the *Journal of the Institute of Navigation,* January, 1952.

The bibliography of the first letter of Christopher Columbus describing his discovery of the New World, compiled by R H Major, 1872, has been reprinted by The Meridian Publishing Company, Amsterdam, together with a facsimile of Columbus' letter published at Basle, 1494.

335 ' Commons and village greens: *a study in land use, conservation and management based on a national survey of commons in England and Wales 1961-1966'* was edited by D R Denman and others (Hill, 1967), financed by the Nuffield Foundation. Management schedules for many types of land are discussed, followed by a more detailed description of individual special areas, such as ' Livestock husbandry on Dartmoor and the commons of Devon '. The text also includes appendices on the conduct of the surveys and a glossary of special terms used, illustrated by photographs and maps.

336 The Commonwealth Agricultural Bureau, Farnham Royal, Buckinghamshire, was established as a co-operative venture providing a scientific information service for agricultural research workers. This is done mainly through abstracts of world literature covering the whole range of agriculture, agricultural economics, food science and technology, animal health and nutrition and forestry. The CAB comprises the following institutes: Commonwealth Institute of Entomology, Commonwealth Mycological Institute, Commonwealth Institute of Biological Controls and the Commonwealth Bureaux of Agricultural Economics, Animal Health, Animal Nutrition, Dairy Science and Technology, Forestry, Helminthology, Horticulture and Plantation Crops, Pastures and Field Crops, Plant Breeding and Genetics and of Soils. The CAB came into existence in its present form in 1929. Most of the bureaux are located at a research institute working in the same field and library facilities are shared. Many overseas contacts are maintained and numerous publications issued, of which those of most value to geographers are probably the eighteen abstract journals, including *World agricultural economics and rural sociology abstracts, Dairy science abstracts, Herbage abstracts* and *Forestry abstracts.*

337 The Commonwealth Geographical Bureau project arose from the discussions of a meeting of Commonwealth geographers attending the 21st International Geographical Congress in New Delhi. The immediate establishment of a small Commonwealth Geographical Bureau and committee management was agreed; the recommended terms of reference included the study and practice of geography at all levels within the Commonwealth, the dissemination of information by means of a *Bulletin* and exchange of geographic staff. The Bureau has given priority to making contact with university geography departments and other higher education institutions.

338 The Commonwealth Institute, London, is concerned with the human aspects of Commonwealth relations and with the promotion of reciprocal knowledge and understanding. Its function is therefore educational and it carries out its ideals by means of the exhibition galleries, the cinema, the library, publications and travelling exhibitions. Talks to school classes are arranged and a Schools Advisory Service provides information and advice of all kinds. Publications include *Pamphlets* and *Papers* on aspects of the Commonwealth; and *Fact sheets on the Commonwealth,* which vary in size from four to twelve pages, giving important information on individual countries, with maps.

339 Commonwealth Scientific and Industrial Research Organisation (CSIRO), Canberra, the principal official scientific research body in Australia, is recognised as of world importance in soil research, pasture problems and basic research for industry. The work of the various divisions, such as the Northern Australia Regional Survey, 1946, and the Land Research and Regional Survey Section and its successors, 1950-, has been of central interest to geographers and the organisation's publications are of the first importance, frequently produced as technical papers in such journals as *Australian journal of agricultural research* and the *Australia journal of applied science*. There are also the special series 'Land research' and 'Soil publication'; most issues are illustrated and include bibliographies. Closely associated with the CSIRO research is the Waite Agricultural Research Institute of the University of Adelaide, founded in 1925. As the organisation has increased the scope of its activities, more specialised divisions have come into being, such as the Wheat Research Unit at Sydney, the Sugar Research Unit, Melbourne, Dairy Research at Melbourne, Fisheries and Oceanography at Cronulla, etc.

340 'Commonwealth survey', a record of United Kingdom and Commonwealth affairs, is a fortnightly factual listing of important developments in British and Commonwealth projects, trade and industry; the index cumulates into a useful source of reference.

341 'Concise encyclopedia of explorations', compiled by Jean Riverain (Collins; Follett, English language edition, 1969, for the Librairie Larousse, 1966), and re-edited by Miss P Bascom, is a mine of information about explorers and early discoveries. Salient facts given in the entries are strictly confined to geographical journeys and discoveries. The translation was made by Thérese Surridge, with an introduction by Sir Vivian Fuchs and, throughout the text are small monochrome illustrations and a few maps.

342 'A concise glossary of geographical terms' was compiled by J C Swayne (Philip, second edition, 1959), honorary secretary of a committee of professional geographers set up by the British Association and the Royal Geographical Society to provide agreed definitions of geographical terms. Pending the completion of the *Glossary of geographical terms* (*qv*) by this committee, this work was issued, containing brief definitions, intended mainly for students and based on terms used in national geographical examination papers and in standard geographical works.

343 '**Connaissance du monde**', an interesting monthly journal begun in 1955 by the Société d'Edition Géographique et Touristique, Paris, includes articles, notices of books, films and records; it is frequently illustrated and is indexed.

344 '**Contemporary China: a research guide**', by Peter Berton and Eugene Wu and edited by Howard Koch, jr (The Hoover Institute on War, Revolution and Peace, Stanford University, 1967; Hoover Institution bibliographical series, XXXI), includes bibliographical and reference works, selected documentary materials and serial publications dealing mainly with the social sciences and humanities; more than two thousand entries concern post-1949 mainland China and post-1945 Taiwan. Each chapter and sub-section is preceded by an introduction placing its bibliographical material in proper context. Entries are annotated in detail and contain descriptive and evaluative comment. Appendices contain a list of publications devoted to the resources of research libraries and institutions in the United States and abroad, and a selected list of directories and theses on contemporary China accepted by American universities. Four categories of works are distinguished: general and subject bibliographies, lists and indexes of newspapers and periodicals; general reference works; and selected documentary materials in law, government, economics, education and foreign relations; selected serial publications from the mainland, Taiwan, Hong Kong, Japan, the United States and the USSR. Two special lists are included: of descriptive accounts and catalogues of research libraries and institutions throughout the world, and of doctoral and masters' theses.

345 '**Contributions to bibliography of Australia and the South Sea Islands**', compiled by Willem C H Robert in five volumes (Amsterdam: Philo Press, 1967-), includes invaluable material arranged upon the following plan. In volume I, printed matter relating to discovery, exploration and travel issued in the Netherlands to 1921, with special emphasis on lesser known material, was made available by the author in 1964 in a limited edition, with a second edition in 1969; volume II consists of an *Index and bibliography of Dutch manuscripts and manuscript charts relating to the discovery,* 1968; volume III—*Printed material relating to discovery, exploration and* *travel, issued in Europe, except the Netherlands, to 1835,* again, first available in a limited edition, 1967, later revised and enlarged; volume IV, *Printed material relating to discovery, exploration and travel, issued in Europe, except the Netherlands, 1836-1921;* and volume V, *Charts and maps relating to Australia and the South Pacific issued in Europe to 1921.* The aim of these reference bibliographies and indexes is to enable the research worker in the field of Pacific history, geography and discovery to trace numerous references and facts pertaining to the subject. Within the volumes, arrangement is by alphabetical order of authors' names, and each concludes with indexes and lists, with cross references.

346 The Copper Development Association, Radlett, Hertfordshire, set up in 1933 to compile and distribute technical information, is typical of the specialist bodies whose work and publications are so vital to economic geographers. Books, pamphlets and brochures are produced at intervals, but of particular importance are *Copper,* issued three times a year, covering all aspects of the use of copper and its alloys, and *Copper abstracts,* a monthly selection of abstracts from over 120 technical and other periodicals and from relevant books, supplemented on occasion by abstracts from other abstracting services.

347 Coradi, G, Limited, of Zurich, is one of the most influential instrument-making firms in the world. Some recent machines include the 'Coradograph', precision coordinatograph for use in surveying and industry; the 'Automatic High-Precision Coordinatograph with Linear Interpolation Unit'; the 'Digimeter' in several forms, a data acquisition instrument developed by Coradi in collaboration with expert surveyors and the 'Electronic Coordinate Digitizing Unit' by which data are transferred in digital form to the data output units using punched cards or punched tape as data carrier. There is also the 'Coradomat' automatic drafting machine and numerous other models, many of them new designs using magnetic or punched tape.

348 Cornish, Vaughan (1862-1948) was a geographical scholar of wide interests, who, by his many papers and published books, as well as through his influence in the course of work with section E of the British Association and with the Geographical Association and other bodies, exerted great influence on the rising generation of young geographers. His first major work, *The great capitals: an historical geography*, published in 1923, aroused considerable interest. He made a special study of wave formations and used his great geographical knowledge to illuminate the appreciation of natural scenery, a theme to which he devoted much of his later life. Notable among his publications are also *National parks*, 1930; *The poetic impression of natural scenery*, 1931; *The scenery of England*, 1932; *Scenery and the sense of sight*, 1935; *The preservation of our scenery*, 1937; *The scenery of Sidmouth*, 1940; *The beauties of scenery: a geographical survey*, 1943. A selected bibliography is appended to a lecture delivered to the Oxford Preservation Trust by Edmund W Gilbert, ' Vaughan Cornish 1862-1948 and the advancement of knowledge relating to the beauty of scenery in town and country ' (The Trust, 1965).

349 The Corona Library (HMSO) is a series of pleasantly produced monographs, each on a country of the Commonwealth—*Sierra Leone*, by Roy Lewis, 1954; *Nyasaland*, by Frank Debenham, 1955; *British Guiana*, by Michael Swan and *Jamaica* by Peter Abrahams, both 1957; *Uganda*, by Harold Ingrams and *North Borneo*, by K Tregonning, both 1960; *Fiji*, by Sir Alan Burns, 1963. The last volume in the series was Austin Coates: *Western Pacific Islands* (HMSO, 1971). They are illustrated with photographs and maps.

350 ' Cosmographei oder Beschreibung aller Länder Stetten ', by Sebastian Münster, 1544, is regarded as the first of this type of geographical encyclopedia, marking the peak of Renaissance geographical knowledge. The work went through thirty-six editions in six languages during the following century. Münster's aim was to provide a ' Compendium and brief description of all the lands of the Earth ', using all available sources—the Bible, classical works, the comments of contemporary travellers and the reports of such explorers as Marco Polo, Columbus and Vespucci. The result was a glorious collection of geographical, topographical, historical, genealogical, ethnographical, anthropological data—both fabulous tale and scientific fact. The numerous maps and illustrations are a delight and the whole production has provided source material of all kinds for later scholars. Theatrum Orbis Terrarum Limited has now published a facsimile of the first edition, Basel, 1550, from the copy preserved in the Sächsische Landesbibliotek, Dresden, with a new introduction in German and English by Professor Dr Ruthardt

Oehme. It is in 'A series of early books on the history of urbanization ', as part of the *Mirror of the world,* and is bound in facsimile leather and boxed.

351 ' Countries of Europe as seen by their geographers ', compiled by E C Marchant (Harrap, 1971), under the auspices of The Council of Europe *Education in Europe* series, the result of conferences sponsored by The Council for Cultural Co-operation, comprises a series of articles by eminent European geographers in which each outlines the manner in which he would have his country's geographical characteristics presented.

352 The Countryside Commission, under the Countryside Act 1968, replaces and assumes the functions of the National Parks Commission. The commission selects and designates national parks and areas of outstanding natural beauty and draws up proposals for long-distance footpaths and bridleways, acts as the central source of advice on the provision of other recreational areas and makes recommendations on the distribution of grants from the Ministry of Housing for amenity purposes; it is also empowered to do research itself or in conjunction with other relevant bodies, such as the Nature Conservancy, Forestry Commission, British Waterways Board, Water Resources Board and others. A central information system, based on feature cards, is maintained and a monthly newsletter, *Recreation news,* is produced.

353 ' County atlases of the British Isles 1579-1703: a bibliography ' compiled by R A Skelton (Map Collectors' Circle; Carta Press, 1964, 1970), was intended to provide a basic bibliography of the maps in the county atlases of the British Isles from Elizabethan to mid-Victorian times, a project interrupted by Dr Skelton's death at the end of 1970. Donald Hodson, collaborator in the continuation of the bibliography after 1703, will, it is hoped, complete the work, which, to a certain extent, replaces Thomas Chubb: *The printed maps in the atlases of Great Britain and Ireland . . . (qv).* The first volume deals with the period from the publication of the *Atlas of England and Wales* by Christopher Saxton to the maps prepared for Bishop Gibson's 1695 edition of Camden's *Britannia (qv),* comprising 124 county atlases. The cartographic history of each item is given, including commentary and bibliographic sources as necessary. Additions to the original work were made on publication:

'Maps of parts of the British Isles in general atlases before 1650: a select list' and 'The London map trade before 1700; with a biographical list of London map-publishers'. The volume is generously illustrated and indexed.

354 'The crossing of Antarctica: *The Commonwealth Trans-Antarctic Expedition 1955-58'*, by Sir Vivian Fuchs and Sir Edmund Hillary (Cassell, 1958; second edition, 1959), presents the full story by the leader of the expedition and Sir Edmund Hillary, leader of the New Zealand support party. There are coloured and monochrome photographs and some sketchmaps in the text.

355 'The culture of cities', an exhaustive survey and a pioneer work, remains probably Lewis Mumford's greatest achievement (Secker and Warburg, 1938); it is eminently readable and the extensive, annotated bibliography and illustrations add to its usefulness as a source book. Other outstanding works by Mumford on similar themes include *The story of Utopias: ideal commonwealths and social myths,* 1923; *Technics and civilization,* 1934; *The city in history: its origin, its transformations and its prospects* (*qv*); and *City development: studies in disintegration and renewal,* 1945, etc.

356 'A cumulation of a selected and annotated bibliography of economic literature on the Arabic-speaking countries of the Middle East, 1938-1960', prepared by the School of Oriental and African Studies, University of London, from the Bibliography issued by the Economic Research Institute, American University of Beirut. Selected, annotated and classified lists of articles, books, reports, official documents etc in English, French and Arabic amount to about 9,600 entries, the emphasis being on English-language material. The work has been reproduced by G K Hall.

357 'Current European directories', compiled and edited by G P Henderson, was published in 1969 by CBD Research Limited, Beckenham, Kent, in English, French and German. Here is a representative selection from the compiler's records of several thousand titles of directories published in Europe, based on personal assessment of contents or inspection of recent editions, response of publishers to questionnaires and the validity of bibliographical information available on those not actually inspected.

358 '**Current geographical publications**' is the record of additions to the research catalogue of the American Geographical Society, begun in 1938 by Elizabeth Platt and still the responsibility of the librarian and library staff. Ten issues a year are published, in which references to periodical articles as well as books are arranged according to the classification system used in the research catalogue; regional placing takes preference over thematic.

359 '**Cyclopedia of New Zealand**' in six volumes (The Cyclopedia Company, 1897-1908) includes industrial, descriptive, historical and biographical information, with figures and other illustrations throughout the text. Each province is treated separately; each volume has four separate indexes and the whole work may be useful in suggesting ideas towards a historical geography of New Zealand.

360 '**Czechoslovakia: a bibliographic guide**', compiled by Rudolf Sturm, is one of the great number of special bibliographies produced by the Library of Congress, Slavic and Central European Division. Published in 1967, the work is in two parts: the first consists of chapters dealing with bibliographies, general and reference works, 'The land', 'Its peoples', 'History', 'Politics and government', 'Law', 'The economy' and 'Social conditions'. The second part consists of a bibliographic listing of items discussed.

361 '**The Daily Telegraph world atlas**', edited by D L Baker, comprises seventy-eight pages of maps provided by the Esselte Map Service of Sweden. The maps are political and physical, showing railways, but only one category of roads, which gives therefore rather an inadequate impression. Included are three pages of statistics and a section 'Space supplement' compiled by Dr A R Michaelis.

362 '**The Dartmoor bibliography: non-fiction**', compiled by J V Somers Cocks and published by the Dartmoor Preservation Association, 1970, is 'an attempt to fill the need for an up-to-date reference to the printed literature of the area', the first reasonably comprehensive bibliography since Rowe's *Perambulation of Dartmoor,* 1896. The first part is a listing of books and pamphlets, arranged under authors' names, the entries appending annotations as necessary. Relevant periodicals, handbooks, guidebooks and directories are named, but not individual articles. The entries in the second section are arranged under subject headings and here more than two hundred additional articles and papers are included. In addition to its intrinsic value, this work has been mentioned as an example of bibliographical studies undertaken by local societies in Britain, providing key documents for the local geographer.

363 David and Charles, publishers, of Newton Abbot, Devon, issued their first complete catalogue in 1969; supplements are circulated at intervals, also special communications, such as ' Christmas news '. The firm publishes books on history, geography, natural history, travel and transport, topography and local history, with special emphasis on borderline subjects, to ' bridge gaps ', such as that between geography and history or between history and industrial archaeology. They are also the European representatives of the Smithsonian Institution Press. Series include ' Industrial archaeology of the British Isles '; ' Railway holiday ', for example, in France, in Switzerland; ' Railway history '; ' Railway histories of the world '; ' Railway history in pictures '; ' Inland waterway history '; ' Industrial history '; ' Industrial history in pictures '; ' Canals of the British Isles '; 'A regional history of the railways of Great Britain' in several volumes; 'Studies in historical geography', edited by Alan Baker and J B Harley; ' Problems in modern geography ', edited by Professor Richard Lawton; 'Industrial Britain'; ' Islands '; ' The many worlds of wildlife '; ' Ports of the British Isles '; ' Old . . .', for example, ' Old Liverpool ', ' Old Mendip '. A special line is naturally concerned with the West Country, such as *Geology explained in South Devon and Dartmoor*. 1969-70 saw the production of paperback editions. Important reprints include Camden's *Britannia; Murray's Handbook for Devon and Cornwall*, 1859, and *Murray's Handbook for Scotland*, 1894; *Baedeker's Russia*, 1914, *with Teheran, Port Arthur and Peking: a handbook for travellers; The Royal English atlas: eighteenth century county maps of England and Wales; Annals of coal mining and the coal trade,* by R L Galloway, in two volumes, 1898 and 1904; the classic, *Dartmoor,* by R Hansford Worth; and a *Reprint of the first edition of the One Inch Ordnance survey of England and Wales.* Journals published include *Industrial archaeology* and *Maritime history.* The firm has recently taken over the book club, Readers Union, with its magazine, *Readers news.*

364 Davis, William Morris (1850-1934), Professor of Physical Geology at Harvard from 1893, exercised one of the greatest influences on modern geomorphology. His conclusions, expressed in some five hundred pages, were strongly criticised both at the time and later, but his theories concerning cycles of denudation and erosion are still fundamental, though modified by later research. *Geographical essays,* 1909, included papers on normal erosion and on desert, coastal and glacial erosion, also papers on education; an unabridged republication of this collection was issued by Dover Publications and Constable in 1954.

365 'The dawn of modern geography: *a history of exploration and geographical science from the conversion of the Roman Empire to AD 900, with an account of the achievements and writings of the Christian, Arab and Chinese travellers and students'*, by C R Beazley, published in three volumes by Murray, 1897-1906 (reprinted in New York by Pater Smith, 1949), was a monumental work dealing with the history of geographical exploration and scholarship from the fourth century to the early fifteenth century. The chapters on the history of medieval cartography remain particularly valuable.

366 de Gruyter, Walter, and Company, Berlin, publish many scholarly works in the fields of geography, geology and cartography and are notable to geographers particularly for the series of monographs which make up the *Lehrbuch der allgemeinen geographie* (*qv*). Other works of considerable importance are R Maack: *Kontinentaldrift und geologie des Südatlantischen Ozeans,* 1969; Richard Finsterwalders and others: *Photogrammetrie,* 1963-1968; Horst Falke: *Die Geologische Karte . . .,* 1969; *Kartographische Geländedarstellung* (*qv*); Martin Schwind: *Das Japanische Inselreich . . .,* in three volumes, 1967-; Hans-Günter Gierloff-Emden: *Mexico: eine landskunde,* 1969; Viktor Heissler: *Kartographie,* 1968. The firm publishes also *Die Erde . . .,* the *Hamburger Geographische Studien . . .* and the *Hamburger Geophysikalische Einzelschriften.*

367 de Martonne, Emmanuel (1873-1955) became the leading French geographer after the death of Paul Vidal de la Blache, whose student and son-in-law he was. His major interests lay in physical and regional geography and, through his academic work at Rennes, Lyon and the Sorbonne, where he directed the Institut de Géographie, 1927-1944, and by his publications, he exerted a permanent influence on these two aspects of geography. He was elected an honorary president for life to the International Geographical Union, to which his support was invaluable. His chief published work was the *Traité de géographie physique* (*qv*); he contributed the *Europe Centrale* volume to the *Géographie universelle* (1930-31) and the volume dealing with the physical geography of France, 1942.

368 Debenham, Frank (1883-1965) led a varied life, both practical and academic; his wide interests in both arts and sciences, his vital personality and his gift for exposition are revealed in his writings and in the even more numerous works which he directed, edited or to which he contributed. He specialised in geology and went as geological expert with Captain Scott to the Antarctic, where he carried out research on the geology of Granite Harbour; he subsequently became a founder-member and, in 1925, first Director of the Scott Polar Research Institute. He was then at Cambridge and in 1931 he became first professor of geography there, exerting great influence on Cambridge academic geography. After publishing his polar researches, he wrote on many subjects. Greatly concerned to encourage cartographical knowledge and use, he wrote *Map making: surveying for the amateur,* which went into a third edition in 1955 (reprinted 1963); and he was one of the first geographers in this country to see the possibilities of the new experiments in three-dimensional cartography, demonstrated, for example, in Harrap's *3-D junior atlas,* which he edited. This interest and his flair for vivid presentation are shown also in *The world is round: the story of man and maps,* first published by the Istituto Geografico de Agostini, Novara, and by Macdonald/ Rathbone Books, 1958; in *The McGraw-Hill illustrated world geography* 1960; *The Reader's Digest great world atlas* (*qv*) and *Discovery and exploration . . .* (*qv*).

369 'Decorative printed maps of the 15th to 18th centuries' (Spring Books, 1952, 1965) is an edition of *Old decorative maps and charts,* by A L Humphreys, 1926, revised by R A Skelton, with a new text and eighty-four reproductions. Humphreys, a connoisseur of fine books, compiled his work from the collector's point of view and with the interests of the collector in mind. The new edition is intended 'both as a specimen book and as an elementary guide to the study of maps printed from wood blocks or copper-plates . . .'. Individual sections deal with the map as a work of art, the printed map, the map trade, map-collecting and aspects of map features and conventional signs; Ptolemy, woodcut maps; Italian, Netherlands, English, German and French cartography. More than half the volume is devoted to reproductions. Bibliographies are included of reference material on the history of cartography and for each of the special sections.

370 'Deep-sea research and oceanographic abstracts', edited by Mary Sears, of the Woods Hole Oceanographic Institution, and Mary Swallow, of the National Institute of Oceanography, Wormley, Surrey, is a bi-monthly journal founded in 1953 (Oxford: Pergamon Press) at the instigation of the Joint Commission on Oceanography of the International Council of Scientific Unions. Papers are concerned with the results of research, improvements in instruments and new laboratory methods. Book reviews are included and there is a separate ' Instruments and methods' section, 'Abstracts' and a bibliography section.

371 Demangeon, Albert (1872-1940) was one of the great French geographers who trained in history and geography under Paul Vidal de la Blache. His monograph on Picardy, 1905, demonstrated his regional method; *Le déclin de l'Europe,* 1920, followed, then *L'Empire Britannique,* 1923, both of which were translated into English. He contributed to the *Géographie universelle* the first two volumes on the British Isles and that for Belgium, Holland and Luxemburg; *The British Isles,* translated and edited by E D Laborde, was published by Heinemann (third edition, third impression, 1961). His *Problèmes de géographie humaine* went into a fourth edition (Colin, 1952), in which was included a section on Demangeon and his published work. Many papers and notes made his ideas and his scholarship widely known, but perhaps his most lasting contribution to geographical studies lay in his direction, as editor, of the *Annales de géographie* (*qv*).

372 'The demographic yearbook', published annually since 1949, is one of the many United Nations publications essential to the geographer. Besides standard tables giving the latest available details for each country, each volume usually specialises in one particular aspect.

373 '**Denmark: collective papers:** *some contributions to the geography of Denmark and other topics discussed by Copenhagen geographers*' was edited by N K Jacobsen and R H Jensen (Københavns Universitets Geografiske Institut, 1968) on the occasion of the 21st International Geographical Congress. The papers, illustrated with photographs, sketchmaps, figures and an end section of folded maps, range over many aspects of the geography of Denmark and of Danish geographers. Each paper includes an abstract.

374 '**A descriptive atlas of New Zealand**', edited by A H McLintock (Wellington: Government Printer, 1959), was the first systematic atlas of New Zealand, which proved so much in demand that a second edition was put in hand immediately and was issued in 1960. The atlas comprises an analysis and assessment of the country's resources on maps of varying scale; the topographical map is on the scale 1:1M, while maps of geology, soils, population and other factors are mainly on 1:3,200,000. Descriptions accompany each map sheet and a comprehensive text, illustrated by diagrams, graphs, maps and photographs, was contributed by members of various government departments. There is a full gazetteer.

375 '**A descriptive atlas of the Pacific Islands, New Zealand, Australia, Polynesia, Melanesia, Micronesia, Philippines**', edited by T F Kennedy (Praeger, 1966, 1968), contains monochrome maps by Julius Petro and Lionel Forsdyke, with textual explanation and statistical information.

376 '**Deserts of the world:** *an appraisal of research into their physical and biological environments*', edited by William G McGuinnies, B J Goldman and Patricia Paylore (University of Arizona Press, 1968, second printing 1970), was a United States contribution to the International Hydrological Decade programme. A massive work, it was intended 'for those seriously interested in planning, managing and executing research or development efforts in the arid parts of the world'. The text is prefaced by 'locator maps' of relevant parts of the world, followed by some general comments pertaining to the major deserts. 'Appraisal of research on weather and climate of desert environments'; 'Appraisal of research on geomorphology and surface hydrology of desert environments'; 'Appraisal of research on surface materials of desert environments' continues with examination of vegetation, fauna and desert coastal zones. An appendix gives a 'General summary of the state of research on ground water hydrology in desert environments'. In addition to subject and author indexes, there is an index of scientific names of plants.

Refer also An inventory of geographical research on desert environments published by the University of Arizona, 1968, for the Office of Arid Zone Studies.

377 'Deutscher generalatlas' at 1:200,000 (Mairs Geographischer Verlag, 1967/68) consists mostly of double spreads, many folded. The hill shading and the deeper blues for hydrographic features are effective and roads, place-names and state boundaries stand out well. Much factual information is incorporated, but there is no sense of overcrowding. The town plans included are useful and the whole work is well documented.

378 'Deutscher Planungsatlas' is a superb atlas in ten volumes, 1956-, by the Academy for Area Research and Land Use Planning in Hanover. Scales range from 1:500,000 to 1:1M. The material was organised mainly by the individual states, bringing together a wealth of information based on the latest census data and other relevant sources, arranged as follows: 1, North Rhine-Westphalia; 2, Lower Saxony; 3, Schleswig-Holstein; 4, Hesse; 5, Rhineland-Palatinate; 6, Wurttemberg-Baden; 7, Bavaria; 8, Federal Republic of Germany; 9, Berlin; 10, Hamburg.

Refer W Witt: 'Deutscher Planungsatlas', *International yearbook of cartography*, 1962.

379 'The development of Australia: *a study commissioned by the Australian Development Research Foundation, Sydney*' (Free Press of Glencoe; Collier-Macmillan, 1964) is a foundation illustrated work, including sketchmaps.

380 'Diccionario Geografico-estadistico-historico de España' provides an essential guide to the historical geography of Spain (Madrid, second edition, 1846), covering in several volumes also the overseas possessions of Spain.

381 'Dictionary catalog of the Department Library, United States Department of the Interior', Washington DC, was reproduced in thirty-seven volumes by G K Hall, 1966. These volumes represent more than 724,000 items, published and unpublished, covering the scientific, technical, economic and social aspects of natural resources, land management, mines and mineral resources, including journals and early works, also a unique collection on the American Indians.

382 ' **Dictionary catalog of the Stefansson Collection on the Polar Regions in the Dartmouth College Library** ', Hanover NH, consisting of 120,000 entries, was published in eight volumes by G K Hall, 1966. Historical coverage is the main emphasis of the collection, with primary concern for the history of Polar exploration, Alaskan history, biography, description and travel with various other specialisations. Resources on the Arctic and Antarctic are included, within chronological and geographical limits, while documentation of the international relations aspect of the Polar Regions is included without regard to period. An extensive subject index is a particularly valuable feature of the catalogue.

383 '**Dictionary catalogue of the Hawaiian Collection in the Sinclair Library, University of Hawaii**', Honolulu, was published by G K Hall in four volumes, 1963, with a first supplement in one volume. Dating from 1908, this catalogue represents the world's largest and most complete collection of Hawaiiana, including books, pamphlets and serials. Since 1915, the collection has been a depository for Hawaiian government documents and all are fully catalogued. Special reports and a newspaper clippings file are both included in the catalogue, also the acquired microfilm items.

384 ' **Dictionary of discoveries** ', compiled by I A Langnas (New York: Philosophical Library, 1959) and revised by Peter Owen, provides an interesting and most useful alphabetical listing.

Refer also L H Parias, *ed*: *Histoire universelle des explorations* (Paris, three volumes, 1955).

Jean Riverain: *Dictionnaire des explorations* (Larousse, 1968).

385 '**A dictionary of geography** ', by F J Monkhouse (Arnold, 1965; second edition, 1970) included in the first edition about 3,400 entries in one alphabetical sequence, together with an analytical list of entries by subject grouping. The ' main criterion for inclusion has been usage '; current geographical textbooks and periodicals were scanned and those foreign words have been included which have been incorporated into English geographical literature. The new edition contains some six hundred new entries, most of them in the fields of town planning and quantitative research; also metric measures and quantities are given, though imperial equivalents are used where appropriate.

Refer also W G Moore: *Dictionary of geography: definitions and explanations of terms used in physical geography* (Penguin Books, 1949, reprinted 1950; new edition, revised and enlarged, 1952, 1954, 1956, 1962, 1963; again revised and enlarged, 1967).

Thesaurus des termes géographiques (Paris, 1971).

386 'A dictionary of mapmakers' is a comprehensive work, in progress, by R V Tooley, which will include cartographers, publishers and engravers from the earliest times to 1900. Part I (Map Collectors' Circle, volume 2, no 16, part 1) is already available to members and the whole work will eventually be made available outside the Circle.

387 ' Dictionnaire de géographie ancienne et moderne à l'usage du libraire et de l'amateur de livres ', one of the most interesting of the early dictionaries of geography, in two volumes, by Pierre Deschamps, forms part of the supplement to Brunet: *Manuel du libraire et de l'amateur de livres,* published between 1660 and 1680. The dictionary includes gazetteer and glossary type entries in one alphabet, with a supplement of additions and corrections and a French index of Latin names; a feature is the references to early printing presses in various towns. The entire work has been reprinted by Rosenkilde and Bagger of Copenhagen.

388 'A dictionary of natural resources and their principal uses ', compiled by Nora Jackson and Philip Penn (Oxford: Pergamon Press, 1966; second edition, 1968), examines a comprehensive number of products, mineral and vegetable, together with conditions of growth, users and potentials. The second edition has been up-dated and contains illustrations.

389 ' Dictionary of rubber technology ', by A S Craig (Butterworth, 1969), includes brief annotations on the meanings of common technical terms used in the natural or synthetic polymer industries.

Sketches and diagrams supplement the explanations and there are references to specialised periodicals. Appendices give sources of information about rubber, notes on the composition of the types of products, grouped according to their uses, and a list of special publications.

Refer also Rubber developments.

390 Directorate of Overseas Surveys, which originated in 1946, in response to the urgent demand for accurate medium-scale maps to assist economic development in many countries, is now one of the units of the Ministry of Overseas Development. The department helps developing countries overseas at their request in basic surveying and mapping of their territories and co-operates with local survey departments, so that maps may be produced as rapidly and as economically as possible and also that work may cross national boundaries if necessary in order to plan geodetic surveys and air photography on a continental basis. In recent years, as more detailed development plans have been required, the department has increasingly produced large-scale maps. The directorate acts also as a centre for the collection and dissemination of technical information on surveying, field survey and mapping; comprehensive map and air photograph libraries are maintained covering overseas territories. Separate sections are concerned with cartography, photography, forestry, land use and resource survey. The Land Resources Division now has eight or more project teams of scientists overseas, while many more projects are at preliminary or final stages. Geology, geomorphology, climatology, soil science, hydrology, irrigation engineering, ecology, forestry, agriculture, livestock husbandry and agricultural economics are among the specialist disciplines represented on the scientific staff. The *Annual reports* are central sources of information on the work of the directorate; it is in four parts: General view of the year's work by region; Report on activities; Details by countries; Establishment and finance. Summaries and lists are presented in Appendices and a number of folded maps are included in a back cover pocket. A *Fact sheet* is revised as required; a quarterly *Newsletter* is issued and a *Catalogue of maps,* compiled in 1960, is supplemented by monthly lists of additions.

Refer also K M Clayton, *ed: Guide to London excursions,* section 28 (20th International Geographical Congress, London, 1964).

D R Warren: 'Surveys for development' in *The geographical magazine*, October 1969.

391 'Director's guide to Europe', a comprehensive business guide to Western Europe and Scandinavian countries, endorsed by the Institute of Directors, London, was compiled by Thornton Cox, 1968, specialists having contributed individual sections. Articles, 'The European Free Trade Association'; the 'European Economic Community'; 'Profit and loss of the Kennedy Round'; 'Doing business with Eastern Europe'; 'The Nordic Council and Nordic co-operation', are followed by chapters on the countries of Europe, each being dealt with in the same sequence and format. A vast amount of factual information has therefore been collected together on raw materials, agriculture, population, trade and exports etc, of practical use to the economic geographer.

392 'Directory of meteorite collections and meteorite research' (Unesco, 1968) was based on information received in reply to two questionnaires; the first was sent by the Unesco Department for the Advancement of Science to government departments and scientific institutions in Member States and the second by the Committee for Museums of Natural History of the International Council of Museums to all museums of natural history affiliated to the International Council. Collections and catalogues of research institutions in forty-nine countries are listed.

393 'Discovery and exploration': *an atlas-history of man's journeys into the unknown,* by Frank Debenham (Hamlyn, 1960), is a profusely illustrated work, for general background interest; a bibliography on explorers and exploration is a useful feature.

394 'The discovery of America', by G R Crone (Hamilton, 1969), in the *Turning points in history* series, sets the main theme against the background of medieval and Renaissance scientific thought and European power politics, including a specially important topic, the Vinland Map controversy and the development of cartographical techniques. An introduction mentions the source material available, following which the characteristics of the medieval world are discussed, Irish and Norse navigation in the Atlantic, the Portuguese achievements and the voyages of Columbus, Amerigo Vespucci and expansion to the recognition of the two Americas. An Appendix,

'Navigation and cartography of the Discovery', provides a useful summary. There are reproductions and sketchmaps throughout the text and a short bibliography, divided within broad subject headings.

395 '**The discovery of North America:** *a critical documentary and historical investigation*', by Henry Harrisse, was reproduced by Theatrum Orbis Terrarum in 1969 from the London-Paris edition of 1892. The work is one of the outstanding contributions to the history of American geography, including a study of the early cartography of the New World, a chronology of one hundred voyages westward between 1431 and 1504 and bibliographical accounts of three hundred pilots who first crossed the Atlantic, with a copious list of the original American geographical names.

396 ' **Documentation on Asia** ' was begun in 1960, edited by Girja Kumar and V Machwe (Allied Publishers, 1963-). It is arranged under regional topics, then again by systematic. Under the auspices of the India Council of World Affairs, an unofficial body founded in 1943 to encourage and facilitate the scientific study of Indian affairs, it is hoped that the work may be kept updated.

397 ' **Domesday geography of England** ' is a most interesting reconstruction, by H C Darby and a team of editors, of the social geography and economic life of early medieval England, by interpreting the entries in the Domesday Book (Cambridge University Press, 1952-). The volumes run as follows: 1, Eastern England, 1952, second edition, 1957; 2, Midland England, 1954; 3, Southeast England, 1962; 4, Northern England, 1962; 5, South-west England, 1967.

398 '**Early charts of New Zealand, 1542-1851**' is a magnificent production, compiled by Peter Bromley Malins (A H and A W Reed, 1969), consisting of plates arranged in thirteen sections, with notes opposite each plate. The charts are grouped as Pre-Tasman, Tasman, Post-Tasman and pre-Cook, Cook's officers, French visitors to New Zealand, Vancouver and Makaspina, 1800-1820, 1820-1830, Dumont d'Urville, 1830-1840, Early sheet maps and two Maori charts. An Appendix lists Admiralty Charts, both British and French and there is a bibliography.

399 Earth Science Editors: the two Associations, European (EDITERRA) and United States, covering North America (AESE), have a Co-ordinating Council, thus providing an international structure for handling questions of common interest. The International Union of Geological Sciences is also represented on the Council. The council is working to extend the system to include those parts of the world not already covered by the present associations and to promote the institution of similar ones, as desirable. Affiliation with similar organisations in biology and other sciences is also under consideration.

400 'Earth science reviews', published by Elsevier, Amsterdam, from 1966, is an international journal, presenting review articles relating to mineralogy, igneous and metamorphic petrology, geochemistry, geophysics, volcanology and economic and applied geology. Four issues make a volume.

401 'Earth science symposium on Hudson Bay', edited by Peter J Hood and others (Department of Energy, Mines and Resources, Geological Survey of Canada, 1969), consists of the proceedings of the symposium held in Ottawa in 1968, under the sponsorship of the National Advisory Committee on Research in the Geological Sciences and the Associate Committee on Geodesy and Geophysics of the National Research Council. There are illustrations, maps and diagrams in the text.

402 'Earthquakes—atlas of world seismicity' (Oxford: Pergamon Press, 1969), edited by C Lomnitz, aims to assemble the basic available data in an accessible form and to ' secure a theoretical foothold for a concept of earthquake risk '.

403 'The East African geographical review' has been published by the Uganda Geographical Association from 1963 as an annual issue; articles and book reviews are the chief features.

Refer also The East African economic review (twice a year, OUP, East African Branch, Nairobi).

404 'East Midland geographer', founded by Professor K C Edwards in 1954, is produced twice a year by the Department of Geography, University of Nottingham, and is an excellent example of the academic publications which have multiplied in Britain during recent years and which are particularly valuable for local studies. A cumulated index is issued every four years.

405 'Economic atlas of Ontario', a fine research atlas, edited by W G Dean (University of Toronto Press, 1969), with the assistance of the Department of Geography of the University, the Government of Ontario and a number of graduate and undergraduate students, is arranged in ten sections—aggregate economy, population, manufacturing, resource industry, wholesale and consumer trade, agriculture, recreation, transportation and communications, administration and reference maps. Data are presented usually in the form of ratios or indices of comparison rather than as absolute numbers. Cartography is varied according to the topic and is generally clear, readily informative and artistically pleasing, under the direction of G J Matthews. Dr Dean's preface describing the evolution and organisation of the project is particularly interesting. Sources and notes are added for each map, usually on the map face or on the reverse. The atlas measures the two principal factors of production and market potential, its major purpose being to provide a useful aid to decision-making in economic affairs. In so doing it presents a record of the current economic geography of the province.

Refer W G Dean: 'An experiment in atlas structure: *The economic atlas of Ontario'*, in *The cartographic journal,* June 1970.

406 'Economic bulletin for Latin America', the excellent publication by the Secretariat of the Economic Commission for Latin America, has been published twice a year since 1956, to provide a résumé of the economic situation of the region designed to supplement and bring up to date the information issued in the commission's annual economic surveys; specialised articles are included on topical issues relative to the economy. Since October 1958, the bulletin has regularly included a 'Statistical supplement'; this subsequently became sufficiently important to warrant separate publication, one issue being prepared for 1960, another for 1961, two for 1962, each bi-lingual. From 1964, the *Statistical bulletin for Latin America* began regular publication.

407 'Economic geography', published quarterly, 1925-, by Clark University, Graduate School of Geography, Worcester, Mass, is the only English-language periodical devoted to the subject; editorial policy is directed towards geographers, economists, teachers, professional and businessmen and for all others interested in the intelligent use of the world's resources, with the aim of increasing the understanding of the world's economic patterns. A small section of careful reviews is included.

408 ' Economic geography of China ', by T R Tregear (Butterworth, 1970), emphasises the political development of China during this century, especially since 1949, as an essential background to the geographical description and analysis of the country and its economy. Dr Tregear has drawn his information from first-hand experience of the country and constant monitoring of Chinese journals and reports, carefully sorting facts from propaganda. Numerous illustrations, tables and maps are included in the text; there is a glossary and a selected bibliography.

Refer also Charles A Fisher: 'Containing China?' *in The geographical journal,* December 1970 and September 1971.

409 ' Economic implications of the size of nations with special reference to Lebanon ', by Nadim G Khalaf (Leiden: Brill, 1971), is an attempt to analyse the economic stability, diversity of production and trade, degree of dependence on foreign trade and economic growth or development of a small nation. Part II of the work applies these theories to the Lebanese economy. A useful bibliography is appended.

410 ' Economic, social and political studies of the Middle East ', edited by C A O Van Nieuwenhuijze, of the University of Guelph, forms one of a new series published by Brill of Leiden designed as a channel of information taking the form of social scientific publications dealing superficially with contemporary problems of the Middle East. The editorial committee includes a number of specialists on the Middle East, social scientists living in that area and others having particular contributions to make to the series. The text is in English or French.

411 ' Economic survey of Latin America ', published by the United Nations from 1964, annually, supplemented by two issues a year of the *Economic bulletin* (*qv*), presents a comprehensive survey of regional and internal economic developments in Latin America and the impact of world events on the trends of economic growth of the areas from the years 1960 to 1963. Prepared in English and Spanish editions, central themes are on agriculture, manufacturing, energy, industry, transport, housing, commodity trade and the balance of payments.

412 'The economy of Pakistan: *a select bibliography, 1963-65'*, compiled by A H Siddiqui, was published by the Pakistan Institute of Development Economics, Karachi, 1967. This bibliography is a continuation of *The economy of Pakistan: a select bibliography, 1947-1962,* published by the institute in 1963. More than 1,100 entries are included referring to books, government documents, reports, conference literature and articles from periodicals published in Pakistan or abroad, arranged according to a special classification devised by the International Committee for Social Sciences Documentation, with some modifications. There is an author index and a detailed list of periodicals examined.

Refer also K S Ahmad: *A geography of Pakistan* (OUP Pakistan, 1964; second edition 1971).

413 'The Edinburgh world atlas' or *Advanced atlas of modern geography,* first published by Bartholomew in 1949, with a second edition in 1957, was issued also with *Everyman's encyclopaedia* in 1958 and 1959. It is a useful general purpose reference atlas, comprising world maps for geology, physiography, vegetation, population and ethnology, climate and oceanography and economic geography, also including regional physical maps. Revised editions have followed since 1959, reaching a seventh edition in 1970, in which the world maps have been redrawn; British temperature charts have been re-styled and contours removed from the general maps, making for improved clarity. Heights and depths have been converted to metres and temperature charts show degrees centigrade.

414 'Eduard Imhof: werk und wirken'; a festschrift volume for Professor Imhof at the age of seventy-five, was prepared by Hans Hauri and others (Zurich: Orell Füssli Verlag, 1970), with contributions from his colleagues and former students. Professor Imhof will be remembered in particular for his unique influence on the science of cartography and his development of the 'Swiss style'. The emphasis in this book is very much on the combination of artistic qualities and scientific accuracy upon which he insisted in all his work. His drawings, maps and relief models are reproduced throughout the text, showing especially his oblique hill-shading technique. Included is a bibliography of Professor Imhof's works.

415 'Education and training in the mapping sciences, a working bibliography', compiled by Harry Steward (American Geographical Society of New York, 1969), consists of sixty pages containing 720 entries on the subject published between January 1955 and December 1969. Surveying and photogrammetry are included.

416 Educational field studies (15 Wulfruna Gardens, Wolverhampton) aims at providing advanced field study courses for students interested in botany, geography and geology. In 1971, field courses were based on Weymouth, for Dorset; on Scarborough, for East Yorkshire; on Malham, for North Pennines; on Llanryst, for Snowdonia; and on Church Stretton, for the Welsh Borderland. A Prospectus is available.

Compare *Geographical field classes in Benelux and the Eifel* (Educational Travel Limited, 1971).

417 'Elements of cartography', by A H Robinson and R D Sale (Wiley, third edition, 1969), has become a 'standard' work since its first publication in 1953. Cartographical teaching and the additions made to the new edition render it even more suitable for modern courses, especially a new chapter 'Compilation from air photography' and much new material bringing the text into line with contemporary thinking. There is a useful bibliography.

418 Elsevier Publishing Company Limited, Amsterdam, has issued many publications of interest to geographers. The House of Elsevier was founded by Louis Elsevier (1540-1619), the son of a printer of Louvain. The family became famous, printing many scholarly scientific works. In 1655, the concern, then directed by Abraham Elsevier III, moved to Amsterdam. For the geographer, the name stands for quality journals, monographs and maps. Periodicals in the geosciences include *Marine geology,* international journal of marine geology, geochemistry and geophysics; *Sedimentology,* journal of the International Association of Sedimentologists, and *Agricultural meteorology,* an international journal with a unique subject content. The *Elsevier oceanography* series and the *Developments in sedimentology* series are particularly valuable. The firm also published the *Reports* of the Soviet Antarctic Expedition, 1955. The looseleaf *Grosse Elsevier atlas,* 1950- is in two folders, with explanatory text containing sketchmaps and diagrams. A recent Elsevier publication has been the *Picture atlas of the Arctic,* compiled by

R Thorén, 1969. Nine chapters deal with 'The Arctic Ocean', 'Drifting ice stations', 'Arctic Alaska', 'Arctic Canada', 'Greenland', 'Iceland' (the northernmost headland), 'The Norwegian Arctic Islands of Jan Mayen and Svalbard', 'Arctic Scandinavia' and 'The Soviet Arctic'. 567 photographs are included, a location map for each section and some diagrams. *Elseviers weekly* is a valuable check on current publications.

419 '**Emerging Southeast Asia:** *a study in growth and stagnation*', by Donald W Fryer (Philip; McGraw-Hill, 1970), aims to assess the role of South East Asia in the modern world: 1, 'The region', dealing with its part in the world economy, land use, urbanisation and industrialisation; 2, 'Progress', considering the natural resources and their use in Thailand, the Philippines, Malaysia and Singapore; 3, 'Stagnation', concerning Indonesia, the Union of Burma, Vietnam, Laos and Cambodia; and 4, 'Prospect', attempting to assess the region's potential trade and regional co-operation. There is a selected bibliography.

420 '**Encyclopaedia of Australia**', compiled by A T A and A M Learmonth (Warne, 1968), contains more than 2,700 entries and fifty special articles, illustrated, sometimes in colour, as in the case of 'Arms and flags of Australia' and the flora and fauna; there are small black and white maps and drawings in the text. Cross-referencing is helpful and there are frequent guides to further reading at the end of entries; collections and special libraries are included.

421 '**Encyclopaedia of Ireland**', edited by Victor Meally and others (Dublin: Allen Figgis, 1968), deals with all aspects of the country under broad subject groupings. The text is illustrated and there are bibliographies—'Books and periodicals', 'Transport' 'The land' and 'Agriculture and fisheries' among others.

422 '**An encyclopaedia of the iron and steel industry**', compiled by A K Osborne and M J Wolstenholme (The Technical Press, 1956, 1967), is a central source of information on the industry. Photographs and diagrams are included throughout the text and references are noted. Particularly useful for reference purposes are the Appendices on 'Conversion tables', 'Weights and measures', 'Properties', 'Signs and symbols' and 'List of scientific, technical and trade societies and other bodies related to the iron and steel industries'.

423 '**Encyclopaedia of Latin-American history**', in one volume, compiled by Michael Rheta Martin and others, first issued in 1952, was revised by L Robert Hughes in 1968 for the Bobbs-Merrill Company. The work is particularly useful for geographers for the statistics included, and for the historical information concerning major cities and industries.

424 '**An encyclopaedia of London**' was first published in 1937, edited by William Kent, revised and reset in 1951. In 1970, it was further revised by Godfrey Thompson (Dent), illustrated with photographs and including references and bibliographies.

425 '**An encyclopaedia of New Zealand**', edited by A H McLintock, is in three volumes (Wellington: Government Press, 1966). Many of the headings are specific, such as 'Anchovies, pilchards and sprats', some longer, such as 'Geology'. The whole work is a mine of information, including plates, monochrome illustrations and diagrams, statistics and endpaper maps.

426 '**Encyclopaedia of Southern Africa**', first published by Warne in 1961, is now in its fourth updated edition, under the direction of Eric Rosenthal and a team of experts. All aspects seem to be covered in this handbook, which contains some five thousand entries, cross-referenced, and illustrated with eleven colour plates, many half-tones and numerous sketchmaps and drawings throughout the text.

427 '**The Endeavour journal of Joseph Banks 1761-1771**', in two volumes edited by J C Beaglehole (Angus and Robertson, 1962), was produced as a memorial to Sir Joseph on behalf of the Government and people of New South Wales; it comprises the entire *Journal*, illustrated by portraits, landscapes and natural history notes. Dr Beaglehole, an acknowledged authority on Cook and Banks, also contributed the introduction.

428 '**Environment and economic life:** *an economic and social geography*', by Howard F Gregor (Van Nostrand, 1963), considers the role of the environment in the economic activities of mankind. The text is illustrated and contains maps and diagrams; there is a bibliography.

429 'Environment and land use in Africa', edited by M F Thomas and G W Whittington (Methuen, 1969), sets out the problems facing geographers and agricultural planners; these topics were the subject of a symposium on the natural and social environments of selected African areas. An introduction considers agricultural geography in tropical Africa; part 1 presents studies generally; part 2, studies of the social environment and in part 3, selected case studies are analysed in detail, from Malawi, Nigeria, East Central Sudan, Swaziland and Kenya. There are photographs and numerous line drawings in the text.

430 'The environmental handbook', edited by Garrett de Bell (New York: Ballantine Books Inc, 1970; 'Friends of the earth' series), was proposed at the first National Environmental Teach-in, April, 1970. Included in this stimulating work are articles, written by students, suggestions towards an ecological platform, a number of reprinted essays by authorities on economic, ecological, cultural and political change, a bibliography of books and one of films.

Refer also Max Nicholson: 'Environment on record' *in The geographical magazine,* November 1971.

H W Helfrich: *Agenda for survival: the environmental crisis* (Yale University Press, 1970).

The environmental handbook: action guide for the United Kingdom, edited by John Barr (Ballantine with Pan), based on the above.

Fieldworker: the environmental studies magazine.

Man's impact on the global environment: assessment and recommendation for action. Report of the Study of Critical Environmental Problems (SCEP) (MIT Press, 1970).

431 'The environmental revolution: *a guide for the new masters of the world',* a monumental work by Max Nicholson (Hodder and Stoughton, 1970), presents a summary of the author's views on the urgent necessity for pursuing a more ecological approach to the problems arising in an overcrowded world; the author's work in establishing the Nature Conservancy and as Chairman of the World Conservation Committee of the International Biological Programme, in addition to his years of study in this field, enable him to give an exceptionally clear analysis of man's use of the earth and the existing relationship between man and his environment. The computer is considered as a tool with which statistics and multiple factual data may be made meaningful. Conservation in Britain and in the United States is discussed, followed by descriptions of the development of international organisations whose aim is to deal with the many problems.

432 Eratosthenes (c276-c194 BC), Greek geographer and scientist, and a librarian at Alexandria, laid the foundations of mathematical geography. He is especially notable for his calculations of the circumference of the earth, which he recognised as a sphere. He compiled a history of geography from the time of Homer, and constructed a map of the world based on a system of parallels and meridians.

433 'Die Erde: *Zeitschrift der Gesellschaft für Erdkunde zu Berlin'*, one of the most scholarly of geographical periodicals, has been published quarterly since 1839 by Walter de Gruyter. The articles, bibliographies and reviews provide an unrivalled source on the development of geographical progress; abstracts are included and the annual index appears in the fourth issue.

434 'Erdkunde: *archiv für Wissenschaftliche geographie'*, one of the periodicals of international stature, is published quarterly by Ferd. Dümmlers Verlag, Bonn. The articles are usually in German, some in English, of a high standard of scholarship and wide range of interest. Full and accurate summaries precede each article and English titles of articles are also listed in the tables of contents.

435 ' Essays in political geography ', edited by C A Fisher (Methuen, 1968), based on papers prepared for the 20th International Geographical Congress, 1964, deal first with the geographical aspects of the internal structure and the external relationships of states; secondly, with individual case studies in decolonization and with aspects of politico-geographical change in the old world. There are extensive notes, references and sketchmaps throughout the text.

436 Esselte Corporation, Stockholm, is a private map firm which has had great influence on Swedish map production. For many years, the firm has maintained an active exchange of technical experiences with the leading map printers in other Scandinavian countries and in Great Britain, America, Germany, Switzerland and France, and has itself taken the lead in glass engraving and the use of plastics in mapping techniques. Scribing at Esselte has been further improved by the design of new engraving tools, which are also exported to other countries. Esselte did much to encourage the formation of the International Cartographic Union and takes its place in co-operative ventures between the foremost map services of the world, as, for example, in the production of the *International yearbook of cartography* (*qv*).

437 '**The European bibliography**' was compiled, in English and French, at the European Cultural Centre, Geneva, under the direction of Hjalmar Pehrsson and Hanna Wulf (Leiden: A W Sijthoff, 1965). The sections 'Europe and the world', 'Economics' and 'Documentation' are of particular interest to geographers; entries are arranged under broad subject headings and there is a name index. Notes are given in the original language in the case of books written in German, Spanish, Italian or Dutch.

Refer also *European research resources* (J Tricart for the Council of Europe).

438 European communities: *Press and information* produces a wide range of publications describing work and policy. Of central importance is the monthly magazine *European community,* an illustrated journal containing short articles and notes. *The European Community in maps,* in a revised edition, 1968, is a folder of twelve four-colour maps depicting the main political, social and economic features of the Community and adjacent countries, including England and Wales. Set in to the folder is a four-page leaflet describing the aims and structure of the Community and a second four-page leaflet gives notes on the maps, with two pages of statistical tables. Sources are named on the inside back cover. 'Economic union and enlargement', 1969, represents 'The European Commission's revised opinion on the application for membership from the United Kingdom, Ireland, Denmark and Norway'; and 'Uniting Europe: the European Community since 1950', an illustrated brochure, describes the achievements of the Community. There are also the booklets 'Farming in the Common Market', 'Research and technology and the European Community' and a smaller summary leaflet, 'Europe's tomorrow'.

439 '**European companies: a guide to sources of information**', in a second edition, 1966, reprinted, 1968, was compiled and edited by

G P Henderson (CBD Research Limited, Beckenham, Kent). Arrangement of entries is alphabetical by countries, with code numbers attached.

440 European Conservation Year was 1970, when twenty-seven countries from Iceland to Turkey, in addition to fifty-seven international organisations, organised under the guidance of The Council of Europe, attempted to focus attention on the way in which human activities affected the natural environment and to discuss policies of conservation. In April, 1963, the United Kingdom held its first National Wildlife Exhibition, the highspot of National Nature Week, which led to the first of the conferences, ' The countryside in 1970 ', in November 1963. When the scope of the work had been fully explored, the European project was prepared. A conference in Strasbourg, in February, 1970, concentrated particularly on urbanisation, industry, agriculture and forestry, and on the increasing demands for leisure. In November, a conference in London summed up the main themes of the year and proposed guidelines for the ensuing decade. The Countryside Commission issued a ' Calendar of events ' in England, ending with a list of the organisations taking part. A booklet, ' Wales in European Conservation Year ', was issued by The Countryside in 1970 Committee for Wales; The Countryside Commission for Scotland published a guidebook ' Scotland's Countryside 1970 ' and a news-sheet setting out Scotland's involvement; a ' Calendar of Events in Northern Ireland ' was issued by the Nature Reserves Committee.

Refer also The whole issue of *The geographical magazine* for January 1970 was devoted to ' Conservation in Europe '.

441 ' The European peasantry: the final phase ', by S H Franklin (Methuen, 1969), describes and discusses the evolution of the people since the war, the economic and social consequences of their incorporation within the dynamic postwar industrial economies, the regional aspects of their evolution and the role of the state in helping to solve their problems. In outline, the scope of the work includes ' Peasants: concepts and context '; ' Bauern: worker-peasants, family farms in Federal Germany '; ' Paysans: property, family and farms in France '; ' Contadini: peasant and capitalist farming in the Mezzogiorno '; ' Peasants in the EEC . . .'; and peasants in the countries of Eastern Europe. There is a section of references, sub-divided according to the chapters, and a glossary of names not widely familiar or generally known by initials. A number of sketchmaps are included throughout the text.

442 '**Everyman's atlas of ancient and classical geography**' was first published in 1907. The latest revision (Dent, 1961) includes a particularly valuable essay by J Oliver Thomson on the development of ancient geographical knowledge and theory.

443 '**Evolution of Australia**', by R M Crawford (Hutchinson University Library, third edition, 1970), examines the policies, events and biographical details which have contributed most to the course of development. Four particularly important chapters, from the geographer's point of view, are devoted to the formative period 1850 to 1900—' The history of Australia is a chapter in the history of migration '.

444 '**The evolution of Scotland's scenery**', by J B Sissons (Oliver and Boyd, 1967), examines the following topics: ' Pre-glacial evolution ', ' Glacial erosion and early ideas about glaciation ', ' Landforms produced by glacial erosion ', ' Glacial deposits: their origin and significance ', ' Glacial deposits and associated landforms ', ' Landforms produced by glacial rivers ', ' The glacial sequence ', ' Changing sea-levels and changing climates ', ' Causes and measurement of sea-level changes ', ' Changing sea-levels in Southeast Scotland ', ' Changing sea-levels in Scotland as a whole ', ' Periglacial and postglacial changes '. The text is accompanied by beautiful and well-chosen photographs, sketchmaps and figures and there is a bibliography of references, mainly to periodical articles.

Refer also Archibald Geikie: *The scenery of Scotland.*

445 '**Experimental cartography:** *report on the Oxford Symposium, October 1963*' (OUP, 1964) includes papers and discussions; pertinent bibliographical references are also appended. The first object of the symposium was to discuss the use of cartography as a research tool, in the mapping of data in industry, geology, demographic distributions, climate, transport, vegetation, flora and fauna, history and archaeology, hydrography and oceanography. The second object was to consider whether the cartographic analysis of different subjects has anything to contribute to topographic mapping.

Refer also D Bickmore, *ed*: *Automatic cartography and planning,* prepared by the Experimental Cartography Unit, Royal College of Art, being a report on a feasibility study commissioned by the former Department of Economic Affairs on behalf of the East Anglia Economic Planning Council (The Architectural Press, 1970).

446 ' Experimental orthophotomap of Camp Fortune Skiing Area of Gatineau Park ', at 1 : 10,000, prepared by the Photogrammetric Research Section of the National Research Council, one of the chief organisations promoting the use of such photomaps, was issued with *The Canadian surveyor,* volume XXII, no 1 (Ottawa: The Canadian Institute of Surveying, 1968). The map was designed by D Honegger under the direction of T J Blachut. In the *Proceedings* of the International Symposium on Photomaps and Orthophotomaps held at Ottawa in September 1967 is an article by T J Blachut, Director of the Research Section, entitled ' Further extension of the Orthophoto Technique '. The author points out that the Camp Fortune map was designed as a tourist ' winter map ' to provide information for skiers. Relief shading was applied, based on contours produced on a conventional stereoplotter.

447 ' Explanation in geography ', by David W Harvey (Arnold, 1969), formulates a methodological framework for geographic thought in the light of recent advances in geographic research and statements in the philosophy of science. In six parts—' Philosophy, methodology and explanation '; ' The methodological background and explanation in geography '; ' The role of theories, laws and models in explanation in geography '; ' Model languages for geographic explanation '; ' Models for description in geography '; ' Models for explanation in geography '—are subdivided; for example, there are separate chapters on definition and measurement, classification, maps and pattern analysis. Figures illustrate the text and there is a list of references.

Note Methuen, from November 1971, announced a new series, ' The field of geography '—' designed to meet the need for inexpensive textbooks which provide a thorough grounding in modern techniques, concepts and principles in the many divisions of geographical study. Twenty volumes are in preparation and more are planned.'

448 'The explorers: *an anthology of discovery'*, compiled and edited by G R Crone (Cassell, 1962), portrays 'a number of explorers in action, either at critical moments in their journeys or when displaying their individual qualities conspicuously', excluding, for example, for the most part, sea voyages and mountaineering expeditions. Actual extracts of texts are presented, each with a brief note of introduction, arranged chronologically within sections dealing with the countries explored: Asia, Africa, North America, South America, Australia and the Polar regions. A map is included in each section.

Note The literature concerned with exploration is vast, either general accounts, summaries of the opening up of individual countries or assessments of the achievements of particular explorers. J R L Anderson, in *The Ulysses factor: the exploring instinct in man* (Hodder and Stoughton) has conceived an interesting theory, presenting evidence of the urge to explore as a combination of recognisable qualities, of which the Greek Ulysses is the type-figure.

449 'Explorers' maps: *chapters in the cartographical record of geographical discovery'*, by R A Skelton (Routledge and Kegan Paul, 1958), reprints, with revisions, a series of fourteen articles written for *The geographical magazine* between July 1953 and August 1956. 'It is designed chiefly as a picture book, and not as a systematic history of exploration', having numerous black and white reproductions throughout. The text is planned as a summary of geographical ideas and events associated with the maps, under the following headings: 'The way of the East'; 'The way of the West'; 'The way of the North'; 'The Spice Islands and Cathay'; 'The South Sea'; 'The continents and the Poles'. There is a short bibliographical note.

450 'Explorers' maps of the Nile sources 1856-1891' is a unique collection of reproductions, from the archives of The Royal Geographical Society, of maps made in the field by Livingstone, Speke, Grant and other great explorers.

451 Faber and Faber, London, are known to geographers chiefly for *The Faber atlas,* which traces its origin to the *Österreichischer Mittelschul-Atlas* (*Kozenn-Atlas*) compiled by the Slovak geographer, Blasius Kozenn, 1864. A long line of atlases developed from the first edition, including Lautensach's *Atlas zur Erdkunde,* the Bordas *Nouvel atlas général* and *The Faber atlas,* published for some time by E Hölzel of Vienna; the English edition was specially designed for the English-speaking market. A new edition edited by D J Sinclair was published in 1970 by GEO Publishing Company, Oxford; this edition was updated to 1965. Intended primarily for teaching, the maps give most areas reasonable coverage for the size of the atlas, with emphasis on the British Isles and Europe. The gazetteer-index contains some twenty thousand entries. The firm's catalogues dealing with specific topics are issued from time to time; an excellent example is 'Faber books on farming', 1969, which includes land use, conservation, forestry, country life and other rural studies.

452 'Facsimile atlas to the early history of cartography, *with reproductions of the most important maps in the XV and XVI centuries',* prepared by A E Nordenskiöld, is a translation from the Swedish original, Stockholm, 1889, indispensable for the study of the Ptolemy editions and for the development of the cartography of the period.

453 Fairey Surveys Limited, Maidenhead, Berkshire, comprises a number of subsidiary companies and the Fairey Group of Companies includes yet more companies in many parts of the world. The whole range of surveying, mapping and photography is carried out, in any kind of terrain. Town planning, railway development, irrigation projects, cadastral and reconnaissance surveys, road construction and aerial surveying have contributed particularly to the development of countries in Africa and Asia. The firm produces pamphlets concerning its work: 'This is Fairey Surveys', for example, and 'Loose print mosaicing', by E L Freeman. Generally speaking, the firm does not manufacture instruments, but there is the Fairey Plotterscope, a viewing aid for stereoplotting machines, with separate co-ordinatograph tables, a compact lightweight instrument designed to aid the machine operator by providing an accurately magnified image of the pantograph pencil together with the surrounding detail which can be viewed from a normal seated position.

454 'The Far East and Australasia, 1969' (Europa Publications Limited) is the first of a new series of surveys annually bringing together factual information of all kinds. A general introduction is followed by a listing of regional organisations, then the main part of the work considers regional areas, South Asia, South-east Asia, East Asia, Australasia and the Pacific Islands. Essays are included on specific topics, reference material is noted and the final section is a 'Who's who' and a list of research institutions, associations and centres of study of the Far East and Australia. There are maps, diagrams and statistics in the text and a bibliography is appended to each part.

455 'Far East trade and development', the monthly independent international for trade expansion, reports in each issue on trends in world production and trade, including reports on new projects, reports on individual countries and on particular industries and commodities; a regular feature is the 'Machinery report'. There are usually some photographs and the advertisements, as always in such a periodical, are valuable indicators. Supplements on some current topic are occasionally issued.

456 'Far Eastern survey', a fortnightly research service (New York: American Council of the Institute of Pacific Relations, 1932-1961), contained factual information and expressions of opinion, mainly on economic topics, excellent background material for the understanding of events and development. Publication ceased in 1961, but the scope and purpose, with certain modifications, of this journal was continued in *Asian survey* (*qv*). The previous volumes were reprinted by Krauss in 1966.

457 'Farming systems of the world', by A N Duckham and G B Masefield (Chatto and Windus, 1970), is intended as a reference book for students and research workers, including, therefore, plenty of maps and diagrams, some photographs and a glossary. The two main divisions consist of 'Analysis of location factors' and 'Farming systems', containing a classification of farming systems and a number of regional reports based on actual surveys. A third part, 'Conclusion', examines, among other topics, world food and population problems and there are end of chapter references throughout.

Refer also H F Gregor: *Geography of agriculture: themes in research* (Prentice-Hall, 1970).

458 Fawcett, Charles Bungay (1883-1952) was one of the leading British academic geographers in the period immediately following the inspiration of Mackinder and Herbertson, especially in Southampton, Leeds and London. His original thought in political and population geography was expressed in numerous articles in geographical and demographic journals and in such published works as *Frontiers: a study in political geography,* a pioneer work, 1918; *The bases of a world commonwealth,* 1941; and *Provinces of England: a study of some geographical aspects of devolution,* 1919, revised, with a new preface by W Gordon East and S W Wooldridge (Hutchinson University Library, 1960). A bibliography of his works was included in the *Transactions of the Institute of British Geographers,* 1952.

459 'The Fenland in Roman times', edited by C W Phillips, is a unique record of Roman settlement and of an ancient landscape, published by The Royal Geographical Society in 1971; twenty-four plates and eighteen figures are included in the text and there are twenty-four maps, some in colour, in a separate envelope.

460 Festschriften: English scholars or groups of scholars have not been so ready to use this form of publication as, for example, those on the Continent—and geographers have been no exception—*Festschrift für Hans Kinzl: celebrating his 70th birthday,* for example, from the Geography Department of the University of Innsbruck and *Essays on agricultural geography: a memorial volume to Dr B N Mukerji,* edited by B Banerjee, from the University of Calcutta, 1969. During recent years, however, several examples of the genre have appeared in Britain, including *Geographical essays in honour of K C Edwards* (qv); *Essays in geography for Austin Miller,* edited by J B Whittow and P D Wood (University of Reading, 1965); *Northern geographical essays* (qv); *Studies of a small democracy* (qv); *Irish studies in honour of E Estyn Evans* (qv); *The allocation of economic resources: essays in honor of Bernard Frances Haley* (Stanford University; OUP, 1959, 1965); *Land use and resources: a memorial volume to Sir Dudley Stamp* (Institute of British Geographers, the first volume in the 'Special' series of publications); *Exeter essays in geography in honour of Arthur Davies,* edited by K J Gregory and W L Ravenhill (University of Exeter, 1971); *Studies in the vegetational history of the British Isles: essays in honour of Harry Godwin,* edited by D Walker and R G West (CUP, 1970). Festschriften with a difference are *Liverpool essays in geography:* a jubilee collection (qv) and *Geography in Aberystwyth:*

essays written on the occasion of the Departmental Jubilee 1917/18 ... (qv). One of the latest to appear is the very valuable volume for *Eduard Imhof... (qv).*

461 Field Studies Council (formerly The Council for the Promotion of Field Studies) issued the first number of *Field studies* in 1959, its contributions 'well spread in interest and locality', illustrated with maps and diagrams. Field work being now an integral part of the training for a university degree in geography, geology, botany, zoology and others of the earth sciences, the field centres organised by the council, with their resident wardens, collections of essential books and maps, simple laboratories and conference room facilities, are an absolutely essential unit in our system of higher education.

Refer also Field Group studies, Nottingham University.

462 ' Field studies in the British Isles ', edited by J A Steers (Nelson, 1964), was published for the 20th International Geographical Congress, 1964; it contains thirty-three studies by specialist scholars. Many kinds of landscape are examined, including coasts, the Weald, scarplands, highlands and lowlands. The second group of essays in the book concentrates on Wales, followed by sections on Scotland and Ireland. There are textmaps and end of chapter references.

463 ' The fitness of man's environment ' prepared by Robert McC Adams and others, was published by the Smithsonian as Annual II (available in Britain through David and Charles of Newton Abbot). The contributors discuss what has gone wrong with man's environment and how our future surroundings might be made more attractive.

464 ' Fiziko-geografičeskij atlas mira ', published by the Academy of Sciences of the USSR, together with the Department of Geodesy and Cartography GGK, SSSR (Moscow, 1964), is the most detailed atlas of physical geography yet produced; an entirely new atlas, compiled by the foremost scientists of the USSR, incorporating much new research, especially regarding the atmosphere, hydrosphere and upper layers of the lithosphere, and including the latest earth science theories. In the section of general geography, some maps from the 1954 *Atlas mira (qv)* have been used in revised form, but all other maps are original and previously unpublished.

The atlas is in three main parts: seventy maps devoted to the natural features of the world, including the Arctic and Antarctic, on relatively small scales of from 1:60M to 1:150M, and a summary world map, 1:80M, of natural landscape zones; maps of individual continents, 1:10M to 1:40M, each series comprising about twenty pages of maps, together with a final zone map for each continent; an extensive study of the USSR, on more than eighty maps on scales of 1:15M, 1:20M and 1:35M, and a summary map of natural zones and provinces. Forty pages of explanatory text in Russian summarise the basic conclusions derived from the content of the maps and discuss the methods used in data selection and map compilation. The whole atlas notably demonstrates new cartographical techniques. A complete translation of legend matter, contents and explanatory text is presented in a combined two-month issue of *Soviet geography: review and translation,* May-June 1965.

465 Fleure, Herbert John (1877-1969), one of the great modern scholar-geographers, was the acknowledged leader among human geographers in Britain. Throughout his professional life he encouraged generations of students, served in various offices, a great number of learned organisations, travelled much and wrote an amazing number of major works, in addition to innumerable reviews and papers: *Human geography in Western Europe,* 1918; *Peoples of Europe,* 1922; *The races of England and Wales,* 1923; *The races of mankind,* 1927; *French life and its problems,* 1942; *The corridors of time,* with H J E Peake, in ten volumes completed in 1958; *The natural history of man in Britain,* 1951, 1959, revised edition by Dr Margaret Davis (Collins *New naturalist library,* 1971); *Guernsey,* 1961. Mrs Sinnhuber compiled a bibliography of his publications, 1954, and The Geographical Association Library in the headquarters building has been renamed the Fleure Library. Among the many tributes to Professor Fleure's work, one of the most informative is that contributed by Professor E G Bowen and Professor Alice Garnett to *Geography,* November 1969.

466 'The flora of Greenland', by Tyge W Bocher and others (Hurst, 1968), is considered a standard reference work; every species of flowering plant and fern is discussed, a total of five hundred. The determination of species is facilitated by keys and a number of clear and realistic line illustrations by Ingeborg Frederiksen. For each species a morphological description, the chromosome number and distribution of the species are given.

467 'Focus' is a monthly (except July and August) periodical published by the American Geographical Society from 1950, 'presenting a brief, readable, up-to-date survey of a country, region or resource, helpful in understanding current world events'. Each issue carries valuable bibliographies on the chosen theme and gives information on the country's place in the world, its physical setting, climate and people, economic activities and resources, referring to any relevant historical developments and adding notes on any special policies or projects in the area. Small photographs are introduced into the text, also at least one page of suitable sketchmaps. In 1971, Praeger published the first of a series of monographs based on the issues of *Focus*.

468 Food and Agriculture Organization of the United Nations sponsors research and produces publications of vital importance to geographers. Initiated by the first Conference on Food and Agriculture among the United Nations, at Hot Springs, Virginia, in 1943, the organization was formally created at Quebec in 1945, with temporary headquarters in Washington. In 1951, the headquarters removed to Rome. Publications cover agriculture, economics, fisheries, forestry and nutrition, in addition to the official records of the functioning and work of FAO. Periodicals such as the *Monthly bulletin of agricultural economics and statistics, Cocoa statistics, Unasylva, World fisheries abstracts* (qv), *Food and agricultural legislation,* the *FAO plant protection bulletin* and the statistical yearbooks of agricultural, fisheries and forestry products constitute a world intelligence service on production, prices and trade that covers almost every commodity in the world. The state and outlook of world agriculture is recorded in a major annual publication, *The state of food and agriculture*. Technical monographs, reports, regional studies and pamphlets such as the *Commodity series* are the result of extensive co-operation between governments and experts throughout the world. *The FAO review* began publication in six issues a year, 1967-, in English, French and Spanish editions; the journal reports on agricultural technology in relation to the food demands made by increasing world population, progress of development projects, international trade and agro-allied industries. Most interesting and valuable of all, from January 1967 the 'FAO Documentation-current Index', issued monthly with bimonthly cumulations, has provided a 'current awareness' index, an excellent example of indexing in depth. It consists of three parts: an analytical index, an index by authors and a bibliographical list, produced automatically by means of special computer programmes. The FAO Documentation Center was also established in 1967. The *Catalogue of FAO publications* appears every two years in English, French and Spanish; it lists all priced publications since 1945 and titles in preparation, under the headings: 'World food situation—basic information', 'Agriculture', 'Economics and statistics', 'Fisheries', 'Forestry and forest industries', 'Nutrition', 'Legislation', FAO official records and basic documents', 'Vocabularies, glossaries, bibliographies, catalogues', 'Periodicals and annuals'. Complete catalogues covering the years 1945-1968 and 1945-1972 are also available; brochures and leaflets describing new publications are circulated from time to time. The *Annual report* should be consulted for details of the work of the organisation.

469 'Foreign maps and landscapes', by Margaret Wood (Harrap, 1968), provides an introduction to the study of foreign maps; selected map sheets demonstrating a variety of landforms, are reproduced, with explanatory notes, from Switzerland, France, Germany, Norway, Netherlands and the United States. Terms are explained when necessary and there are fine half-tone plates, map extracts and diagrams in the text, also a short bibliography.

470 'Forest resources of the world', by Raphael Zon and William N Sparhawk (McGraw-Hill, 1923), prepared under the authority of the Secretary of Agriculture of the United States and in co-operation with The National Research Council, in two volumes, remains a classic. The text, illustrated by folded maps and diagrams, begins with 'The general forest situation in the world', followed by a series of chapters analysing forestry in various parts of the world and ending with an examination of 'Forest resources other than timber', such as resins and rubber. To its date, it provides a wealth of information, including many statistical tables and references.

Refer also L R Holdridge: *Forest environments in tropical life zones: a pilot study* (Oxford: Pergamon Press, 1971), as one of the most recent examples of research on forestry environment.

471 'Forestry and timber technology', the title of the catalogue of Stobart and Son, Limited, London, includes government publications, Forestry Commission Publications, Forest Products Research Publications, Timber journals, timber specimens and forestry and timber trades accessories; for example, Hough's *Encyclopaedia of American woods; Technical and commercial dictionary of wood*, by G Trippodo; *World timber: trends and prospects*, by T Streyffert; the *Concise encyclopaedia of world timbers*, by F H Titmuss; *Tropical forestry, with particular reference to West Africa*, by C J Taylor.

472 The Forestry Commission, London, was established by Act of Parliament in 1919, with the primary object of providing wood for industry, but with the declared intention of conserving and controlling wildlife, safeguarding the natural landscape and of meeting reasonable requirements for access and other recreational activities. The publications of the Department are set out in Sectional List no 31, revised in 1970; these comprise reports, Acts, regulations, bulletins, leaflets and guides. New booklets include 'The Forestry Commission and conservation', 'Britain's new forests', 'Forestry in England', 'Forestry in Wales' and 'Forestry in Scotland', all amply illustrated.

Refer also Check list of Forestry Commission publications, 1919-1965 (HMSO, 1966).

H L Edlin: 'The Forestry Commission in Scotland, 1919-1969', *in The Scottish geographical magazine,* September 1969.

George Ryle: *Forest service: the first forty-five years of the Forestry Commission of Great Britain* (David and Charles, 1969).

473 'France: a geographical survey', by Philippe Pinchemel, translated by Christine Trollope and Arthur J Hunt (Bell, 1969), is based on the second French edition, 1964, by Colin of Paris, with the inclusion of some of the author's revisions for the third French edition. The aim has been to give ' as coherent a picture as possible of the land of France, its organisation . . .'; the words of Paul Vidal de la Blache—' By what means did a fragment of the earth's surface, neither a peninsula nor an island, which no physical geographer could ever consider a whole, become a political unity and finally reach the status of a nation?' may be considered the keynote of this later study, by another of the leading geographers of France.

With this philosophy in mind, the approach is traditional: an introduction treating of general characteristics; followed by ' The role of nature '; ' The role of man '; ' Industry '; ' The towns and cities '. A carefully selected bibliography is appended, in addition to bibliographical end of chapter notes. The sketchmaps and diagrams, though small, are evocative, in the best tradition of French geographical scholarship.

474 France: its geography and growth ', by Jean Dollfus, one of the *Geography and growth* series (Murray, 1969), surveys France under such headings as physique, climate, history and other general factors, before going on to consider the regions individually. Splendid photographs demonstrate the reaction of man and geographic factors throughout the area. The book is available in a students' edition and a library edition.

475 Freytag-Berndt has been issuing educational maps since 1894, including general, physical and thematic maps. Methods of cartography were completely revised after the last war and the latest techniques are now used. The maps are striking, using contours, distinctive hypsometric tints and spot heights for relief representation and clear drawing of river systems, roads and other main features. The firm is noted for the series *Touristen-Wanderkarten,* 1:100,000, of Central Europe, the *Auto-Atlas Österreich mit kartenteil Mitteleuropa* and other series such as the *Landerkarten,* 1:600,000 and the *Strassenkarten* 1:2M.

Refer Fritz Aurada: ' Entwicklung und Methodik der Freytag-Berndt Schulwandkarten ' *in The international yearbook of cartography,* 1966.

476 ' From geography to geotechnics ', by Benton MacKaye, edited by Paul T Bryant for the University of Illinois Press, 1968, is a collection of thirteen essays arranged in three sections: ' Geography to geotechnics '; ' Control of the landscape '; and ' Uses of the wilderness '. MacKaye's thesis on ' new science ', the philosophy of homesteading, valley authorities, ' folkland ', watershed management, the Tennessee Valley Authority, the Appalachian Trail and ' outdoor culture ' reveals the American bias of the essays and shows echoes of his acquaintance with original thinkers such as W M Davis, Patrick Geddes, Lewis Mumford, Stuart Chase, Henry Gannett and Isaiah Bowman.

477 'Frontiers in geographical teaching', the Madingley Lectures for 1963, edited by Richard J Chorley and Peter Haggett (Methuen, 1965, 1967; second edition, 1970), is divided into three parts: concepts, techniques and teaching. Summaries and analyses are presented of many topics in the forefront of current geographical thinking, with useful lists of references accompanying each section.

478 'Fundamentals of economic geography: *an introduction to the study of resources'* by William Van Royen and Nels A Bengtson, was published in 1935, 1942, 1950, 1956 and in a fifth edition, 1964, extensively re-written. The first section considers 'Economic activities and their physical background'; the second, 'Distribution of mankind: its significance and its dynamic aspects' and the third, 'The field and function of economic geography', dealing with topics such as 'The earth as the habitat of man', 'Factors responsible for climatic differences', 'Climates of the world: classification and distribution', 'Climate, soils and agriculture', 'The regions of humid tropical climates: characteristics of the three types', 'Major agricultural products of the humid tropics', 'The regions of dry tropical climates: areas of agricultural potential' and so on, carefully analysing the main kinds of resources in their environments, including, finally, the metals and manufacturing industries. An appendix examines maps, sketchmaps, map projections and time zones. Throughout the text are photographs, maps and diagrams.

479 Gallois, Lucien (1857-1941) was one of the leading figures in the French school of regional geography after the death of Paul Vidal de la Blache, following him at the Sorbonne. He carried out the great project, *Géographie universelle* (*qv*) envisaged by Blache, and himself published a number of regional monographs, among the best known being *Les géographies allemands de la Renaissance,* 1890 and *Régions naturelles et noms de pays,* 1908.

480 'Gazetteer of the Persian Gulf, Oman and Central Asia', by John Gordon Lorimer, was published in two volumes; volume II was issued in 1908 and volume I, in three parts, with introduction and table of contents, was completed by L Birdwood, in 1915, after Lorimer's death. The Irish University Press reprinted the work in 1970 in six parts, two volumes. In the first are twelve chapters geographically arranged, each of the chapters being sub-divided according to chronological periods, plus a number of appendices.

The second volume is a geographical dictionary, presenting, in a series of alphabetically arranged articles, a detailed account of the physical and political conditions of the Persian Gulf and the surrounding countries, including also information concerning natural products and agriculture, livestock, population, communications and trade, internal and external.

481 Geddes, Sir Patrick (1854-1932) was a pioneer thinker in planning and sociology, whose ideas have exerted considerable influence on geographers interested in urban geography. His association with French ideas and, in particular, with the 'place-folk-work' concept of Frédéric Le Play led to the foundation of the Le Play Society (*qv*). Arthur Geddes, son of Sir Patrick, who died in 1969, was also a scholar, specialising in geography, population problems in relation to environment and France.

Refer also Jacqueline Tyrwhitt, *ed : Patrick Geddes in India* (Lund Humphreys, 1947).

482 Geikie, Sir Archibald (1835-1924) was one of the great scholar geologists and director-general of the Geological Survey of Great Britain. Of his publications, *Earth sculpture and the origin of land forms,* 1898, *Textbook of geology,* in two volumes, 1903, and *Structure and field geology for students of pure and applied science,* 1905, have remained the most useful. He also wrote *Founders of geology,* 1905 (Dover Editions, second edition, 1963).

483 ' Geoderma', *quarterly* 1967-, 'international journal of soil science' (Elsevier Publishing Company), covers all aspects of soil research; text is in English, French and German, with English summaries.

484 ' Geo-ecology of the mountainous regions of the tropical Americas', edited by Carl Troll (Bonn: Ferd Dümmlers Verlag, 1968), is, in effect, the proceedings of the symposium organised by Unesco in conjunction with the Latin American Regional Conference of the International Geographical Union, held in Mexico City, August 1966. The symposium may be considered the result of the lifetime involvement of Professor Troll and his far-encompassing interest in the tropical mountains throughout the world, a subject in which his publications have attracted international acclaim. Professor Troll himself contributed the introductory paper on climatic, phytogeographical and agrarian ecology of tropical American mountains. The whole volume must surely remain a basic reference work for those interested in this group of subjects.

485 'Geografia dell' Africa', the first known atlas of Africa, by Livio Sanuto, published in Venice, 1588, has been reprinted in facsimile by Theatrum Orbis Terrarum Limited. This, Sanuto's main work, was left incomplete at the time of his death; the three indices were made by Giovan Carlo Saraceni and the bibliographical note was contributed by R A Skelton.

> Compare J D Pearson and R Jones: *The bibliography of Africa: proceedings and papers of the International Conference on African Bibliography,* Nairobi, December 1967 (Cass, 1970).

486 'Geo-Forum', a new earth science quarterly, published by Vieweg and Sohn of Braunschweig and Pergamon Press, Oxford, 1969-, aims to offer close evaluation of research in the earth sciences, methodology and application.

487 'The geographer and urban studies', by David Thorpe (Department of Geography, University of Durham, 1966), is number 8 in the *Occasional papers* series, consisting of thirty-five pages, illustrated with sketchmaps, plans, diagrams and including a bibliography. 'The main text is a broad statement of the role of the geographer in urban studies, which has as its aim an outlining of the most profitable approach which he can adopt to the study of towns. The major conclusions of this statement are considered to be of significance to the methods and content of geographical teaching at all levels . . .'. An Appendix deals with some specific aspects, under the headings 'Urban field studies: some suggestions', 'The growth of urban areas', 'Shopping centres', 'Housing estates' and 'Functioning regions', each section having a select bibliography of British works.

488 'The geographer as scientist: *essays on the scope and nature of geography',* by S W Wooldridge (Nelson, 1956), brings together a number of papers written during twenty-five years of active teaching and research, each a classic of its kind, revealing Dr Wooldridge's systematic methods in studying places—the physical study of landforms, geomorphology and soil science and their significance in determining the patterns of settlement and farming, in addition to wide reading and understanding. Other topics include 'Geographical science in education', 'On taking the "Ge" out of geography', 'The status of geography and the role of field work', 'The role and relations of geomorphology', 'The changing physical landscapes of Britain' and further studies of Britain, and 'The conservation of natural resources'. Sketchmaps and diagrams illustrate the text.

489 'The geographer's craft', by T W Freeman (Manchester University Press; New York: Barnes and Noble, 1967), deals with the work of a selected number of geographers who have adopted distinctive approaches to the field of geographical research: Francis Galton, Paul Vidal de la Blache, Johan Cvijić, Ellsworth Huntington, Sten de Geer, P M Roxby and A G Ogilvie.

490 'A geographer's reference book', edited by C H Saxelby, was prepared and published by the Geographical Handbook Committee in 1955. The work presents a mixture of source material and factual information for use in teaching and draws attention to some new developments in world resources. References are given under broad subject groupings.

491 'Geographia' of London publish a great variety of maps and atlases, mostly for quick-reference, commercial or tourist use. Many maps, for different purposes, have been published covering the London area, including the famous *Atlas of London and suburbs*. A series of world maps, both general and special, are frequently revised; more recent ventures include marketing and sales maps of Europe and various marketing and industrial survey maps of Great Britain. The Geographia marketing maps of the conurbations, published in association with the London Press Exchange, began with that for the West Midlands, followed by a similar survey of Liverpool and Merseyside, both prepared in conjunction with the Department of Geography, University of Birmingham; a similar survey of Glasgow and Clydeside was made with the Department of Geography, University of Glasgow. A wide range of plans of towns in all parts of Britain is available and the *Commercial gazetteer of Great Britain* is a useful guide. The firm is also agent for a number of European cartographic houses.

492 ' Geographia ', the first atlas known to be printed in the Italian language, by Francesco Berlinghieri (Florence, 1482), was reprinted by Theatrum Orbis Terrarum in 1966, with an introduction by R A Skelton. The atlas consisted of thirty-one copper engraved maps, the only edition known to be prepared on the original Ptolemaic projection, with equidistant parallels.

493 ' Geographia Polonica ', a journal published by the Polish Academy of Sciences, Institute of Geography, in English, French and Russian, is intended to inform foreign geographers on the achievements of Polish geographers. It contains translations or summaries of comprehensive studies published originally in Polish, as well as specially commissioned articles. There is an emphasis on methodology; and the journal includes also reports prepared by Polish geographers for international meetings and conferences held in Poland.

494 ' Geographical abstracts ' developed from *Geomorphological abstracts* (*qv*). Four separate series are now compiled by a team of experts under the direction of K M Clayton, who also edits series A, and the whole work is based on the department of Geography, London School of Economics. Six issues a year are produced in the following series: A, Geomorphology; B, Biography, hydrology and water resources, meteorology and climatology, oceanography, pedology; C, Economic geography; D, Social geography, cartography. (Early series were arranged rather differently.) Author and regional indexes for each volume are published in the last issue of the year and a comprehensive annual index to all four volumes is in hand. The service is unique in the English language and is of obvious value.

Refer also British Geomorphological Research Group.

495 ' Geographical analysis: *an international journal of theoretical geography'* began quarterly publication by the Ohio State University Press in January 1969, edited by Professor L J King in the Department of Geography. The concept of the periodical is to encourage 'significant research aimed at the formulation and verification of geographical theory through model development and mathematical and statistical analysis '.

496 The Geographical Association is one of the two organisations of professional geographers in England, founded in 1893 to further the study and teaching of geography in all categories of educational institutions in the United Kingdom and abroad. Conferences and courses are arranged throughout the year and the annual meeting has become an event of geographical consequence, based on a main theme chosen each year; a valuable exhibition is also staged, which brings together the work of all kinds of organisations and of commercial publishers interested in geographical subjects. The annual report, including the activities of the Branches, appears in the January issue of *Geography,* together with the financial and publication reports. The Spring conference, held at different centres, also takes a special theme of current importance as its point of central interest, talks, discussions and field excursions being related to it. Local branches and sections exist at a number of provincial centres, each of which conducts its own programme of meetings, discussions, lectures, excursions and field work. There are also the Standing Committee for Teaching Aids, the Standing Committee for Field Studies and the Standing Committee for Sixth Form/University Geography. The quarterly journal of the Association is *Geography* (*qv*). Other publications include *Geography in secondary schools,* revised in 1960 by E W H Briault and D W Shave and several pamphlets, handbooks and reprints on many aspects of school teaching and the equipment of geography rooms; a completely new handbook, by Dr N J Graves, was issued in the Autumn of 1971. A valuable series is *British landscapes through maps,* designed for use with Ordnance Survey maps and with the *Exercises on OS maps* series prepared by the Secondary Schools Section Committee; there are also the *Sample studies, Asian sample studies, The geography room, Geography in primary schools, Air photographs—man and the land,* reprints, air photo packs and L Dudley Stamp: *Hugh Robert Mill: an autobiography with introduction.* A history of the association was published on the occasion of the Diamond Jubilee, *The Geographical Association 1893-1953* ... A library of more than ten thousand volumes is maintained, including British and foreign geographical periodicals, atlases and maps and monographs of all kinds. The library catalogue is available in parts, Asia, Africa, North and Latin America, Australia and New Zealand, and the oceans. The special classification scheme devised for the library uses a numeral notation for regional placings and a literal one for subject placings, each notation being capable of detailed subdivision. Regional placing takes precedence, with added entries in the catalogue as required.

See also ' Teaching geography '.

497 ' Geographical digest ', an annual publication first issued as an experiment by the Research Department of George Philip, under the editorship of Harold Fullard, in 1963, has amply justified the initial aim ' to provide, in a concise form, information on recent changes in the world of interest to geographers and especially to provide such information as is difficult to obtain without consulting many sources '. The eighth edition, 1970, was enlarged to include a detailed analysis of United States trade. Other sections deal with significant political changes, new place-names, population and census details, world production statistics, new sources of raw materials, United Kingdom imports and exports, world trade figures, new projects in progress, communications, developments in exploration, discovery and surveying, including space and major geographical catastrophes in 1969.

498 ' Geographical essays ', by William Morris Davis, 1909 edition, was republished unabridged in the main in 1954 by Dover Publications, edited by D W Johnson. Selections of Professor Davis' most important geographical essays are brought together in this production, representative of his contributions to the science of geography. Minor alterations have been made to the original text, but the originals may be found by consulting the citations given in the Table of Contents. The essays are in two groups—Educational and Physiographic. Within each part, the essays are grouped together according to subject matter rather than in the order of original publication. A few figures have been retained in the text, but most of the original illustrations have not been reproduced.

499 ' Geographical essays in honour of K C Edwards ', edited by R H Osborne and others, have been published by the Department of Geography, University of Nottingham. Twenty-seven papers fall into two groups: those on the East Midlands, which were included simultaneously in *The East Midlands geographer,* 1970; and others dealing with a wide variety of subjects and places—New Zealand, the Soviet Far East, Wales, African pre-history, British seaweeds. In a Foreword, Professor Edwards' career and the influence he has had on geography, particularly in Nottingham and the East Midlands, are described, including a list of his published work.

500 'Geographical essays in memory of Alan G Ogilvie', edited by R Miller and J W Watson (Edinburgh: Nelson, 1959), contains ten contributions by former students and colleagues on the American Geographical Society of New York, on human geography, and on regions such as Scotland, Canada, South Africa and Spain. The volume was originally planned as a Festschrift to mark his retirement, but was converted, by his death in 1953, into a posthumous tribute.

501 'Geographical essays on British tropical lands', edited by R W Steel and C A Fisher (Philip, 1956), makes a useful introduction to the problems involved. Factual material is dated up to 1949-1950. Chapters by six eminent geographers, in addition to the editors themselves, are concerned with ' Geography and the tropics: the geographer's contribution to tropical studies '; ' Some problems of population in British West Africa '; ' The transport pattern of British West Africa '; ' Soil erosion in Nigeria '; ' The trade of Lake Victoria and its marginal lands '; ' Land-use and settlement in Jamaica '; ' Rainfall and water-supply in the dry zone of Ceylon ' and ' The problem of Malayan unity in its geographical setting '. The sketchmaps, figures, notes and references in the text are valuable additions, but there is no index.

502 Geographical Field Group owed its origin in the inter-war years to the Student Group of the former Le Play Society, which, in turn, developed from the Le Play Society (qv). The group has been outstanding in the promotion of field studies both at home and abroad. The headquarters are at the Department of Geography, University of Nottingham, where an annual conference and meeting are held.

> *Refer* P T Wheeler: ' The development and role of the Geographical Field Group ', *in East Midland geographer,* 1967, part 3.

503 'Geographical fieldwork: a handbook ', edited by K S Wheeler and M Harding for Blond Educational, comprises the work of practising teachers in the exchange and demonstration of their experience in taking classes out to look at, enjoy and understand various types of landscapes. Among the specific topics are 'Geography teaching and fieldwork'; 'Fieldwork with schoolchildren'; 'Fieldwork and maps'; 'The visual recording of fieldwork'; 'A farm study'; 'The traverse'; 'Fieldwork in an urban area'; 'Making a study of communications'; 'A parish study'; 'The study of a river'; 'A study of coastal features'; 'The study of a mountain area' and 'Studying a quarry'. Appendices give information on fieldwork preparation and fieldwork centres. There is a bibliography.

Refer also J E Archer and T H Dolton: *Fieldwork in geography* (Batsford, 1968).

M F Cross and P A Daniel: *Fieldwork for geography classes* (McGraw-Hill, 1968).

504 'Geographical handbook' series was compiled by the Geographical Section, Naval Intelligence Division of the Admiralty, for information during the first world war and was therefore of restricted circulation. Recognising their value, Kegan Paul, Trench, Trubner issued a second series in fifty-eight volumes, 1942-, which comprised entirely new works prepared by the Naval Intelligence Division sub-centres at Oxford and Cambridge by trained geographers and scholars in other relevant fields, under the direction of Lt General K Mason and H C Darby. The bibliographies and maps are particularly valuable. Among the especially useful volumes are those for *China proper*, in three volumes, 1944-45; *Jugoslavia,* in three volumes, 1944-45; and *Spain and Portugal,* in four volumes, 1941-45.

505 'Geographical interpretation of historical sources: *readings in historical geography'*, edited by Alan R H Baker, John D Hamshere and John Langton (Newton Abbot: David and Charles, 1970), brings together twenty papers of fundamental significance to the study of historical geography. In the preface, the editors state that the choice of papers 'is organised around the theme of source materials and their interpretation. The primary criteria used in making the selection were that the essays should illustrate the range of historical source material that exists, the types of problems that these present to geographical analysis and some of the methods that have been employed to overcome these problems.' The sample studies cover England and the Welsh borderland; Domesday woodland; population trends and agricultural development shown in the

Warwickshire Hundred Rolls of 1279; the market area of Preston in the sixteenth and seventeenth century; the combination and rotation of crops in East Worcestershire; family limitation in preindustrial England; agricultural changes in the Welsh borderland; locational change in the Kentish hop industry and the analysis of land use patterns; the Lancashire cotton industry in 1840; the population of Liverpool in the mid-nineteenth century and moated settlements in England. There are illustrations, sketchmaps and diagrams in the text; also notes and references.

506 'A geographical introduction to history', by Lucien Febvre, in collaboration with Lionel Bataillon (Routledge and Kegan Paul, 1924, 1949, fourth impression, 1966), was translated into English by E G Mountford and J H Paxton. In the foreword, by Henri Berr, the effect of environment on man and man's exploitation of the earth are discussed, stating Professor Febvre's opinions thus— ' Lucien Febvre has set definite limits to his subject out of regard for scientific accuracy. He does not deny the direct action of the environment on the physical and psychical nature of man; but he holds no brief for it, and leaves the subject severely alone.' The book rouses a wealth of ideas; it is at once objective and personal, revealing an enthusiasm for his predecessors in this field, Paul Vidal de la Blache, F Rauh and J Michelet. The plan of the book was complete in 1912-13, but the work on it was interrupted by the First World War; taken up again in 1919, much had to be revised and it was not until 1924 that the first edition appeared. An interesting bibliography is arranged under headings related to the body of the text.

507 'The geographical journal', the quarterly organ of the Royal Geographical Society (*qv*), began publication in 1893, successor to the *Journal of the Royal Geographical Society*, 1830-1880. The articles are varied in content and international in scope, including accounts of recent exploration and travel, original contributions to geographical research and papers read at the meetings of the society. From time to time, assessments of developments are most valuable, such as, in the issue for June 1964, ' British geography in the twentieth century ', ' The RGS and British exploration: a review of recent trends ' and ' British maps and charts: a survey of development '; and the annual presidential addresses are invaluable as summaries of work in progress. Folded maps and maps in the text are usually original, frequently drawn from an author's survey by the draughtsmen of the society. Plates and text figures also add to the value of the articles. An extensive review section, usually preceded by one or two review articles, includes the most significant works in all aspects of geography, also the best of travel literature, with a high proportion of overseas publications, arranged by broad regional and systematic headings. These reviews are contributed by experts and are frequently in themselves scholarly contributions to their subjects; a 'Books received' section has recently been added, arranged regionally by continent, also an 'Expedition reports' section and notes on University journals. The feature 'Cartographical survey' ('Cartographical progress' until 1967) is a valuable contribution to the scattered information on maps available. In addition, shorter notices and news items, 'The society's news', 'University news' and 'The record' are all essential sources of current information. The obituary paragraphs have through the years accumulated much biographical information. In the journal for January 1943 is contained the article 'The first hundred volumes'. The annual index is included with the December issue and cumulative indexes are prepared approximately every ten years. The seventh general index covers volumes 121 (1955) to 130 (1964), and includes personal, topographical and subject references to all papers, articles, reviews and notes.

508 'The geographical lore of the time of the Crusades: *a study in the history of medieval science and tradition in Western Europe'*, by John Kirtland Wright (American Geographical Society of New York, 1925; new unabridged edition, with introduction by Clarence J Glacken, Dover Publications; Constable, 1965), remains a classic of scholarly thought and geographical application. Two sentences from the preface seem to summarise Dr Wright's approach to his life's work; referring to 'geographical concepts', he writes, 'they have also sprung from the accumulated learning and lore of preceding ages and to no small extent from unfettered flights of the imagination. The history of geography, therefore, leads its students into many fields, affording them a key by means of which they may gain a sounder understanding of the extensive ranges of human activity and of the evolution of important phases of intellectual life.' This work was an enlargement of the thesis offered to Harvard

University, 1922. The thirteen main sections, in two parts, are concerned with 'The contribution of the ancient world', 'The contribution of Western Christendom before 1100 AD', 'The contribution of the Moslems', 'The sciences from the period 1100-1250 AD', 'The place of geography in the medieval classification of knowledge'—and secondly, under the main heading 'The substance and character of the geographical lore of the time of the Crusades', 'Cosmogony, cosmology and cosmography', 'The atmosphere', 'The waters', 'The lands', 'The astronomical geography of the known world', 'Cartography', 'Regional geography', Conclusion', followed by extensive notes, a bibliographical note and a bibliography. There are some illustrations, mainly of early maps.

509 'The geographical magazine' was launched in 1935 on the initiative of Michael Huxley, as an independent illustrated magazine presenting the geography of the modern world and its relation to history, politics and economics. Editorial policy has changed from time to time; some special issues remain in the memory for their particular usefulness, such as the April 1960 'Atlases and mapmaking' number and the superbly illustrated issue for October 1969, devoted to 'Cartography for the 1970s'. A recent innovation has been the occasional issue of map supplements. The features 'World news' and 'Current affairs', illustrated with sketchmaps and diagrams, and 'Shorter news', as well as 'Notes by Ptolemy' have been recently introduced and are particularly useful in drawing attention to new developments and topical issues in all parts of the world. The book reviews section, 'The world in books', is a short feature, but of high standard, followed by 'Short reviews', notes of new periodicals and a classified 'Book list'. As Spring approaches, advertisements and notes about travel, guides, youth hostels, etc, are particularly relevant. The 'slide service' has expanded through the years; early in 1966 the service was reorganised to provide even more varied items, on 2 inch by 2 inch mounts, suitable for all the usual makes of 35 mm projectors, with notes for commentary.

510 'The geographical review' has been the chief periodical publication of the American Geographical Society since 1916. Original studies in depth cover a broad range. Especially useful is the 'Geographical record' section, in which are noted new developments, particularly important monographs, surveys, reports, new periodicals and personal news concerning the world's leading geographers; followed by a section devoted to 'Geographical reviews'. The journal is one of the most reliable sources of information about new maps and atlases, especially those prepared in the United States. A recent feature has been the inclusion of brief abstracts of the articles. A cumulative index is prepared every ten years. With the September 1970 issue, a list of new books was added, arranged alphabetically by author, with prices when available.

511 The Geographical Society of Ireland was founded in 1934 for the promotion of geographical studies in Ireland. *Irish geography*, published once a year, concentrates on Irish material; specialist articles are usually illustrated by text maps and are well documented. A section of practical book reviews is included and 'Geographical literature relating to Ireland' is to be an annual feature. The society's library is housed in the Geography Department of Trinity College, Dublin, and consists mainly of periodicals obtained in exchange for *Irish geography*, review copies of books, and bequests. The society's collection of film-strips was recently donated to the Association of Geography Teachers of Ireland.

512 'Géographie de la population' is a unique work on this subject, in two volumes, by Mme Jacqueline Beaujeu-Garnier of the Sorbonne (Paris: Librairie de Médicis, 1956, 1958). Volume I includes general considerations, West and Southern Europe, the United States of America, Canada, Australia and New Zealand, and Latin America; volume II covers Africa, the Middle East, Monsoon Asia and socialist lands. Translated by S H Beaver (Longmans, 1966), it forms one of the *Geographies for Advanced Study* series. The illustrations, maps and diagrams and the bibliography are component parts of the work.

513 'Géographie d'Israel', by Efraim Orni and Elisha Efrat (Israel Universities Press, 1970), is one of the volumes in the Israel Program for Scientific Translations Limited. The text is in French, including many photographs, maps and diagrams; there is a bibliography and a folded map in a pocket.

514 'La géographie française au milieu du XXe siècle', edited by G Chabot, R Clozier and Mme J Beaujeu-Garnier (Paris: L'Information Géographique et J-B Baillière, 1957), is a collection of forty-two essays on bibliographical and theoretical topics concerning the development and organisation of French geography, including some brilliant new studies on the leading interests of French geographers. It is one of a growing number of scholarly analyses by geographers of the state of the subject in their respective countries.

515 'La géographie humaine', a pioneer work of Jean Brunhes (*qv*), 1910, had a profound influence and so continued an interest that its publishing history has been complex. Among the most interesting of subsequent editions have been a fourth edition, revised by M Jean-Brunhes Delamarre and Pierre Deffontaines, published by the Librairie Félix Alcan in 1934, in two volumes, with a third volume of plates and maps; an abridged edition by the same two editors, published by PUF in 1956; an abridged and edited translation by E C Le Compte, Isaiah Bowman and R E Dodge, under the title, *Human geography,* in 1920; and a new translation and abridgement by E F Row (Rand McNally; Harrap, 1952; third edition, 1956).

Refer also, one of the many outstanding works on human geography produced during the past twenty years or so: M Chisholm: *Research in human geography: Social Science Research Council reviews of current research* (Heinemann Educational Books; the tenth volume in the series, 1971).

516 'Géographie régionale de la France', by Georges Chabot (Paris: Masson, 1966; revised edition, 1969), is not only a factual examination of France, but an outstanding example of the methodology of regional geography. Arrangement is straightforward and double page maps accompany the description of each region, in addition to some sketchmaps in the text.

517 'Géographie universelle' (Paris: Colin, 1927-1955), the vast project to produce a world regional geography at a uniformly high standard, was conceived by Paul Vidal de la Blache before the first world war. After his death in 1918, Lucien Gallois renewed the project and the first volumes appeared in 1927; all except the volumes for France were finished before 1939, but the second war slowed production again and the last French volume did not appear until 1955, completing a work of twenty-three volumes and the *Atlas historique et géographique* by the leading

French geographers; each volume is illustrated with plates, maps, text-maps, figures and bibliographies, and has its own index.

518 ' **Géographie universelle des transports** ', in four volumes, runs as follows: 1, French railways, including those of North Africa and French colonies; 2, European railways, excluding France and the USSR, in four parts; 3, America, Africa and Asia and the land areas of Oceania; 4, Road, sea and air transport. Illustrations, maps and textmaps are generously provided.

519 ' **Géographie universelle Larousse** ', under the direction of M Jean-Brunhes Delamarre and P Deffontaines (Paris; Larousse, 1958-60), is in three volumes on a plan which groups together countries touching the same ocean; *L'Europe péninsulaire,* 1958; *Afrique, Asie péninsulaire, Océanie,* 1959; *Extrême Orient, Plaines Eurasiatiques, Amériques,* 1960. Within these groupings, the unit of treatment is political. The production is sumptuous, with superb illustrations; there are short bibliographies and an index of placenames for the whole work is in the third volume. An English edition was planned, of which the volume for Europe was published by Paul Hamlyn in 1961.

520 ' **Die geographischen Fragmente des Eratosthenes Neu gesammelt, geordnet und besprocken** ', 1880, ' the best commentary and critical edition of the fragmentary geographical work of Eratosthenes (275-195 BC), was reproduced in 1964 by N Israel, Meridian Press and Theatrum Orbis Terrarum.

521 ' **Geographisches Jahrbuch** ' is a remarkable bibliographical series produced since 1866 by various publishers, mainly by Perthes of Gotha and later by VEB Hermann Haack, Geographisch-Kartographisch Anstalt. Each issue includes books and periodical articles covering a different geographical area, preceded by an introductory survey. Indexes for the period 1866 to 1936 are included in volumes 40 and 52; a guide to contents up to volume 58, part 1, is given in *Aids to geographical research* (*qv*); and in A J Walford: *Guide to reference material,* the subjects covered in volumes 60 and 61 are summarised.

522 ' **Geographisches Taschenbuch and Jahrweiser für Landeskunde** ', an annual or bi-annual begun by Franz Steiner Verlag in 1949, presents a guide to German regional geography and considerable information also regarding geographical documentation in the rest of the world, under the present editorship of Professor E Meynen. Articles, bibliographies, lists of organisations and institutes responsible for the promotion of geographical and related work in Germany, geographical societies and their publications, statistical sections, notes on methodology and cartographical techniques, reviews of new official map publications and a complete bibliography of works on German geography make this a most substantial and helpful work. An index includes references to previous as well as to the current issue and, on occasion, to other bibliographical publications also. A special supplement to the 1960/61 volume constituted the 1960 *Orbis geographicus* (*qv*), covering the International Geographical Union and cartographical societies, university professorships and institutions, with lists of geographers from eighty-eight countries.

523 ' Geographisches Zeitschrift ' was launched by Alfred Hettner in 1895. Publication ceased with the second world war, but revived in 1963, now continued quarterly by Franz Steiner Verlag. Original studies, extensive reviews and detailed discussions of important works are included, on an editorial policy which still closely follows the ideals of Hettner, his belief in the unity of geographical studies, in the importance of methodology and the relevance of the subject to everyday life. A cumulated index to volumes 1 to 50 is in preparation.

524 ' Geography ', the quarterly journal of the Geographical Association (*qv*), was begun in 1901. Illustrated articles of geographical and topical interest and on aspects of the teaching of geography are followed by the ' This changing world ' feature, which draws attention to current developments of interest to geography teachers; this feature was redesigned and increased in size in 1971. The articles are preceded by abstracts and the latest feature, indicative of adaptation to automation, is the ' Key words to articles in this issue for information retrieval systems '. Announcements concerned with the status of the subject, new teaching methods, materials and aids, an extensive section of informative reviews and ' Geographical articles listed from periodicals received in the library ', together with ' books and publications received ', both arranged under broad subject groupings, make this an invaluable publication for checking either subject of bibliographical knowledge, particularly from the teaching angle. The index is included with the November issue.

525 'Geography and education', Ministry of Education pamphlet no 39 (HMSO, 1960), is a key document, summarising 'The heritage of geography'; 'The nature of geography'; 'The development of geography as a school subject'; and various aspects of geography teaching at successive standards. The following are included as appendices: 'Letter of the Royal Geographical Society to the universities of Oxford and Cambridge, 1871'; 'Memorial of the Royal Geographical Society to the universities of Oxford and Cambridge, 1874'; 'Geography rooms and their accommodation and equipment'.

526 'Geography and man: *a practical survey of the life and work of man in relation to his natural environment',* edited by W G V Balchin, in three volumes (New Era Publishing Company, second edition, 1955), 'The heritage of man: Europe', 'The British Isles, Asia, Africa' and 'The Americas, Australasia, Man's work in industry', which includes agriculture and fisheries, is a most readable as well as evocative work. The preface deals with general matters, exploration and maps.

527 'Geography and planning', by T W Freeman (Hutchinson University Library, third edition, 1967), considers planning at local, regional and national levels—'The planner and the geographer'; 'The physical landscape'; 'Climate and weather'; 'Rural land use'; 'Aspects of town geography'; 'Some problems of industrial location'; 'National parks'; 'The changing scene'. Notes and references for each chapter are grouped at the end and there are some figures in the text.

Refer also G H J Daysh and A C O'Dell: 'Geography and planning'; paper read at the Royal Geographical Society, 15 April, 1946, published in *The geographical journal,* July 1947 (January-March, 1947).

E C Willatts: 'Planning and geography in the last three decades', the Eva G R Taylor Lecture, 1971, printed in *The geographical journal,* September 1971.

Department of Environment: *Atlas of planning maps,* two volumes, 1970.

528 '**Geography as human ecology:** *methodology by example*', edited by S R Eyre and G R J Jones (Arnold, 1966), ' is defined in this volume as the interaction between human activity and natural circumstances '. The book comprises the essays of eleven scholars, writing with geography teachers particularly in mind.

529 '**Geography at Aberystwyth:** *essays written on the occasion of the Department Jubilee, 1917-18 - 1967-68*', edited by E G Bowen, Harold Carter and James A Taylor (Cardiff: University of Wales Press, 1968), comprises a series specially commissioned and dedicated to Professor H J Fleure. The evolution of geographical studies at Aberystwyth is traced, following which are seventeen chapters by individual scholars on a variety of topics, especially physical and human, illustrated with sketchmaps, figures and with end of chapter references. Subjects covered include ' The role of the sea in the evolution of the British landscape '; ' Raised shore platforms of the western islands of Scotland '; ' The periglacial landscape of the Aberystwyth region '; ' Climatology and the geographer '; ' Scale problems in hydrology '; ' Reconnaissance vegetation surveys and maps '; ' Factors in soil formation in Wales '; ' Surveying techniques in coastal geomorphology '; ' Photogrammetric techniques in geography '; ' The seas of Western Britain: studies in historical geography '; ' Mid-Wales: prospects and policies for a problem area '; ' Geography and political problems '; ' The rural community in Central Wales: a study in social geography '; ' Urban systems and town morphology '; ' New techniques in urban analysis '; 'A multivariate grouping scheme: association analysis of East Anglian towns ' and ' Regional geography at Aberystwyth '.

530 '**The geography behind history**', by W Gordon East (Nelson, 1938; revised edition, 1965), aims to show how the majority of events and developments are caused or conditioned by geographical factors. The outline runs as follows: ' Geography as an historical document '; ' Geographical position '; ' Climate and history '; ' Routes '; ' Towns '; ' Frontiers and boundaries '; ' Habitat and economy '; ' The dawn of civilisation '; ' The dawn of civilisation in the Americas '; ' Europe and China '; ' International politics '. Sketchmaps and diagrams figure in the text and there is a brief reading list, subdivided according to the chapter topics.

531 '**Geography from the air**', by F Walker (Methuen, 1953), was written when the use of aerial photographs as an aid in geographical studies was a comparatively new technique. Beginning with a general explanation, the text proceeds to examine 'Geological information on air photographs'; 'The study of erosion on air photographs'; 'Minor relief features and soils'; 'Coasts and shorelines'; 'Human geography' and 'Economic conditions and the formation of settlements'. A number of excellent aerial photographs are included, also some diagrams, and there is a short technical explanation of photogrammetry.

Refer also D R Harris and R U Cooke: 'The landscape revealed by aerial sensors' in *The geographical magazine,* October 1969.

532 '**Geography in education**': *a report prepared by the Education Committee of the Royal Geographical Society,* 1955, and *Geography and technical education: a memorandum prepared by the Education Committee of the Royal Geographical Society,* published with the approval of the Council of the RGS and the Executive Committee of the Geographical Association, 1958, are documents of central importance in the establishment of geographical studies in the context of contemporary education; in the words of the members responsible for the second report, 'to place on record what they regard as the important place which geography should have in such developments'.

533 '**Geography in the field**', by K S Wheeler (Blond Educational, *Teachers' handbook* series, second edition, 1970), is a revised edition of *Geographical fieldwork: a handbook,* by K S Wheeler and M Harding, which evolved from the work of a group of Surrey teachers, who, under the auspices of the County Inspectorate, formed the Surrey Fieldwork Society in 1958. It is one of the most helpful texts in the methodology of fieldwork, dealing with many different aspects and amply illustrated with sketchmaps and diagrams. The work ends with 'A selected list of fieldwork centres and bibliography' and an 'Index of fieldwork techniques'.

534 '**Geography: an outline for the intending student**', edited by W G V Balchin (Routledge and Kegan Paul, 1970), begins with a Preface and first chapter by the editor—' The nature and content of geography', also an Appendix at the end giving details of relevant organisations and publishers. Between are four sections: 'The basic skills of geography'; 'The organization of geographical material'; 'Pure and applied geography'; and 'Information for the intending geographer', within which specialist geographers have contributed the individual chapters. Short lists of 'further reading' are appended to each chapter and there is a brief index.

Refer also T W Freeman: *The writing of geography* (Manchester University Press, 1971) written for the geography student.

535 'Geography in Great Britain, 1956-60', a report submitted to the XIXth International Geographical Congress, Stockholm, 1960, by K C Edwards and G R Crone (Royal Geographical Society, 1960, also reprinted in *The geographical journal,* December 1960), is a systematic study, covering research and publications.

536 'Geography in the secondary school, *with special reference to the secondary modern school',* a report by E W H Briault and D W Shave (Geographical Association, 1952; second revised edition, 1955), stresses the importance of geography in the balanced education of all children, both in the systematic and regional approaches.

537 ' Geography in the twentieth century: *a study of growth, fields, techniques, aims and trends'* (New York: Philosophical Library; Methuen, 1951; second revised and enlarged edition, 1953; third enlarged edition, 1957; in Methuen's *Advanced geographies* series) is the work of twenty-two specialist authors, under the editorship of Griffith Taylor. Chapters are grouped in three main sections: 'Evolution of geography and its philosophical basis'; 'The environment as a factor'; and 'Special fields of geography'. 'A concise glossary of geographical terms' is appended.

538 ' Geography in world society: *a conceptual approach'* is a monumental work by Alfred H Meyer and John H Strietelmeier (*Lippincott geography* series, 1963). In six parts, with subdivisions —' The nature of the discipline of Geography'; 'Man's planetary domain'; 'The geographic aspects of technology and ethics in societal and national life'; 'The life of man as viewed in its areal setting'; 'The regional and resource factors of nation-states'; and ' Summary observations on man's relations to space in time '—the text was in preparation at the time when exploration of space and the planets was beginning to become a reality. It was designed as an introduction to university geographical studies and to develop intelligent awareness of the critical role of geography in the world. There are photographs, maps and diagrams throughout and references at the ends of chapters. A folding plate depicts the climates of the earth and an appendix considers the 'significance of climate identification—classification and its application to geographic understandings'.

539 'The geography of economic activity', by Richard S Thoman, Edgar C Conkling and Maurice H Yeats (McGraw-Hill, 1962, 1968), is intended to be an introduction to economic geography: 'The human being and his economies'; 'The natural environment'; 'Patterns and theories of production and exchange'; 'The sources and application of energy'; and, particularly detailed, 'The roles of selected commodities'. Throughout the text are photographs and diagrams and, as an appendix are special red and blue maps pertaining to agriculture designed from the latest United States Census of Agriculture and on manufacturing, using data from the 1963 United States Census of Manufactures. Endpaper maps show 'Density of world population' and 'The world's political units'.

540 'A geography of Europe', by Jean Gottmann (Holt, Rinehart and Winston, fourth edition, 1969), as in the previous editions, reviews geography as 'dedicated to the description and analysis of constantly changing facts and situations'. The opening chapters view Europe as a whole, the following revised sections concern the changes in the geography and economy of Western Europe, the rapid evolution of the Mediterranean countries and their ties with the West, Central Europe, the marginal Soviet Republics and the USSR. Special emphasis is placed on such organisations as the Common Market, urban growth and its consequences, the use of resources and the need for planning. In conclusion, 'The diversity and unity of Europe' is discussed; there is a bibliography.

541 'The geography of greater London: *a source book for teacher and student'*, prepared by the Standing Sub-Committee in Geography of the Institute of Education, University of London, under the general editorship of R Clayton (Philip, 1964), was conceived as a joint enterprise, bringing together the results of recent research by university teachers and the knowledge and experience of teachers of the institute engaged in the training of geography teachers for schools, to provide for all teachers interested in urban geography up-to-date information and guidance. Authoritative sections are included on all aspects, such as 'Maps of Greater London', 'Museums, libraries, record offices, and other sources of information', 'Some statistical sources on Greater London', 'Teaching aids'. An extensive bibliography by P M Wilkins comprises entries, annotated as necessary, under the following headings: A 'Main section': Physical aspects and the land; Topographical-historical surveys; London's growth and history; The face of London; The government of London; London's economy; The demographic background; The metropolitan community: its geographical character; Planning: problems and administration; B 'Local reference section'; C 'Books of general interest'; D 'Books for the junior library'.

542 'Geography of international trade', by Richard S Thoman and Edgar C Conkling (Prentice-Hall, 1967, *Foundations of economic geography* series), examines trade data, interpreting the pattern and structure of international trade in terms of current monetary and economic blocs. The result provides one of the first analyses of the basic types of international relations. The series as a whole is designed so that each volume focuses on a major aspect of economic geography. A number of sketchmaps and diagrams are included in the text, which contains also a glossary of terms and a short bibliography.

Refer also B J L Berry: *Geography of market centres and retail distribution* (Prentice-Hall, 1967).

543 'The geography of modern Africa', a major work by W A Hance (Columbia University Press, 1964, 1965), has as its purpose the presentation of the chief features of the economy, the handicaps and some of the potentialities that affect the future of economic growth. With this aim in mind, the arrangement of material is regional—featuring an introduction, 'Northern Africa', 'West Africa', 'Equatorial Africa', 'Eastern Africa', 'South Central Africa', 'Southern Africa', 'Madagascar and the Mascarenes'. Illustrations and sketchmaps accompany the text and there is a regionally arranged bibliography.

544 'The geography of Norden: *Denmark, Finland, Iceland, Norway, Sweden'*, edited by Axel Sømme (Oslo: F W Cappelens for the Norwegian National Committee of Geography, 1960; Heinemann, 1961), was prepared for the Nineteenth International Geographical Congress, Stockholm, 1960, by thirteen eminent geographers from the leading universities and colleges of the five countries. It is an excellent example of co-operation in such work, skilfully co-ordinated and produced. There are fine plates of aerial and ground photography and a section at the end of coloured general and thematic maps, in addition to maps and diagrams in the text.

545 'The geography of religion in England', by John D Gay (Duckworth, 1971), begins with an examination of the approaches adopted by other geographers of religion in a comprehensive survey of research in this topic already completed, following with an analysis of the source material available, attaching special importance to the 1851 Census. The remaining chapters discuss in detail ' The Church of England '; ' The Roman Catholics '; ' The Nonconformists ', ' The Baptists, the Presbyterians and the Congregationalists '; ' The Methodists '; ' The smaller Christian groups '; ' Quasi-Christian groups and Eastern religions ' and ' The Jews '. Appendices are given to tables, chapter by chapter references and bibliography, with a section of explanatory maps.

Refer also H J Fleure: ' The geographical distribution of the major religions ', *in The bulletin of the Egyptian Royal Society of Geography.*

546 'The geography of Romania', by Dr Tiberiu Moraru and others (Bucharest: Meridiane, 1966), concentrates particularly on physical and economic aspects. There are a number of photographs and sketchmaps throughout the text and a folding physical map in a pocket.

547 'The geography of towns', by Arthur E Smailes (Hutchinson University Library, 1953; fifth edition, 1966, paperback edition), traces the origin and bases of towns, the setting of towns and their cultures, their morphology, urban regions and the development of the town structure. There are maps and town plans throughout the text to demonstrate points and suggestions are included for further reading, mostly in English.

Refer also, among a rapidly growing literature M Ash: *Regions of tomorrow: towards the open city* (Evelyn, Adams and Mackay, 1969).

R G Putnam *et al*: *A geography of urban places: selected readings* (Toronto: Methuen Publications, 1970).

548 'A geography of the USSR: *the background to a planned economy'*, by J P Cole and F C German (Butterworth, second edition, 1970), presents much information necessary to the study of recent economic developments and policies in the USSR. The second edition was completely re-written, taking into account the 1966-70 Five Year Plan and the increasing influence of Comecon. Tables and maps pin-point details in the text.

549 Geological Society of London, founded in 1807 for the investigation of the mineral structure of the earth, has become one of the great geological organisations of the world, the *Quarterly journal,* 1845-, being of central importance. Annual *Proceedings* and occasional *Memoirs* are also published. The library, holding some hundred thousand volumes and over three thousand geological maps, is available to Fellows and on the introduction of a Fellow.

550 Geological Survey of Canada, founded in 1841, has been responsible for much of the exploration of Canadian territory, especially in the west and, during this century, detailed mapping has been systematically carried out, together with studies of mineral resources. The publication programme includes *Annual reports, Memoirs, Bulletins,* the *Economic geology series* and *Geophysical papers.* In addition to the map-making division, the survey maintains a museum, library and laboratories.

551 Geological Survey of Great Britain, instituted as the Geological Ordnance Survey in 1935, was foreshadowed in 1832, when De la Beche, secretary of the Geological Society, was authorised to add geological colouring to the one inch ordnance maps of Devon, with adjoining Somerset, Dorset and Cornwall. Since that time, a national programme of mapping and publication has been gradually formulated, comprising the map series, *Palaeontological memoirs, Mineral statistics, Iron ore memoirs, Stratigraphical monographs, Water supply memoirs* and special reports and memoirs. Memoirs accompany the sheet maps. The geological maps of the British Isles are now in five main series: the twenty-five miles to one inch, 1:1,584,000, fourth edition, 1957; the ten miles to one inch, 1:625,000, second edition, 1957; the quarter-inch to one mile, 1:253,440, in various editions; the one inch to one mile, 1:63,360, new series reprints with the national grid; and the six inches to one mile, 1:10,560. There are also special maps of the coalfields and other areas of significance. The Survey is now within the Institute of Geological Sciences.

Refer Sir Edward Bailey: *Geological Survey of Great Britain* (Allen and Unwin, 1952); Sir J S Flett: *The first hundred years of the Geological Survey of Great Britain* (HMSO, 1937), which includes a bibliography.

John Challinor: *The history of British geology: a bibliographical study* (David and Charles, 1971).

552 The Geologists' Association was formed in 1858 to encourage interest and research in geology and allied subjects through discussion, visits and field work. More recently, the Association has directed most of its work to the discovery and demonstration of new geological knowledge, whilst still providing facilities for students and amateurs to gain practical knowledge of the subject. Individual groups have developed in north-east Lancashire, the Midlands, North Staffordshire and South Wales. The *Proceedings*, issued in four parts each year, contain papers embodying the research of members, reports of the geology demonstrated at the field meetings and in museums and, from time to time, résumés of current knowledge of particular subjects; the volumes are fully indexed, with ten-yearly cumulated indexes, and form an important source of reference in geological literature, well illustrated by plates, maps and text figures. *Monthly circulars* contain announcements of meetings and papers to be read, with full particulars and lists of maps, books or other source material suggested for consultation; they form excellent guides to the geological features of the districts. In addition, there are special publications, descriptive geological pamphlets and the *Geologists' Association guides,* of which more than thirty have been published concerning individual areas. The library, incorporated with that of University College, London, contains most of the standard geological works of reference, the publications of the leading geological and natural history societies at home and abroad and of the British and colonial geological surveys.

Refer G S Sweeting: *The Geologists' Association 1858-1958. a history of the first hundred years* (The Association, 1958).

553 'The geometrical seaman: *a book of early nautical instruments'*, by E G R Taylor and M W Richey (Hollis and Carter, 1962), provides a unique, illustrated monograph on the instruments available to early seamen and therefore of central interest to the geographer interested in the progress of exploration and discovery. Following an introduction and a fascinating chapter on 'the old instrument-maker's shop', the apparatus described includes lead and line, the azimuth compass, the cross-staff, quadrant, mariner's astrolabe and the 'kamal', the back-staff, Hadley's quadrant, sextant, the reflecting circle, the helmsman's traverse board, the sinical quadrant and traverse board, marine chronometer, Gunter's scale and sliding gunters, the vernier, bubble level, station pointer, telescopes, globes, the marine chair, rutter and pilot books, tide-tables, nautical tables and catalogues of nautical instruments.

554 'Geomorphological abtracts' has been published by the Department of Geography, London School of Economics, since 1960, first in four issues a year, then in six issues yearly from 1964, when the scope was enlarged. Coverage is international; abstracts are mostly informative, arranged by a broad subject grouping. Author and regional indexes have been prepared and a cumulative index is available covering 1960-1965; from 1966, an author index and subject index of more than 22,000 entries has been prepared on a computer. In 1966, the service formed the basis of a fully developed abstracting service, *Geographical abstracts* (*qv*).

Note from the numerous monographs on geomorphology, it is perhaps invidious to refer to only one:

F Machatschek: *Geomorphologie,* translated into English by D J Davis (Oliver and Boyd, 1969), a standard text in Germany.

555 'Geomorphology of cold environments', by Jean Tricart, translated into English by Edward Watson (Macmillan, 1969, 1970), presents in a condensed form the material contained in the two volumes of *Traité de géomorphologie* (SEDES, Paris, 1963), with a revision of text and bibliography by Professor Tricart. While the treatment is original, the outline and approach are traditional enough—'Extent of frost climate phenomena'; 'Periglacial processes and landforms'; 'Glacial processes and landforms'. The few photographs included are superb.

Compare A M Harvey, *ed: Geomorphology in a tropical environment* (Occasional Paper no 5, British Geomorphological Research Group, May 1968).

556 'Geophysikalische Bibliographie von Nord-und Ostsee', a comprehensive work by Fritz Model, published for the Deutsches Hydrographisches Institut of Hamburg by the firm of Gebrüder Borntraeger, Berlin, 1966, covers a clearly defined area between the west coast of Ireland and the eastern limit of the Baltic, in two parts, to 1961. Books and articles are included, in a wide range of languages, including Russian. The second part consists of an alphabetical list of authors and of periodicals, plus other sections designed to facilitate the use of the bibliography; the introductory matter to the first part is in English as well as in German.

557 The Geoscience Information Society, United States of America, organised as an independent non-profit professional body in 1965, aims to promote the exchange of information in the earth sciences through the mutual co-operation of its members. Literature guides and bibliographies are planned, including a *Bibliography of theses in the United States and Canada.* The annual meeting is held in conjunction with the annual convention of the Geological Society of America.

558 GeoServices, of London and Alberta, a current awareness service for all interested in the earth sciences, prepares three periodicals (published by Lea Associates, 1968-). *Geotitles weekly* lists in English the titles of all new publications within the field; publications scanned include journals, trade magazines, books, theses, patents and standards, conference announcements, trade literature and broadcasts. Searching is done in London and by correspondents in different parts of the world and the time-lag between publication and inclusion in the index is claimed to be no more than ten to fourteen days, distribution being by air mail. Titles are arranged by what is called a UDC-linkable GeoServices Decimal Classification and they are also available on library catalogue cards. The weekly editorial draws attention to significant trends. *Geoscience docu-*

mentation, 1970-, consists of a world list of current serial publications, giving details of frequency, publisher and changes in publication; new serials will be added, as will analyses of geoscience literature, bibliographic news and information on data handling projects. The monthly *Geocom bulletin,* 1968-, concentrates on news of research methodology and exploration, with emphasis on mathematical and computer studies. Regular cumulative indexes are planned.

559 The Geo-stat system for geographical statistics, which includes transparent map overlays, so that items can be placed and identified according to geographical location as well as by other factors, uses national grid co-ordinates. Geo-stat works with any metric scale and a map can contain one or a hundred national grid squares; different scales can be used for different jobs, so long as the scale gives a reasonable spread of holes.

560 'A German and English glossary of geographical terms', compiled by E Fischer and F E Elliott (American Geographical Society, 1950), includes those terms which are of central importance to the subject.

561 The German Copper Institute, Berlin, founded in 1927 for the purposes of research, industry and scientific studies in copper, is mainly concerned with the dissemination of information on copper to a wide public. A library is maintained, in which a comprehensive card index is kept. The institute works closely with the New York Copper and Brass Research Association and with the British Copper Institute. A monthly *Kupfer-Mitteilungen* contains essays and articles on German and foreign publications in the field of copper research and about ninety technical journals are regularly evaluated. In 1961, the institute introduced a teaching service including lectures, demonstrations and film presentations for technical colleges and private firms.

562 ' Geschichte der Kartographie ' was completed in 1943 by Leo Bagrow and published by Safari-Verlag, Berlin, in 1951; an English edition, with some rearrangement of the text, was undertaken by R A Skelton, with the title *History of cartography* (Watts, 1964). This erudite work is concerned with maps as ' craft products ', presenting an account of the principal groups of maps to the mid eighteenth century, especially in their aesthetic and cultural aspects. The majority of the illustrations, line and half-tone, are printed from the original blocks. A useful 'List of cartographers to 1750' and a 'Select bibliography', together with references throughout the text, make this the most definitive English-language history of the period to date.

Refer also A Libault: *Histoire de la cartographie* (Paris: Chaix, 1959).

563 Gesellschaft für Erdkunde zu Berlin, founded in 1828, is one of the great geographical societies of the world. Particularly important is the quarterly journal *Erde,* 1839-, which contains abstracts, bibliographies and detailed reviews, in addition to original articles of consistent scholarship. Another important bibliographical work is the *Bibliotheca geographica: jahres-bibliographie der geographischen literatur,* edited by Otto Baschin, which covers the years 1891-1912 in nineteen volumes (Berlin: Kühl, 1895-1917).

564 ' Glacial and periglacial geomorphology ', by Clifford Embleton and Cuchlaine A M King (Arnold, 1968), emphasises particularly the landforms and the processes of erosion and deposition; features actively being created by ice today are considered alongside features relict from the Pleistocene, which combine to form the present environment. The first section of the book describes the characteristics of contemporary ice-sheets and glaciers, including their physical nature, régimes and mode of flow, the causes and effects of glaciation on other land features. The second part deals with the erosional work of glacier ice and meltwater, with chapters on cirques and U-shaped valleys, followed, in part three, by studies of special forms, including till fabrics, moraines, drumlins and glacial lakes. Finally, such topics as weathering processes and permafrost are discussed. The whole work is well documented and there are 125 maps and diagrams, as well as thirty plates, illustrating the text.

565 ' Glacier sounding in the Polar regions ' was the title of a Symposium held at the Royal Geographical Society on the afternoon of 17 February, 1969. The papers, subsequently published in the December 1969 issue of *The geographical journal,* consisted of five specialist reports: ' The VHF radio echo technique ', by Dr S Evans; 'Airborne radio echo sounding of the Greenland ice sheet ', by P Gudmandsen; 'Airborne radio echo sounding by the British

Antarctic Survey', by Dr Charles Swithinbank; 'Results of radio echo sounding in Northern Ellesmere Island, 1966', by Dr G Hattersley-Smith; 'Long-range echo flights over the Antarctic ice sheet', by Dr G De Q Robin. The discussions are reproduced, as well as some magnificent plates, maps and figures in the text, and references follow each section.

566 Glaciological Society, founded in 1936 to encourage research on and to stimulate interest in the practical and scientific problems of snow and ice, is based on the Scott Polar Research Institute, Cambridge, and is open to all who have a scientific, practical or general interest in any aspect of the subject. Chief among the society's publications are the *Journal of glaciology* (*qv*) and *Ice* (*qv*); literature on polar expeditions and applied aspects of glaciology is noted in each issue of the *Polar record* (*qv*).

567 Globes: The oldest extant globe is probably that in the Royal Museum at Naples, the *Atlante Farnesiano,* which is thought to date from about 300 BC. Both celestial and terrestrial globes were used by Greek scholars and many are the references to the use of celestial globes and armillary spheres by Persian and Arab astronomers. Crates, the Greek philosopher (*c* 150 BC), asserted that the only way of portraying accurately the true shape of the various parts of the world was to show them on a globe; he built a globe on which were shown three imaginary land masses in the regions of the Americas and Australasia, to balance the land masses he knew.

The earliest known modern terrestrial globe is that made by Martin Behaim (*qv*), which is richly coloured and decorated and based probably on a map by Henricus Martellus of about 1490. During the following years, many more were built, some of great size, elaborate and fairly accurate for those parts of the earth which were at all well known. In 1507, Martin Waldseemüller began to make gores, lens-shaped flat sections, which could be pasted on a globe with a minimum loss of accuracy. This made possible the 'mass production' of globes and they continued to be used in conjunction with the increasing variety of maps and charts by explorers, cosmographers and liberal scholars alike. The earliest examples of English globes are known by reference only or by their presence in portraits, as in the picture of Sebastian Cabot when an old man and in the Holbein painting 'The ambassadors'. From about the 1540s, the globes of Gemma Frisius were held in the greatest esteem in England; John Dee, for instance, who studied under Gemma Frisius and another of his pupils, Gerard Mercator, in 1547 brought back to Cambridge 'two great globes of Gerardus Mercator's making'. The Frisius globes were gradually superseded by those of Mercator, whose 1541 terrestrial globe and 1551 celestial globe had a very wide circulation. There are many references in the literature of the time, also in records of acquisitions in such libraries as the Bodleian. Several of the Mercator engraved gores have survived, still in sheet form. Jodocus Hondius and Willem Blaeu also made globes as well as maps. In the Netherlands, the Van Langrens (or Florentius) family established a famous globe-making business. Through the seventeenth century, globes became even more fantastic. The great cartographer, P Vincenzo Maria Coronelli (1650-1718), made a globe fifteen feet in diameter for Louis XIV, which people could actually climb into; in 1688, he produced engraved gores for globes three feet six inches in diameter, the largest engraved gores to that time, and in 1696, he made one also for William III, to show the expanding world. A modern globe featured by Georama reproduces the original Coronelli stand. Fine globes were made also by Gerhard Valck and his son between 1700 and 1750, and in England, by John Senex, publisher of atlases, maps and globes. The Adams family, also, in their Fleet Street publishing house, set a high standard for English globe-making, with their well-known eighteen inch diameter globes. With the globes of William and John Cary and such European globe-makers as Rigobert Bonne and Joseph Jérôme le Français Lalande in France and Giovanni Maria Cassini in Italy, the great period of decorative globes came to an end.

A facsimile of the rare second edition of *Libro dei Globi,* 1701, volume X of Coronelli's thirteen-volume *Atlante Veneto,* 1693, was issued by TOT, 1969, together with a portrait of Coronelli and a bibliographical note by Dr Helen Wallis; included are the gores of Coronelli's largest pair of engraved globes, $3\frac{1}{2}$ feet in diameter. There is a Coronelli World League of Friends of the Globe; a third international conference was held in September 1969, for the first time together with the International Conference on the History of Cartography.

The present-day renewed interest in globes has resulted in considerable experiment in printing, finishing and mounting, and a number of magnificent examples are available, though expensive. For teaching purposes at various levels, Philip and Georama offer perhaps the most interesting range, from the six inch 'London' Library Globe to the seventy-two inch Pilkington-Jackson Orographical Globe. Between the two is a great variety of political, physical and special-purpose models. The latest developments include relief model globes, plastic inflatable globes and illuminated globes. The Istituto Geografico de Agostini, Novara, has specialised in globes of all kinds, for which full, often illustrated, catalogues are available. There are now several important collections of globes in various parts of the world, of which the Jagellonian University Museum, Cracow, holds one of the finest; a special catalogue of the collection was prepared in connection with the XIth International Congress of the History of Science, 1965.

A catalogue of early globes, made prior to 1850 and conserved in the United States, compiled by Ena L Yonge (American Geographical Society of New York, 1968; *Library series* no 6), includes armillary spheres, planetariums, orreries and astronomical clocks, also globe gores. The items are grouped by types, thereafter alphabetically by name of maker. A list of globemakers is a useful feature.

Refer also D R McGregor: ' Geographical globes ', *in The cartographic journal,* June 1966.

E L Stevenson: *Terrestrial and celestial globes: their history and construction including a consideration of their value as aids in the study of geography and astronomy,* in two volumes (Yale University Press for the Hispanic Society of America, 1921).

George Goodall: *The globe and its uses* (Philip, second edition, 1948).

W Bonacker: *Das Schrifttum zur Globenkunde* (Leiden: Brill, 1960).

Oswald Muris and Gert Saarmann: *Der globus im Wandel der Zeiten: eine Geschichte der Globen* (Berlin: Columbus Verlag Paul Oestergaard KG, 1961).

Raymond Lister: *How to identify old maps and globes: with a list of cartographers, engravers, publishers and printers concerned with printed maps and globes from c 1500 to c 1850* (Bell, 1965) (*qv*).

Helen M Wallis: ' The first English globe: a recent discovery ' *in The geographical journal,* September 1951.

Helen M Wallis: ' Further light on the Molyneux globes ' *in The geographical journal,* September 1955.

568 ' Glossary of geographical names in six languages, *English, French, Italian, Spanish, German and Dutch',* compiled and arranged by Gabriella Lana, Liliana Isabez and Lidia Meak (Elsevier, 1967), forms one of the series *Glossaria Interpretum,* supervised and co-ordinated by M Jean Herbert and sponsored by many of the world's linguistic organisations. The languages chosen were according to the frequency of their use at international conferences on the subject.

569 'A glossary of geographical terms ' was prepared by a committee of the British Association for the Advancement of Science (Longmans, 1961) under the editorship of Sir Dudley Stamp. Definitions are drawn, with acknowledgements, from leading authoritative works, covering physical, human, economic and political geography. A second edition appeared in 1966.

570 ' The golden encyclopaedia of geography ' was adapted and edited by Theodore Shabad and P M Stern from the *Westermann Bildkarten Welt-Lexikon* (Macdonald, 1960). An introduction dealing with physical, economic and climatic features is followed by a series of articles, alphabetically arranged, on the main regions and towns, typical plants and animals and economic products, the whole being profusely illustrated.

571 ' Goode's world atlas ', previously *Goode's school atlas,* went into an eleventh revised edition by E B Espenshade, jr (Rand McNally, 1960), using improved cartographic and reproduction techniques. A pronouncing index of more than thirty thousand entries is included.

572 ' Gough map of Britain ', by an unknown cartographer, probably between 1347 and 1366, was first noticed by Richard Gough in his *British topography,* 1780, and the map itself was bequeathed with his collection to the Bodleian Library in 1809. The Bodleian has made a reproduction, with a matching transparent overlay showing the corresponding modern locations.

573 ' Gran atlas Aguilar ' is an outstanding atlas in three volumes, produced in Madrid in 1969. The method of relief representation is particularly fine, showing contours and clear layer colouring; a full blue is used for hydrographic features and the continental shelves and ocean troughs are well depicted. The town plans are another feature to note.

574 ' Le grand atlas ou cosmographic Blaviane ', the work of Johan Blaeu (Amsterdam, 1663), has received revived notice through the reproduction in 1967-68 of all the twelve volumes by N Israel, Meridian and TOT, with an introduction by Professor Dr Ir C Koeman in a separate volume. It is difficult to over-praise this exciting atlas and the words of the facsimile publishers may be taken literally—' The complete facsimile in twelve volumes of the most splendid atlas ever published. Extensive historical, geographical and cartographical descriptions of Europe, Africa, Asia and America . . . The engraved maps are not only surprisingly accurate, they are also wonderful works of art. The complete work is printed on specially made paper. The maps are mounted on guards.' All the frontispieces are included in the reproduction.

575 ' Grande atlante geografico ' is a beautiful atlas, edited by L Visintin (Istituto Geografico de Agostini, Novara, in a fifth revised edition, 1957). 232 plates of maps on astronomy, geology, physical, economic and political geography, climate and population are on scales varying from 1:50M to 1:1,500,000. Numerous city plans are most useful. Special features include fine distribution diagrams and block diagrams and the clear, varied symbols used throughout. Brief text notes accompany the maps and there is an index of about one hundred thousand entries.

576 Gresswell, Dr Ronald Kay (1905-1969) specialised in the study of coastlines and glaciation in Britain and Scandinavia, his work being recognised not only by his colleagues, but by Lancashire CC, the Nature Conservancy and by various government ministries. In his fine teaching work, he emphasised the value of field work and his interest in the training of young geographers is shown in his school textbooks: *The physical geography of glaciers and glaciation,* 1958; *The physical geography of rivers and valleys,* 1958; *Geology for geographers,* 1963; *Physical geography,* 1967; in addition to a number of articles for periodicals, notably the *Transactions* of the Institute of British Geographers.

577 'The Griffin', journal of the Department of Geography, North-Western Polytechnic, London, began in April 1970; in form, it is mimeographed, otherwise it is a typical example of the numbers of excellent journals which are prepared now by almost every Geography Department in the country. It covers a wide range of subjects, the articles being contributed by students, past and present, and by staff.

578 'Der Grosse Bertelsmann Weltatlas', compiled by Dr W Bormann at the Cartographical Institute, Bertelsmann Verlag, 1961, 1963 and 1964, is an excellently produced atlas, the result of a most significant publishing venture. A special cartographic institute was established for the purpose, official information was gathered from all over the world and editions were published simultaneously in Denmark, Finland, Germany, the Netherlands and Sweden. Subsequently, French and Spanish editions were published and the *McGraw-Hill international atlas* (*qv*) was based on it. In the original, a valuable preface in German and English discusses 'The atlas—its development and possibilities', followed by a detailed introduction explaining the plan of this atlas itself, the scales employed, projections, contents and principles of nomenclature adopted.

579 'Der Grosse Brockhaus Atlas', accompanying the sixteenth edition of *Der Grosse Brockhaus* and *Der Neue Brockhaus* 1960, is a useful reference atlas, including nearly four hundred coloured maps, town plans and statistical data. The index-gazetteer comprises some sixty thousand entries.

580 'Ground water year book', the first, 1964-66, prepared by the Water Resources Board (HMSO), contains ground water statistics for aguifers in England and Wales, together with related rainfall, evaporation and soil moisture deficits for the period January 1 1964 to December 31 1966. The index of observation wells and table of selected ground water level observations are prefaced by introduction, explanatory notes, other specialised notes and a few references. Folded diagrammatic maps are grouped at the end. The

Water Resources Act 1963 requires the publication, inter alia, of information relating to ground water resources.

581 'Guide bibliographique d'hydrogéologie', compiled by Jean Margat (Paris: Bureau de Recherches Géologiques et Minières, 1964), contains references to the more important and most recent books and articles in the French language, mainly later than 1950. It is planned to produce a companion guide to the more important works in other foreign languages.

582 'A guide to historical cartography: *a selected annotated list of references on the history of maps and mapmaking'*, by W W Ristow and Clara E LeGear (Library of Congress, Map Division, Reference Department, 1960, second revised edition, 1961), is an important contribution to the definitive documentation of this subject; in addition to all usual reference sources, map catalogues are included, also such comprehensive monographs as those of Beazley and Bunbury.

583 'A guide to Irish bibliographical material, *being a bibliography of Irish bibliographies and some sources of information'*, by Alan R Eager (The Library Association, 1964), was compiled as 'an exploratory volume' to make a beginning in the indexing of Irish enumerative bibliography and to serve as a quick reference guide to those interested in Irish studies. The arrangement of entries is based generally on the DC, with author and subject indexes; no annotations are included.

584 'A guide to Latin American studies', by M H Sable (Los Angeles: University of California, Latin American Center, 1967), was intended mainly for undergraduate and graduate students in all fields relevant to Latin America; all aspects of Latin American civilisation and all disciplines in the humanities, natural, applied and social sciences are covered. About five thousand entries, annotated in English, include monographs and reference books, pamphlets, periodical articles, government documents and conference proceedings, mainly in English and Spanish, but also in French, German and Portuguese.

Refer also Irene Zimmerman: *Current national bibliographies of Latin America: a state of the art study* (University of Florida Press).

585 '**A guide to literature on Iraq's natural resources**', by Abu Ghraib of Baghdad, 1969, was prepared by A P Srivastava within the framework of the Unesco/Special Fund Project, Iraq 12; the work lists studies and reports from 1833 to 1968 broadly grouped under hydrology and allied areas, plants, climate, irrigation, soil and drainage.

586 '**Guide to London Excursions**', edited by K M Clayton, was prepared for the 20th International Geographical Congress, London, 1964. Usefully collected in this publication is a vast amount of information on the geographical features of interest as far distant as the radius of Oxford, with effective maps and diagrams. Chapters are included on the Meteorological Office Headquarters, the Ordnance Survey, the Royal Greenwich Observatory and other centres of advanced work of interest to geographers.

587 '**A guide to manuscripts relating to America in Great Britain and Ireland**', edited by B R Crick and others, was published for the British Association for American Studies by the Oxford University Press, 1961. It is comprehensive, giving annotations and locations, dealing by counties with England, Wales and Monmouthshire, Scotland, Northern Ireland and the Republic of Ireland.

588 '**Guide to New Zealand reference material and other sources of information**', compiled by John Harris (New Zealand Library Association, 1946; second edition, 1950), includes all subjects in their application to New Zealand to be found in monographs, official publications, serials and journals of societies and listing library collections. The sections are numbered and classified according to subject, following generally the classification of the Otago University Library, which uses the Bliss Bibliographic Classification. The first supplement appeared in June 1951 and the second in 1957, both prepared by A G Bagnall. Items are printed on one side of the page only, so that libraries may cut out and mount, if desired, the numbers referring to place in the main work.

589 '**Guide to Russian reference books**' in six volumes, of which the second deals with history, auxiliary historical sciences, ethnography and geography, has been published by the Hoover Institution on War, Revolution and Peace (Stanford University, 1962-). It lists and annotates 1,560 reference tools relating to Soviet and Russian history in general as well as to the histories of various specific events or geographical areas. Each section is preceded by an introductory survey.

590 ' Guide to South African reference books ', compiled by Reuben Musiker (Cape Town; Amsterdam: Balkema, fourth edition, 1965), arranges the items cited according to broad category headings or subject groupings, such as Bibliographies, Periodicals, Statistics, Economics, Commerce and trade, Agriculture, Geography. Brief annotations are appended as considered desirable.

Refer also Reuben Musiker: *South African bibliography* (Crosby Lockwood, 1970).

591 Guides. *Note*: A short article on *Guides* was included in the first edition; the items named as examples need updating, otherwise the only comment that requires mention here is to emphasise the increasing numbers and variety of this genre of publication.

592 Gutkind, Dr Erwin E (1896-1968) directed his chief scholarship to the study of urban planning and rural settlement. Of his many written works, the one most interesting to the geographer is *Our world from the air: an international survey of man and his environment,* 1952. In 1956, he was appointed by the University of Philadelphia to undertake the project ' International History of City Development ' under the auspices of the Institute for Environmental Studies; the work was planned in ten volumes and was completed by his daughter, a research associate at the University.

593 Gyldendal of Copenhagen produce from time to time works of great interest to geographers. Perhaps their most significant recent innovation has been the introduction of graded series of atlases for systematic study; *Atlas 1, Atlas 2* and *Atlas 3* were designed for progressive use in Danish schools. The idea has been continued in this country by Collins and Longmans.

594 ' Habitat ', the illustrated journal published bi-monthly by the Central Mortgage and Housing Corporation, is now mainly concerned with urban centres and the problems of city life, on an international scale.

595 Hakluyt, Richard (c 1552-1616) graduated at Oxford in 1574 and lectured for a while on maps, globes and geography generally. He travelled widely, making exhaustive notes concerning geographical exploration and travels, which he subsequently prepared for publication. Notable among his writings was the *Principall navigations* . . . ' 1589, which was enlarged to three volumes, 1598, 1599 and 1600. The best complete modern edition is that published by the Hakluyt Society and MacLehose, Glasgow, in twelve volumes, 1903-5, with the ' Essay on the life and work of Hakluyt ' by Professor Sir Walter Raleigh. A considerable literature has grown on Richard Hakluyt and his brother and on the society (*see* below) which bears his name; especially interesting is *The original writings and correspondence of the two Richard Hakluyts*, published by the Hakluyt Society in 1935, with introduction and notes by E G R Taylor.

596 Hakluyt Society, founded in 1846, is a publishing society only, having for its object the publication of original narratives of important voyages, travels, expeditions and other geographical records. Many of them, especially the accounts and translations of the Elizabethan and Stuart periods, are admirable examples of contemporary prose styles. In the case of a foreign original, the work is rendered in English, either in a new translation, or in an earlier translation, unabridged, providing that this is accurate and suitable. Two volumes are published annually, each edited by an expert, including notes, maps, portraits and other illustrations, and facsimile reproductions whenever possible of original plates, woodcuts or drawings. In the first series, a hundred volumes were issued between 1847 and 1898; the second series began in 1899. Indexes of the society's publications are included in *Richard Hakluyt and his successors,* edited by Edward Lynam, published to mark the centenary of the Society in 1946.

Refer also Alison M Quinn: ' The modern index to Richard Hakluyt's *Principall Navigations' in The indexer,* Spring, 1967.

David B Quinn: *Richard Hakluyt, editor: introduction* to the complete facsimiles of *Divers voyages* (1582) and 'A shorte and briefe narration of the two navigations to neue Fraunce ' (1580) (TOT, 1967, *Mundes novus* series, translated by John Florio), introduced by a survey of the sources and bibliography of the books, with a modern index.

George B Parks: *Richard Hakluyt and the English voyages* (New York: American Geographical Society, 1928; Frederich Ungar Publishing Company, second edition, with introduction by James A Williamson, 1961); the author refers to the younger Hakluyt as ' the first professional geographer in England's history ' and to the *English voyages* as ' the flower of his career, the planned and purposeful career of the geographer '. An appendix sets out 'A list of English books of travel overseas and geographical description to 1600, from 1496 ', arranged chronologically. Another appendix brings together Hakluyt's writings, letters, *mss* and published works, also studies about him.

Janet Hampden: *Richard Hakluyt: voyages and documents: selected with an introduction and a glossary* (OUP, 1958).

597 Hallwag maps are a familiar feature of contemporary life. The political maps are fully coloured to show de facto boundaries, relief is shown by hill shading, with spot heights, and such features as railways and shipping routes, roads, pipelines, ruins and oases are depicted with the clarity demanded for road and tourist users. Town stamps are graded according to size of population.

598 ' Handbook for geography teachers' was first edited by Miss D M Forsaith and produced by Goldsmith's College in 1932. A new edition was prepared by G J Cons, chairman of the University of London Institute of Education Standing Sub-Committee in Geography, published by the committee in 1955; a third edition followed in 1957 and a fourth, edited by R C Honeybone, in 1960. A fifth edition by M Long, in 1964, was reprinted in 1965. Sections deal with geography in schools of various kinds; geography both out of doors and indoors; geographical societies. Booklists are included for successive levels of study and for teachers.

599 ' Handbook of American resources for African studies ', prepared by Peter Duignan (Stanford (California): Hoover Institution on War, Revolution and Peace, 1967), mentions ninety five manuscript and library collections, also missionary and church, ethnographic and business archives.

600 'Handbook of Hispanic source materials and research organizations in the United States', edited by Ronald Hilton (Stanford University Press, second edition, 1956) was first issued in 1942 by the University of Toronto Press, under the title *Handbook of Hispanic source materials,* compiled by the same editor. This is a comprehensive work, covering Spain, Portugal and Latin America of the pre- and post-Columbian periods; Florida, Texas and California until their annexation by the United States. Arrangement is regionally, then alphabetically. An index map and a list of published sheets of 'The Map of Hispanic America' 1 : 1M are included.

601 'Handbook of Latin American studies' has been published approximately annually by Harvard University Press since 1936; arrangement is by subject, subdivided according to country or region.

Refer also to the works of George Pendle, the acknowledged expert on Latin American countries, including *Argentina* and *Paraguay: a riverside nation* (OUP for The Royal Institute of International Affairs, 1955, 1961, 1963; and 1954, 1956, 1967, respectively); *The land and people of Peru* (Black, 1966).

602 'Handbooks to the modern world' series, published by Blond, London, provide essential information and comment about every country, by means of essays, statistical tables, maps and half-tone illustrations, compiled by a panel of experts. *Western Europe . . .,* edited by John Calmann, covers the area from Iceland to Turkey; part I provides basic information on each country, part II deals with important questions affecting all or much of Western Europe and part III with the Common Market and various aspects of European integration. *Africa . . .,* edited by Colin Legum, went into a second, enlarged edition to include the developments of the new states; *Asia . . .,* edited by Guy Wint, is in four main parts—the usual basic information by country, surveys of the history and development of each country, essays on all aspects concerning the continent as a whole and an appendix giving selected texts of important treaties and agreements. *Latin America and the Caribbean . . .,* edited by Claudio Véliz, 1967, follows the general pattern; volumes on Australia and New Zealand, the Soviet Union and North America complete the series.

603 'The haven-finding art: *a history of navigation from Odysseus to Captain Cook',* by E G R Taylor, with a foreword by Commodore

K St B Collins, Hydrographer of the Navy (Hollis and Carter, 1956), sets out in a readable narrative that wears its scholarship easily, the gradual improvements in apparatus and instruments that made accurate navigation possible. In five parts, the first is devoted to a general introduction, followed by 'Navigation without magnetic compass or chart'; 'With compass and chart'; 'Instruments and tables'; 'Towards mathematical navigation'. A number of reproductions and figures are included throughout the text and there is a select bibliography of original documents and secondary works. A new augmented edition, published in association with the Institute of Navigation, contains an appendix on Chinese Medieval Navigation based on the research of Joseph Needham.

604 'The heart of the Antarctic: *being the story of the British Antarctic Expedition 1907-1909, with an introduction by H R Mill'*, by Sir E H Shackleton, is in two volumes, with maps in a pocket. Shackleton and his scientific helpers were the first to establish the position of the South Magnetic Pole; quite apart from the discoveries and important scientific results of his expeditions, Shackleton left his influence on Antarctic exploration by his adoption of a new technique in sledge travelling and the abandonment of many of the older traditions of polar exploration.

605 W Heffer and Sons Limited, Cambridge, new and secondhand booksellers, hold stocks of publications on all general subjects; a journal and periodical subscription service and a reasonable search service are maintained and recently developed have been the individual services. Heffer's Paperback Shop, Heffer's Penguin Shop, Heffer's Children's Bookshop, Heffer's Stationery Shop and Printing Works. *The book news* is circulated every other month. Of special interest to geographers are the individual catalogues issued from time to time, including, for example, 'Geographical and earth sciences', with entries grouped under the headings 'Geography', 'Planning and urbanisation', 'Atlases and cartography', 'Polar', 'Oceanography', 'Geology', 'Meteorology and Climatology', 'Periodicals'. Other catalogues have been compiled on 'Latin America', 'India and Pakistan', 'Africa and Asia', 'Oriental and African Studies', 'Modern Europe' and 'Soviet Union'.

606 Herbertson, Andrew John (1865-1915) was one of the scholar-geographers who helped to lay the foundations of modern academic geography. He studied in Germany and France, developing theories over a broad field, his early work being in physical geography, especially meteorology and oceanography, from which he proceeded to the study of botany and worked as demonstrator in botany under Sir Patrick Geddes at Dundee. In 1899 he was appointed lecturer in regional geography at Oxford as assistant to H J Mackinder, to whom he succeeded as reader and director of the School of Geography in 1905. He introduced the theory of the major natural regions of the world, based on structure, relief and climate; his paper to the Royal Geographical Society in 1905 on 'The major natural regions' was not greeted with enthusiasm at the time, but many of his ideas now seem surprisingly up-to-date. *Man and his work: an introduction to human geography,* written with F D Herbertson, also reveals his modern outlook. Herbertson's meteorological knowledge is shown particularly in his contribution to the *Atlas of meteorology* (*qv*), the third volume of *Bartholomew's physical atlas.* He was a founder-member of the Geographical Association, which has established a Herbertson Memorial Lecture in his honour. In a special issue of *Geography,* November 1965, an appreciation of his life and work is given, with lists of his published works, both books and periodical articles, together with an outline of his editorial activities.

Refer also E W Gilbert: 'Andrew John Herbertson 1865-1915' in *The geographical journal,* December, 1965, which includes a bibliography.

607 'Hereford mappa mundi', now in Hereford Cathedral, was probably made between 1285 and 1295. Cartographically, it is of the greatest interest; it reveals also considerable evidence concerning secular life in the Middle Ages. *The world map of Richard Haldingham in Hereford Cathedral* was reproduced by The Royal Geographical Society in 1954, with a memoir by G R Crone (Reproductions of early manuscript maps, III), in nine overlapping sheets, with a sheet of diagrams, at a scale of about nine-tenths the original. Richard Gough said of the map in 1780—' In the library of Hereford Cathedral is preserved a very curious map of the world, inclosed in a case with folding doors, on which are painted the Virgin and the Angel. It is drawn with a pen on vellum, fastened on boards, and is six feet four inches high to the pediment and five feet four inches wide.'

Refer also ' Key to the Photographs of the Ancient Map of the World preserved in Hereford Cathedral ', available at the Cathedral.

W L Bevan and H W Philliott: *Mediaeval geography: an essay in illustration of the Hereford Mappa Mundi,* 1873, reprinted by TOT, 1969.

A L Moir: *The world map in Hereford Cathedral,* a paper read to the Woolhope Naturalists' Field Club, Hereford, on 25th November, 1954; Malcolm Letts: ' The pictures in the Hereford mappa mundi ', *Notes and queries,* January, 1955; both reprinted as a pamphlet available from the Cathedral, Hereford.

G R Crone: ' New light on the Hereford map '. *Geographical journal,* December, 1965.

608 Hettner, Alfred (1859-1941) was a physical and regional geographer and one of the founders of modern German geography. He travelled widely and his influence reached into many spheres, but his main work was achieved as professor at Heidelberg. One of his chief published works was *Die Geographie: ihre gueschichte, ihre Wesen und ihre Methoden,* 1927; and his ideals and geographical theories were embodied in the great journal *Geographisches Zeitschrift (qv)* which he founded. Hettner's *Surface features of the earth,* a key work in geomorphology first published in 1921, has been translated into English by Philip Tilley and published by Macmillan.

609 Hinks, Arthur Robert (1873-1945) exerted a widespread influence through his many geographical activities. He was a meticulous scholar, particularly in astronomy, mathematics and all aspects of surveying and cartography, and a stimulating lecturer, communicating his formidable knowledge with an infectious energy. He also wrote many articles and books on maps and surveying, which have proved constantly valuable to students. His other main contribution to the status of geography arose from his appointment to the staff of the Royal Geographical Society in 1912; he improved the production, especially the typography, of *The geographical journal,* revised the *Hints to travellers* series and initiated the *Technical* series. Between 1914 and 1918, he directed the production of more than a hundred sheets of the provisional 1:1M Map of the world and some sheets of the 1:2M Map of Africa; and from 1919, as secretary to the Permanent Committee on Geographical Names, he added greatly to the success of the project.

Refer T S Blakeney: 'A R Hinks and the first Everest Expedition 1921' *in The geographical journal,* September, 1970.

610 Hipparchus (*c* 160-125 BC), the most celebrated astronomer of the ancient world, evolved the concept of map projections. His power of thought was in advance of technical developments, but he divided the great circle into three hundred degrees and greatly improved the spherical astrolabe.

Refer H Berger: *Die Geographischen Fragmente des Hipparch,* 1869.

The geographical fragments of Hipparchus, edited, with introduction, by D R Dicks (Athlone Press, 1960).

611 'Histoire des villes de France avec une introduction général pour chaque province', in six volumes by M Aristide Guilbert and others (Paris: Furne et Cie, Perrotin, H Fournier, 1845), is a magnificent work, including superb steel engravings.

612 'A historical atlas of Canada', edited by D G G Kerr, was published by Nelson of Canada, undated, but about 1961, sponsored by the Canadian Historical Association. Cartography is by Major C C J Bond and the maps were drawn by Ellsworth M Walsh and others. The maps are mainly in two colours, with brief textual entries, beginning with ' Environment and prehistory ', and a selected bibliography.

613 'An historical atlas of China', in the new edition, is based on the *Historical and commercial atlas of China,* compiled by Albert Harrmann and published by the Harvard-Yenching Institute in 1925 (Monograph series, volume 1). A new edition was first published in 1966 by the Aldine Publishing Company, Chicago; Edinburgh University Press were agents for Great Britain and the Commonwealth. The work has now been extended and re-drawn under the supervision of Professor Norton S Ginsburg, with a prefatory essay by Paul Wheatley (Djambatan/Morton). Ten new maps have been added on the main characteristics of Chinese agricultural land use, density and patterns of population, distribution of minerals and industries and of the transportation network. Endpapers present a physical map of modern China and one of the political and administrative divisions, making altogether seventy five six-colour maps and town plans. The bibliography has been updated; a list of Chinese characters has been included and there are two indexes of geographical and proper names.

614 'The historical atlas of the holy land', edited by E G Kraeling (Rand McNally, 1959), is one of the many works of scholarship inspired by modern archaeological discoveries. The atlas comprises a brief outline of the historical geography by maps, photographs and text, which begins with an account of the finding of the Dead Sea Scrolls.

615 'Historical atlas of Latin America: *political, geographic, economic, cultural'*, edited by A Curtis Wilgus (New York: Cooper Square Publishing, Inc, 1943; new edition, enlarged, 1967), consists of sketchmaps, with facing commentary, on the following topics: 'The geographical background'; 'The ethnological background'; 'The European background'; 'The colonial period'; 'The revolutions for independence'; 'The national period'; 'Latin American boundary controversies'; and 'Latin American relations with the United States and Europe'.

616 'The history and use of diamond', by S Tolansky, published by Methuen, examines the strange and scientific characteristics of diamonds, together with their history from Biblical times, their folklore and anecdotes connected with them and the techniques of mining and cutting. The final chapters cover the importance of diamonds in contemporary industry and their synthetic production in large quantities. A bibliography is included.

617 'A history of ancient geography among the Greeks and Romans from the earliest ages till the fall of the Roman empire' is a classic study in two volumes by E H Bunbury (Murray, 1879; second edition, 1883). Dover Publications issued an edition in 1932, with a second edition in 1959.

618 'A history of cartography: *2,500 years of maps and mapmakers'* (Elsevier, 1968; Thames and Hudson, 1969) is a fine display of the art of the cartographer, of 'those old maps that fascinate collectors'. The maps were chosen by R V Tooley, the text contributed by Charles Bricker and the preface by G R Crone. An introduction discusses the mapmaker's skills, following which each section deals with the history of cartography in each continent—'Europe: the rise of map publishing'; 'Asia: seaway to the Indies'; 'Africa: the mysterious continent'; 'The Americas: a new world'; 'Australia: the southern land'. Throughout are illustrations of the instruments and other engravings. In some cases, the detail of the maps is reproduced, in others the whole of a sheet or plate in colour—'The New Description of America, or the New World', for example, from Ortelius's *Theatrum Orbis Terrarum* and the enchanting plan of seventeenth century Goude from Jan Blaeu's *Stedenboek*. There is a short bibliography 'for further reading'.

619 'A history of geographical discovery in the seventeenth and eighteenth centuries', by Edward Heawood (CUP, 1912; reprinted New York: Cass; Octagon Books, 1965), was one of the *Cambridge geographical* series; it dealt with a period of discoveries which has 'met with less attention, perhaps, than it deserves', using primary sources of information as much as possible, from the Arctic, 1550-, through Oceania, 1605-, Africa, 1600-, Russian discoveries in the North-East from 1700 and all areas to 1800. A brief, neat summary concludes the text, with supplementary notes and a really useful index. Folding maps and illustrations are in the text.

620 'The history of geography: *papers by J N L Baker'* (Oxford: Blackwell, 1963) comprises eighteen papers selected by his former students to mark his retirement; they are prefaced by an appreciation of his achievements in geographical studies, a list of his published works and a portrait. The subjects of the papers are mainly historical, some of the most interesting and valuable drawing attention to individual contributions of earlier, some little-known writers —'Nathanael Carpenter and English geography in the seventeenth century'; 'Mary Somerville and geography in England'; 'The geography of Bernhard Varenius'; 'The history of geography in Oxford'; 'Major James Rennell and his place in the history of geography'; 'The geography of Daniel Defoe'; 'Geography in the Essays of Elia'; 'The earliest maps of H Moll'; and 'Some geographical factors in the campaigns of Assaye and Argaon'.

621 'A history of land use in arid regions', a composite work edited by Sir L Dudley Stamp (Unesco: Arid Zone Research, 1961), was prepared with the practical aim of the improvement of arid and semi-arid lands. Following an introduction by Sir Dudley Stamp, a number of sections, each by an expert, deal with ' Climatic change in arid regions since the Pliocene '; ' Evolution of irrigation agriculture in Egypt '; ' The arid zone of India and Pakistan '; ' Land use development in the arid regions of the Russian plain, the Caucasus and Central Asia '; ' Development of land use in northern Africa (with references to Spain) '; ' Land use in the Sahara-Sahel region '; ' Land use in pre-Columbian America '; ' Post-Columbian development in the arid regions of the United States of America '; ' Land utilization in the arid regions of Southern Africa and South West Africa '; ' The problems of arid Australia '—and ending with a consideration of health problems, to which Jacques M May contributed. There are a number of sketchmaps in the text and end-of-chapter bibliographies.

622 ' The history of the study of landforms: *or the development of geomorphology'*, by R J Chorley and others, will when completed, be a definitive work. Volume I, *Geomorphology before Davis* (Methuen, 1964), is a work of exact scholarship, including many excerpts from original sources, and is well documented with references and bibliographies.

623 ' How to find out in geography: *a guide to current books in English'*, by C S Minto (Oxford: Pergamon Press, 1966), includes principally books published within the previous five years. The author states ' This book is not regarded as a bibliography: its purpose is limited to assisting the undergraduate student and the adult general reader in selecting books on general and special aspects of geography. It is further limited in scope by being concerned with books in English readily available in bookshops and libraries.' For these reasons also the entries are arranged according to the Dewey Decimal classification, that most used in British public libraries; with this arrangement, the index should really be more detailed to be helpful. There are some reproductions of map extracts, title pages or covers, some being too much reduced for clarity.

624 ' How to identify old maps and globes, *with a list of cartographers, engravers, publishers and printers concerned with printed maps and globes from about 1500 to c1850'*, by Raymond Lister (Bell, 1965), is illustrated by fifty nine plates of reproductions. Following 'An outline history of maps and charts ', ' Celestial maps and charts' are considered—'Methods of map reproduction', 'Decoration and conventional signs', 'Terrestrial and celestial globes and armillary spheres'. The appendices give information otherwise scattered: 'The use of watermarks in dating old maps and documents', a 'select bibliography' and 'A list of cartographers, engravers . . . etc'. A second book on early maps by Raymond Lister was published by Bell in 1970: *Antique maps and their cartographers* (*qv*).

625 Hudson's Bay Record Society, London, was constituted in 1938 for the purpose of issuing in book form the unique records of the Hudson's Bay Company; the intention was to select for publication material of various dates between 1670 and 1870 rather than to print in historical sequence. The aim is two-fold—to make available to members of the society those records in which public interest already exists, such as Ogden's journals, Rae's letters and Isham's ' Observations'; secondly, to direct attention to less well known topics or regions to the study of which the company's archives seem able to make a worthwhile contribution; for example, with volume XXIV, 1963, Labrador and Northern Quebec are concerned, about which few records of any kind had previously been published. Each document chosen for publication has covered a vital topic in the history of the company, of the fur trade and of Canada. Each has been published in full, accompanied by an introduction by a chosen scholar and supported by extracts from other documents. The total archives comprise more than thirty thousand documents, covering every branch of the company's activities throughout the years. The *History of the Hudson's Bay Company,* by E Rich, in two volumes dealing with the periods 1670-1763 and 1763-1870 respectively are numbered volumes XXI and XXII.

626 ' Human geography from the air ', by R M Minshull, published by Macmillan, provides practice in the interpretation of types of air photographs, based on a carefully chosen selection of oblique aerial photographs illustrating the major aspects of human geography in Britain. Each photograph is followed by a set of exercises and questions constructed to encourage students to find out as much as possible for himself by the study and analysis of each picture. A variety of work is involved, such as tracing, sketching, the drawing of sketchmaps and cross-sections.

627 'Human nature in geography: *fourteen papers, 1925-1965'*, by John Kirtland Wright (Harvard UP; OUP, 1966) represents a 'selection from the professional papers of a lifelong student of the history of geography'. The papers, some of which have been previously published, others not, are concerned with many periods and personalities. The book is without doubt important in itself and it is also significant of a growing number of works embodying the reflections of mature geographers, revealing deep knowledge of and devotion to their subject.

Refer also Two new books presenting current thoughts on different aspects are D C Money: *Patterns of settlement: human geography in colour* (Evans, 1972); Peter Toyne: *Techniques in human geography* (Macmillan, in paper edition or boards).

628 Humboldt, Friedrich Wilhelm Karl Heinrich Alexander, Baron von (1769-1859) brought a scientific inspiration to modern geography. Educated at Göttingen and for several years a mining geologist, he travelled widely and on his first-hand observation and experience many of his published works are based. During his lifetime, he became noted as a scholar in social and political as well as in scientific fields. His major work, *Kosmos: a sketch of a physical description of the universe,* in five volumes, 1845-62, is an attempt at a physical description of the earth, a work of superb scholarship and stylistic excellence; in it, Humboldt attempted to demonstrate the inter-relation of all creation, dominated by physical geography. He made a notable contribution to geographical methodology, and emphasised the importance of maps, inspiring Berghaus (*qv*) and Justus Perthes. The Humboldt University of Berlin, Institute of Political and Economical Geography, founded in 1906 and reconstructed in 1950, stands as a monument to his influence on academic geography. His many-sided activities and his publications are examined in *Alexander von Humboldt: Studien zu seiner universalen Geisteshaltung,* a volume of essays published under the editorship of Professor J H Schultze, by the Berlin Geographical Society in 1959, as part of the commemoration of his death.

A vast literature has grown up around him and his work. In addition to the above, Hanna Beck, in *Alexander von Humboldt, 1959,* completed the first full length study of his work and his American journeys; and published an anthology of his work on the American tropics, in 1969. A special exhibition, 'Alexander von Humboldt and his world 1769-1859' was held in Berlin to mark the two hundredth anniversary of his birth and a film, 'Alexander von Humboldt —from his life and work', was made under the auspices of Inter Nationes. The Humboldt Foundation, established in 1860 and reorganised in 1953, awards scholarships to young teachers at universities, research and other academic institutes abroad for study at universities and institutes in the Federal Republic. Each year approximately three hundred such scholarships are awarded.

Refer also L Kellner: *Alexander von Humboldt* (OUP, 1963), which contains a bibliography both of Humboldt's own works and of works contributing to his biography.

A H Robinson and Helen M Wallis: ' Humboldt's map of isothermal lines: a milestone in thematic cartography ', *in The cartographic journal,* December, 1967.

629 'A hundred years of geography', by T W Freeman (Duckworth, 1961, in *The ' Hundred years'* series), traces progress during this period under broad headings; useful ' Notes and references ' and ' Short biographies of geographers ' sections are included.

630 ' Hungarica ' contains references to English books, prints, maps and periodicals relating to Hungary, collected by Béla Iványi-Grünwald and published by Mrs Jocelyn Iványi (Bures, Suffolk, 1967); entries are arranged alphabetically by author's name, and are numbered, with a subject index referring to the numbers.

631 The Hunting Survey Companies comprise Hunting Surveys Limited, Hunting Geology and Geophysics Limited and Hunting Technical Services Limited, for which the holding and management company is Hunting Surveys and Consultants Limited, Boreham Wood, Hertfordshire. Hunting Surveys Limited carries out panchromatic and colour photography for surveys and interpretation; auxiliary aids include airborne profile recorder, Doppler navigator, statoscope and horizon camera. Laboratory photography includes the making of photo mosaics, rectified enlargements, electronic printing and precision mapping and holds a record library of worldwide photographic cover. Topographic and planimetric mapping by photogrammetry are carried out, including profiles, cross-sections and capacity tables, electronic computations and alignment programmes for road and engineering calculations. Land surveys and models are also undertaken. Hunting Technical Services Limited maintain a group of scientists providing a consultancy service in the land use and agricultural aspects of natural resources development, specialising in regional and project planning, land use and land capability studies, soil surveys and land classification, irrigation and drainage studies, watershed management and forestry, and agricultural and project economics. Hunting Geology and Geophysics Limited provides all forms of airborne geophysical survey, overland and offshore for mineral and oil exploration and for regional geological studies; geology, geophysics and geochemistry applied to groundwater, mineral and construction material resources evaluation and to civil engineering projects; sea, lake, harbour and river bottom and sub-bottom surveys for mining, construction materials and civil engineering work. Aerofilms Limited and Hunting Photographic Limited are associated companies. Hunting Surveys Limited and Hunting Geology, and Geophysics Limited both issue *Information sheets*. The *Hunting Survey review* contains descriptions of new projects, illustrated by photographs, and individual booklets are produced from time to time on specific topics, for example 'A modern approach to road design' and the more comprehensive and finely produced 'In a changing world'. Hunting Technical Services also publishes illustrated booklets showing the wide range of services in the 'investigation and development of natural resources'. The quarterly *Hunting Group review* is the trade journal.

632 Huntington, Ellsworth (1876-1947) made a unique contribution to geographical thought with his challenging theories, which stimulated much thought and discussion. From geological studies, he turned to climatology, heredity and eugenics, striving increasingly in his publications to analyse the influences of biological inheritance and physical environment in shaping the course of history. He published nearly thirty books, parts of a number of other books and innumerable articles, notably *The pulse of Asia,* 1907; *Palestine and its transformation,* 1911; *Civilization and climate,* 1915; *Character of races,* 1924; *The human habitat,* 1927; and *Mainsprings of civilization,* 1945.

633 Hutchinson university library series includes a sub-series devoted to geographical subjects, notably G R Crone: *Maps and their makers* ...(*qv*); S W Wooldridge and W G East: *The spirit and purpose of geography* (*qv*); F W Morgan: *Ports and harbours;* Norman

J G Pounds: *The geography of iron and steel;* James Bird: *The geography of the Port of London.* The policy of publishing in hard cover and paperback simultaneously began with the new edition of G R Crone: *Maps and their makers.*

634 Hydrographic Department of the Admiralty, founded in 1795, has the duty of publishing and distributing Admiralty charts to the Merchant Navy, the Fleet and the general public. Publications are world-wide in geographical scope and constantly under revision; many special charts are also prepared. Other nautical publications are concerned with the safety of navigation, such as the *Admiralty list of lights,* the *Admiralty list of radio signals* and the *Admiralty notes to mariners.* The *Annual reports* are central documents; also the *Admiralty manual of hydrographic surveying* and the *Admiralty manual of tides. British Admiralty charts and other hydrographic publications: their use and correction* lists every chart and includes at the back a series of maps showing the limits of each available chart. An annual *Catalogue of Admiralty charts and other hydrographic publications* is issued.

Refer also A R Boyle: 'Automation in hydrographic charting' *in Canadian surveyor,* December 1970.

F A Pielou: ' Special purpose navigation charts' *in The cartographic journal,* June 1971.

635 Ibn Battuta (1304-1378) is undoubtedly one of the most outstanding of all land travellers who have recorded their experiences. Samuel Lee first translated the *Travels* into English, from an abridged text, in 1829. Professor Hamilton Gibb translated selections for the *Broadway travellers* series, 1929, and the full text for the Hakluyt Society in 1958.

636 'Ice', the news bulletin of the Glaciological Society, has been issued twice a year since 1958. General information is published here, of more personal interest than in the *Journal of glaciology* (*qv*), concerning expeditions and the activities of members; also reviews and short articles of less enduring value.

637 ' The Icefields Ranges Research Project ' scientific results began publication in 1969. The first volume, edited by Vivian C Bushnell and Richard H Ragle (published jointly by the American Geographical Society of New York and the Arctic Institue of America), consists of nineteen papers on natural, climatic, glaciological and quaternary studies, illustrated, and with figures and maps.

638 Idrisi *see* Al-Idrisi.

639 'Imago mundi: *a review of early cartography'* was founded by Leo Bagrow in 1935 to serve as a vehicle for the publication of scholarly work on early maps and mapmakers and as a forum for discussion and exchange of information in this field. It is the only international review covering the early history of cartography, of the highest scholarship and superbly produced. The journal's publishing history has been complex. Bagrow himself edited thirteen volumes before his death; Edward Lynam was co-editor for the second and third volumes, which were published in London. At one time, the journal was sponsored by the trust 'Humanistiska Fonden', thereafter under the patronage of King Gustav of Sweden; it is produced now under the direction of an editorial committee by N Israel of Amsterdam, with Professor E M J Campbell as editor. *Imago mundi* sets out to communicate the results of original research, to inform those interested by comment, reviews of recent publications, correspondence and an annual bibliography. Each issue contains a number of superb map reproductions. A series of supplements, published from time to time, under separate subscription arrangements, present larger facsimiles and studies of early maps too extensive for the regular issues; Bagrow's own *Anecdota cartographica* was the first. Volume XXII contains the papers read at the symposium on the History of Cartography held in London at the time of the 20th Congress of the IGU (1967). Cumulative indexes were published in volumes X and XX.

640 'Index India' began publication with the January to March 1967 issue, from the Rajasthan University Library, Jaipur. Articles, editorials, notes and letters deal with India and Indian matters, selected from periodicals and newspapers published in English throughout the world; about 375 periodicals are scanned for each issue.

641 India Office Library and Records: after 1947, when the India Office Library and Records passed into the custody of the new Commonwealth Relations Office, the printed, manuscript and other resources grew rapidly, the use of the material for research purposes expanded, much additional space was needed for an enlarged staff and for modern technical facilities such as photocopying and conservation. The Library and Records Conservation Departments were amalgamated at the end of 1967 and at about the same time the India Office Library moved to Blackfriars Road, London, following one hundred years residence in the India Office building in King Charles Street, Whitehall. The East India Company's archives, now included within the India Office Records, accumulated steadily from 1600 and the Library was founded by the Company in 1801 in the East India House in Leadenhall Street. The Official Publications amount to more than 100,000 volumes and the Map Collection of maps, plans and drawings, includes topographical, subject and archival material. The only printed catalogue of the latter was published in 1878.

Refer Sir William Foster: *Guide to the India Office Records, 1600 to 1858* (London, 1919), reprinted in 1961 and again in 1966.

India Office Records, Report for the years 1947-1967.

Annual Report of the India Office Library and Records.

642 'India: regional studies', edited by Dr R L Singh, was published by the Indian National Committee for Geography in 1968 on the occasion of the 21st International Geographical Congress. The contributors, all academic geographers, analyse the geography of thirteen regions of India, including some photographs and a number of small maps, drawings and tables. End of chapter references are included.

643 'Industrial activity and economic geography: *a study of the forces behind the geographical location of productive activity in manufacturing industry'*, by R C Estall and R O Buchanan (Hutchinson University Library, second edition, 1966), is a succinct text working out the detailed title ideas, illustrated by sketchmaps and including references.

644 'Industrial archaeologists' guide 1969-70', illustrated with photographs, figures and diagrams in the text, is published by David and Charles of Newton Abbot; it contains information on such as the National Record of Industrial Monuments, scientific and technological museums, science and technology collections, individual sites, the use of photography in industrial archaeology, relevant local societies, journals and newsletters.

645 'Industrial archaeology: *the journal of the history of industry and technology'*, 1964-, published quarterly by David and Charles, aims, by original papers, comprehensive reviews of new projects, notes, news and readers' correspondence, to keep industrial archaeologists informed of new developments; the journal is illustrated with plates and drawings in the text.

646 Industrial Diamond Information Bureau, London, publishes the monthly *Industrial diamond abstracts,* which covers current scientific and technical literature, including patents, on industrial diamonds and associated subjects, frequently illustrated with diagrams; also *Industrial diamond review.* A library is maintained, which acquires world literature in this field, especially technical and scientific pamphlets. The *Industrial diamond trade name index,* 1961-, includes an Index of Manufacturers, Agents and Sales Organisations, an Index of Proprietary and Trade Names and an Index of Available Products, together with much information on diamonds and diamond tools. Selected bibliographies are prepared, as are specialist monographs, such as 'Diamonds in industry', 1961, in English, French, German, Dutch and Italian and 'Industrial diamonds made in South Africa', 1962.

647 'Industrial minerals', published monthly by Metal Bulletins Limited, London, is an illustrated journal having world coverage. Usually, one long article deals with one mineral in detail throughout the world or with the mineral industries of a certain country, followed by a section 'World of minerals', which contains news of developments, usually accompanied by statistical tables. Current topics are expertly noted and the 'Company news' section and advertisements are equally useful to those concerned with current markets.

Refer also Herbert Wöhlbier, *ed: Worldwide directory of mineral industries.*

648 'Influences of geographic environment on the basis of Ratzel's system of anthropo-geography', by Ellen Churchill Semple (Constable, 1932), was an attempt to interpret and explain Ratzel's theories to the English-speaking world. The work, which was meticulously documented and detailed, caused considerable discussion at the time and opinions still ebb and flow on the whole subject. Harriet Wanklyn in *Friedrich Ratzel: a bibliographical memoir and bibliography* (CUP, 1961) makes some comments on Miss Semple's work; and in J K Wright: *Human nature in geography* (*qv*), section 12 is ' Miss Semple's *Influences of geographic environment:* notes toward a bibliography '.

649 ' Information bulletin of the Soviet Antarctic Expedition ', 1958-, was published, in three volumes, by Elsevier, 1964-65; from 1964, a bi-monthly translation, illustrated with maps and charts, was issued by Dr M M Somov of the Russian Arctic and Antarctic Scientific Research Institute, sponsored by the Geophysical and Polar Research Center of the University of Wisconsin with support from the National Science Foundation, with the intention of bringing together all the data obtained from the Expedition, especially in glaciology and meteorology. Russian work in the Antarctic programme of the International Geophysical Year and during the years following, has been particularly valuable.

650 L'information géographique ', ' revue illustrée paraissant tous les deux mois pendant la période scolaire ', is an excellent service for schools, compiled by educationists in Paris and circulated by Baillière five times a year since 1936. Included are articles, well illustrated by sketch-maps, diagrams and graphs; notes and statistical information; selected abstracts; bibliographies; and notes of developments and projects useful for teachers to know about. Practical notes on new publications are usually included.

651 ' Information Hungary ', edited by Ferenc Erdei and others (Oxford: Pergamon Press, 1968-, ' Countries of the world information series '), is arranged according to broad subject groupings, beginning with ' Land and people '. It is a monumental work, in which two other sections of particular interest to geographers are 'Atlases and maps ' and ' Historical maps '. There are photographs in the text.

652 ' Information USSR: *an authoritative encyclopaedia about the Union of Soviet Socialist Republics',* edited by Robert Maxwell (Oxford: Pergamon Press, 1962), is an expanded and updated English-language version of the Soviet official handbook, in its second edition, 1957. Four appendices have been added: ' Statistical data to 1960 '; a directory of establishments of higher learning in the USSR; a select bibliography of recent books in English on the USSR; and ' Data on trade with the Soviet Union '.

653 Institut Français d'Afrique Noire at Dakar is noted for scholarly research work and for informative, frequently erudite, publications. The *Bulletin d'IFAN,* 1939-, has been divided into two series since 1954, of which series A is devoted to 'sciences naturelles' and series B to 'sciences humaines'; this bulletin replaced the former *Bulletin* of the Comité d'Etudes Historiques et Scientifiques de l'Afrique Occidentale Française. *Mémoires* have been issued since 1940; also *Notes africaines* and a number of issues of *Catalogues et documents.*

654 Institut Géographique National, founded in Paris in 1940, is the civilian body which replaced the Service Géographique de l'Armée and is the national agency responsible for the topographic maps at scales of 1:10,000 or smaller. The central task of the institute is the preparation of the 1:20,000 base map of France. The majority of maps are prepared by the stereo-plotting of aerial photographs, the Photogrammetric Department of the IGN being one of the largest in the world. Mechanographic punched-card equipment is used for geodetic documentation, capable of undertaking a wide range of search and classification functions. *Exposé des travaux de l'IGN* and the catalogue *Cartes en service,* comprising ' cartes terrestres ', ' cartes en relief ' and ' cartes aéronautiques ' are annual publications. The *Lexique anglais-français des termes appartenant aux techniques en usage à l'Institut Géographique National* was published in three volumes, 1956-58.

655 The Institute of Agricultural Economics, University of Oxford, publishes the *Farm economist* and the *Digest of agricultural economics and marketing,* now incorporated in *World agricultural economics and rural sociology abstracts,* published by the Commonwealth Agricultural Bureaux, Farnham Royal, Buckinghamshire (Amsterdam: North Holland Publishing Company). Books published by the institute include—K E Hunt and K R Clark: *Poultry and eggs in Britain . . .,* 1963; and by the same authors, *Poultry and eggs . . .,* 1967; H Frankel: *Economic changes in British agriculture 1959/64* (The Agricultural Register), 1964; G T Jones: *Analysis of Census data for pigs,* 1964; D Wood: *Economic aspects of pig-*

meat marketing, 1965; K E Hunt and K R Clark: *The state of British agriculture . . .,* 1966; K S Woods: *The development of country towns in the south-west Midlands during the 1960s,* 1968; E Neville-Rolfe: *Economic aspects of agricultural developments in Africa,* for the US Department of Agriculture, 1969.

656 Institute of Australian Geographers was established in 1959 at the Department of Geography, University of Sydney, as a result of a growing conviction among professional geographers that a national organisation was necessary to promote the study of geography in Australia. *Australian geographical studies* has been issued twice a year since 1963; the original contributions concentrate mainly on Australia and neighbouring regions, but include wider geographical studies from time to time, and the reviewing service is world-wide in scope.

657 The Institute of British Geographers, a graduate body, was formed in 1933 for the study, discussion and advancement of geography, primarily at academic level. From 1965, the secretariat has been centred on the Royal Geographical Society headquarters. The annual conferences, held at different university centres, and the summer field meetings, promote valuable research and discussion. The Institute now issues three publications each year, incorporating papers by leading geographers on all aspects of geography. The *Transactions* include the results of members' research, 1935-; abstracts are now appended. *Area (qv)* is published four times a year and the first volume of the Institute's Special Series of Publications was *Land use and resources: a memorial volume to Sir Dudley Stamp (qv).* Among the separately published works, two of the most notable have been S W Woolridge and D L Linton: *Structure, surface and drainage in south-east England,* 1939 (Philip, revised edition, 1955), and Henri Baulig: *The changing sea-level,* 1935, re-issued 1956. Lists of publications are available from the institute.

658 Institute of Documentation Studies was established on a new site in Frankfurt, in 1969, together with the Automatic Documentation Centre. The Institute forms part of the Max Planck Society and will, with the newly founded branch office in the United States, help in the co-ordination and promotion of state, scientific or industrial information and documentation projects, act in an advisory capacity and assist in the training of documentation specialists. The

Automatic Documentation Centre is concerned with the application of computers in documentation, possessing initially a fully equipped IBM 1460 computer and an IBM 113 for computer type-setting.

659 The Institute of Marine Engineers, founded at the London Docks in 1889, encompasses specialist sections—Marine Electrical Engineering; Education; Materials; Ocean Engineering, among others. Each section arranges its own meetings and reports of these meetings are published in the *Transactions*. The library is one of the most complete in the world in marine engineering and associated subjects. A punched card information retrieval system, operated in conjunction with Lloyd's Register of Shipping, enables the library to provide advice about published material on a wide range of marine developments. The *Transactions* are available in paper or microfiche editions. *Marine and shipbuilding abstracts,* formerly included in the *Marine engineers review,* is now a quarterly information service covering more than seventy journals, also available in paper or microfiche editions, as is the *Marine engineers review,* which presents a monthly survey of developments throughout the world. A news sheet, conference papers and some half-dozen textbooks are also among the Institute's publications.

660 The Institute of Pacific Relations is an unofficial international organisation founded for the study of the social, economic and political relations of the Pacific area; it is composed of independent National Councils in a number of countries, including the Royal Institute of International Affairs and the University of British Columbia.

661 The Institute of Transport, London, founded in 1919, was incorporated by Royal Charter in 1926 'to promote, encourage and co-ordinate the study and advancement of the science and art of transport in all its branches'. Meetings, lectures and discussions are held during the winter in London and at branches throughout Great Britain and Ireland, Australia, New Zealand, Nigeria and South Africa. There are graduate and student societies, separately constituted but working closely with the parent body. The principal papers presented to the institute are reprinted in the *Institute of Transport journal,* published every other month. The library at headquarters contains a substantial collection of books, periodicals and reports on transport and allied subjects; some of the branches hold smaller collections. A biennial handbook sets out all information concerning the institute.

662 The Institute of World Economics (Kiel) Library is one of the largest of its kind in the world. The German Research Association has passed a resolution that its holdings shall be increased still further and it will have special information retrieval services in its fields of coverage.

663 The Institution of the Rubber Industry was founded in 1921 and now has many active sections in Great Britain and other parts of the world. Its main purpose is to further technical training and to provide a means of communication between those engaged in various parts of the industry, to promote the dissemination of knowledge of rubber science and technology by meetings and the publication of *Transactions* and other literature. The *Transactions and proceedings* are published six times a year; the *Annual report on the progress of rubber technology,* the *Proceedings of international conferences* are key documents and also published are monographs on specialised subjects from time to time. No separate library is maintained; the Library of the Rubber and Plastics Research Association is available for use by courtesy.

Refer also The *International rubber digest,* published monthly by the Secretariat of the International Rubber Study Group, a comprehensive listing of statistics and news items concerning rubber production throughout the world.

664 Intergovernmental Oceanographic Commission, under the sponsorship of Unesco, co-ordinates the activities of regional and national research in marine sciences. The *Five years of work* (Unesco, 1966) contains the history and description of current operations, international expeditions and co-operating groups; a list of documents pertaining to oceanographic activities; and the ' Radio communication requirements for oceanography: a short outline for oceanographers' needs for radio frequencies to be used for oceanographic data transmission ', appended, 1967. The *Manual on international oceanographic data exchange,* second edition, 1967, assembles data in a convenient form for the instruction of practising oceanographers; the various documents concerned with the exchange of oceanographic data; and a list of national oceanographic data centres with information concerning their mode of operation.

Perspectives in oceanography was published in 1968, 1969 and there are also various monographs on specific methodology. The *international marine science* series comprises a quarterly newsletter, published jointly by Unesco and FAO, 1963-; collected reprints, such as the reports on the International Indian Ocean Expedition, published by Unesco in three volumes to 1966 and continuing as desirable; the Technical Series, 1965-; reports of sessions, 1961-, and special publications, including the reports of individual symposia and the reports from the Japanese Oceanographic Data Center, Hydrographic Division, among others, in four or five issues a year, 1965-, and *A manual of sea-water analysis*, by J D H Strickland and T R Parsons, published by the Commission and the Unesco Office of Oceanography, 1965.

665 'International African bibliography', published by the International African Institute, London, from 1971, has incorporated the bibliographical section from the journal *Africa*. With the expansion of Africanist literature, it has doubled in size over the past few years and the number of titles of books and articles from periodicals has now reached an annual level of about three thousand, making the separate publication necessary. The scope of the bibliography remains as before.

Refer also Bibliography of African bibliographies (South African Library, 1961; new edition, 1971).

International conference on African bibliography, *Papers,* 1967.

J M D Crossley: 'Notes on Africana in the Yale University Library' *in South African libraries,* April 1971.

Reuben Musiker: 'The bibliographical scene in South Africa' *in South African libraries,* April 1971.

Hans E Panofsky: 'The African Studies Library of Northwestern University' *in South African libraries,* April 1971.

666 International African Institute was established in 1926 primarily to promote the serious study of Africa and its peoples through research, publication and information services. The Institute's official organ, *Africa,* quarterly since 1931, is an erudite journal containing articles of wide interest, notes of publications, careful book reviews and until 1971 a 'Bibliography of current publications' prepared in co-operation with other organisations interested in African studies. *African abstracts,* published quarterly, contains abstracts of all significant articles in European languages on ethnographic and social aspects, in the widest sense. Two further series of publications are *The handbook of African languages* and *The ethnographic survey of Africa,* begun in 1945, which present a conspectus of up-to-date information, with comprehensive bibliographies. The value of all this work for the geographer lies particularly in reports and reviews of agriculture and land use, population analysis, distribution and migration. The Institute's library holds some five thousand books and a large collection of pamphlets, manuscripts, periodicals and government reports; a classified bibliographical card index is maintained.

667 International Association of Physical Oceanography was founded in 1919, based on Göteborg, Sweden, as the Oceanographic Section of the International Union of Geodesy and Geophysics, to promote the study of scientific problems relating to the oceans, to initiate and co-ordinate research and to provide for discussion and publication. *Publications scientifiques* have been issued since 1931, usually annually; *Procès-verbaux* of the General Assemblies appear every third year.

668 International Association of Quaternary Research was founded in Copenhagen in 1928 as the International Association for the Study of the European Quaternary Period; the scope of the work was enlarged and the present name adopted in 1932. Congresses are held every four years and the papers relating to them are included in the *Transactions.*

669 International Association of Scientific Hydrology is a constituent association of the International Union of Geodesy and Geophysics, formed to promote the study of hydrology, to encourage and facilitate research necessitating international co-operation and to ensure the dissemination of information. Separate commissions have been set up on continental erosion, surface waters, subterranean waters, snow and glacier study. The annual *Bibliographie hydrologique* and the *Bulletin* are central publications; in addition, irregular bibliographies are prepared on specific topics and reports of research are issued from time to time.

670 International Association of Sedimentology, founded in 1952, is based on the Geological Institute, Wageningen. The *Annual reports* are valuable records and in 1959 the *Bibliographie internationale* began publication.

671 International Association of Seismology and Physics of the Earth's Interior was established in 1901 to develop studies in the economic, social and scientific aspects of seismology. Among its important published works are the *Travaux scientifiques, Bulletin mensuel* and the *International seismological summary*.

672 'The international atlas' (Philip, 1969), not to be confused with Philip's international atlas of 1931-1951, is one of the few new major world atlases to be published during the past fifty or so years; the product of co-operation between Rand McNally, George Philip, Cartographia, Esselte and Teikoku-Shoin of Tokyo, under the leadership of Russell Voison of Rand McNally. The editorial board was chosen from thirteen countries and cartographers from six countries were responsible for the maps; glossaries are in four languages, English, French, German and Spanish, as are the Preface and Table of contents. The work reflects the knowledge now available by means of new tools in mathematics, electronics, photography, radar and under-sea sounding. Five distinctive series of maps give a balanced coverage of the world, with particular attention given to areas of growing importance. Six principal scales, 1:24M, 1:12M, 1:6M, 1:3M, 1:1M, and 1:300,000 are followed through with different styles of cartography and colouring. Twenty nine pages of large scale plans enable comparison of sixty of the most important cities in the world. Hill shading is the most outstanding relief feature; all signs and symbols are standardised; placenames and physical features have been given in national spelling and every map page carries its own glossary in addition to the complete glossary preceding the index. A lavishly illustrated essay by Marvin W Mikesell, 'Patterns and imprints of mankind', deals with population, settlement, land use, urbanisation and resources and several extra factual features add to the reference value of the work. Although the atlas was so long in planning and creation, continuous revision kept all information as accurate as possible. Smyth sewn binding enables the double pages to lie flat and provides extra strength.

673 'International atlas of West Africa', prepared under the auspices of the Organisation of African Unity, Scientific, Technical and Research Commission, and with the assistance of the Ford Foundation, began its appearance in 1969-, with eight of the total intended forty-two plates. In the main, scales range between

1:10M and 1:20M, with four maps on smaller scales and a two-page spread at 1:5M, showing administrative and political boundaries. Layer colouring or shading is used to show relief and symbols are mostly those in standard use. Twenty-three pages of text are in French and English and sources used are quoted.

674 International Bank for Reconstruction and Development sponsors much high-level enquiry of interest to geographers. In particular, the economic development series, published by Johns Hopkins Press, furnishes basic surveys, frequently of the lesser-known regions of the world; typical are those for Colombia, 1950; British Guiana, 1953; Venezuela, 1961.

675 'International bibliography of vegetation maps', edited by A W Küchler (University of Kansas Libraries, Library series, 1965-1970, sponsored by the National Science Foundation), was agreed at the International Colloquium on Vegetation Mapping held at Toulouse in 1960. In four volumes, the first covers all of North America, compiled by Dr Jack McCormick and A W Küchler; the second, 1966, covers Europe; the third, 1969, the USSR, Asia and Australia; and the fourth, 1970, Africa, South America and world maps generally. Entries are arranged geographically, grouped in regions, sub-regions and countries. World maps at scales down to 1:20M are included. Detailed definitions and plan of the Bibliography are set out in the Introduction, which explains the individual entries and their arrangement of information.

676 International Cartographic Association had its origin in a meeting of cartographers from both national and commercial mapping organisations, near Stockholm, in 1956, under the sponsorship of the Swedish Esselte organisation. A small committee of representatives from France, Germany, Great Britain, Sweden, Switzerland and the United States of America was formed to examine possibilities, and in 1958 a second international conference was held, based on the house of Rand McNally and Company, Chicago. The German Cartographic Society organised a conference at Mainz in 1958, at which the name of the International Cartographic Association was adopted and an Executive Committee appointed. At the IGU Congress, Stockholm, 1960, a Special Commission on Cartography was set up to study the affiliation of the ICA to the IGU; the first General Assembly of the Association was held in Paris in 1961, the second in London in 1964, arranged to coincide with the meetings of the XX International Geographical Congress in the hope of facilitating liaison between the two organisations. The objects of the ICA are the advancement of the study of cartography and its problems, the encouragement and co-ordination of cartographic research, the exchange of ideas and documents, the furtherance of training in cartography and the dissemination of cartographic knowledge. ' Cartography ' is understood to include not only the design and study of maps but also their evaluation, and the compilation, drafting and all stages of reproduction.

Refer International yearbook of cartography, 1965, which contains a selection of papers and discussions from the 1964 technical conference.

The cartographic journal, December 1964, for an outline of the progress and objects of the ICA.

677 International Civil Aviation Organisation developed in 1947 from the provisional ICAO founded at San Francisco in 1944. At successive conferences, the convention on international civil aviation has been established. The sixth World Map Conference of the ICAO was particularly important, when proposals were made for new chart requirements and also modifications for the geographical map series. In 1960, at the Stockholm International Geographical Congress, proposals were discussed for the production of a common base map from which both the 1:1M International map of the world and the ICAO charts might be derived. The standard air charts and maps are produced at scales of 1:1M, 1:500,000 and 1:250,000. An *Aeronautical chart catalogue* was begun in 1951, presenting a world list of charts conforming to the ICAO standards, with details of price and how they may be obtained; supplements keep this catalogue up-to-date.

678 International Commission for Aeronautical Charts was founded in 1907 at the Third Congress of the Fédération Aéronautique Internationale and was responsible for the first practical air maps, in 1909, on a scale of 300,000.

679 International Commission for the Northwest Atlantic Fisheries was established during the years 1949-50, with a membership of twelve countries. The *Annual reports,* issued from 1951, trace the work of the commission. Other publications include an annual abstracting service covering statistical and technical articles, and annotated papers and newsletters from time to time. A small reference library is maintained at the secretariat at Halifax, Nova Scotia, and here are collected scientific and statistical works, research reports and maps relating to the fisheries and all documents bearing on the commission's work.

680 International Commission for the Scientific Exploration of the Mediterranean Sea was founded in 1919. *Annual reports* and other irregularly timed documents are issued reporting on the work of the commission; one of the most important publications undertaken has been the *Fiches, faune et flore de la Méditerranée*.

681 International Council for the Exploration of the Sea was established in 1902 as a result of the discussions at the First International Conference of the Study of the Sea, held in Stockholm in 1899, and the Second Conference at Oslo in 1901. A vast publishing programme is undertaken, including the *Journal du conseil, Rapports et procès-verbaux*, the *Bulletin hydrographique*, the *Bulletin statistique* and the *Annales biologiques*. A *Herring atlas*, edited by W C Hodgson, was published in 1951. The library holds some fifteen thousand volumes and a hydrographic card index is maintained.

682 International Demographic Symposium held in Zakopane in 1964 was especially important; the *Proceedings* were published by the Polish Academy of Sciences, Committee for Demographic Sciences, in English, French, German and Russian, illustrated and including drawings. The papers deal with the regularity of population development in socialist countries, with changes in population structure from several aspects; the discussions and the conclusions reached at each session are included.

683 ' International dictionary of stratigraphy ', recommended at the International Geological Congress in Mexico, 1956, was undertaken jointly by the Comité Français de Stratigraphie and the Centre National de la Recherche Scientifique. A collection of about 120 fascicules, containing contributions from some hundred specialists, covers nearly all countries, grouped according to continent—Europe, USSR, Asia, Africa, Latin America, Oceania, North America, with a final section, 'Major stratigraphic terms'. Each fascicule contains in alphabetical order the different stratigraphical terms used, description of section type and bibliographical references. Maps show locations and the whole is indexed. In addition, in each of the fascicules in the eighth section, the special significance of a term, for example, 'Pennsylvanian', in any one country, can be studied, with the description of the formations attributed to each continent; this section was published in collaboration with the Commission on Stratigraphy of the IUGS, Sub-Commission on Stratigraphic Terminology, complementing the previous series by giving a synthesis on a world scale for each major stratigraphic term.

684 The International Federation of Library Associations recognised the special interests of map librarianship in August 1969 when a new Sub-section for Geography and Map Libraries was incorporated with Dr Walter R Ristow, of the Library of Congress, as its first chairman. The compilation of a world directory of map collections was agreed as a first priority. Papers presented at meetings (alternate years) are to be published in *INSPEL*, the journal of the Special Libraries Section, edited by Dr Baer. Dr Helen Wallis is secretary of the new section.

685 International Geographical Congress, the assembly of world geographers which has been held every three or four years since the first was inaugurated in Antwerp in 1871, has been now established at four-yearly intervals. A complete list of the congresses is to be found in *Orbis geographicus* (*qv*). Since 1922, the International Geographical Union (*qv*) has provided the Executive, the organising committee of the congress, which was previously the function of each of the host countries. The details of the programmes and overall organisation are carried out by the host country; individual sections participate in lectures, symposia and study tours. English and French are now the working languages. The body of publications prepared for each congress is invaluable and *Abstracts of papers* are published, systematically arranged according to appropriate subject content and subdivided as necessary. In the 1964 volume, additional sections for political geography, teaching of geography, methodology and bibliography were added to bring together abstracts of papers submitted for symposia not covered by sections of the Congress. *Proceedings of congress meetings and symposia* have been published since 1871. Full details of congress history and procedure are to be found in *Orbis geographicus* and in the issues of the *IGU newsletter.*

686 International Geographical Union was founded in 1923 to encourage the study of geographical problems, to initiate and co-ordinate researches requiring international co-operation and to provide for their scientific discussion and publication, to provide for meetings of the International Geographical Congress and to appoint commissions for the study of special matters during the intervals between congresses. Each of the countries belonging to the union establishes a national committee formed on the initiative of the principal academy of the country or through its national research council or some similar institution or scientific society. Full details concerning the union are to be found in *Orbis geographicus* (*qv*) and in the issues of the *IGU newsletter*, published in English and French once or twice a year since 1950. Publications, mainly bibliographical, are numerous, issued from the headquarters office or from the various Commissions, such as *Bibliography 1959-1963,* issued by the Geographical Institute, University of Copenhagen under the direction of Axel Schou, 1964, for the Commission on Coastal Geomorphology, and the *Bibliography 1955-1958,* issued by the Coastal Studies Institute, Louisiana State University, for the Commission on Coastal Sedimentation, 1960, with financial assistance from the Geography Branch, Office of Naval Research and the co-operation of the Graduate School, Louisiana State University.

687 International Geological Congress, held every three or four years in different countries, was established in 1878. It is the occasion for conferences and meetings of many related bodies and much bibliographical work is produced. *Congress proceedings and papers* are published, while irregular *Circulars* make known the details of work in progress. Special commissions sponsor specific projects, such as the International geological map of the world, 1:5M, the International geological map of Europe (*see* below) and the Geological map of Africa.

688 'The international geological map of Europe' is one of the oldest examples of international scientific co-operation; the idea was first promoted at the Second International Geological Congress at Bologna in 1881, when the commission for the project was created. At the Third International Congress at Berlin in 1885, the editorial work was entrusted to the Preussische Geologische Landesanstalt in Berlin and the first edition was published between 1893 and 1913. Work began on the second edition in 1933 and was delayed by the second world war. The map will be completed in forty-nine sheets, on a scale of 1:1,500,000, the work jointly of the Bundesanstalt für Bodenforschung and UNESCO.

689 International Geophysical Year 1957-58 was the occasion for co-ordinated scientific research by many countries under the aegis of the International Union of Geodesy and Geophysics. A vast literature is accumulating on the researches and all the analyses arising from them; the IGY collection of Arctic and Antarctic research in the IGY at the National Academy of Sciences, for example, has already reached some ten thousand books, articles, reports, conference proceedings, manuscripts and memoranda. The *Annals of the IGY* have been published, 1957-, by Pergamon Press, Oxford. The IGY organisation includes an Artificial Satellite Sub-Committee.

>   Refer *The United Kingdom contribution to the International Geophysical Year, 1957-58* (Royal Society, 1957).
>
>   Werner Buedeler: *The International Geophysical Year* (Unesco, 1957).
>
>   Sir Archibald Day: *Guide to IGY World Data Centres.*
>
>   A Hayter: *The year of the Quiet Sun* (1968).
>
>   D C Martin: 'The International Geophysical Year', *in The geographical journal,* March 1958.
>
>   Walter Sullivan: *Assault on the unknown: the International Geophysical Year* (1961).

690 International Hydrographic Bureau was founded by Prince Albert I of Monaco in 1921 following the Hydrographic Conference of 1919, to establish close association between national hydrographic offices, to encourage co-ordination and uniformity in research and to ensure publication. The *Annual reports,* issued since the Bureau's inception, trace the development of its work; a *Yearbook,* published since 1928, forms a guide to the hydrographic services of the world. *The international hydrographic review,* including a continuing bibliography, has been published twice a year since 1923, and the *International hydrographic bulletin* monthly since 1928. The library collects all hydrographic reports, periodicals, standard technical and scientific works, hydrographic catalogues and index charts published throughout the world and maintains a documentation service. Of the mapping programme, one of the most important achievements has been the *General bathymetric chart of the oceans,* published in 1903 and 1912, with subsequent revisions.

Refer J R Dean: 'The International Hydrographic Bureau', *The geographical journal,* December 1963.

D W Newson: 'The General Bathymetric Chart of the Oceans —seventy years of international cartographic co-operation', *in The cartographic journal,* June 1971.

691 The International Hydrological Decade came into being in 1965, following endorsement by the Unesco General Conference two months previously; within a short time, some seventy committees had been set up throughout the world to carry out relevant research projects. A quarterly periodical, *Nature and resources,* replaced the *Unesco Arid Zone newsletter,* which finished in December 1964. This new journal, considered as the *Bulletin* of the International Hydrological Decade, covers all the different aspects of the programme of the Division of Natural Resources Research, in the fields of hydrology, geology, soil sciences, ecology, conservation of nature and arid zone research. Extracts from the resolutions adopted by the General Conference are included in the first issue, 1/2 combined issue, June 1965. In addition to articles and news items, the journal is invaluable for the information it imparts concerning national and international meetings and conferences, new research projects, United Nations and Unesco sponsored publications; the feature ' Publications received ' contains annotated references.

692 ' International journal of comparative sociology ' was founded in 1960 by the editor, K Ishwaran (Brill of Leiden), appearing at first biennially, every second issue being devoted to a special theme; since 1968, the journal has become a quarterly. Each issue contains usually at least one article of direct interest to geographers, the others providing desirable background reading.

693 ' International library of sociology and social reconstruction ', a series published by Routledge and Kegan Paul, contains many regional economic studies of direct interest to geographers, also, for example, R E Dickinson: *City, region and regionalism: geographi-*

cal contribution to human ecology, 1947, followed by his *City and region: a geographical interpretation*, 1964.

694 'International list of geographical serials', edited by C D Harris and J D Fellmann (University of Chicago, Department of Geography, research paper no 63, 1960), includes both current and retrospective titles. This work followed *A union list of geographical serials*, by C D Harris and others (second edition, 1950), and *A comprehensive checklist of serials of geographic value*, by Harris and Fellmann, 1959. Over a thousand numbered titles are listed and arranged first internationally, then by region and country. Complete holdings are noted for the Library of Congress, the American Geographical Society, the University of Chicago and the Royal Geographical Society. A revised edition now in preparation contains an additional 635 titles. Chauncy Harris has also compiled *Major foreign language geographical periodicals with English summaries of articles*.

See also *An annotated world list of selected current geographical serials in English* . . .

695 'International map of the world on the millionth scale', first proposed by Professor Albrecht Penck in 1891, became a reality during the course of successive discussions. At a meeting of the VIIth Assembly of the International Geographical Union in Lisbon, 1949, a new commission on the IMW was established and fresh terms of reference were adopted; and at the UN Technical Conference on the IMW in Bonn, 1962, the specifications were again revised. For details of the history of the map and the complete recommendations, Professor Penck's article ' Construction of a map of the world on a scale of 1:1 million ' (*Geographical journal*, 1893) should be consulted; also the *Report of the Commission on the International Map of the World 1:1,000,000* (American Geographical Society, 1952) and the *Report* of the Bonn Conference, 1962. During the early years, the Central Bureau was centred on the Royal Geographical Society, but was transferred to the Cartographic Office of the United Nations in 1953, as part of the programme for co-ordinating cartographic services. The Central Bureau has published progress reports since 1921 and *Annual reports* have been issued from 1954- by the Department of Economic and Social Affairs, New York. Bibliographies of selected official documents covering the IMW appeared with the

1954 *Report* in *World cartography,* volume IV, 1954, and a first supplement to the bibliography appeared with the *Report* for 1961. An *International bibliography of the 'Carte internationale du monde au millionième'* was compiled by E Meynen as a special publication of *Bibliotheca cartographica,* no 1, on the occasion of the Bonn Conference, 1962; this included titles of all books, papers and reports since 1891, together with an abstract of the history of the map series, a register of all published sheets and several index maps showing the different issues of the IMW and their derived and related editions. New sheets are noted in the ' Recent maps ' section of *The cartographic journal.*

Refer ' The International map of the world on the millionth scale in the field of cartography ', *World cartography,* no 13, 1955.

R A Gardiner: ' A re-appraisal of the International map of the world (IMW) on the millionth scale ', *International yearbook of cartography,* I, 1961.

G R Crone: ' The future of the International million map of the world ', *Geographical journal,* March 1962.

696 ' International marine science ', a newsletter prepared jointly by Unesco and FAO, is designed to help marine scientists, administrators and government officials to be better informed on international, regional and national activities of significance to marine science. It contains information on organisations, national oceanographic programmes, international projects, meetings, training facilities, miscellanous news items and publications.

697 International Nickel comprises The International Nickel Company of Canada, Limited, Copper Cliff and Toronto; the International Nickel Company Inc, New York; and The International Nickel Company (Mond) Limited and Henry Wiggin and Company Limited, in London. From the publicity departments of these companies issues a continuous stream of information of general and educational importance. Two regular publications are the *Nickel bulletin,* a monthly abstracting service of current information on nickel and nickel alloys; and *Inco-Mond nickel,* an illustrated journal featuring applications of nickel-containing materials. The *Annual report* is naturally a central source of information. Technical films, filmstrips, specimens, samples and photographs are available on loan, and educational publications, including many of considerable length and erudition, may be had free, covering aspects of the steel industry, non-ferrous alloys, nickel-iron alloys, cast irons, corrosion-resisting materials, nickel plating, precious metals, Mond chemical products and iron powders.

698 The International Road Federation, a non-profit service organisation established in 1948 to encourage the development and improvement of highways and highway transportation, has offices in London, Paris and Washington, co-ordinated by a joint committee. Each office holds a library and archives. Technical, economic and statistical information and documents are distributed and the Federation's own publishing programme is extensive. The *Annual reports* provide the body of current information, illustrated with photographs and maps concerning road developments in all parts of the world; the excellently produced quarterly *Road international* (*qv*) is the chief journal of the Federation, but there is also the *World highways* newsheet, 1950-, circulated monthly in English, German and Spanish; the *World directory of highway officials* has been issued annually since 1954 and *World road statistics* from 1950 in English and French. *Proceedings,* 1952- and *Reports,* 1951- are published as necessary.

699 The International Road Research Documentation Scheme, 1965-, began as a regular international exchange of abstracts of published and unpublished literature, sponsored by the OECD. The abstracts, in English, French and German, are distributed through the Road Research Laboratory, United Kingdom, the Laboratoire des Ponts et Chaussées, France and the Forschungsgesellschaft für das Strassenwesen, in the Federal Republic of Germany.

700 International Society for Photogrammetry has, since 1910, provided an important forum for the discussion of the latest developments; there are seven autonomous technical commissions and an International Congress of Photogrammetry is organised every four years under the direction of the committee of the Society. Publications, which are nearly all published in English, French and German, include *Photogrammetria,* from 1938; each volume of four quarterly parts contains the Proceedings of the Conferences, national reports, reports of the technical commissions and original papers.

701 The International Society of Soil Science, located at the Royal Tropical Institute, Amsterdam, dates back in effect to 1911, when the *Internationale Mitteilungen* was initiated; in 1924, the society was reconstituted under its present title and its journal became the *Proceedings of the International Society of Soil Science*. The society meets at intervals for discussion at various centres and its findings are published as *Congress transactions*, in English and French, usually including bibliographies.

702 International Soil Museum, joint project of Unesco and the International Society of Soil Science, was established in January 1966 under the direction of F A van Baren, Professor of Soils at the State University of Utrecht and Secretary General of the International Society of Soil Science, the first meeting of the Advisory Panel taking place at The Hague in September 1967. Contacts are maintained with FAO and with institutions and soil scientists in all parts of the world.

703 ' International studies ', quarterly journal of the Indian School of International Studies, New Delhi, published by the Asia Publishing House, usually concentrates in each issue on one country. Original articles, notes and memoranda, bibliographies and book reviews are included.

704 The International Tin Research Council, founded in 1932, with the object of developing the consumption of tin, is financed by the major tin producers of the world. With headquarters at Greenford, Middlesex, the council controls organisations for the technical development of tin in other countries, centred on the Centre d'Information de l'Etain, Brussels; Technical Service Centre for Tin, Montreal; Centre d'Information de l'Etain, Paris; Zinn-Informationsbüro, Düsseldorf; Technisch Informatie Centrum voor Tin, The Hague; Centro d'Informazioni dello Stagno, Milan and the Tin Research Institute Inc, Columbus, Ohio. The whole group is active in the documentation of tin throughout the world; research findings are published in the journals of scientific societies, in the technical and trade press, in practical handbooks and in *Tin and its uses*, quarterly journal of the Tin Research Institute. Library facilities are maintained at each centre, including reference, bibliographic, loan, translating, photocopying and abstracting services.

705 International Union for Conservation of Nature and Natural Resources was founded in 1948 to facilitate international co-operation, to promote scientific research, to disseminate information and to conduct programmes of conservation education, especially in Africa and the Middle East. The maintenance of a documentation centre is also important. A quarterly *Bulletin* and *Annual report* are published, also special technical reports from time to time.

706 International Union of Geodesy and Geophysics, founded in 1919, is a member of the International Council of Scientific Unions, its function being to promote the study of problems relative to the figure and physics of the earth, to initiate, facilitate and co-ordinate research and to provide for discussion and publication. The union consists of seven international associations: the International Association of Geodesy, the International Association of Geo-magnetism and Aeronomy, the International Association of Meteorology and Atmospheric Physics, the International Association of Oceanography, the International Association of Scientific Hydrology, the International Association of Volcanology and the International Association of Seismology and Physics of the Earth's Interior. It is thus one of the great co-ordinating world organisations and the publishing programme, both of the main body and of the constituent associations, is vast. *Procès-verbaux* of meetings have been published since 1922, in English and French; the initial quarterly *Bulletin d'information de l'UGGI* was superseded by *Chronique* from 1957, issued monthly in English, German, French and Russian, with bibliographical supplements on geodesy, hydrology and oceanography from 1958.

Refer The International Union of Geodesy and Geophysics, its scientific role, its international character and its organization (Paris, 1960).

A description of the Union and its constituent associations and commissions with an account of their scientific activities (Bureau of the IUGG, Richmond, Surrey, 1947; revised edition, 1961).

707 'International vegetation map 1:1M' is in progress under the direction of Henri Gaussen. The first sheet, of Tunisia, by Henri Gaussen and A Vernet, was completed in 1958; some sheets of India began to appear from 1961. Gaussen's system for the 1:1M map, first shown in the vegetation map in the *Atlas de France,* is being used.

708 The International Wool Secretariat, London, is a non-governmental, non-profitmaking concern, dating from 1937. The Library, established in 1947, holds books, pamphlets, periodicals, films and filmstrips relating to economics, history, science and technology of wool. Official publications and statistical bulletins from Australia, New Zealand, South Africa and Canada, in addition to other material relating to wool producing and manufacture in those countries and in the United Kingdom, India and Pakistan are received; loan is made to other libraries, but photocopies are readily supplied to individual enquirers. The classification used is a modified UDC; periodical articles are indexed and issued, arranged by UDC. Quarterly accessions lists are issued, also subject bibliographies, lists of periodical holdings, the *IWS news service,* 1946-; semi-monthly, *Wool knowledge,* the journal of wool education, 1948-, which includes bibliographic references; *Wool science review,* irregularly from 1948; *The world of wool;* an *Annual review of IWS and the Wool Bureau; World wool digest,* 1950-, semi-monthly, and numerous other handouts and visual aids.

Refer also The wool trade directory of the world (Skinner, London).

709 'International yearbook of cartography' has resulted from a most imaginative and valuable international partnership, beginning in 1961, between some of the leading cartographical agencies, first under the editorship of Professor Eduard Imhof of Zurich, in collaboration with experts from Austria, Great Britain, Italy, Germany, Sweden, the Netherlands, France and the USA, published by George Philip, together with Armand Colin, C Bertelsmann Verlag, the Esselte Map Service, Freytag-Berndt und Artaria, the Istituto Geografico de Agostini, Art. Institut Orell Füssli AG and Rand McNally. Articles are in German, French or English, with synopses in the two languages not used. A variety of topics is covered in each volume regarding the compilation, design and methodology of maps and on new productions. The reproductions of map sheets are particularly valuable. The volume for 1965 contained a selection of papers and discussions from the 1964 Technical Conference of the International Cartographic Association. There is so far no index, but each volume contains contents lists for the previous volumes.

710 'An introduction to human geography', by J H G Lebon (Hutchinson University Library, 1952; sixth edition revised, 1969), defines his purpose as 'the comparative study of major societies in the areas of their characterisation', proceeding, in the first chapter, to justify this definition 'and to examine its connections both with the ideas of earlier authorities and with modern doctrine on the nature of geography'. Further chapters are devoted to 'Climate and man'; 'The foundations of the human economy'; 'The major human societies'; and 'Civilisations in confrontation'. Bibliographies are appended to each chapter and there are a few clear sketchmaps and figures in the text. A folding chart—'Centres of origin of the chief cultivated plants' follows the text. The book is available in both hard cover and paperback editions.

711 'An introduction to quantitative analysis in economic geography', by Maurice H Yeates (McGraw-Hill, 1968), was prepared to demonstrate the use of new techniques in the study of economic geography. With the aid of graphs, figures and tables, individual theories are explained, including some regional examples. A selected bibliography is included.

712 'An introduction to the study of map projections', by J A Steers (ULP, 1948; thirteenth edition, 1962), has remained a 'standard' work for students, being not so technically presented as A R Hinks: *Map projections*. The text includes explanations of latitude and longitude, scale and the nomenclature of projections, followed by chapters on the main groups of conventional projections, interrupted and orthoapsidal projections, with final chapters on the Ordnance Survey map projection and grids, the choice of projection and notes on the history of projections. Appendices set out relevant tables and a short list of references. Explanatory diagrams and figures accompany the text.

713 'Irish forestry', journal of the Society of Irish Foresters, is published twice a year. Articles, notes, reviews and miscellaneous news items are the main features; in the Autumn issue is included a report of the Society's activities during the previous year. The

Autumn 1963 issue is especially useful for the article by T Clear: 'A review of twenty-one years of Irish forestry'.

714 'Irish geographical studies', edited by N Stephens and R E Glasscock, was produced by the Department of Geography, The Queen's University of Belfast, in honour of Professor E Estyn Evans, following his forty years of teaching, 1928-1968. Twenty-three essays representative of recent geographical research, with the geography of Ireland as a common theme, are grouped in three parts—the evolution of the physical landscape, historical aspects of the cultural landscape and contemporary geographical problems. There are twenty plates and numerous line drawings in the text.

715 Irish University Press (International) Limited, London, is to be noted by geographers for its planned series of Area Studies—USA (*qv*), China and Japan (*qv*), Russia, Central and South America, being reprints of British Parliamentary Papers, each multi-volume sets for which detailed catalogues are available. Many other regional sets have been issued, also some based on some special subject, for example, *Mining districts, Shipping, The Industrial Revolution, Factories* and *Agriculture*. T P O'Neill wrote *British Parliamentary Papers,* a monograph, in 1969. The Irish UP issues a *Bulletin,* check lists and announcements, either as leaflets or single sheets. The *Bibliotheca Hibernicana* catalogue, compiled by William Shaw Mason, originally published in 1823, is one of the most important of the reprints (1970).

716 'The Isle of Man: *a study in economic geography'*, by J W Birch, published by CUP for the University of Bristol, 1964, considers all aspects of the economy and life of the island, accompanied by photographs, sketchmaps and figures. 'The fascination of the Isle of Man is that you can study everything in miniature there—history, geography, climatology, economics, political evaluation...' can stand as the keynote of the book; there is a selected bibliography.

717 Istituto Geografico de Agostini of Novara was founded in Rome in 1901 and has become one of the great cartographic agencies of the world. Staff are trained at the institute in all the techniques of modern cartography. Atlases, road maps and special-purpose maps of all kinds are planned, constructed and printed.

Most famous is the *Grande atlante geografico* (*qv*); other fine atlases include the *Atlante mondiale,* the *Nuovo atlante geografico moderno* and many regional atlases, mostly under the direction of L Visintin. Plastic relief maps and globes are also specialities, and wall maps for school use are produced in effective new relief styles.

Refer Umberto Bonapace: ' La production cartographique de l'Institut Géographique de Agostini: buts et problèmes actuels ', *International yearbook of cartography,* 1963.

718 ' James Cook: the opening of the Pacific ', written by Basil Greenhill, Director of the National Maritime Museum (HMSO, 1970), is a most useful record of Cook's life and achievements, illustrated with photographs, maps and engravings. Illustrations include contemporary instruments and a plan of the Britannia Otaheite war canoe.

719 ' Japanese geography: *a guide to Japanese reference and research materials',* compiled by Robert B Hall and Toshio Noh (University of Michigan Press, 1956; Center for Japanese Studies, Bibliographical series, no 6), is concerned with twentieth century sources—bibliographies, encyclopedias, etc, yearbooks, collections of statistics and censuses, sets and collections, the history of Japanese geography, physical geography, historical and cultural geography, economic geography and regional descriptive geography —followed by a list of publishers with a special interest in Japanese affairs.

720 ' Japanese studies on Japan and the Far East: *a short biographical and bibliographical introduction',* prepared by Teng Ssu-Yü and others (Hong Kong University Press, 1961), includes, among other more general topics, Eastern anthropology and ethnology, economics and economic development, Far Eastern and Southeast Asian geography and geology. Items are numbered, titles are given in Japanese, usually with English translations and annotations are in English; there is an index to subject matter and authors.

Refer also Japanese geography, 1966: its recent trends (Special paper, no 1, The Association of Japanese Geographers, Tokyo).
Nippon, 1936- (Tsuneta Yano Memorial Society).
Japanese cities: a geographical approach (Special paper, no 2), The Association of Japanese Geographers, Tokyo, 1970).

721 'Jena review', the bi-monthly journal of VEB Carl Zeiss Jena, is finely produced, including excellent technical articles and notes, with illustrations, on new equipment; a work of central importance in practical geography.

722 'Jerusalem studies in geography', a new journal published 1970- by the Department of Geography, The Hebrew University of Jerusalem; the aim is to publish the results of research both in Israel and abroad. The journal is at present cyclostyled, including illustrations, maps and diagrams.

723 'Jewish history atlas', compiled under the direction of Martin Gilbert, with cartography by Arthur Banks (Weidenfeld, 1969), traces the history, world-wide migrations, achievements and life of Jewish people from the civilisation in ancient Mesopotamia to the present day. Maps portray early Jewish migrations, the Promised Land, the kingdom of David and Solomon, the destruction of Jewish independence, the rise of the Roman empire, Palestine at the time of Jesus Christ, European Jewry from 500 to the emancipation of European Jewry 1789-1918; Zionism 1860-1939; and modern movements which have affected the Jews.

724 'John Speed's atlas of Wales', Part II of 'The Theatre of Great Britaine', published in 1676, has been reproduced in facsimile, with a bibliographic introduction, by SR Publishers Limited, East Ardsley.

725 Johnston, Alexander Keith (1804-1872), a great cartographer and a scholar much in sympathy with the work of leading continental geographers, whose ideas he helped introduce into Britain, founded the firm of W and A K Johnston, which still operates in Edinburgh. His *Physical atlas* (second edition, 1856) was inspired by the *Physikalischer Atlas* of Berghaus (*qv*). Another notable atlas was the *Royal atlas*, 1859.

726 Jones, Llewellyn Rodwell (1881-1947) was one of the generation of geographers following Sir Halford Mackinder who, by their scholarship and energy, established British academic geography; his chief academic work was at the London School of Economics, where he contributed greatly to the new Joint School of Geography of King's College and the LSE between 1921 and 1945. His regional studies on Northern England, North America and East Africa have become established texts; he had a great interest in port geography and his *Geography of London River,* 1931, in particular, became a classic. *London essays in geography, the Rodwell Jones Memorial Volume,* was published in 1951.

727 'Journal of Asian and African Studies' is issued from the Department of Sociology, York University, Toronto; there are editors in twenty-four countries, including Great Britain, and a Board of scholars representing many parts of the world (Brill, Leiden). It is an interdisciplinary journal covering the whole area and is described as ' the first attempt to answer a need in the relevant field, in that it unites contributions from anthropology, sociology and related social sciences into a concerted emphasis upon building up systematic knowledge and using the knowledge derived from pure research for the reconstruction of societies entering a phase of advanced technology '.

728 'The journal of Christopher Columbus ', translated by Cecil Jane, was published by Blond in 1968, including ninety illustrations from prints and maps of the period, notes and a bibiography. Particularly informative is the chapter by R A Skelton, ' The cartography of Columbus' first voyage ', and the ' Letter of Columbus describing the results of his first voyage '; there are extensive notes on both the Journal and the letter.

729 'The journal of economic abstracts ', an international journal published quarterly by the contributing journals under the auspices of The American Economic Association, which acts as agent, has, since 1963, aimed to assist economists around the world to become acquainted with methods and conclusions currently being reported in the general economic periodicals. In most cases, abstracts are prepared by the author of the original article.

730 ' Journal of the economic and social history of the Orient ' was founded by N W Posthumus in 1957 and is now edited by Claude Cahen and W F Leemans, with the help of an international board of Orientalists. The *Journal* has proved a forum for all kinds of studies on the economic and social conditions of Asia and North Africa from the earliest times to the nineteenth century. Originally designed to supplement the thirty-volume *Economic and social history of the Orient,* the *Journal* has developed independently, published by Brill of Leiden; three issues make a volume.

731 ' **The journal of geography** ' has been published since 1902, in nine issues a year, by the National Council for Geographic Education; edited by A J Nystrom, it has been devoted primarily to the place of geography in education, but includes also substantive geographical articles.

732 ' **Journal of glaciology** ', published three times a year from 1947, contains the papers read at meetings of the Glaciological Society (*qv*), together with the discussions, articles and short notes of current interest, review articles, abstracts and a bibliography of recent glaciological work. The longer articles are preceded by abstracts in German and French.

733 ' **Journal of hydrology** ', first issued in March 1963, marked the need for a journal devoted exclusively to the subject; published quarterly by the North Holland Publishing Company, Amsterdam, its scope has from the beginning been international, including surface and ground waters, agro-hydrology, the hydrology of arid zones, hydraulic engineering, meteorology and climatology.

Another journal with the same title was begun by the Hydrological Survey, New Zealand Hydrological Society, Ministry of Works in 1968; there is also the *Annuaire hydrologique,* Québec, Ministère des Richesses naturelles du Québec, 1965— and similar works in many other countries, during the past decade.

734 ' **Journal of Latin American studies** ' (CUP, 1970) has been sponsored by the centres and institutes of Latin American studies in or attached to universities, covering the continent from the point of view of all the social sciences.

Refer also C Furtado: *Economic development of Latin America: a survey from Colonial times to the Cuban Revolution* (CUP, 1970, Cambridge Latin American studies, no 8).

735 ' **Journal of Nigerian studies** ', a monthly periodical published by the African Book Company Limited, aims to disseminate knowledge of all aspects of the life of the country by means of articles, news items, bibliographical assessments and information on collections of Nigerian literature.

Refer also R K Udo: *Geographical regions of Nigeria* (Heinemann Educational Books, 1970).

736 '**Journal of Pacific history**' (OUP, 1966-) is an annual publication, its scope extending beyond history in all its aspects to anthropology, archaeology, geography, economics and bibliographical features concerning documents and archives. It is sponsored by the Research School of Pacific Studies of the Australian National University, though not to be considered a journal of the School.

737 '**Journal of regional science**', published by the Regional Science Research Institute in co-operation with the Department of Regional Science of the Wharton School, University of Pennsylvania, in April, August and December, contains articles, usually with appended references and including maps and diagrams as necessary, also reviews and a 'Books received' feature.

738 '**Journal of South-East Asia and the Far East**', published twice a year since 1967 by the Institut de Sociologie de l'Université de Bruxelles, contains articles on a wide range of topics, with special studies in English and French on the current social, economic and political problems of India, Pakistan, Indonesia and the Philippines, China, Japan and Korea.

Refer also George Allen and Unwin, ' Books on Asian Studies' catalogue, 1971.

Barry Cohen: *Monsoon Asia: a map geography* (Heinemann, 1971).

K Ishwaran *ed*: *Contributions to Asian studies* (Brill, 1971).

739 '**The journal of transport history**', published twice yearly by Leicester University Press, 1953-, is the only journal to be devoted to the history of transport as a whole, in all its branches, ancient and modern. The scope of the articles is wide-ranging. The *Journal* is also very much concerned with the literature of the subject, a continuing feature, 'Sources of transport history' being particularly useful. In each May issue appears a list of the articles on transport in British periodicals during the preceding year. Book reviews are a regular feature, normally including some overseas publications.

740 '**Journal of tropical geography**', semi-annual from 1953, is the publication of the Departments of Geography, University of Singapore and the University of Malaya, Kuala Lumpur. Useful bibliographies are included.

741 'Journal of West Midlands regional studies', published annually since December 1967 by the Wolverhampton College of Technology, to provide information on research being done in the College and to encourage the preparation of material in a form suitable for publication. The foreword states—' The emergence of the journal is a natural development stemming from the work of interested people from many professional fields who, whether natives of the West Midlands or not, find this region a rich source of study. Originally concerned with certain aspects of West Midlands industrial archaeology, investigations led inevitably to projects about the lives and conditions of the people. The studies therefore cover fields such as economic and political history, geology, geography, sociology, as well as the scientific and technical aspects of industrial developments.' There are photographs occasionally and sketchmaps and figures in the text.

742 'Kansas geographer' is now revived following a seven-year suspension, published by the Kansas Council for Geographic Education, in co-operation with the Division of Geography at Kansas State Library.

Refer also Kansas in maps, prepared by R W Baughman for the Kansas State Historical Society (McCormick-Armstrong, 1960).

743 'Kartographie: Kartenaufnahme, Netzentwürfe, Gestaltungsmerkmale, Topographische Karten', by V Heissler and G Hake (Berlin: Walter de Gruyter, 1970) is a most comprehensive and technical book on all aspects of cartography. There are diagrams in the text and eight map sections in a cover pocket.

744 ' Kartographische Geländedarstellung ', by Eduard Imhof (Berlin: Walter de Gruyter, 1965), a major work on relief representation, is in German, but the illustrations would in any case be intelligible. The whole process of portraying the geographic landscape is considered, from the topographic, artistic and technical points of view.

745 'Kartographische Nachrichten', prepared by the Deutsch Gesellschaft für Kartographie, 1951, first quarterly then bi-monthly (Bertelsmann Verlag), contains erudite articles, reviews and bibliographies.

746 Keltie, Sir John Scott (1840-1927) had a varied career in publishing before joining the staff of the Royal Geographical Society in 1883. His life's work lay in his administration of the Society, as librarian between 1885 and 1892, then as secretary until 1915; but he is chiefly remembered for his research into geographical education on the continent and in America, which he described in the influential report of 1885 which bears his name and which led directly, in conjunction with the work of Sir Halford Mackinder, to the establishment of modern academic geography in this country and the foundation of the Oxford School of Geography.

Refer ' Geographical education. Report to the Council of the Royal Geographical Society ' (RGS *Proceedings,* Supplementary Papers, volume 1, part IV, 1882-1885).

747 Kendrew, Wilfred George (1885-1962) was an outstanding influence in British geography, especially at Oxford, and a pioneer in climate geography, as director of the Radcliffe Meteorological Station and subsequently reader in climatology. His *Climates of the continents* (*qv*) in its successive editions became known throughout the world; *Climate,* 1930, which later became *Climatology,* 1949, went into a second edition in 1957; *Weather: an introductory meteorology for airmen,* 1942, and *The climate of central Canada,* with B W Currie, 1955, are among his other most interesting works, in addition to innumerable contributions to books and periodicals.

748 ' Kinship and geographical mobility ', edited by Ralph Piddington (Leiden: Brill, 1965), in the ' International studies in sociology and social anthropology ' series, examines the effects of migration, urbanisation, industrialisation and acculturation on a system of closely-knit kinship bonds outside the ' primitive ' or ' folk ' societies with which social anthropologists were in the past largely concerned. Experts have as yet not reached agreement and further observations over a long period will be necessary.

749 Kinvig, R H (1893-1969), a much travelled geographer, will be remembered especially for his great achievement in founding the Department of Geography at the University of Birmingham, which established a Chair of Geography in 1948. His interests lay in human geography, in the development of geography and in maps. In 1930, he published an essay on the North West Midlands and encouraged local study in the department, particularly land use and physical planning. He also studied the Atlantic side of Britain and wrote a *History of the Isle of Man;* also, works on Eastern Europe and Latin America.The Geographical Society of the Department was named the Kinvig Geographical Society in 1958, as a mark of esteem, on his retirement.

750 Köppen, Vladimir Peter (1846-1940) developed an early interest in climate and vegetation. His first version of the world climatic regions, in 1900, was based largely on plants as indicators; in 1918, this was modified to make it freer from botanical geography. He edited the great *Handbuch der Klimatologie* and wrote many papers, also the climatic section of *Die Klimat der Geologischen Vorzeit* with his son-in-law, Alfred Wegener. The Köppen climatic regions were used by A J Herbertson and J G Bartholomew in the *Atlas of meteorology (qv).*

751 'Kultura' *see* Cartactual.

752 'Kulturgeografisk atlas', compiled by J Humlum (Copenhagen: Gyldendal, 1955), is a re-issue in two volumes of Humlum's well-known atlas of economic geography. In this publication, with maps and text separated, the length of text has been increased, also the number and variety of the maps. All maps are in red and black and one of the features of the work is the skill with which symbols have been used to such effect; notes on the map plates are in Danish, German, French and English. The maps on population, crops, mineral production and trade are especially well conceived. The text volume is in Danish, lavishly illustrated by photographs and diagrams. The whole work emphasises the trends in world affairs, especially in production and trade.

753 Laborde, Edward Dalrymple (1909-1962) exerted a great influence on the teaching of geography in British public schools, especially through his work at Harrow. He edited many texts and, in particular, his translations made a number of French geographical works available in English, notably Georges Jorré: *The Soviet Union: the land and its people* and A V Perpillou: *Human geography.*

754 Lafreri, Antonio is noted particularly for the collections of maps he is known to have made up into 'atlas' form, as required by his customers. For one such collection, about 1570, he engraved an elaborate title-page, using the figure of Atlas, and this is the first occasion known when this figure was used. Later, it was adapted by Rumold Mercator and thereafter gave the name to such a collection of maps brought together according to some unifying principle.

755 'Land and water', by Patrick Thornhill (Methuen, 1968), forms the first of a series of transparency atlases intended for use with a particular overhead projector, either as base maps or as a means of highlighting specific data. The maps are made of durable plastic, enclosed in a binder. A political map is included and the base map on which the whole map is built. The land maps contain data such as relief features, rock formations, glaciation and drainage; the Polar regions and oceans are treated separately.

756 'Land evaluation' was the collective title given to the thirty-two papers of a CSIRO Symposium organised in co-operation with Unesco, at Canberra, August 1968; they were edited by G A Stewart, Chief of the CSIRO Division of Land Research (Macmillan, 1969). The papers include some general comments on land evaluation, by G A Stewart, followed by papers on the ' Principles of land classification ', ' Land evaluation reviews and case studies ', ' Data handling and interpretation ', ' Land parameters ' and ' Sensors for land parameters ', illustrated in half-tone and line.

757 'Land from the air: *a photographic geography',* by G Dury and J A Morris, published by Harrap, has the intention of explaining the reading and interpretation of geographical aerial photographs showing a wide variety of topographical types, illustrated with twenty-one half-tone plates and a number of figures in the text. Scope is limited to the British Isles and each photograph demonstrates a specific theme—'A Welsh mining valley ', ' Orchard and marsh in North Kent ', among others. A section 'Books for further reading ' is appended.

758 'Land of the 500 million: *a geography of China',* by George B Cressey (McGraw-Hill series in geography, 1955) has remained a recommended book for background reading, dealing with the people, physical features, especially climate, crops and resources, commerce and industry, followed by regional assessments and a chapter on ' China's prospects '. There are photographs, sketchmaps and figures throughout the text and an extensive bibliography, grouped under subject headings.

Refer also G B Cressey: *China's geographic foundations* and *Asia's lands and peoples.*

759 '**Land use and resources:** *studies in applied geography*' was created by the Institute of British Geographers in 1968 as a memorial volume to Sir Dudley Stamp, forming the first of the Institute's series of *Special publications.* Chapters were contributed by a number of leading geographers in this country and abroad, including ' The man and his work ', which contains a bibliography of his publications; ' The land of Britain '; ' The developing world '; and ' Sir Dudley Stamp and his life and times ', by M J Wise.

760 '**Land use information:** *a critical survey of US statistics, including possibilities for greater uniformity*' was prepared by Marion Clawson and Charles L Stewart and published by The Johns Hopkins Press and OUP. The title is self-explanatory; the authors point out that United States data have developed each in isolation and the purposes of this work are to present information and ideas and to develop a general overall system for handling land-use data, suggesting the establishment of an organisation that would assist in tabulating, cross-referencing and collating the gathered data.

761 '**Land utilisation survey of Great Britain**' was initiated in 1930 by L Dudley Stamp. The field work was carried out mainly between 1931 and 1933 by teams of voluntary workers, every acre of land in England, Wales, Scotland, the Isle of Man and the Channel Islands being recorded on six inch Ordnance Survey maps. The results were reduced to the scale of one inch to the mile and published in 150 sheets for England and Wales and the more populous parts of Scotland. The work was organised on a county basis and the findings were eventually co-ordinated and published in a series of *Reports,* one for each county, under the title of *The land of Britain.* The whole undertaking was summarised by Dudley Stamp in *The land of Britain: its use and misuse* (Longmans, 1948); a shorter summary had been issued for the British Council in the previous year, with the title *The land of Britain and how it is used.* This monumental work represented Britain's land use at the time of the second world war. A *Second Land Use Survey of Britain* on the scale of 1:25,000 was begun in 1960 by the Isle of

Thanet Geographical Association under the direction of Alice Coleman of King's College, London. Full details are to be found in the *Land Use Survey handbook,* by Alice Coleman and K R A Maggs and in ' The first twenty-four published land use maps ', a special issue of *Panorama* (Isle of Thanet Geographical Association, 1964).

Refer Alice Coleman: ' The Second Land Use Survey: progress and prospect ', *Geographical journal,* June 1961.

H C Darby: ' Domesday Book—the first land utilization survey ' *in The geographical magazine,* March 1970.

762 ' Landforms and life: *short studies on topographical maps ',* by C C Carter (Christophers, 1931, reprinted 1933, 1936, 1938), made a great impression on geographical thought when it first appeared, a pioneer work and the forerunner of many similar texts. In the introduction, the author could still say, what is now taken for granted —'The power of the topographical map, as an essential tool, is being increasingly recognised . . .' The text is divided into parts, each including several sub-sections: ' The map '; ' The surface expression of denudation '; ' The surface expression of structure '; ' Land-forms and life '; ' Coast forms '; ' Coast life ', giving examples taken from the relevant parts of Britain. Throughout the text are photographs, sketchmaps and diagrams, showing some surprisingly ' modern ' concepts, such as the effect of industrial conurbations on natural life.

Refer also Roy Millward and Adrian Robinson: ' Landscapes of Britain ' series (Macmillan, 1971-).

763 ' Landscape ', published three times a year from 1951 in Santa Fe, New Mexico, covers all aspects of landscape forms—cultural geography, human ecology, urban geography, planning and conservation. It is well illustrated; there are articles, brief notes and comments, and book reviews.

764 ' Landscape studies: *an introduction to geomorphology ',* by K E Sawyer (Arnold, 1970), deals with the study of the processes of landscape formation and precise quantitative measurement in field work, discussing actual land forms with the aid of specially adapted Ordnance Survey map sections and of photographs and line diagrams. The standard of the work has been designed for students in college of education or first year university.

765 'The large scale county maps of the British Isles 1596-1850: *a union list'*, compiled in the Map Section of the Bodleian Library by Elizabeth M Rodger (The Library, 1960), includes all printed county maps issued separately during the period specified on a scale between half an inch and three inches to one mile, with a few exceptions. Arrangement of entries is first by county alphabetically, within the four divisions, England, Wales, Scotland and Ireland, then by date of publication of the first edition, each entry bearing a running number. Details include the existence of index maps or gazetteers. An introduction explains the approach to the work and the characteristics of each map are given in the entries.

766 'Larousse agricole: *encyclopédie de l'agriculture'* is a monumental work prepared under the direction of Raymond Braconnier and Jacques Glandard, consisting of entries of varying length on all aspects of agriculture, domestic plants and animals, forestry, modern agricultural techniques and a comparison of French agriculture with that in other countries. More than a thousand photographs, diagrams and figures illustrate the text and there are fifty-six separate plates, many in colour.

767 'Larousse encyclopedia of the earth', translated by R Bradshaw and M M Owen from *La terre, nôtre planète,* was published by Paul Hamlyn in 1961, with a second edition, 1965. The *Larousse encyclopedia of world geography* (*qv*) was adapted from *Géographie universelle Larousse* (*qv*). Both are lavishly illustrated.

768 'Larousse encyclopedia of world geography', adapted from *Géographie universelle Larousse,* edited by Pierre Deffontaines, was first published in three volumes in France by Augé, Gillon, Hollier-Larousse, Moreau et Cie (Librairie Larousse, Paris) and Paul Hamlyn, 1964; The Western Publishing Company published the work in one volume in 1965. Arrangement of the text is by regional grouping of countries and there are illustrations and maps throughout.

769 'Late Tudor and early Stuart geography 1583-1650: *a sequel to Tudor geography, 1485-1583'*, by E G R Taylor (Methuen, 1934; New York: Octagon Press, 1968), comprises a 'Bibliography of English geographical literature 1583-1650', first arranged under authors' names, with running numbers, then grouped chronologically, prefaced by explanatory chapters setting out the geographical features of the period under review—'Richard Hakluyt the younger'; 'The first edition of the *Principal voyages,* 1589'; 'The second edition of the *Principal voyages,* 1598-1600'; 'Regional geography, or chorography: 1583-1625'; 'Samuel Purchas: 1612-26'; 'Mathematical geography, navigation and surveying'; 'The realm of nature'; 'Economic geography; 1625-50'; 'The urbane traveller', and 'Colonial geography'. Some reproductions are included and the text is rich in quotations and notes.

770 'Latin America: *an introduction to modern books in English concerning the countries of Latin America'* was prepared by The Hispanic and Luzo-Brazilian Councils in 1960, with a second edition, revised, by the Library Association in 1966. The later edition incorporates works published in the previous five years, placing greater emphasis on economic and social conditions and omitting guide books and children's books. A section on 'General books' is followed by the body of the work, which deals with the literature on individual countries, arranged alphabetically. Brief annotations are added to the entries as necessary and there is an author index. A useful list of British-Latin American organisations in London is appended.

Refer also George Pendle: *South America, a reader's guide* (National Book League, 1957).

771 'The Latin American markets: *a descriptive and statistical survey of 30 markets made up of almost 173 million people'* was compiled by J Walter Thompson Company and published by McGraw-Hill in 1956, with maps and a bibliography. For most of the countries information is provided under the following headings: the land and its climate; population and characteristics; mineral and vegetation sources; energy and power; agriculture, livestock and fisheries; manufacturing; transport and communications; imports and exports; income and standard of living; market classification by administrative division.

772 'Latin American Publications Fund handbook' was first published in London under the auspices of the Fund, in 1969. Papers include 'Cities in a changing Latin America: two studies of urban growth in the development of Mexico and Venezuela' and 'Urbanization and economic development in Mexico', by David J Fox;

and 'The city as centre of change in modern Venezuela', by D J Robinson.

773 'Laxton: the last English open field village', by J D Chambers (HMSO, 1964) is a well-produced brochure illustrated with reproductions, photographs, figures and sketchmaps. The text deals with the village, farming system, the Manorial Court and other aspects of the Manor, of interest to historical geographers and those interested in agricultural or settlement geography; an Appendix sets out 'Rules and regulations for the grazing of the open fields, Laxton Estate in the County of Nottingham'.

774 Le Play, Frédéric (1806-1882) exercised a continuing influence on British geography, largely through Professor Herbertson, Sir Patrick Geddes and other geographers who contributed to the work of the Le Play Society. His insistence on the importance of fieldwork and the relation of communities to environment led the way to the acceptance of this aspect of geographical studies and to the founding of such organisations as The Geographical Field Group.

Refer M Z Brooke: *Le Play, engineer and social scientist: the life and work of Frédéric Le Play* (Longman, 1970).

775 Le Play Society was founded in 1930 by Sir Patrick Geddes, Professor C B Fawcett and Miss M Tatton for the promotion of international studies involving field work in geography and geology by groups visiting various countries. Pioneer work was accomplished before the Society ceased in 1960. The studies of 'place, work, folk' had followed closely the methods of the French sociologist, P G F Le Play and linked the work of geographers, historians and sociologists.

Refer S H Beaver: 'The Le Play Society and field work' *in Geography*, 1962, part 3.

776 Lebon, Dr J H G (1909-1969) was an outstanding geographer who, during his academic career, passed on his inspiration to generations of students in this country and abroad. His chief interests lay in physical, particularly in climatological, geography and in regional work in the Middle East and Sudan, but he was learned and experienced in many facets of the subject and stressed always the essential unity of geographical studies. He wrote *An introduction to human geography*, 1952, which made new contributions to the subject, also many articles and *Land use in the Sudan,* no 4 in the *World Land Use Survey;* and at the time of his death he had almost completed a new geography of the Middle East. A bibliography of his writings is included in the *Transactions of the Institute of British Geographers,* November 1970.

777 'Lehrbuch der Allgemeinen Geographie' is a comprehensive survey of world geography, published by Walter de Gruyter, Berlin, in ten volumes, of which the first appeared in 1961. Unlike the *Géographie universelle (qv),* the approach is thematic, the plan of the complete work being as follows: Herbert Louis: *Allgemeine Geomorphologie;* Joachim Blüthgen: *Allgemeine Klimageographie;* Fritz Wilhelm: *Allgemeine Hydrogeographie;* Josef Schmithüsen: *Allgemeine Vegetationsgeographie;* Hans Bobek: *Allgemeine Sozial-und Bevölkerungsgeographie;* Gabriele Schwarz: *Allgemeine Siedlungsgeographie;* Erich Obst: *Allgemeine Wirtschafts-und Verkehrsgeographie;* Martin Schwind: *Geographie der Staaten;* Josef Schmithüsen: *Allgemeine Landschaftskunde;* Eduard Imhof and others: *Karte und Luftbild: als Arbeitsmittel des Geographen.* Individual volumes have already run into more than one edition.

778 'Leicestershire landscapes: *case studies in local geography for the CSE pupil',* prepared by the Leicestershire Association for Local Geographical Studies (Blond Educational, 1970), is a particularly successful, well illustrated example of many such studies that have recently appeared, demonstrating the renewed interest of geographers in field work and local geography. Thirteen contributors consider the characteristics of the Soar Valley, Charnwood Forest, the City of Leicester, Market Harborough, East and West Leicestershire, the Vale of Belvoir and farming in the county, with a foreword by Professor J F Kerr.

779 'Leland's itinerary in England' was reprinted in 1964 from the version based on Hearne's second edition (1744 and 1745) with a foreword by Thomas Kendrick (Centaur Press Limited, five volumes), edited by Lucy Toulmin Smith first in 1907, then in 1964. The reprint is now available in paperback.

780 'The library atlas', by Harold Fullard and H C Darby (Philip, 1938; eighth edition, 1965) is in two parts, general and economic. The general section, on scales ranging from $1:2\frac{1}{2}$M to $1:8$M, was published previously as *The university atlas*. Thirty-two pages of economic maps and diagrams include distribution maps of major products throughout the world; diagrams and maps of world occupations, transport, imports and exports; economic development and production, land use and agriculture, minerals and industry for Great Britain and for other selected areas. Climatic graphs are included for more than two hundred representative stations throughout the world. The index comprises over fifty thousand entries, conforming in general with the rules of the Permanent Committee on Geographical Names and the US Board on Geographic Names.

781 'The Library catalogue of the School of Oriental and African Studies' was printed by G K Hall in twenty-eight volumes, 1963; entries dealing with the whole of Asia, Oceania and Africa are available in the complete set or in fourteen sections. The author index and title index are followed by the subject index—General, Africa, Middle East, South Asia, South-East Asia and Pacific Islands and the Far East. Manuscripts and microfilms are included, also periodicals and serials; there are also entries in Chinese and Japanese.

782 'Library guide for Brazilian studies', compiled by William Vernon Jackson (Pittsburgh: University of Pittsburgh Book Center, 1964), is a guide to the major United States research collections of Brazilian materials; the holdings of seventy-four libraries are described and evaluated. Summary chapters analyse particular collections as a whole and sketch co-operative projects for collecting current materials. Appendices form important parts of the book, comprising an annotated list of ninety-eight Brazilian periodicals in the humanities and social sciences, and a breakdown of the Library of Congress classification numbers for Brazilian history and literature. Emphasis throughout is on materials for advanced study and research.

Note Most of the country's bibliographic activities emanate from the Instituto Brasileiro de Bibliografia e Documentaçao. Subject bibliographies have been issued on agriculture, botany, zoology and social sciences.

783 Library of Congress Map Division aims to keep at least one copy of each edition of every map, atlas, globe and other form of cartographic publication with any reference value. The basic collection to date amounts to some three million maps and more than twenty thousand atlases, and the collections are growing at the rate of some ninety thousand to a hundred thousand maps and up to fifteen hundred atlases a year. A *Bibliography of cartography* has been maintained, of which microfilm copies are available. The library has issued printed cards for maps and atlases from the beginning of the card printing programme; a more comprehensive cataloguing of maps began in 1946, though still on a selective basis. Cards for maps and atlases printed before 1953 were included in the *Library of Congress author catalog;* from 1950, entries were included also in the *Subject catalog*. With the re-organisation of the catalogue in 1953, a separate part, *Maps and atlases,* was issued and this policy continued until 1955 (1956), when entries were again included in the main sequences, *Books: authors* and *Books: subjects*. Bibliographical publications of central interest to geographers are frequently published, notably the *List of geographical atlases* . . . below-mentioned.

784 'Lighthouse of the skies: *The Smithsonian Astrophysical Observatory: background and history 1846-1955',* written by Bessie Zaban Jones and published by David and Charles, presents an analysis of its contributions in the study of solar radiation, long-range weather forecasting, etc, including a list of publications issued by the Institution.

See also The Smithsonian Institution . . .

785 The Linmap system, the first commercial British computer mapping system, was originally developed by Computation Research and Development for the Ministry of Housing and Local Government; in its version 1 form it has the ability to process up to forty-two data items for each map produced and a capacity of four thousand points of information per map. These capacities will be increased as the scope of the system is enlarged. Linmap has been designed to accept magnetic tape input and to work for the Co-ordinate Reference System of data identification which uses Ordnance Survey National Grid Co-ordinates. 'Colmap' is a colour map system based on Linmap. Symap (*qv*) has been developed at Harvard University and Mapit is another programme for the production of flow maps, dot maps and graduated symbols; and a new design by CalComp is a high-speed, high-resolution computer-output-microfilm system.

Refer to recent issues of *The cartographic journal,* especially that for December 1971; the Oxford system of automatic cartography (*qv*) and, from recent monographs, G R P Lawrence: *Cartographic methods* (Methuen, 1971), which includes notes on computer-produced maps.

786 Linton, David Leslie (1906-1971), a notable scholar and teacher, was particularly interested in geomorphology. In association with S W Wooldridge, he wrote *Structure and surface drainage in South-East England;* and he passed on his knowledge and inspiration to generations of students in numerous addresses, papers and edited works. He was Chairman of the British Geomorphological Research Group, editor of *Zeitschrift für Geomorphologie* and, for some years, editor of *Geography.*

787 'A list of American doctoral dissertations on Africa' is compiled and published by the Library of Congress. A comparable work published by SCOLMA is *Theses on Africa accepted by universities in the United Kingdom and Ireland,* kept up to date by annual volumes of *United Kingdom publications and theses on Africa.*

See also next item, 788.

788 'List of French doctoral dissertations on Africa, 1884-1961', compiled by Marion Dinstel, former librarian of the African Document Center, Boston University, was sponsored by the Boston University Libraries and reproduced in 1966 by G K Hall. The work is arranged by country and/or area, and alphabetically within each area by author; each title is numbered and there is an author index, also a limited subject index. Forty African states are represented, as well as material on the continent in general, amounting in all to nearly three thousand entries.

789 'List of geographical atlases in the Library of Congress', the only extensive bibliography of atlases and an indispensable reference work, is especially useful to the student of early cartography. The basic work was in four volumes, 1909-1920, by Philip Lee Phillips; a fifth volume, with bibliographical notes, was compiled by Clara Egli LeGear, 1958, containing references to over two thousand world atlases acquired by the library between 1920 and 1955; a further supplement by the same compiler, 1963, on the same plan, contained also some eight hundred oriental atlases and a valuable list of the contents of Yousuf Kamal's *Monumenta cartographica Africae et Aegypti*. A seventh volume comprises a consolidated author list and the eighth volume an index. Author lists and topographical lists are also included in the first four volumes. Theatrum Orbis Terrarum Limited published a reprint in 1970, with bibliographical notes.

790 'A list of maps of America in the Library of Congress *preceded by a list of works relating to cartography'*, compiled by Philip L Phillips, published in Washington, 1901, was reprinted in 1967 by Theatrum Orbis Terrarum Limited. About fifteen thousand maps of America are arranged chronologically, then according to subject and the bibliography on cartography lists some 1,200 books on mapmaking and its history.

791 ' Die Literatur über die Polar-Regionen der Erde bis 1875 ', prepared by J Chavanne, A Karpf and F Le Monnier (1878), comprises more than six thousand items, including works on adjacent regions; it was reprinted in 1962 by N Israel, Meridian and Theatrum Orbis Terrarum Limited.

792 ' Lithofacies maps: *an atlas of the United States and Southern Canada'* was the work of graduate students at Northwestern University engaged in regional stratigraphic studies under the editorship of L L Sloss, E C Dapples and W C Krumbein (Wiley, 1960). Together with the accompanying text, it presents a unique source of information concerning the thickness and lithology of the sedimentary rocks of the USA.

793 ' Liverpool essays in geography: *a jubilee collection'*, edited by Robert W Steel and Richard Lawton (Longmans, 1967), contains contributions from 'over thirty authors who, between them, span successive generations of Liverpool-trained geographers from the 1920s to the 1960s '; a festschrift, not to a single scholar, but to a Department of Geography. Each author has written on his own speciality, within the unity which shows the compass of geographical studies at Liverpool University. There are maps and figures throughout the text and appendices present the staff of the Department of Geography at Liverpool from 1902 and graduates of the department; but there is no index.

794 'Locational analysis in human geography', by Peter Haggett (Arnold, 1965), was based on a series of lectures given at Cambridge. Part I, 'Models of locational structure', discusses the geographically significant aspects of the main theoretical models—both classical (Thünen, Weber, Christaller and Lösch) and modern (Isard, Garrison, Berry and Hägerstrand), attempting to weld them around the theme of the geometrical symmetry detectable within regional systems. Part II, 'Methods in locational analysis', includes a particularly interesting section on 'Region-building'. There are numerous diagrams throughout the text, references and a bibliography.

795 'London 2000', by Peter Hall (Faber, 1963; second edition, 1969, 1971), is, to quote the author, 'an exercise in academic polemic'. In the second edition, the text remained unchanged, but postscripts were added to each chapter setting out such facts as were necessary to bring the content up to date. The five main parts: 'Posing the problem', 'Guiding growth', 'Building the new London', 'Running the new London', 'Living in it' are subdivided into chapters, and an appendix to chapter VI is 'Twenty-five new towns for London' arranged alphabetically by name and giving the basic facts concerning each. Sketchmaps and photographs illustrate the text.

796 Longmans, Green and Company Limited, founded in 1724, has published many serious monographs and bibliographical works in geographical subjects, such as the UNESCO *source book for geography teaching* (qv), *Longman's dictionary of geography* (qv) and *A glossary of geographical terms* (qv). The *Geographies for advanced study* series, edited by Professor S H Beaver, is a reliable series of monographs on individual countries or regions, including, for example, Georges Jorré: *The Soviet Union: the land and its people;* F J Monkhouse: *Western Europe;* R J Harrison Church: *West Africa;* C O'Dell: *The Scandinavian world;* and thematic studies such as B W Sparks: *Geomorphology,* all well produced and illustrated by photographs and sketch-maps. For school work, there are *Europe in maps,* 1969; the *Concepts in geography* series, 1969-; *Longman's geography paperbacks; Sketchmap geographies; First lessons in human geography;* and *New geography in the classroom,* edited by Rex Walford. Longman's loops in geography, produced by Halas and Batchelor Animation Limited, are edited under the supervision of H J Clarke and W J Allen; a new series of 8 mm film loops uses colour and animation to elucidate and interpret some of the more complex topics in meteorology, edited by Dr R G Barry.

797 'Longman's dictionary of geography', 1966, edited by L Dudley Stamp, brings together in one alphabetical sequence information on societies and their awards, brief explanations of geographical terms and physical features and notes on countries and cities of the world, commodities and international trade. Biographical entries are quoted from the *Dictionary of national biography*. A 'Selected bibliography of geographical books published in Britain' is included, based on one prepared by the editor for the National Book League in 1964, with later additions, intended as a guide in the setting up of a school or college library; books published in North America are also included, but limited to those in the English language.

798 'The lost villages of England', by Maurice Beresford (Lutterworth Press, 1954), remains a classic work, well documented, including a county index to lost village sites mentioned in the text, notes and a list of works frequently cited. There are pertinent sketchmaps and illustrations.

799 'The Lower Swansea Valley project', edited by K J Hilton (Longmans, 1967), presents a comprehensive survey of the area made by members of the University College of Swansea, with support from the Nuffield Foundation; it provides also an excellent example of a recent case study of an area, isolating the factors that have been inhibiting the social, physical and economic use of the land and estimating its potential. The text, as it stands, including several sketchmaps and diagrams, is based on twelve reports—'The human ecology of the Lower Swansea Valley', by Margaret Stacey; 'Report on transportation and physical planning in the Lower Swansea Valley', by R D Worrall; 'Report on the hydrology of the Lower Swansea Valley', by D C Ledger; 'Report on the geology of the Lower Swansea Valley', by W F G Cardy; 'The soil mechanics and foundation engineering survey of the Lower Swansea Valley project area', by H G Clapham and others; 'The prospects for industrial use of the Lower Swansea Valley: a study of land use in a regional context', by Susanne H Spence; 'Lower Swansea Valley: housing report' and 'Lower Swansea Valley: open space report', by Margaret Stacey; 'Plant ecology of the Lower Swansea

Valley'; 'Soil biology of the Lower Swansea Valley', by P D Gadgil; 'Afforestation of the Lower Swansea Valley', by B R Salter and 'Tips and tip working in the Lower Swansea Valley', by G Holt.

800 Lundqvist, Gösta, who died in 1968, was one of Sweden's foremost cartographers, working especially with the Esselte Company in Stockholm and the Generalstabens Litografiska Anstalt. Many original atlases and map series owe their existence to him. In 1962, he became head of all the map production departments in Esselte. His wide travel enabled his reference atlases, wall maps and tourist maps to be really practical and helpful as well as of cartographical excellence.

801 Lyell, Sir Charles (1797-1875) became the first professor of geology at King's College, London, where the Lyell Club is named after him. His influence on geological thought in academic circles was very great, also in the Geological Society, of which he was president 1835 to 1837 and 1848 to 1850; his publications have perhaps been even more influential. The *Principles of geology* was published in 1830, having a second edition in 1832; a third edition, in four volumes, followed in 1834, of which the fourth volume, rewritten as *Elements of geology,* 1838, went into six editions and was followed by *The student's elements of geology.*

Refer Sir Edward Bailey: *Charles Lyell,* 1962. F. J. North: *Sir Charles Lyell: interpreter of the principles of geology, 1965.*

L G Wilson, *editor: Sir Charles Lyell's scientific journals on the species question* (1970, one of the Yale *Studies in the history of science and medicine).*

802 ' McGraw-Hill Encyclopedia of Russia and the Soviet Union ', edited by Michael T Florinsky and others (Donat Publishing Company, 1961), contains a mass of useful information on the principal developments in Russia before and after the revolution of 1917, presented in alphabetical order; some entries are single or two-line or longer, for example 'Agriculture '. There are maps and diagrams throughout the text, also small black and white portraits. References are given at the ends of articles.

803 ' McGraw-Hill international atlas ' was based on *Der Grosse Bertelsmann Weltatlas (qv)* edited, in English, French and German, under the direction of Dr W Bormann, 1964 (dated 1965). Information gained during the IGY researches is incorporated, as well as other recent developments in roads, railways, water and power projects.

804 McGraw-Hill Publishing Company Limited is particularly interesting to geographers for the *McGraw-Hill series in geography*, which includes such standard works as V C Finch and others: *Elements of geography, Physical elements of geography* and *The earth and its resources;* O W Freeman and H F Raup: *Essentials of geography;* G B Cressey: *Asia's lands and peoples, Land of the 500 million: a geography of China* and *China's geographic foundations;* Erwin Raisz: *General cartography* and *Principles of cartography;* G T Trewartha: *An introduction to climate. The McGraw-Hill illustrated world geography,* edited by Frank Debenham and W A Burns, was published in 1960. *See* also *The McGraw-Hill international atlas,* and the *McGraw-Hill encyclopedia of Russia* . . . Other series include the ' Biological sciences ', ' Population biology ', 'Agricultural sciences ', 'American forestry ', ' International development ' and a ' Field guide series '. A complete catalogue is issued annually, with information sheets and handouts as necessary. The House of Grant publications were acquired a few years ago.

805 ' Mackenzie Delta bibliography ', edited by Mary Jane Jones, was sponsored by the Northern Science Research Group, Department of Indian Affairs and Northern Development (Queen's Printer, 1969). The bibliography brings together references to monographs and articles showing the academic and other interest in the project since 1964.

806 Mackinder, Sir Halford John (1861-1947) is notable particularly for his achievements in establishing academic geography in England. In many ways, he was foremost among British scholar-geographers, exerting the greatest influence not only on those aspects of the subject in which he was specially interested, such as regional and political geography, but on succeeding generations of geographers. His academic work was centred on Reading, London and Oxford, but his theories and geographical beliefs reached a wider public through his papers to the Royal Geographical Society, his lectures given throughout the country and his British

Association address of 1905. Among the highlights of his career was his address to the 1895 International Geographical Congress and his presidential address to the Geographical Association on the theme 'Modern geography: German and English', which led to the foundation of the first School of Geography in Oxford in 1889, with himself as director and A J Herbertson as his second. Two papers to the Royal Geographical Society which achieved far-reaching results were 'The scope and methods of geography' and 'The geographical pivot of history', reprinted separately by the society in 1951, with an introduction by E W Gilbert. Among his published books, *Britain and the British seas* (*qv*) is probably the most remarkable.

> *Refer* E W Gilbert: 'The Right Honourable Sir Halford J Mackinder, PC 1861-1947', *Geographical journal*, March 1961, which includes a bibliography.
>
> E W Gilbert and W H Parker: 'Mackinder's democratic ideals and reality after fifty years', *in The geographical journal*, June 1969.
>
> J F Unstead: 'H J Mackinder and the new geography', *in The geographical journal*, June 1949.

807 Macmillan and Company Limited, London, issues monographs, booklets and visual aid materials of great use to geography students and teachers. The *Landscapes of Britain* series, for example, 1970-, beginning with *The South-West peninsula, The West Midlands* and *South-East England,* describes and discusses in each volume the physical, economic and historical elements of the landscape of the area, lavishly illustrated with photographs, maps and diagrams. Frequent catalogues are issued of both books and non-book materials. Since the overhead projector became universally accepted as a teaching aid, Macmillan, with many other publishers, have been increasingly preparing transparencies; the geographical series cover the world, with an emphasis on the British Isles. Macmillan Educational include geography as one of their lines—new titles being *Topics in geography, London* and *Transport* by Oswald Hull, *Towns,* by M Turner, *Food, clothing and shelter,* by L W Steven, *A map book of the Benelux countries,* by A J B Tussler and A J L Alden and *Beginning practical geography,* by Arthur Bray.

808 Mair Geographical Publishing House, Stuttgart, founded in 1950, specialises in road maps; also, since 1960, a number of road atlases of European countries has been issued. The Deutsche Generalkarte 1:200,000 was published between 1952 and 1957, in a map series of uniform size, based on a new survey. The *Grosse Shell-Atlas of Germany and Europe*, 1960-, revised annually, aims at supplying, for any given need, the proper map at the most convenient scale. Arranged in systematic order, the maps are at scales of 1:4,500,000, 1:1,500,000, 1:500,000 and 1:200,000, in addition to a number of maps at larger scales. Brief comments below the map faces provide compact sources of information.

Refer Volkmar Mair: ' Strassenkarten aus Mairs Geographischen Verlag ', *International yearbook of cartography.* 1963.

809 ' The major seaports of the United Kingdom ', by James Bird (Hutchinson, 1963), describes the history, layout and development of the chief groups of British ports, illustrated by fine aerial photographs, sketchmaps and diagrammatic plans. Following an introduction, ' Industrial and commercial estuaries ' includes Newcastle, Glasgow and Clydeside, Belfast, Hull and other Humber ports, Southampton and Southampton Water and the Bristol complex. South Wales ports are considered separately, especially Swansea; ' Packet ports ', Dover, Harwich and Holyhead; ' The inland port ', such as Manchester; ' General cargo giants ', Liverpool, Merseyside, the Thames and the Port of London. A final section is headed ' Ports in perspective ', dealing with port hinterlands and the port of the future; an appendix comments on the Report of the Rochdale Committee. Notes and references are given at the end of each section and there is a short list of ' selected further references ' published since 1953.

810 ' The makers of modern geography ', by Robert E Dickinson (Routledge and Kegan Paul, 1969), was an endeavour ' to trace the development of modern geography as an organised body of knowledge in the light of the works of its foremost German and French contributors ', with a first chapter entitled ' From Strabo to Kant '. Leading names are Alexander von Humboldt, Carl Ritter, Friedrich Ratzel, Ferdinand von Richthofen, Albrecht Penck, Alfred Hettner, Otto Schlüter, Frédéric le Play and Vidal de la Blache.

811 ' Man and his world ', the Noranda Lectures, Expo 67 (Toronto University Press, 1968), with an introduction by Helen S Hogg, comprise essays by scholars from many countries. ' The development of earth science ', by Dr J Tuzo Wilson, is of special interest to geographers, illustrated with photographs, sketchmaps and diagrams; references are included.

812 ' Man and the land: a cultural geography ', by George F Carter (Holt, Rinehart and Winston, 1964; second edition, 1968), was conceived as an introduction to geography, especially to the question why people in different lands live differently from one another. Beginning with ' The origin of man and the problem of race ', the author continues to explore the characteristics of peoples living in arid environments, wet tropics, Mediterranean settings, forests and grasslands, mountains and polar regions, concluding with ' The role of physical environment and culture '. There are plenty of photographs, diagrams and maps in the text and suggestions for further reading at the end of each part.

813 ' Man and nature: *or physical geography as modified by human action'*, by George Perkins Marsh, originally published in 1864, is an acknowledged classic, forward-looking work, republished from the first edition by the Belknap Press of Harvard University Press, 1965, edited by David Lowenthal (second printing, 1967) for The John Harvard Library series. Referred to by Lewis Mumford as ' the fountainhead of the conservation movement ', Marsh attempted to reveal the menace of man's waste of the earth's resources, showing how man has changed the earth, suggesting means of conservation and reform. The second chapter deals with the ' Transfer, modification, and extirpation of vegetable and of animal species ', the third, fourth and fifth successively with ' The woods ', ' The waters ', ' The sands ' and ending in chapter six with ' Projected or possible geographical changes by man '. The style is discursive, deeply sincere and revealing an intensive study of the subject as well as powerful vision.

814 ' Man and society in Iran ', by A Reza Arasteh, in collaboration with Josephine Arasteh, first published in 1964, was reprinted, with some corrections, in 1970. The authors' main preoccupation in this study was to evolve a comprehensive frame of reference with which to analyse man and culture in Iran, which might also serve as a model for the study of other countries. To suit the various topics, a variety of methods have been used, essentially those of historical, statistical, descriptive and speculative analysis. The three main aspects chosen were 'Man in traditional Iranian society'; 'Man in contemporary Iranian society' and 'A measure for the future', each being subdivided.

815 'Man and wildlife', by Dr C A W Guggisberg (Evans, 1970), an account of the relationship between man and animals, examines the progressive destruction that man the hunter, the farmer and the industrialist have wrought on his environment. Photographs, many in colour, are included, also maps. There is a comprehensive survey of national parks throughout the world.

816 'Manitoba historical atlas: *a selection of facsimile maps, plans and sketches from 1612 to 1969'*, edited, with introduction and annotations, by John Warkentin and Richard Ruggles (The Historical and Scientific Society of Manitoba, Winnipeg, 1970), was prepared for the centenary of the province. The first part includes exploration and mapping 1612 to 1800 and the imaginative mapping of Manitoba to 1731, followed by the first stage of scientific mapping, to 1681, and the second stage from 1681 to 1731 and onwards in greater detail to the French mapping and the English maps of the remaining years of the century. Part II deals with early European settlement and scientific exploration 1801 to 1869, early settlement maps and scientific surveys. In part III, geographical patterns begin to appear, 1870 to 1969, land surveys, topography and hydrography representation, land settlement and rural settlement patterns, village, town and city plans, the portrayal of resources, cultural features and economic and population changes. Reproductions and maps are in black and white, using appropriate techniques; there are facing notes throughout, extensive text between the topics and a bibliography.

817 'Man's rôle in changing the face of the earth: *an international symposium'* under the Chairmanship of C O Sauer, Marston Bates and Lewis Mumford (University of Chicago Press for the Wenner-Gren Foundation for Anthropological Research and the National Science Foundation, 1956), in two volumes, edited by W L Thomas and others, is a monumental presentation of all aspects of conservation and regional planning from antiquity to modern concepts, illustrated with maps in the text. There is a bibliography. The work is prefaced by an essay contributed by E A Gutkind: 'Our world from the air: conflict and adaptation'. Some of the main themes concern 'Man's tenure of the earth', 'Through the corridors of time', 'Man's efforts on the seas and waters of the land', 'Alterations of climatic elements', 'Slope and soil changes through human use', 'Modifications of biotic communities', 'Ecology of wastes', 'Urban-industrial demands upon the land', 'Limits of man and the earth' and 'The rôle of man'. The symposium discussions and summary remarks are included and many interesting photographs, sketchmaps and figures add force to the text.

818 'Manual of map reproduction techniques', by E D Baldock, published by the Surveys and Mapping Branch, Mines and Technical Surveys, Ottawa, describes in methodical detail the procedures that make it possible to print any number of copies of maps, each sharply precise and delicately coloured. Illustrated with photographs, diagrams and map specimens, the manual considers 'Drawing techniques', 'Negative scribing', 'Subsequent colour tints', 'Positive scribing', 'Photo colour proof', 'Map editing' and 'Lithography'.

819 'Map collections in the United States and Canada: a directory', edited by Marie C Goodman (Special Libraries Association, Geography and Map Division, 1954), includes 527 collections, excepting the following, because of their comprehensiveness: the American Geographical Society, the University of California at Los Angeles, Canada Department of Mines and Technical Surveys, Geographical Branch; the University of Chicago; Harvard College Library; New York Public Library; the Free Library of Philadelphia; Stanford University; Yale University and the US government departments. Arrangement is by state. There is an appendix listing United States Government repositories and a folding map is attached to the inside back cover.

820 Map Collectors' Circle was formed in 1963 to stimulate interest in and publish material on early maps and atlases and on cartographers and map publishers. The *Map collectors'* series, edited by R V Tooley, is issued on a present programme of ten works in every two years; these are available only to subscribers in the first instance, some of the more comprehensive works being subsequently reproduced in book form. Each volume contains an introductory essay, carto-bibliographical descriptions and full-page or double-spread plates of maps, many of which have not previously been reproduced. While every work is of specific interest, two in particular are of outstanding importance. *County atlases of the British Isles, 1579-1850: a new bibliography*, by R A Skelton, in co-operation with the staff of the British Museum Map Room (part I, no 9, part II, no 14), includes full descriptions of every county map in atlases, beginning with Christopher Saxton, and the complete set will largely supersede Chubb's monumental work. *A dictionary of mapmakers*, by R V Tooley, is also being issued in parts, beginning with no 16, 'A to Callan'; the concept of the work is comprehensive, including cartographers, engravers, geographers and publishers from the earliest times to 1900. Reports of meetings are included in *The cartographic journal*.

821 ' **Maps and air photographs** ', by G C Dickinson (Arnold, 1969), is divided into four sections: ' Thinking about maps ' considers their development in all parts of the world, the evolution of national map series, including the Ordnance Survey; ' Working with maps ' examines all the practical aspects of the making of maps; ' Looking at maps ' describes the interpretation of various landscapes from maps and 'Air photographs ' completes the text, which is generously illustrated by photographs, sketchmaps and diagrams.

822 ' **Maps and charts published in America before 1800:** *a bibliography* ', compiled by James Clements Wheat and Christian F Brun (Yale University Press, 1969), a major contribution to the bibliography of early American cartography, describes 915 maps, from the crude wood-cut map of New England, ' White Hills ', 1677 to 1799, by which time forty-one maps had been recorded. The contents are arranged geographically with the world maps first, followed by those of the continents and oceans, with a natural emphasis on America. Eighteen reduced facsimiles of the more important maps are included and there is a comprehensive bibliography.

823 ' **Maps and diagrams:** *their compilation and construction* ', by F J Monkhouse and H R Wilkinson (Methuen, 1952, in a ' University paperback ', 1963; third edition, 1971; *Methuen's advanced geographies* series), is a comprehensive summary of the materials and techniques employed in plotting data relating to relief, climate, economics and population. An appendix by R G Barry deals with numerical and mechanical techniques.

824 'Maps and map-makers', by R V Tooley (Batsford, 1949; third edition, 1971), in hardback and paperback, is one of the most comprehensive studies, in English, of the development of map-making from the earliest times to the mid-nineteenth century, excellently illustrated with reproductions. The second edition incorporated additions to the chapter bibliographies, and a new chapter on the maps of Scandinavia. In the third edition, a new bibliography was prepared, the lists of authorities at the ends of chapters was extended and numerous minor corrections and additions were made.

825 'Maps and their makers: *an introduction to the history of cartography'*, by G R Crone (Hutchinson University Library, 1953; second revised edition, 1962, reprinted 1964; third edition, in hard cover and paper-back simultaneously, 1966; fourth edition, 1970), provides a most useful succinct account. Lists of general works on cartography and of reproductions of early maps and charts are given in appendices. Considerable re-arrangement and additions of material were made in the fourth edition.

826 'Maps and survey', a classic work by Arthur R Hinks (CUP, 1913, 1923, 1933, 1942, 1944, 1947), ' was designed as an introduction to the study of maps and the processes of survey by which they are made '. Beginning with 'A brief history of early maps ', accompanied by some reproductions, the modern map in all its complexity is discussed and explained, including British official maps and international maps, maps of the countries of Europe and other foreign maps. The various kinds of survey are explained, with photographs of instruments used and other diagrams in the text, ending with a section of additions and corrections.

827 'Maps for books and theses', by A G Hodgkiss (Newton Abbot: David and Charles, 1970), with illustrations by the author, concerns those maps which are composed to illustrate and to amplify the written or spoken word. A chapter on ' Cartographic illustration ' opens the text, followed by ' Map compilation'; ' Drawing instruments and equipment '; ' Lettering '; ' The visual presentation of statistical data '; ' Map design and layout '; ' Specialised maps ' and ' Reproducing illustrative maps '. Suggestions for further reading are included and appendices deal with the effects of reduction, scale changing equipment, book and paper sizes and copyright.

828 '**Maps of Costa Rica:** *an annotated bibliography*', by Albert E Palmerlee (Lawrence: University of Kansas Libraries, 1965), is one of the University of Kansas Publications, Library series, no 19. The compiler has attempted to be comprehensive, including also maps found in books and other publications. In five major divisions— ' General maps of the whole of Costa Rica ', ' Subject maps of Costa Rica ', such as climate, agriculture, industry and population, ' Provincial, cantonal and district maps ', ' Regional maps ' and ' City plans ', whole sets of topographical maps have been incorporated, such as those of the Instituto Geográfico Nacional 1:25,000 and 1:50,000, and the Aeronautical Charts 1:1M, prepared by the United States Hydrographic Office and the United States Aeronautical Chart and Information Service. References stress the work by Luis Dobles Segreda: *Lista de mapas, parciales o totales de Costa Rica* (San José, 1928) and Jorge A Lines: *Bibliografía aborigen de Costa Rica* (San José, 1944).

829 '**Marco Polo: Venetian adventurer**', by Henry H Hart (University of Oklahoma Press, 1967), is the successor and, in a sense, a revised edition of *Venetian adventurer* (Stanford University Press, 1942) which went into several editions in English, translations into Spanish, German and Polish and a paperback edition. Excerpts from the narrative itself have been taken from the variorum edition prepared in English and from the version of Giovanni Batista Ramusio, also from the edition in French and Latin published by the Société de Géographie, Paris, 1824. A substantial bibliography represents works used in the course of preparing this volume, there are notes and references throughout the text and 'A note on Ramusio, translator and annotator of the first important collection of accounts of discovery and exploration . . .'. Several reproductions taken from manuscripts and early maps have been included.

830 '**Marine cartography in Britain:** *a history of the sea chart to 1855*', by A H W Robinson (Leicester University Press, 1962), was sponsored by The Royal Society, The Pilgrim Trust and The Crompton Bequest. A major section of the book is devoted to a group of reproductions, with notes on each. The early manuscript charts of the sixteenth century and the sea atlases are noted, followed by a section on ' The development of an accurate marine survey technique ', tracing the improvement of instruments, official hydrography, the private chart publisher and on to the Survey of the

British Isles. Appendix A contains useful biographical notes on some of the sixteenth century surveyors and chart-makers; appendix B lists 'Manuscript charts of the Tudor cartographers' and C is a listing of 'The charts of Greville Collins and John Adair'; D, 'Seventeenth century charts'; E, 'Surveys and charts of Murdoch Mackenzie (Senior)'; F, 'Charts and surveys of the "amateur" hydrographers'; G, 'Other published charts and manuscript surveys of the eighteenth century'—ending with detailed lists of the charts and surveys of Admiralty surveyors.

Refer also I D Kember: 'Some distinctive features of marine cartography', *in The cartographical journal,* June 1971.

G P Britton: *Marine meteorology and oceanography* (University of Reading, 1969).

C H Andregg: 'The scientific art of mapping and charting', *in Surveying and mapping,* June 1969.

V T Miscoski: 'Marine charting today', *in Surveying and mapping,* June 1969.

A Thunberg: 'Hydrographic surveying and data processing', *in The cartographic journal,* December 1968.

831 'Marine climatic atlas of the world' *see under Atlas of meteorology.*

832 'Marine geophysical researches', produced quarterly from January 1970 by the Reidal Publishing Company, Dordrecht, an international journal for the study of the earth beneath the sea, will be published mainly in English. Papers in French will also be accepted if accompanied by an abstract and an extensive summary in English; the text figures and captions are bilingual. Included will be original research papers, research notes and preliminary papers, papers on instrumentation and methods, letters and concise reports. Work is based largely on the Vening Meinesz Laboratorium, University of Utrecht.

833 'The mariner's mirrour' of Lucas Jansz Waghenaer was reprinted by Theatrum Orbis Terrarum, 1966, with an introduction by R A Skelton; the maps are of particular interest as some of the earliest copper engravings to be done in England.

Refer G R Crone: 'The Mariner's Mirrour, 1588', *in The geographical journal,* December 1955.

834 'Maritime history', April 1971-, is published twice a year by David and Charles of Newton Abbot, Devon. Its emphasis is on international merchant shipping and subjects such as ports, naval architecture, maritime law and insurance. Illustrated articles, based on current research, will include many from overseas; and reviews, notes, news on research in progress, and a current bibliography complete the text.

835 'Marketing and management: *a world register of organisations'*, in English, French and German, is edited by I G Anderson in co-operation with the Institute of Marketing (CBD Research Limited, Beckenham, Kent). Indexed are the publications of organisations, national and international, devoted to the dissemination of information on marketing in all parts of the world.

836 Markham, Sir Clements R is known chiefly for his *General sketch of the history of Persia* (Longmans, 1874) and the *Report on the Geographical Department of the India Office 1867-1877*.

Refer A H Markham: *Life of Sir Clements R Markham,* 1917, which includes a list of his publications.

Donovan Williams: ' Clements Robert Markham and the Geographical Department of the India Office 1867-77', *in The geographical journal,* September 1968, with a portrait.

837 Marsh, George Perkins (1801-1882) exerted great influence on geographical thought in the United States, notably through his *Man and nature: or physical geography as modified by human action,* 1864 (*qv*).

838 Maury, Matthew Fontaine (1806-1873) inspired the international study of oceanography, notably at the Brussels conference he originated in 1853 to establish a co-ordinated plan for the observation of winds and ocean currents. Maury's *Wind and current charts* and *Whale charts* showed considerable advance on previous work. He produced the first systematic textbook of oceanographic physics, *Physical geography of the sea and its meteorology,* 1855 (reprinted under the editorship of John Leighly, by Belknap Press, 1963). His contributions to geography are revealed also by his voluminous official and semi-official correspondence with scientists throughout the world which is in the archives of the US Naval Observatory.

Refer G E R Deacon: 'Matthew Fontaine Maury' *in USN journal of the Royal Naval Scientific Service,* May 1964.
F L Williams: *Matthew Fontaine Maury, scientist of the sea* (Rutgers UP, 1963).

839 Mawson Institute of Antarctic Research was inaugurated by the University of Adelaide in 1961 to foster polar studies and research. The collections include the library of Sir Douglas Mawson, also his equipment and specimens.
Refer Paquita Mawson: *Mawson of the Antarctic: the life of Sir Douglas Mawson* (Longmans, 1964).

840 ' Mélanges de géographie, physique, humaine, économique, appliquée ', was prepared as a festschrift to Professor M Omer Tulippe, edited by José A Sporck (Editions J Ducolot, 1967). In two volumes, it includes a bibliography of the works of Professor Tulippe and a survey of his career. The essays, by scholars of several nationalities, are on many aspects of geography, chiefly in English, French, German and Russian, with some photographs, maps and diagrams in the text.

841 ' Mémoires et documents ', an invaluable series begun in 1949 by the Centre de Documentation Cartographique et Géographique, Centre Nationale de la Recherche Scientifique, Paris. Each issue is devoted to a particular area: volume I, 1949, Canada; II, 1951, Belgium; III, 1952, British Isles and so on. Section A consists of high level articles; sections B and C are invaluable bibliographies, broadly classified as necessary, ' Documentation cartographique ' and ' Documentation bibliographique '.

842 ' Memoirs of hydrography, *including brief biographies of the principal officers who have served in HM Surveying Service between the years 1750 and 1885 ',* compiled by Commander L S Dawson RN, is in two parts; part I, 1750 to 1830; part II, 1830 to 1885 (Eastbourne: Henry W Keay, The ' Imperial Library ', 1885). The work was issued in facsimile reprint by the Cornmarket Press, London, 1969. Quoting from the preface to this reprint—' These Memoirs originally published in 1885 have until recently been the only comprehensive work on the activities of the oldest department of the Admiralty, now the Ministry of Defence, and of the naval surveying service acting under its direction. In recent years a continuation of the Memoirs has been increasingly in need and in the early 1950s it was agreed that it should be an official undertaking. Partly owing to the magnitude of the task by reason of the large mass of original material buried in the department's records as well as at the Public Record Office, no author was then found. The next step was the employment of Messrs G B Stigant and L A Luff, both in retirement after a lifetime in the department to index and digest the department's letter and minute books which date back to 1804. Then in 1961 Vice-Admiral Sir Archibald Day was invited to continue the history from 1885 to the present day. It was decided for various reasons to start at 1795 when the department was formed and to end at 1919 . . . With the then available digest of departmental papers and after a search at the Public Records Office, it was sensible to recapitulate Dawson's work up to 1885, particularly as to the activities of the department from its inception; there being at the time no reprint of Dawson in prospect . . . a research worker in the subject must still turn to Dawson for his immensely detailed account of the worldwide surveys undertaken . . .' Details of Commander Dawson's career are given, followed by a list of errata; the books are part of a series of fifty being reprinted in facsimile from originals in the National Maritime Museum Library on naval history, naval strategy, biography, history of the mercantile marine, voyages and hydrography.

Refer also ' The Admiralty Hydrographic Service 1795-1919 ' (HMSO, 1967).

Admiral G S Ritchie: *The Admiralty chart (qv).*

M Chriss and G R Hayes: *An introduction to charts and their use* (third edition, 1964).

Per Collinder: *History of marine navigation,* 1954.

843 ' Men and meridians: *the history of surveying and mapping in Canada',* by Don W Thomson, is in two volumes, ' Prior to 1867 ' and ' 1867 to 1917 ' (Queen's Printer, Ottawa, 1966 and 1967), sponsored by the Department of Mines and Technical Surveys, Ottawa and published in English and French editions. It is the first comprehensive, non-technical story of the surveyors of Canadian land and waters and of the maps and charts resulting from their work. A bibliography and reference notes end the work and there are also endpaper maps as well as numerous diagrammatic maps and reproductions throughout the text.

844 'Mendelssohn's South African bibliography' is in two volumes, being the 'Catalogue raisonné of the Mendelssohn Library of works relating to South Africa . . .' with a descriptive introduction by I D Colvin (Kegan Paul, Trench, Trubner, 1910). The references, annotated, include periodical articles from periodical literature throughout the world, there is a section on the cartography of South Africa and twenty-six reproduced engravings.

845 'Mercantile marine atlas', first published in 1904 by George Philip and Son and constantly revised since, presents full information on the ocean highways of the world from the point of view of trade and commerce; it is not intended for navigational purposes, and the land areas are of secondary importance. Large-scale charts, on the Mercator projection, tables of distances between ports and much other information, in addition to 135 port plans, make this a unique and invaluable atlas. The index comprises some twenty-five thousand entries.

846 Mercator, Gerard (1512-1594) was a globe and instrument-maker, at first in association with Gemma Frisius, who in the course of his life made great advances in geographical concepts and in all aspects of map design, especially in the construction of atlases. He was a cartographer of great integrity, his maps being compiled and drawn by himself and his family after much careful study. His first world map of 1538 was widely copied or used as a base by Ortelius and others. Another world map on the projection which has since borne his name, was published in 1569. This projection, on which every compass direction is shown as a straight line, was that most used by navigators from the seventeenth century onwards. He produced an edition of Ptolemy's *Geography* and made plans himself for a complete *Geographia*. His concept of geography was threefold: ancient, Ptolemaic and modern, and the *Geographia* was intended to show this progression, but only a few parts of the modern atlas were completed before his death. The parts of the 'Gallia' and 'Germania', however, printed in 1585, and of 'Italia, Slavonia et Graecia', in 1589/90, pointed the way for the modern concept of regional and national atlases. **Rumold Mercator,** who died in 1600, continued his father's work. In 1595, he published a collection of maps embodying his father's genius, the *Atlas sive cosmographicae meditationes de fabrica mundi et fabricati figura*. Mercator's introduction of the italic script for map lettering was not so well known until the publication of *Mercator: a monograph on the lettering of maps, etc., in the sixteenth century in the Netherlands with a facsimile and translation of Ghim's Vita Mercatoris*. Dr A S Osley explains how this work came about (Faber, 1969); the facsimile is of the first edition of Mercator's *Literarum Latinarum . . . scribendi ratio,* 1540, from the only complete copy, with an English translation by Dr Osley, followed by his translation of Walter Ghim's biography of 1595. *Gerhard Mercator und die Geographen unter seinen Nachkommen* (Gotha, 1914) was reproduced in facsimile by N Israel, Meridian and Theatrum Orbis Terrarum, 1969.

> *Refer* J van Raemdonck: *Gérard Mercator: sa vie et ses oeuvres,* 1869. F van Ortroy: *Bibliographie sommaire de l'oeuvre mercatorienne,* 1920.
>
> E G R Taylor: ' Gerard Mercator: AD 1512-1594' *in The geographical journal,* June 1962.

847 Meridian Publishing Company, Amsterdam, reproduce almost entirely early out of print works of topographical, bibliographical or cartographical interest, frequently in collaboration with Theatrum Orbis Terrarum and N Israel. Notable are *The Cambrian bibliography* (*qv*); W Englemann: *Bibliotheca geographica,* in two volumes, 1857; E Favenc: *The history of Australian exploration* . . ., 1888; L Gallois: *Les Géographes Allemands de la Renaissance,* 1890; and A Neubauer: *La géographie du Talmud,* 1868. General catalogues are issued usually twice a year and frequent advance notices are circulated.

848 'Metallogenic map of Europe', 1969-1971 compiled by the Drafting Committee on the Metallogenic Map of Europe, under the auspices of the Sub-Commission for the Metallogenic Map of the World of the Commission for the Geological Map of the World, has been co-ordinated by Professor Pierre Laffitte of the Ecole nationale supérieure des mines, Paris, and published jointly by Unesco and the Bureau de Recherches géologiques et minières, Paris. Nine sheets have been printed in thirty to thirty-six colours, with legends in French; and nine lists of deposits printed separately refer to each of the sheets. An explanatory brochure of about fifty pages is available in English and French (1969) and *The metallogeny of Europe* (1971) is a companion volume. Foremost European specialists in the geology of mineral deposits have combined their efforts to produce this map, which represents a big step forward in metallogenic studies, especially as regards the methodology of prospecting. It offers a variety and a wealth of information such as has never yet been assembled on a map at this scale, *ie* 1:2,500,000.

849 '**Meteorite research:** *Proceedings of the International Symposium on Meteorite Research, Vienna, Austria, 7-13 August 1968*', edited by Peter M Millman, was published by the D Reidel Publishing Company, Amsterdam, 1969. The symposium was sponsored by Unesco, the IAU, the International Union of Geological Sciences, the International Association of Geochemistry and Cosmochemistry, the Meteoritical Society and the International Atomic Energy Agency. The subjects presented covered a wide field—early history of meteorites, their composition and structure, isotope studies and chronology and ' orbits '. The discussions are also included.

850 Methuen and Company Limited, London, has a special line of monographs of geographical interest, including T W Freeman: *Ireland* . . .; E G Bowen: *Wales* . . .; Monica Cole: *South Africa;* and revised editions of such thought-provoking works as M I Newbigin: *Southern Europe* . . . in a third edition, 1949. More recently have appeared R J Chorley and Peter Haggett, *editors*: *Models in geography* (*qv*); and by the same team, *Frontiers in geographical teaching* (*qv*); Paul Wagret: *Polderlands;* B W Hodder: *Economic development in the Tropics;* R G Barry and R J Chorley: *Atmosphere, weather and climate.* Particularly useful is the entire *Methuen's advanced geographies* series, which includes both regional and systematic titles.

851 '**Mexico atlas;** *atlas géographique et physique de la Nouvelle-Espagne*', comprises the maps, diagrams and sketches contained in the three works on Mexico written by Alexander von Humboldt between 1811 and 1834, reproduced by Hanna Beck and Wilhelm Bonacker. There are twenty-eight folding and other plates, thirty four pages of editor's introduction in German and ninety-two pages of author's introduction in German and French—issued to mark the bicentenary of Von Humboldt. Particular attention was given in the work to the existing conditions and the potentialities of the Mexican state, using population and production statistics of its constituent administrative divisions to make comparisons between the regions.

852 'Meyers Neuer Geographischen Handatlas', prepared by the Mannheim Bibliographisches Institut (Harrap, 1967), comprises seventy-one terrestrial maps, with a few pages of astronomical charts and photographs. Scales range from 1:1M to physical and political maps of the world on 1:80M. Contours and layer tinting are used, relief being emphasised by hill shading; the variety of landforms and hydrological features is stressed, including sand fields, lava fields, coral reefs and glaciers. Maps of the oceans incorporate new research completed during the International Geophysical Year. Some maps tend to look rather crowded, especially in areas of dense population, where more complex cultural features are indicated.

853 Michelin: André Michelin wrote his first travel guide book in 1900 and together with the first Michelin road map for car travellers, in 1913, began the tourist service department of the Michelin Tyre Company, Services de Tourisme du Pneu Michelin. About seven million maps are now published each year and are among the most widely used maps in the world. The Michelin road map at 1:200,000 is now completely revised; there is also the road map at 1:1M and maps of various parts of Africa are a new feature. The *Guides rouges* and *Guides verts* contain many maps of topography and roads, together with excellent city plans. For the 1970 edition of the French Guide, composition was mechanised on 'Monotype' machines and the style was rationalised. The text is now divided into six sections under the following headings: general information and places of interest, tourist offices, distances to principal cities, hotels and restaurants (fixed information), hotels and restaurants (variable information) and garages.

> *Refer* ' Les services de tourisme du Pneu Michelin: histoire et évolution des publications cartographiques Michelin ', *International yearbook of cartography, 1963*.

854 'The Middle East: *a physical, social and regional geography*', by W B Fisher (Methuen, 1950; sixth edition, revised, 1971), aims ' to present a factual and reasoned statement on all the principal elements, physical and human, that influence environment, ways of life and development within the area . . .'. The pattern of the work is therefore straightforward—' Physical geography of the Middle East ', ' Social geography of the Middle East ', ' Regional geography of the Middle East', ending with Appendices, 'Outline of the geological history of the Middle East'; 'Temperature distribution in the atmosphere'; 'The origin of the Mesopotamian plains'; 'The racial and cultural affinities of the people of Israel'; 'Glossary of geographical terms, mainly of Arabic origin'; and a bibliography arranged according to the subjects of the chapters. Sketchmaps and diagrams, though not numerous, are helpfully placed throughout the text and there are folding maps at the end.

Refer also C A O Van Nieuwenhuijze: *Sociology of the Middle East: a stocktaking and interpretation,* the first volume in the series *Social, economic and political studies of the Middle East* (Brill, 1971).

855 'A million years of man', by Richard Carrington (Weidenfeld and Nicolson), presents a non-technical account of the evolution of man from his animal ancestors to present-day intellectual and spiritual achievements, viewed throughout against the background of his environments. In five parts, illustrated in half-tone, the topics stressed include the geological history of the earth and the origin and progress of life forms, man-like creatures, *Homo sapiens,* the contributions of individual civilisations and the human 'adventure' in relation to the whole of nature.

856 'Models in geography: *the second Madingley Lectures for 1965',* edited by R J Chorley and Peter Haggett (Methuen, 1967), comprises seventeen contributions on the theme of geographical generalisation, or model-building.

Refer also R J Chorley and Peter Haggett, *ed*: *Integrated models in geography* (Part II of *Models in geography* in university paperback—*Physical and information models in geography* (Methuen paperback 1969).

857 'Modern Egypt: *a list of references to material in the New York Public Library',* compiled by I A Pratt under the direction of Dr Richard Gottheil (The Library, 1929), was reprinted by Kraus in 1969. The work contains references to bibliographies, periodicals and society publications, followed by sections divided under broad subject headings, such as 'Census and vital statistics' and 'Geography'; essential bibliographical information is given on each, with a brief annotation when desirable.

858 'The modern encyclopaedia of Australia and New Zealand', in one volume (Sydney: Horwitz-Grahame, 1964), contains brief, unsigned articles ranging from a few lines in length to one or two columns; maps, coloured plates and drawings are included. A concise chronology precedes the main alphabetical section; a 'Quick reference section', tables and statistics and a list of abbreviations follow. There is no index.

859 'Modern France: *a social and economic geography'*, by I B Thompson (Butterworth, 1970), presents France as a nation in a state of rapid change. The book incorporates the most recent data available, working on the 1968 census of population, and among its most interesting features is a synthesis of development problems on the basis of the offical planning regions. The text, illustrated by seventy-four maps and diagrams, runs as follows: 'Patterns of social development'; 'Human resources'; 'The geography of population'; 'Rural settlement'; 'Urban development'; 'Patterns of economic activity'; 'State and regional economic planning'; 'The production of energy'; 'Transport'; 'Agriculture'; 'Manufacturing'; 'Regional disparities in economic and social development'.

860 'Modern geographers: *an outline of progress in geography since AD 1800,* by G R Crone (Royal Geographical Society, 1951, reissued 1960, new and enlarged edition, 1970), is a pamphlet reproducing, with revision, seven articles originally contributed to *The geographical magazine* between 1949 and 1951, together with a final chapter on Isaiah Bowman and American geography and some references for further reading.

861 'The Moon', by Zdeněk Kopal (Dordrecht: D Reidel Publishing Company, 1969), attempts to cover almost the entire field of lunar studies, each section being accompanied by bibliographical notes—'Motion of the Moon and dynamics of the Earth-Moon system'; 'Internal constitution of the human globe'; 'Topography of the Moon'; 'Lunar radiation and surface structure'. An outline of the exploration of the Moon is included and there are coloured skeleton maps of near and far sides of the moon at the scale of 1:6M.

Note Reidel publish also *The Moon: an international journal of lunar studies.*

862 'Morskoy atlas' is a major work from the point of view of both oceanography and cartography, compiled in three volumes under the editorship of I S Isakou and produced by the Soviet Naval Cartographic and Hydrographic Staff and the Chief Administration for Geodesy and Cartography in 1950. The Soviet Admiralty and Naval General Staff contributed largely to the work, also the geographers and the administration of the North Sea Route and the Soviet Arctic Institute. General maps of the oceans are on a scale of 1:50M, smaller areas on scales of 1:2M to 1:100,000; port plans are included except for Russian ports. Coast and shore features reveal particularly meticulous cartography.

863 'The mountain geologist', quarterly publication of the Rocky Mountain Association of Geologists, began in 1964. Scope is confined to the Rocky Mountain region and adjoining areas and the emphasis is on the short papers containing original material not normally published in national journals.

864 'The mountain world', published bi-annually by the Swiss Foundation for Alpine Research, Zurich, is translated into English by Michael Barnes and made available in Britain by Allen and Unwin Limited. Descriptions of new discoveries, expeditions and scientific research on mountains throughout the world are magnificently illustrated.

865 'Mountains and rivers of India', edited by Dr B C Law, with a foreword by Professor S P Chatterjee, was published by the National Committee for Geography, Calcutta, 1968, on the occasion of the 21st International Geographical Congress. The text, interesting in itself, brings together much information hitherto unavailable, including also references to the mountains and rivers of early Indian literature.

866 'National atlas of Britain' has been the subject of discussion and proposals for many years. In 1935, a committee appointed by the Council of the Town Planning Institute recommended a Commission to direct a national survey of the natural and economic resources of the country, and the National Atlas Committee of the British Association also submitted proposals in 1939. In 1941, an advisory committee was established under the chairmanship of the director-general of the Ordnance Survey, as a result of which two map offices were set up, in London and Edinburgh, under the aegis of the Ministry of Local Government and Planning, with the Department of Health for Scotland, to examine the mass of official data to be represented on successive sheets of such a national survey. Large-scale maps, maintained in continuous revision, are available for official use. From these manuscript maps a national series has been compiled on a scale of 1:625,000 by the Ministry of Housing and Local Government, with other departments and research organisations, and published by the Ordnance Survey, from whom the current list of published sheets and prices may be obtained. Each map is in two sheets, Scotland and north England, and the rest of England and Wales. Explanatory texts are available for the following sheets: land classification; average annual rainfall; population; limestone; vegetation, the grasslands of England and Wales; local accessibility; vegetation, reconnaissance of the Survey of Scotland.

Refer S W E Vince, revised by W A Payne: ' Towards a national atlas ', in *Government information and the research worker*, edited by Ronald Staveley and Mary Piggott (Library Association, second revised edition, 1965).

867 ' The national atlas of Canada ', in its original form, published by the Department of the Interior of the Canadian Government, 1906, was revised and enlarged in 1915. Revised again by J E Chalifour, in 1957 (1958), the work was a superb achievement, in looseleaf format, covering all aspects of the physical background and economic development of the nation at mid-century. Particularly useful was the series of urban land use maps, linking with the World Land Use Survey of the International Geographical Union. The maps demonstrate a variety of cartographic techniques and are clear and aesthetically pleasing, with notes on the backs of the plates. The gradual opening-up of the country is shown in sections—' Routes of early explorers 1534-1870 ' and ' Mapping the coasts, 1492-1874 '. The first maps from Stephanius to Zaltieri illustrate the emergence of the concept of a new continent of America; Behaim's globe indicates the general belief that Europe and Asia were separated principally by water; Ruysch's map shows the discoveries of Columbus, Cabot, the Corte Reals and Vespucci. The fourth edition of the Atlas constitutes a fresh geographical study of the nation, consisting of three main sections: Physical geography; Human geography and

Economic geography. The vast water resources of the country are emphasised, as are the other resources, both natural and man-made. A new map incorporates the census divisions and sub-divisions.

868 'National atlas of China', edited by Chang Chi-Yun, in five volumes (Yang Ming Shan, Taiwan; National War College in co-operation with the Chinese Geographical Institute, 1959-1962) is the latest in a long line of Chinese 'national' atlases, beginning in 1718 with the *Atlas of the empire* commissioned by the Emperor K'ang Hsi. In Chinese and English, the five volumes are arranged as follows: 1, Taiwan, 1959; 2, Tibet, Sinkiang and Mongolia, 1960; 3, North China, 1961; 4, South China, 1962; 5, General maps of China, 1962. The political maps reflect the Nationalist Chinese viewpoint.

869 'National atlas of disease mortality in the United Kingdom', completed in 1962 by G Melvyn Howe on behalf of the Royal Geographical Society in association with the British Medical Association (Nelson, 1963), was the first project of its kind. The first volume covers the period 1954 to 1958; a second edition (Nelson, 1970) was revised and enlarged to include data covering 1959-1963 in Part II, using the 1961 census material. Maps of population density and mortality for all causes are followed by fourteen maps of the chief causes of death. The historical introduction is valuable, and each map is accompanied by descriptive text.

Refer G Melvyn Howe: 'A national atlas of disease mortality in the United Kingdom' *in The geographical journal,* March 1964.

870 'National atlas of Hungary' (Budapest: Cartographia, 1967) was prepared with the co-operation of the Geographical Committee of the Hungarian Academy of Sciences, edited by Dr Sándor Radó. The work gives a comprehensive survey of natural, social and economic conditions in the country, maps and data being contributed by individual institutes, scientific research organisations and societies; on the basis of the recommendations of the Commission on National Atlases, work began in 1959, with two aims: to promote the work of economic management and planning and to form a historical document for further research. The main divisions of material, all sub-divided, are 'Natural conditions', 'Population and settlement', 'Agriculture', 'Industry', 'Transport and communication', 'Retail trade, tourism, foreign trade and international relations', 'Cultural and social standards'. A supplement, inset in the back cover, is 'Administrative map of the Hungarian People's Republic' at 1:500,000, using graded lettering. There are diagrams where necessary, especially in the section on climate, a fine use of symbols, proportional circles, etc, and text introduces each major division.

871 'National atlas of India' was issued in a preliminary Hindi edition, under the direction of S P Chatterjee, in 1957 (Ministry of Education and Scientific Research, Dehra Dun), with text in Hindi and English and translations of map-keys in English on the backs of the sheets. An English edition, also edited by Professor Chatterjee, was published by the Ministry of Education, National Atlas Organization, 1968, comprising fourteen selected maps from the original atlas. Several regional atlases have also been published, which will eventually constitute a national atlas in even greater detail; *An atlas of resources of Mysore State,* in two volumes, 1961 and 1962, is an excellent example.

872 'The national atlas of the United States of America', a concept of John Wesley Powell some hundred years ago, was actually begun by the National Research Council, National Academy of Sciences and the first sheets available in 1957; in 1962, the National Atlas project was transferred to the United States Geological Survey for re-planning and co-ordination. The editors state in the introduction that it is 'designed to be of practical use to decision makers in Government and business, planners, research scholars, and others needing to visualise country-wide distributional patterns and relationships between environmental phenomena and human activities'. It is therefore an eminently practical work, containing 756 maps and a substantial index, bound into a single volume; it is in two main sections, a reference section, containing general maps mainly at 1:2M, also maps of the twenty-seven largest cities at 1:500,000, and, secondly, 281 pages of special subject maps.

> *Refer* A C Gerlach: 'The national atlas of the United States of America', *in The cartographer,* May 1965.
>
> *Refer also* C O Paullin: *Atlas of the historical geography of the United States* (Carnegie Institution and American Geographical Society of New York, 1932).

873 The National Book League, London, is now associated with Artsmail, a selective information service for arts activities. The main value of the League for geographers lies in the regional bibliographies produced at intervals—on China, for example, the USSR, Greece, Spain—which follow the policy of an introduction to the literature, followed by a bibliography arranged under suitable subdivisions. ' Man and environment ' was selected by The Nature Conservancy, the ' Readers' guide to the Commonwealth ' by the Commonwealth Institute and the ' Readers' guide to Scotland ' (*qv*) as major introductions to the respective source material.

874 ' National geographic atlas of the world ' originated from January, 1958, when a new series of map supplements was introduced for the *National Geographic magazine,* uniform in size and designed to be assembled in a loose-leaf binder (The society, 1963; second enlarged edition, 1966). In the second edition each of the fifty-six double-plate maps has been brought up to date. The map of Antarctica was completely redrawn, incorporating the findings of recent surveys; also boundaries were checked and settlement patterns and towns adjusted, especially in those countries which have gained independence since 1963. The three-dimensional quality of many of the maps reflects the style of the Swiss cartographer, Paul Ulmer, who passed on his method to Henri A DeLanghe, the present cartographer of the Society. Changes in place-names were recorded, the most recent population figures possible were consulted and the index contains some 139 thousand entries. There are two editions: the Standard Edition, a flexible, leather-grained plastic cover, and the De Luxe Edition, hard-bound in a cloth cover with matching slip-case. Emphasis is on the USA, but the rest of the world is covered with reasonable balance. Arrangement is by geographical area, each section being preceded by a short description of the country. The aim was to interest laymen as well as scholars, students, travellers and businessmen. A useful list of addresses of information centres is included.

Refer Athos D Grazzini: ' Problèmes que présente la préparation de l'Atlas Mondial de la National Geographic Society ', *International yearbook of cartography,* 1965.

875 The National Geographic Society was founded in Washington in 1888 as a non-profit making scientific and educational organisation for the diffusion of geographical knowledge and for promoting research and exploration. The Society has sponsored more than two hundred major expeditions and scientific projects and has greatly assisted the widespread appreciation and use of maps. The first map supplement of the *National Geographic magazine* was issued in 1899; a new series of supplements formed the basis of the *National Geographic atlas of the world* (see above). The magazine itself is the major periodical of the Society, published monthly since 1888. The articles are of high standard, though not usually technical, and are profusely illustrated, with a lavish use of colour. The cumulative index is complete to 1963, and includes references to both subjects and illustrations. A *Geographic school bulletin* has been issued weekly, October to May, since 1922. The Cartographic Department of the Society has made a considerable contribution to cartographic history. The society also publishes books from time to time and maintains a News Service Division. Two recently established publications make known the research undertaken by the Society: *Abstracts* includes reviews of research and explorations sponsored by the Society during the previous year, edited by Paul H Oehser under the direction of the Committee for Research and Exploration; and *Research reports,* by the same editor, 1968-, summarise these activities of the Society since 1890.

Refer G H Grosvenor: ' The National Geographic Society and its magazine ', originally in the index volume 1947-1956, also reprinted separately.

876 The National Maritime Museum catalogue of the Library is being published (HMSO, 1968-); under the heading *Voyages and travels,* chapters are allotted to geographical area or type of voyage and within this section the list is chronological. An important publication of the Museum was *Man is not lost: a record of two hundred years of astronomical navigation with the Nautical Almanac 1767-1967* (1968).

877 National Institute of Oceanography was set up in 1949, incorporating the staff, collections and data of the Discovery Committee (1924-1949) and the National Oceanographic Council was established by Royal Charter in 1950. The headquarters is at Wormley, in Surrey, and the *RRS Discovery* is based on Plymouth. Research is concerned with long-term investigations into the interchange of energy between the atmosphere and oceans, with the response of the sea surfaces to wind and pressure changes and with the general circulation of the oceans; much research is conducted in the North Sea and English Channels and in far distant waters, such as the Antarctic. During the International Geophysical Year, a series of oceanographic sections was made across the Atlantic Ocean, in collaboration with the Woods Hole Oceanographic Institute, the results being published by the latter Institute in an *Atlantic Ocean atlas*. The *Annual reports* summarise the work of the Institute, while *Collected reprints* and the *Discovery reports,* published by the Cambridge University Press, give accounts of research in greater detail.

878 The National Library of New Zealand Map Room acquires about five hundred maps a year, of which a large number are catalogued for the *New Zealand national bibliography*. Lack of full-time staff has hampered services, but reference enquiries are dealt with, material provided for the preparation of research work and the demand for photocopies has been on the increase.

879 National Library of Scotland Map Room was opened in 1958, but the collection itself dates from the early eighteenth century, comprising a fine set of early maps of Scotland, original work by leading cartographers, official surveys and a representative collection of more modern maps and atlases.

880 National Library of Wales Department of Prints, Drawings and Maps maintains a large collection of early geographies, atlases, maps and charts, especially county maps of Wales. A complete catalogue is not available, but ' Some cartographical works in the National Library ', an article by M Gwyneth Lewis in *The National Library of Wales journal,* volume v, 3 1948, has also been issued separately.

881 National Oceanographic Data Center, Washington DC, was set up in 1961 as a division of the US Naval Oceanographic Office, supported by the Bureau of Commercial Fisheries, the National Science Foundation, the Coast and Geodetic Survey, the Department of the Navy, the Atomic Energy Commission, the US Weather Bureau and other government departments. The NODC is equipped to provide all types of data pertaining to oceanography, including the systems devised for geological and geophysical data storage and retrieval; the archives already contain millions of readings from bathy-thermograph observations taken by monitoring stations and research vessels and the Center will eventually provide indexes to quantitative and qualitative data relating to, for example, magnetism, seismicity, radio activity and heat flow.

882 National Parks Commission, now the Countryside Commission, was set up in 1949 to select and label as national parks the most suitable extensive areas in England and Wales ' for the purpose of preserving and enhancing their natural beauty and promoting their enjoyment by the public '. The Commission maintains close liaison with The Forestry Commission and The Nature Conservancy. A descriptive and illustrated booklet, *The National Parks of England and Wales,* was issued in 1967 (HMSO).

Refer ' Our National Parks ' *in The geographical magazine,* January 1971.

W M Condry: *The Snowdonia National Park* (Collins, second edition, 1967).

883 National Research Council of the National Academy of Sciences is the national committee representing the United States of America on the International Geographical Union; the council works through special boards and committees, of which the Division of Earth Sciences is the body of greatest interest to geographers. Section E of the American Association for the Advancement of Science is devoted to geology and geography.

884 ' Nature and resources: *newsletter about scientific research on environment, resources and conservation of nature',* is the *Bulletin* of the International Hydrological Decade, June 1965-, issued by UNESCO four times a year. The programme of the UNESCO Division of Natural Resources Research in the fields of hydrology, geology, soil sciences, ecology and conservation of nature is covered by general and regional articles, reports of symposia, research and meetings and news items. A section devoted to publications received is very useful. Beginning with the March 1971 issue, the editorial policy was enlarged to include activities connected with intergovernmental programmes and, in addition to the original sub-title, ' Bulletin of the Man and Biosphere Programme ' was added.

885 The Nature Conservancy was founded by Royal Charter in 1949 as the official government agency responsible for the conservation of wild life and natural features in Great Britain and for the conduct of research work relating to the conservation of nature and natural resources. Since 1965, it has been a component of the Natural Environment Research Council. Nearly 130 nature reserves come within its jurisdiction and there are major research stations at Monks Wood, Huntingdon, Bangor, Edinburgh, Merlewood, Furzebrook and Norwich. A series of ' Habitat teams ' are each concerned with one main type of land, such as mountain and moorland, woodland, coasts or wetlands. The experimental stations include many departments, such as ' geographical sciences ' branch, ' climatology section ', etc. The Biological Records Centre originated as the distribution maps scheme of the Botanical Society of the British Isles, which produced the *Atlas of the British flora (qv)* in 1962 and the Critical Supplement to the Atlas in 1968. Publications include the *Nature Conservancy handbook; Nature Conservancy progress 1964-1968* and for 1968-1970; the *Monks Wood Experimental Station Report;* and *Nature conservancy: the first twenty one years,* 1970. Nature conservancy in Northern Ireland is encouraged by the Amenity Lands Act (Northern Ireland) 1965, by which the Ministry of Development is responsible for matters relating to nature, advised by the Nature Reserves Committee, a statutory body of scientists and laymen. *Nature trails in Northern Ireland* (Belfast: Nature Reserves Committee, 1970) consists of eighteen individual leaves in a pocket folder, each giving details of a nature reserve, numbered on a map.

> *Refer also The countryside in 1970: proceedings of the study conference,* November 1963, edited by The Nature Conservancy (HMSO, 1964).
>
> *The countryside in 1970:* second conference, 1965, *Proceedings: reports of study groups* (Royal Society of Arts, 1965).
>
> *Man and environment* (National Book League in association with The Nature Conservancy, 1970).
>
> Robert Arvill: *Man and environment: crisis and the strategy of choice* (Penguin, revised edition, 1969).
>
> Robin Feddon: *The continuing purpose: a history of The National Trust, its aims and work* (Longmans, 1968).
>
> Garth Christian: *Tomorrow's countryside: the road to the 'seventies* (Murray, 1966).

Nan Fairbrother: *New lives, new landscapes* (Architectural Press, 1970).

C A W Guissisberg: *Man and wildlife* (Evans, 1970).

John Hillaby: *Nature and man* (Phoenix House, 1960, *Progress of science* series).

Joyce Joffe: *Conservation* (Aldus Books, 1969, *Interdependence in nature* series).

J A Lauwerys: *Man's impact on nature* (Aldus, 1969, *Interdependence in nature* series).

Max Nicholson: *The environmental revolution: a guide for the new masters of the world* (Hodder and Stoughton, 1970).

Keith Reid: *Nature's network* (Aldus Books, 1969, *Interdependence in nature* series).

J Rose, *ed*: *Technological injury: the effect of technological advances on environment, life and society* (Gordon and Breach, 1969).

W M S Russell: *Man, nature and history* (Aldus Books, 1967, *Modern knowledge* series).

Sir L Dudley Stamp: *Nature conservation in Britain* (Collins, 1969).

Mary Anglemeyer, *comp: Natural resources—a selection of bibliographies* (second edition, 1970).

886 ' Nature in focus ', bulletin of the European Information Centre for Nature Conservation, Council of Europe, is published in English and French; articles, illustrated with photographs, are international in scope and news of projects in all parts of the world is presented non-technically.

887 ' The nature of geography: *a critical survey of current thought in the light of the past ',* by Richard Hartshorne (Association of American Geographers, 1939; reprinted, 1956) is a scholarly assessment, which will surely remain a classic work. In 1959, the author published *Perspective on the nature of geography,* which brought up to date both his and others' thinking on the subject.

888 ' Nederlands Historisch Scheepvaartmuseum catalogus der bibliotheek, 1960, is distributed by N Israel, Amsterdam, in two volumes, with fifty-four reproductions of title-pages, plates, maps and globes. The library is strong in material relating to the early and later voyages, travels, history and art of navigation, atlases, maritime history and law, ship construction, cartography, including the important collection of early globes. The catalogue was compiled by V Cannenburg, including bibliographical descriptions of nearly thirteen thousand items, together with full indices of personal and geographical names.

889 '**Netherlands India: a study of plural economy**', by J S Furnivall (CUP, 1939, reprinted, 1967), begins with comments on the geography, political geography and the peoples of the area, including a chapter on the East and notes on the India Company, 1600-1800. There is a 'List of general references' and a glossary.

890 '**Netherlands journal of economic and social geography**' (*Tijdschrift voor economische en sociale geografie*), is published by the Royal Dutch Geographical Society Editorial Committee in six fascicules a year, illustrated with many figures and sketchmaps.

Refer also Frank E Huggett: *The modern Netherlands* (Pall Mall, 1971).
Audrey M Lambert: *The making of the Dutch landscape: an historical geography of the Netherlands* (Seminar Press, 1971).
A compact geography of the Netherlands (The Hague: NV Cartografisch Instituut, Bootsma, 1970), prepared by Utrecht State University, with a wall map.

891 '**Network analysis in geography**', by Peter Haggett and Richard Chorley (Arnold, 1969), was designed as supplementary reading to *Locational analysis in human geography* (*qv*), forming the first of a trilogy of volumes dealing with spatial structures of direct concern to geographers. In the first part, three main topics are examined— topological and geometrical characteristics, evaluation of network structures and problems of the growth and transformation of networks. Part II, 'Networks as regional system' discusses the interactions between network structure and regional environments with emphasis on concepts of hierarchy, growth and system interactions. Throughout the text the emphasis is on common research problems rather than the specific features of the empirical networks, as, for example, stream patterns or traffic flows. The text is accompanied by more than a hundred figures, specially drawn, and there is an extensive research bibliography.

892 'The new Cambridge modern history atlas', edited by H C Darby and Harold Fullard, was a new compilation, using the most up-to-date cartographic techniques. With special emphasis on North America, Latin America, the Far East and Australasia, particularly economic and social conditions, double spreads are used to great effect. Maps are arranged on an area basis, as far as possible to the same scale and chronologically within each group. Hill shading is used where especially applicable.

893 'The new contour dictionary', 1971, compiled by J B Goodson and J A Morris, replaces *The contour dictionary*. More than a hundred excerpts from Ordnance Survey maps in full colour are included, six full-page maps in colour and extensive use of small excerpts with reference to contour patterns. Exercises and simple mapping techniques have been introduced, incorporating metric units of length and area.

894 'New geographical literature and maps', begun by the Royal Geographical Society in 1951 and now issued twice a year, in June and December, is a classified list of accessions to the Society's Library and Map Room, including all new atlases and maps. From 1958, the scope was increased to include all articles contained in twenty of the most important geographical periodicals in English, French and German, and a selection is made from 150 other titles. An annual list of completed geographical theses has been added since 1960. A reading list is included on a topical subject, also notes on books not reviewed in *The geographical journal*.

895 'The new Israel atlas: *Bible to present day'*, compiled by Z Vilnay (H A Humphrey Limited, 1968), was published to mark the twentieth anniversary of the establishment of the State of Israel. Three thousand years of history are represented in maps, almost all with explanatory texts by Dr Vilnay recording the development of the people, settlement patterns and economic activities. Arranged in four sections, the maps, by Carta of Jerusalem, cover the geography of modern Israel, the struggle for independence, history from Biblical times to 1918 and proposals for a Jewish state. Appended is a gazetteer of settlements, with foundation dates. Photographs and drawings are included and the presentation of the material is remarkably objective.

896 'The new large shining sea-torch' (*De Nieuwe Groot Ligtende Zeefakkel*), prepared by Johannes and Gerard van Keulen, 1716-1753, was reproduced in three volumes, 1969-1970 by Theatrum Orbis Terrarum Limited, Amsterdam. This atlas is considered by experts as the most attractive, reliable and informative of sea atlases. The original folio volumes contain more than 240 double-page charts showing in detail all the navigable waters, coasts, river estuaries, inlets and harbours known in the first half of the eighteenth century. The charts are decorated with ships, sea-gods and mermaids, as well as with representations of local trades and industries in the ports to which the charts refer. Successive editions of the work show many alterations and improvements. The six original sections were 'The Eastern and Northern navigation', 1728; 'The Western navigation', 1728; 'The Mediterranean', 1716; 'The West-Indies and North America', 1728; 'The Atlantic navigation', 1728; 'The Indian Ocean and South-East Asia', 1753. An introduction to the reproduction was written by Dr C Koeman; the paper for the atlas was manufactured especially to match the tint, texture and thickness of the paper used for the original edition.

897 The New Product Centre, London, was a unique but short-lived venture opened in April 1964 to act as a clearing house of information about new products and processes throughout the world, their arrival, manufacture and use, agenting and licensing; information was assessed, summarised, classified, stored and retrieved as required. Still published, however, from 13 Homewell, Havant, Hants, is the monthly *Product licensing index,* which lists products and processes for which manufacturing or sales licences are available; it also lists new products, industrial and consumer, as they come on the market. Articles are included.

898 'New towns of the Middle Ages: *town plantation in England, Wales and Gascony* (Lutterworth Press, 1967), by Maurice Beresford, deals in detail with this topic illustrated with sketchmaps and diagrams. Chapters 15 to 17 are annotated gazetteers for the three countries. There are frequent references in the text.

899 'The new world': *problems in political geography,* by Isaiah Bowman (*qv*), stemmed from the Versailles Peace Conference in 1919, at which Bowman headed the American delegation (World Book Company, 1921; fourth edition, 1928). The work presents a balanced and broad outlook and, in several editions, made a great impression throughout the world.

900 New York Public Library, Map Division of the Research Libraries of the Library has maintained a dictionary catalogue of holdings. Some 175 thousand cards list maps and other cartographical material from early American and European rarities to up to date representations of all parts of the world, including recently explored areas of the universe. The Map Division holds about 280 thousand sheet maps, being a depository for the United States Army Map Service and for extensive series of maps issued by foreign governments. G K Hall reproduced the catalogue in ten volumes in 1970.

901 'New Zealand contemporary dictionary', (Whitcombe and Tombs, 1968) published in association with Collins, contains more than sixty thousand references. A special feature is a supplement giving distinctive New Zealand and Australian terms.

902 'New Zealand geographer', official journal of the New Zealand Geographical Society, has witnessed the expansion of geographical studies in the Dominion, from 1944; edited from the University of Otago, it is intended to meet the requirements of university students and teachers, also with the wider aim of 'satisfying the common curiosity' in the human geography of New Zealand. It has appeared half-yearly since 1945 and each issue has contained articles contributed by specialists on geographical subjects mainly dealing with New Zealand, Australia and the Pacific Islands. A feature, 'Geographic notebook', provides a valuable commentary on themes of current interest, while generous space is usually given to reviews of relevant publications.

903 New Zealand Geographical Society, as a project, began in 1939 at Christchurch, when a group of enthusiasts formed what in due course became the Canterbury branch. The Society was founded in 1944 at the University of Auckland, with branches at Auckland and Manawatu. In addition to the chief periodical, *New Zealand geographer (qv)*, a *Record of proceedings* is issued as required, *Special publications* series from 1950 and a *Reprint* series, comprising selected articles from the *New Zealand geographer*. A *Research* series is planned to include not only original research, but special lectures and studies.

904 Newbigin, Marion Isabel (1869-1934) came to a geographical studies through biology, succeeding J Arthur Thomson as lecturer in biology and zoology in the extra-mural medicine school for women at Edinburgh. Thereafter, as writer and teacher, her influence on geographic thought and methodology was very great, especially in biogeography and in the regional study of Mediterranean lands. From 1902 to her death, she edited *The Scottish geographical magazine* (*qv*), and made it a leading national periodical; an annual Newbigin Prize is offered by the Royal Scottish Geographical Society for the best essay in Scotland suitable for publication in the magazine. Among the most outstanding of her published books were: *An introduction to physical geography,* 1912; *Animal geography: the faunas of the natural regions of the globe,* 1913; *Geographical aspects of Balkan problems in their relation to the Great European War,* 1915; *Mediterranean lands: an introductory study in human and historical geography,* 1924; and *Southern Europe: a regional and economic geography of the Mediterranean lands,* of which a third edition was prepared by Harrison Church in 1949.

905 'Newsletters on stratigraphy' is a new journal published by Brill of Leiden, designed as a forum for the exchange of information on topics of general stratigraphical interest. It includes articles, preferably in English, but also, on occasion, in French, German or Spanish, on such subjects as descriptions of new stratotypes, the subdivision of stratigraphical units and discussions of stratigraphical boundaries; it provides an opportunity also for the publication of comments on and critiques of articles of international interest. The chief editor is Gerd Luttig, assisted by a distinguished team of associate editors.

906 'Nigeria: a guide to official publications', compiled by S B Lockwood (Library of Congress, General reference and bibliography division, 1966), contains 2,451 entries, mainly references relating to the period 1861-1965 by both the British administration in Nigeria and the Nigerian federal and regional governments; also to selected works on Nigeria issued in the United Kingdom. About a hundred other entries refer to the Cameroons, published by the British administration, the League of Nations and the United Nations.

907 Norden, John (1548-1625 or 1626), topographer and cartographer, contributed greatly to contemporary knowledge of the country. His county maps of England are most informative, especially those for Middlesex, Hertfordshire, Essex, Surrey, Sussex and Hampshire; they were the first to show roads. His estate surveys remain particularly important. The *Speculum Britanniae* was an ambitious project, of which only parts on Middlesex and Hertfordshire were actually published in his lifetime, but his notes, distance tables and 'thumb-nail' maps continued to form the basis for further work for a long time afterwards. His maps were used in the 1607 edition of Camden's *Britannia* (*qv*) and later also by Speed in his *Theatre of Great Britain*.

908 'Nordic hydrology', published by the Joint Committee for a Nordic Hydrological Journal, was created by the National Committees for the International Hydrological Decade in Denmark, Norway and Sweden and the Geophysical Society of Finland. The scope includes surface-water hydrology, groundwater hydrology, hydrometeorology, snow and ice and hydromechanics, with an emphasis on the study of the hydrological cycle and the process involved. One issue is being published during the year, 1970-.

909 Norge, in four volumes, by J W Cappelens, 1963, is an encyclopedic work, magnificently illustrated and containing a superb atlas, the result of co-operation between the Norwegian Geographical Institute, the Army Map Service in Washington and the Esselte AB Kartografiska Institutet, Stockholm. Volume 1 treats of Norway regionally and systematically; volumes 2 and 3 are a geographical encyclopedia of Norway and volume 4 is a historical survey entitled 'Norway in maps', in which are several reproductions of early maps of the country and a sixty-four page atlas. The whole work is well documented, with indexes, a gazetteer, list of sources and a glossary of Lapp terms.

910 'Northern geographical essays in honour of G H J Daysh' (Newcastle-upon-Tyne, Oriel Press, 1966, for the Department of Geography), contains twenty-two essays, a festschrift for Professor Henry Daysh on his retirement. Emphasis is on northern regional geography, but eight of the essays deal with overseas topics.

911 'Northernmost Labrador mapped from the air' was prepared by Alexander Forbes and others (American Geographical Society, Special publications, no 22, 1938); the work, with six sheet maps of Northernmost Labrador and navigational notes on the Labrador coast, is published in a slip-case. The text includes photographs with explanatory text on narratives and expeditions carried out in the area, the mapping, geology and physiography and phytogeographical observations in Labrador, with an appendix 'An impression of Northernmost Labrador as viewed from the air' and notes on placenames.

912 'Norway exports', 1957-, the journal of The Export Council of Norway, established by Royal Decree in 1945 to promote Norwegian exports, is a well produced, fully illustrated production containing feature articles and notes about Norwegian products and suppliers. The advertisements are valuable to those interested. Other publications, in English, German and French circulate information on new products and surveys; and a Directory of Norwegian products and exports, in five languages, gives 2,500 firms and their products.

913 'Nuevo atlas geográfico de la Argentina', produced by Ediziones Geográficas Peuser, Buenos Aires, 1969, was compiled by José Anesi and is now in an eighth edition. Maps are printed on the recto only, with notes; some are large, folded maps. The layer colour, with contours, shows up particularly well on the coastline, with its islands, and the complex relief of the Andes.

914 Obruchev, V A (1863-1956) specialised in geology and physical geography, playing a leading part in the development of interest in the earth sciences in Russia. His work on the geomorphology of sand deserts and on the alluvial origin of the Kara Kum sands, also his classification of sand relief into four sub-divisions—barchans, sand mounds, sand ridges and sand steppe—are still accepted. In addition to academic teaching, he carried out a number of surveys and expedition work; his three-volume book, *Frontier Dzungaria,* 1912-1940, embodied the results of research, as did *The geological map of the Lena gold-bearing region,* completed in four volumes. The *Field geology,* in two volumes, ran to four editions, and *Geologie von Sibirien,* later expanded to three volumes, has remained a classic. He wrote also *The history of the geological exploration of Siberia,* in five volumes, completed in 1950. He was an editor of

Izvestiya, Geological Series. During the later years of his life, he continued to write—*Eastern Mongolia,* for example, accounts of his travels and some novels based on his experiences, such as *Gold seekers in the desert.* It has been estimated that his geological surveys covered some forty-six thousand square miles and his contribution to the exploration of Asia is incalculable. There is a substantial body of literature, in Russian, on his life and work.

915 ' Ocean wave statistics: *a statistical survey of wave characteristics estimated visually from Voluntary Observing Ships sailing along the shipping routes of the world'* was edited by N Hogbem and F E Lumb (HMSO, 1967 for the Ministry of Technology, National Physical Laboratory), to provide systematic information about environmental conditions for use in research on the sea-going qualities of ships. Nearly two million sets of observations on sea conditions were reported over a period of eight years, 1953 to 1961. Several interested organisations helped in the task of preparing and presenting the material. Data Tables are prefaced by introduction and explanation and there is a short list of references.

916 O'Dell, Andrew Charles (1907-1966) was a leading academic geographer, especially in Scotland, his particular achievement being the establishment and direction of the Department of Geography in the University of Aberdeen, from 1945. He wrote constantly on the geography of Scotland, notably *The highlands and islands of Scotland,* published with Kenneth Walton, in 1962; he also edited the British Association Aberdeen volume in 1963. Another abiding interest was Scandinavian geography, revealed in his early work on the Land Utilization Survey of Zetland, in his MSC thesis, *The historical geography of the Shetland Islands,* 1933, published in 1939, and in *The Scandinavian world,* 1958. *Railways and geography,* 1956, made a substantial contribution to the growing literature on transport geography.

917 Office de la Recherche Scientifique et Technique Outre-Mer (ORSTOM) issues valuable series of publications for geographers, for example, Pierre Vennetier: *Pointe-Noire et la façade maritime du Congo-Brazzaville,* a study in depth, illustrated with photographs, maps and diagrams and having a sound bibliography.

918 ' Official map publications: *a historical sketch, and a biblio-*

graphical handbook of current maps and mapping services in the United States, Canada, Latin America, France, Great Britain, Germany and certain other countries', by Walter Thiele (American Library Association, 1938), is a comprehensive guide to its date.

919 Ogilby, John (1600-1676), one of the great cartographers who helped to improve topographic mapping in Britain, gained official recognition with the title King's Cosmographer and Geographic Printer. Following on John Norden's work, he established roads as important features of maps. Among his publications were a series of atlases and maps and a road book, which describes the chief roads in England and Wales, from original surveys, published in 1675 as *Britannia, or an illustration of the Kingdom of England and Dominion of Wales* . . .

920 Ogilvie, Alan Grant (1887-1954) made great contributions to academic geography, particularly at the School of Geography, University of Edinburgh, and exerted a widespread influence on standards of geographic scholarship and map production in the course of his work with the Royal Scottish Geographical Society, the International Club in Edinburgh, the National Committee for Geography, the British Association, the Institute of British Geographers, the Geographical Association, the American Geographical Society and the IGU. He wrote extensively, edited *Great Britain: essays in regional geography,* 1928, and contributed to the foundation work on the Map of Hispanic America 1:1M. *Geographical essays in memory of Alan G Ogilvie,* edited by R Miller and J Wreford Watson (Nelson, 1959), contains 'Essays on Scotland' and 'Essays oversea'.

921 'On the structure and distribution of coral reefs: *also geological observations on the volcanic islands and parts of South America visited during the voyage of HMS Beagle'*, Charles Darwin's classic account, was published by Ward Lock in 1890 (Minerva Library of Famous Books) and reproduced in paperback in 1963, with a foreword by H W Menard; the text includes illustrations and maps.

Refer also The Darwin reader, edited Marston Bates and Philip S Humphreys (Macmillan, 1957).

F Wood-Jones: *Corals and atolls: their history, description, theories of their origin both before and since that of Darwin* (1910).

W M Davis: *The coral reef problem* (New York: American Geographical Society, 1928, special publication no 9 in the Shaler Memorial series).

J S Gardiner: *Coral reefs and atolls: being a course of lectures delivered at the Lowell Institute at Boston,* February 1930 (Macmillan, 1931).

922 ' Orb ', the journal of the Geographical Society of the University of Aberdeen, was founded primarily as a platform for undergraduate comment, covering a wide variety of topics within the subject of geography.

923 ' Orbis geographicus ', a world directory of geography, was compiled and edited on behalf of the IGU by E Meynen in cooperation with the National Committees and published as a special supplement to the *Geographisches Taschenbuch,* 1960/61 (Franz Steiner Verlag, Wiesbaden). The work, which is in English and French, with the introduction also in German, includes information on the IGU and National Committees, on geographical societies, cartographical societies, geographical chairs and institutes at university level, official agencies, hydrographic offices, cartographic and topographic surveys, important map collections, national authorities and committees on geographical names; also a list of professional geographers, grouped according to their country of residence, with their academic qualifications and appointments. At the International Geographic Congress in Stockholm, it was proposed to publish *Orbis geographicus* every eight years; more frequent revision has proved desirable, however; part I was reissued in 1964 and part II, the ' Who is who of geographers ', in 1966. A further revised edition, also in two volumes, covers 1968 to 1970.

924 Ordnance Survey of Great Britain, the official cartographic agency, was established as the Trigonometrical Survey in 1791, being the outcome of survey operations for the linking of England and France by Cassini and William Roy in 1787. The first task of the Survey was the making of a one-inch map of the country, beginning with the south-eastern counties of England. After the Napoleonic Wars, civil uses of maps became increasingly important and in 1825 the maximum effort was transferred to Ireland to make a six inch to the mile map of the whole of that country for evaluation purposes; this survey was completed about 1840. The subsequent publication history of the various series of topographical maps has been complex, sheetlines and symbols of relief representation being revised to suit changing circumstances and advances in technique.

The Ordnance Survey is responsible for the official surveying and mapping of Great Britain, including geodetic surveys and the associated scientific work, topographical surveys and production of maps at appropriate scales from these surveys. At present, emphasis is on restoration, which includes bringing the nineteenth and early twentieth century 1:2,500 surveys up to date and compiling them on a national sheet line system; also making entirely new surveys of the major towns at 1:1,250 scale and of mountain and moorland areas at six inches to one mile. All the new and revised maps produced in this way are kept up to date by a system of continuous revision which ensures that changes on the ground are surveyed soon after they occur. Once National Grid plans at 1:1,250 or 1:2,500 scales have been completed in a locality, the Ordnance Survey appoints surveyors to keep a master copy, called 'field sheets', of each plan up to date. Copies made directly from these sheets may be bought by members of the public who require survey information in advance of the publication of revised editions. The standard of completeness of an Advance Revision Information Sheet is not the same as that of the printed map; but all important changes will have been entered. Shortly to be offered are the principal roads, town and coastline at the 1:625,000 scale.

In 1968, the Ordnance Survey moved into a new building, functionally designed for it, officially opened in May 1969, on the outskirts of Southampton, thus bringing together the whole of the department except for the regional surveying organisation. An ICL 1902 computer has been installed, replacing the smaller punched card system used previously; it carries out calculations from aerial surveys, minor control and levelling and produces management statistics. The Survey is now able also to supply magnetic tape for drawing a map of the United Kingdom coastline digitised at the scale of 1:1,250,000. The Survey continues to study the possible application of automation to cartography. Experimental work is now being concentrated on digitising contours directly by encoders attached to the output shafts of stereo-plotting machines with the object of compiling a contour data bank which could be used, with automatic plotting equipment, to produce contours for 1:25,000 and smaller scale maps. The Survey has sought the co-operation of the resources of the Experimental Cartography Unit of the Royal College of Art to investigate some of the difficulties involved in automation. A microfilm service has been introduced, so that copies of 1:1,250 and 1:2,500 national plans can be supplied in the form of 35 mm Diazo negative microfilms mounted in aperture cards; they are produced from the glass negatives of the Plans, not from the paper copies. Reductions of the 1:1,250 plans at 1:2,500 scale can be supplied on either film or electro-photographic paper. Normally matt film is used, but clear film is available if required. These reductions are intended for assembly on a gridded film base supplied by the Department in a 2-kilometre format gridded at five hundred metre intervals, at 1:2,500 scale. The assembly of reductions provides an up-to-date master film from which a normal transparency can be produced for the printing of paper copies. The Survey has since 1969 been adjusting to metrication, but it will be many years before the conversion is complete.

The state of the mapping programme at present is that the whole of Britain is covered by the one-inch series, which is under continuous revision; the quarter-inch series, derived from the one-inch, covers Great Britain in seventeen sheets, reprinted with revision every three years. The six-inches-to-one-mile series is the largest scale covering the whole of the country and plans at twenty-five-inches-to-one-mile cover the country with the exception of some areas of mountain and moorland. The fifty-inches-to-one-mile plans cover most of the large towns and will eventually cover all towns with a population of over 20,000 and other areas of dense population. There are also a number of series at intermediate scales and several special maps such as the Route Planning Map, first issued in 1964. The set of historical and archaeological maps has proved very successful, including 'Britain in the Dark Ages', 'Southern Britain in the Iron Age', 'Ancient Britain', 'Roman Britain', 'Monastic Britain' and 'Hadrian's Wall'. Three pamphlets issued by the Directorate General of the Ordnance Survey provide a guide to understanding the map series, *A description of Ordnance Survey small-scale maps,* 1947, reprinted with an addendum, 1951; *A description of Ordnance Survey medium-scale maps,* 1949, reprinted with corrections, 1951; and *A description of Ordnance Survey large-scale plans,* 1947.

The Survey publishes geological maps on behalf of the Institute for Geological Sciences and soil maps for the Soil Survey of England and Wales and the Macaulay Institute for Soil Research (Scotland). New maps, especially at a large scale, are in constant demand by local authorities, architects and the legal profession. Large-scale maps have been provided from the 1971 Census; the provision of maps and copies of the field plots make it possible for the first time to record a National Grid reference for every dwelling. Outline maps are issued as well at various small scales; they are printed in grey or black and make useful bases for planning or recording purposes. From January 1971 the wholesale distribution of Ordnance Survey maps in England and Wales was undertaken by the Survey direct from Southampton; Thomas Nelson and Sons Limited, of Edinburgh, continues as the distribution centre for Scotland. The standard copyright regulations and licence conditions apply to the reproduction of Ordnance Survey materials.

From 1969, in collaboration with the Ordnance Survey, David and Charles of Newton Abbot have been republishing in exact facsimile the later printings of the first edition of the One Inch Ordnance Survey maps, covering England and Wales in ninety-seven sheets; later the Scottish sheets will be done. The reprint is edited by Dr J B Harley, of the Department of Geography, University of Liverpool, who is providing introductory notes for each sheet, designed to assist in the dating and interpretation of the maps. The sheets will be available either flat or folded, within covers; an index map is available, showing the arrangement of the reprint. Also in facsimile reproduction is 'Ancient map of Kent', drawn by Philip Symonson, 1596, as published by Stent about 1650; size and detail of this map was superior to any English county map of its time. A note on the history of the map is printed at the foot of the sheet. The 'Bodleian map of Great Britain' (14th century) has been reproduced in single colour, with the ancient names transcribed in red.

From 1885, details of the progress of the Ordnance Survey can be traced through the *Annual reports*. The 1968-69 issue was the first to incorporate the new 'house style' designed by the Central Office of Information. Plates are included showing the programme and progress of publication of map series. Much historical and technical information appears in the *Reports of the departmental committee on the Ordnance Survey,* 1935-38. Early map publications are listed in the *Catalogue of maps,* 1904 and 1920 and a full description of the six-inch and twenty-five-inch maps was given by

H St J L Winterbotham in *The national plans,* in the Ordnance Survey *Professional papers,* new series, 16, 1934. Each month, a *Publication report* is distributed to those interested, giving exact information on new publications and revisions at all scales, and the ' General information and price list ' brochure is frequently revised.

The atlas of Great Britain comprises seventeen quarter-inch maps in a pull-out style; and the *Ordnance Survey gazetteer of Great Britain,* 1969, lists all the names that appear in this atlas, giving National Grid references to all features named.

Refer also History of the retriangulation of Great Britain, 1935-62 (HMSO for the Ordnance Survey, 1968).

The historian's guide to Ordnance Survey maps, published for The Standing Conference for Local History by The National Council of Social Service, 1964, from a series of articles originally in *The amateur historian.*

Ordnance Survey, its history, organisation and work (OS, 1969).

John Aylward: ' The retail distribution of Ordnance Survey maps and plans in the latter half of the nineteenth century: a map-seller's view ' *in The cartographic journal,* June 1971.

R V Clarke: ' The use of watermarks in dating old series one-inch Ordnance Survey maps ' *in The cartographic journal,* December 1969.

Col Sir Charles Close: *The early years of the Ordnance Survey,* 1926; reprinted, with an introduction and index by J B Harley, by David and Charles.

R C A Edge: ' Ordnance Survey at home ' *in The geographical magazine,* October 1969.

Lt Col W R Taylor: ' The Ordnance Survey of Northern Ireland: an outline of its history and present mapping tasks ' *in The cartographic journal,* December 1969.

Ian Mumford and Peter K Clark: ' Engraved Ordnance Survey one-inch maps—the methodology of dating ' *in The cartographic journal,* December 1968.

J B Harley: ' Error and revision in early Ordnance Survey maps ' *in The cartographic journal,* December 1968.

925 Organisation for Economic Co-operation and Development (OECD) was formed in 1948 between the following countries: Austria, Belgium, Canada, Denmark, Finland, France, Germany, Greece, Iceland, Ireland, Italy, Japan, Luxembourg, Netherlands, Norway, Portugal, Spain, Sweden, Switzerland, Turkey, the United Kingdom and the United States of America; Australia participated in the work of the Development Assistance Committee and Jugoslavia was a full member for confrontation of economic policies, scientific and technical matters, agriculture, fisheries questions, technical assistance and productivity and had observer status in other matters. The present body, with the title The Organisation for European Economic Cooperation (OEEC), came into being to allocate Marshall Plan aid and to work together for post-war recovery; the scope of the organisation was widened to include the dissemination of relevant information in 1961 and to provide a forum for the exchange of ideas and experiences. In the course of its work, a vast number of publications of central importance are produced. The Organisation's Publications Office in Paris and the Center in Washington produce a Catalogue, which is the key to a mine of information. Ten main groups of publications are distinguished: Economics; International trade and payments; Statistics; Development; Agriculture, food, fisheries; Energy; Industry, transport, tourism; Manpower and social affairs; Education and science; General information. Each section is preceded by a short introduction and full annotations to the entries are given as necessary. The *Catalogue* is not published every year, supplements being issued in the intervals. For particulars relating to the entire range of publications between 1948 and 1968, the 1958, 1966 and 1968 catalogues should be consulted.

926 ' Oriental and Asia bibliography: *an introduction with some reference to Africa'*, compiled by J D Pearson (Crosby Lockwood, 1966), is concerned with books, whether written in indigenous languages or in European, which relate to this enormous region. Three parts deal with institutions producing literature, the bibliographical apparatus available for control and use of this literature and the libraries and archives where the literature is stored. A vast amount of information is concentrated herein, together with commentary and discussion of problems involved. Appendices set out ' Booksellers in Asia ' and a ' List of works referred to in the text '.

Refer also J D Pearson: ' Oriental and Asian bibliography ' *in Progress in library science,* 1967.

927 ' The origin of continents and oceans ', by Alfred Wegener, was translated into English by J G A Skerl from the third German edition (Methuen, 1924) and from the fourth German edition by

John Biram in 1968 (New York: Dover Publications). The results of *The theory of continental drift* symposium on the origin and movements of land masses both inter-continental and intra-continental, as proposed by Alfred Wegener, were published in 1928 by the American Association of Petroleum Geologists.

Refer also A L Du Toit: *Our wandering continents: an hypothesis of continental drift* (*qv*).

American Philosophical Society: *Gondwanaland revisited: new evidence for continental drift,* 1968.

A symposium on continental drift, held by the Royal Society in 1965 (*Philosophical Transactions,* no 1088).

E G R Taylor: 'The origins of continents and oceans: a seventeenth century controversy' *in The geographical journal,* December 1950.

S K Runcorn, *ed*: *Continental drift* (New York: Academic Press, 1962).

J T Wilson: 'Continental drift' *in The scientific American, April* 1963.

D H and M P Tarling: *Continental drift* (Bell, 1971).

928 Ortelius, Abraham (1527-1598) became a cartographer and map publisher, probably under the influence of Mercator. His earliest known work dates from about 1564, but his greatest achievement was the *Theatrum orbis terrarum,* which was the first collection of maps known to have been brought together on a scholarly method. The seventy maps were carefully selected, as the long list of 'acknowledgements' shows. Before his death, about thirty editions had been published, increasing in size; an English edition was published in 1606 and his final Latin edition in 1612.

929 'Our developing world', the well-known work by Sir Dudley Stamp (Faber, 1960), appeared in a Faber paperback edition in 1969, with the statistics brought up to date by Audrey N Clark.

930 'Our wandering continents: *an hypothesis of continental drift'* by Alex L du Toit (Oliver and Boyd, 1937, 1957), was dedicated to Alfred Wegener and considers the controversial theory he propounded under the following headings: 'Current theories versus Continental drift', 'Historical and geological principles', 'Gondwana', 'Comparisons between the fragments of Gondwana', 'Laurasia', 'The tertiary history of the lands', 'The oceans', 'The paramorphic zone and its import', 'Application of the paramorphic principle', 'Past climates and the Poles', 'Biological relationships', 'Geodetic evidence', 'The pattern of the earth's orogenies' and 'Causes of continental drift'. There are a number of small sketch-maps and diagrams throughout the text, a glossary of technical terms relating to the subject and a substantial bibliography.

Refer also G D Garland, *ed*: *Continental drift* (University of Toronto in co-operation with The Royal Society of Canada, 1966).

See also ' The origin of continents and oceans '.

931 Overseas Geological Surveys, founded in 1947 by the Colonial Office, became in 1961 part of the Department of Technical Co-operation, now the Ministry of Overseas Development. Work includes the assessment of mineral resources by geological mapping and other scientific techniques, in close co-operation with national governmental geological surveys, in Commonwealth and other countries. Regular publications include the progress bulletins, *Overseas geology and mineral resources* and *A statistical summary of the world mineral industry,* which records world production, imports and exports.

932 ' Overseas railways ', a Railway Gazette publication since 1961, is divided into six sections, concerning the Far East, Africa, Europe, the Americas, Australia and New Zealand. Illustrations and sketch-maps in the text usefully support the text.

933 'The Oxford atlas', 1951, fifth reprint with revision 1963, further printing, 1971, was conceived after the second world war when the need for a new atlas became pressing; a fresh approach and new layout were necessary to illustrate the altered emphases in the strategic, political and economic spheres. There are 112 pages of six-colour maps and a 90-page gazetteer. Uniformity of scale sequence was aimed at in the atlas, to enable comparison between the countries of each continent, and much thought was given also to projections. A valuable feature of each map is a footnote giving full particulars of the projection used and indicating where to look for scale errors and the corrections to be applied to them. In the revised reprint of 1952, edited by Sir Clinton Lewis and J D Campbell, a section of distribution maps was included, by Professor Linton, of Sheffield University. In these, the more detailed maps are confined to areas where the topics chosen are known with sufficient accuracy to warrant representation at large scales. Contour lines have been omitted and the layer tints are limited to those which could be produced by four colours, light blue, yellow, brown and red. The atlas attempts to distinguish between all-weather and fair-weather roads. English forms of placenames have been used when these are in common use, and geographical terms are translated except when the local form is the more familiar. Derived from *The Oxford atlas* are *The concise Oxford atlas, The Oxford economic atlas of the world* (*qv*) and *The Oxford junior atlas* (*qv*). *The Oxford atlas* is also available in a school edition.

934 'The Oxford economic atlas of the world', first published in 1953, went into a second edition in 1959, a third in 1965 and a fourth in 1971. The latest editions of this atlas were entirely revised. They present in map form a selection of world physical, political, social and economic topics that is significantly larger than before and in greater detail; the maps illustrate world patterns and a complementary statistical index gives detailed figures for each country. Data have been obtained in such detail as to allow the portrayal of production both by centre of activity and in precise quantitative terms; for many products the maps now show major world trade flows. The demographic section covers population change, birth and death rates, life expectancy, birth control, migration, accidents, disease, medical services, education and employment; there are also maps showing foreign aid, bilateral trade, political structure and economic alliances. Supplementary tables, notes and economic commentary accompany each section. The regional economic atlases are in series with the general atlas, the complete series comprising *The Middle East and North Africa; The USSR and Eastern Europe; Africa; The United States and Canada; Western Europe; Latin America; India, China and Japan; South East Asia; Australia and New Zealand*. In these, general reference maps are followed by topical maps, notes and statistics, bibliographies and gazetteers. *The shorter Oxford economic atlas of the world,* in a second edition, with revisions, 1961, contains all the pages of the parent atlas, in paper covers, but omits the statistical index.

935 'Oxford economic papers' (new series), of which eight issues appeared at intervals between 1938 and 1947, became a periodical in 1949 and is published three times a year by The Clarendon Press.

Originally intended as a channel for publication of articles by Oxford authors, contributions from elsewhere are also welcomed. Most of the articles are concerned with economic and allied subjects. Review articles are included and references are added at the ends of articles as necessary.

936 'The Oxford junior atlas', when published in 1964, was an entirely new atlas; it was reprinted with revision in 1967. The atlas contained forty-eight pages of seven-colour maps and seven pages of gazetteer. Simplicity and clarity were the aims throughout; most of the maps are double-page spreads, without insets. The series of regional maps are useful; Britain is covered by six regional maps at sixteen miles to the inch. Careful instructions are included for using the atlas and both the planning and selection of content and the imaginative three-dimensional relief representation make this an evocative atlas for use by children. Short descriptions of places are added to the gazetteer entries.

937 'Oxford New Zealand encyclopaedia' is the fourteenth volume of the *Oxford junior encyclopaedia,* reproduced as a self-contained one-volume work, edited by Laura E Salt and John Pascoe (OUP, 1965). The work includes accounts of all aspects of the country and of the great men who have made it, types of farming and the industries arising from farming, the Maori way of life and culture and New Zealand natural features, flora and fauna. There are figures in the text, some four-colour plates and maps of the North and South Islands and of the distribution of the main types of farming.

938 'Oxford plastic relief maps' are an imaginative series, particularly designed for educational purposes. Of the world series planned, the map of Great Britain and Northern Ireland, 16 miles to one inch, was the first available, 1964-. The maps are available in two series, series 1, an outline edition, and series 2, fully coloured.

939 'The Oxford school atlas', in the third revised and enlarged edition, 1960, reprinted in 1963 and reprinted again, with revision, in 1970, contains 112 pages of six-colour maps and a 32-page gazetteer. In this new edition, the 1:1M maps of Britain show roads as well as railways; new maps include the Mediterranean and Southern Europe, the Middle East and China, and the section of world maps has been extended, incorporating a series of economic maps based on the second edition of *The Oxford economic atlas of the world*. *The Oxford home atlas of the world*, in a third edition with revision, 1963, is the general edition of the school atlas. *The shorter Oxford school atlas* is an abridged version of this atlas, giving the broad essentials of world geographical and topographical information; the third edition, revised in 1960, reprinted 1963, contains sixty-four pages of six-colour maps and short index gazetteer. *The little Oxford atlas*, second edition with revision, 1963, is the general edition of *The shorter Oxford school atlas*, giving world coverage, but with a high proportion devoted to Britain. The hill shading technique of the Cartographic Department of the Clarendon Press has produced some striking three-dimensional relief maps. Special regional editions are available for Australia, New Zealand, Canada and Pakistan.

940 Oxford system of automatic cartography, developed by D P Bickmore, head of the Cartographic Department of the Clarendon Press, and Dr A R Boyle of Dobbie McInnes (Electronics) Ltd, Glasgow, translates map compilations into high quality reproduction material, through the media of magnetic tape and punched cards. This eliminates repetitive plotting and drawing, expensive camera work and checking; it also enables information from different sources to be accurately co-ordinated. Thus maps can be prepared with greater speed, versatility and accuracy, so that the skilled cartographer is able to concentrate on original preparation and experimental work.

941 Oxford University Exploration Club published from 1930 to the second world war, a series of *Annual reports,* and, from 1948-, *Bulletins* reporting on Oxford University expeditions, the majority to Africa or the Far East.

942 Oxford University Press, in addition to cartographic works, produces frequent monographs and series, at various grades, of interest to geographers. 'Oxford books', geography section, is a newsheet issued from the Education Department, presenting a selection of books on applied geography for pupils in middle and upper forms of secondary schools. Catalogues, including several sections of direct interest, are issued two or three times a year, in addition to special subject catalogues, such as the 'Some Oxford books 1970-

'71' on Africa. The OUP also publishes books for specialist organisations such as the International African Institute. The ' Oxford social geographies' series is in three sections: *Work and leisure* introduces basic concepts and ideas which, in the subsequent series are built up and expanded by means of sample studies taken from Britain and the rest of the world. *The changing world* selects major environmental themes—urban growth, industry, transport and the changing countryside—which, although applied to Britain, are also of relevance to the rest of the world; while *Regions of Britain* provides detailed local studies which vividly illustrate on a small scale both the geographical method and the problems and concepts illustrated in previous volumes. The Oxford series devoted to fauna and flora are all of interest, particularly *The Oxford book of food plants,* by S G Harrison and others. The series *Oxford in Asia historical reprints* will eventually build up a considerable history on Asia, beginning with Thomas Forrest's *A voyage to New Guinea and the Moluccas 1774-1776* (1779; second edition 1780); the second was the text used for reproduction, introduced by Dr D K Bassett.

See also Clarendon Press, Oxford.

943 Oxford University School of Geography was established in 1899 under the influence of Sir Halford Mackinder, with a Chair of Geography in 1932. The library contains more than twenty-six thousand volumes, some forty thousand maps and atlases, including colonial survey maps and a large collection of Canadian and Australian material, world gazetteers, including a particularly fine set for India, and large collections of pamphlets, reprints and periodicals, making a fine collection for study purposes. For a note on the Oxford special classification for geography, *see* the article on classification above, The map curator, Miss E Buxton, has compiled a list of geological, climatic, economic, population and other thematic maps published after 1940 available in the Oxford libraries.

Refer Juliet Williams: 'The first department' *in The geographical magazine,* January 1972.

944 'The Pacific Basin: *a history of its geographical exploration',* edited by Herman R Friis (American Geographical Society of New York, 1967), was sponsored by the National Science Foundation as 'an outgrowth of a symposium on "Highlights of the history of scientific geographical exploration in relation to the development of the Pacific map ", held as part of the Tenth Pacific Science Congress meeting at the University of Hawaii in Honolulu, August 21-September 6, 1961'. The text is an expanded and revised version of the original papers and there are extensive notes and references for each section. All aspects are considered, including 'The art and science of navigation in relation to geographical exploration before 1900', map compilation, explanation of all parts of the region and, finally, 'The intellectual assumptions and consequences of geographical exploration in the Pacific'. There are a number of reproductions and maps throughout the text.

945 'A Pacific bibliography; *printed matter relating to the native peoples of Polynesia, Melanesia and Micronesia',* compiled by C R H Taylor (Polynesian Society, Wellington, 1951; revised and enlarged edition, Oxford, Clarendon Press, 1965), has been accepted as the standard bibliography. The scope of the work embraces the most important writings on the peoples of the Pacific islands, including New Zealand, entries being classified by island group and by subject, with appendices and exhaustive index. In general, entries are complete to 1960, references amounting to more than sixteen thousand. The work is well guided and with a comprehensive index.

946 ' Pacific island bibliography ', compiled by Floyd M Cammack and Shiro Saito (New York: Scarecrow Press, 1962), covers, with sub-divisions, ' Oceania ', ' Melanesia ', ' Micronesia ' and ' Polynesia'. The work is based on the materials in the Pacific Collection at the University of Hawaii, Gregg M Sinclair Library, alphabetically within the sections.

947 ' Palestine exploration quarterly ', formerly the *Quarterly statement,* 1869-, has been the journal of the Palestine Exploration Fund, founded in 1865. The issues, many of them now out of print, provide a wealth of information on the archaeology, topography, geology, physical geography and peoples, in addition to excellent maps and photographs. Dawson Reprints, in collaboration with the Palestine Exploration Fund, have reprinted the necessary issues to make the whole series again available; the volume for 1869-70 includes the whole of the Warren Reports on the surveys and explorations of Major-General Sir Charles Warren, 1867-1870.

948 Pan American Institute of Geography and History was founded in 1929 for the encouragement and co-ordination of cartographic, geographic and related work in the western hemisphere. Separate committees deal with geodesy, geo-magnetism and aeronomy, seismology, topographic maps and air photographs, aeronautical charts, hydrography, tides, special maps and urban redevelopment. The mapping and publications programme is vast; among the most central are the *Revista geografica,* 1941-, *Revista cartografica,* 1952-, and the *Bibliographical bulletin of American oceanography and geophysics,* 1958-, all issued twice a year. A library was established in 1930 at the headquarters in Mexico; the stock includes some seventy thousand books and pamphlets, a valuable collection of periodicals and more than four thousand maps and charts, covering all aspects of the Institute's work, with emphasis on the western hemisphere. Reference and bibliographical services are maintained, and abstracts, reviews and accessions lists are prepared for inclusion in the Institute's publications.

949 'The passing of tribal man in Africa', edited by Peter C W Gutkind (Leiden: Brill, 1970), brings together nine original papers on changing Africa, in the series ' International Studies in Sociology and Anthropology '. Following an editorial preface, the titles of the papers run thus: ' The passing of tribal man: a Rhodesian view '; ' Reflections on the African revolution: the point of the Biafran case '; ' The illusion of tribe '; ' The passing of tribal man: a West African experience '; ' Rural-urban communications in contemporary Nigeria: the persistence of traditional social institutions '; ' Tribe and social change in south central Africa: a situation approach '; ' Tribal survival in the modern African political system '; ' Political hygiene and cultural transition in Africa '.

950 Paris, a regional atlas, prepared under the direction of Madame Beaujeu-Garnier of the Sorbonne; with the similar project for Berlin, volume 9 of the *Deutsche Planungsatlas* (*qv*) and that in progress for London, it is significant of this new trend in atlas production.

951 Penck, Albrecht (1858-1945) was trained as a geologist and subsequently had a distinguished academic career. As director of the Geographical Institute, Berlin, he exercised considerable influence on geographical thought in Germany. His most lasting achievement was probably the initiation of the ' International map of the world on the millionth scale ' (*qv*).

952 'The Pergamon world atlas', published by the Pergamon Press and printed in Poland by Wojskowe Zaklady Kartografizzne, 1968, was based on the *Atlas Swiata,* 1962, produced by the Polish army and annotated in English for the United States and Canada. The atlas is in loose-leaf form, in a soft leather binding. One feature of special value is the inclusion of maps of Eastern Europe made from sources not readily available in the West. Forty-four pages of thematic maps cover the world and from two to four pages are allotted to thematic maps of each country or group of countries. Some pages unfold to give extra width, but scales are mostly small, especially noticeable in the city plans. Pergamon Press issued a general world atlas in 1966, with S Knight as chief editor, in which were ninety-five maps of areas and topics of international interest. Great Britain is shown on a series of 1 : 1M maps. The metre system is used throughout, with cross-references to feet; names are in the vernacular form, but there are many English equivalents in the gazetteer. The *Pergamon general historical atlas,* edited by A C Cave and B S Trinder, contains sixty-four pages of maps printed in five colours, covering the 'Ancient world'; 'Medieval Europe'; 'Modern Europe'; 'Asia, Africa and America'; 'Ancient and medieval Britain'; 'Modern Britain'; and the 'British Empire, Commonwealth and United Nations.

953 'The periglacial bulletin', organ of the Periglacial Geomorphology Commission of the International Geographic Union, published by the Polish Scientific Publishers, 1924-, brings together notes on the results of research, both in Poland and abroad. Tables of contents are in English and Russian; articles were first in Polish, French, German and Russian, later in English also, with summaries in other languages.

954 'The periglacial environment', edited by Troy L Péwé (McGill-Queen's University Press, 1969), is the printed collection of papers read at the Symposium on Cold Climate Environments and Processes held at the University of Alaska, August 1965; some were especially prepared for the volume and all are by acknowledged experts.

955 Periodicals: some specific periodicals have been mentioned in this text; others of interest to geographers may be located in *An annotated world list of selected current geographical serials in English* (*qv*) or, for example, in *Ulrich's periodicals directory.* They stem from the organs of learned and professional societies, from nearly all geographical societies throughout the world, from university geographical departments and societies based on these departments, from industrial concerns and trade organisations; they are international, national or local in scope and may include original source material or report on achievements completed or projected. Some are highly technical, others, generously illustrated, are more popular in appeal.

956 Permanent Committee on Geographical Names for British Official Use was set up in 1919 at the suggestion of the Admiralty, to study and advise on the problems of geographical nomenclature. It is an advisory body composed of representatives of the Admiralty, the Colonial Office, the Foreign Office, the Ministry of Defence, the War Office, the Ordnance Survey, the Ministry of Transport and Civil Aviation, the Post Office, the Royal Geographical Society and the Royal Scottish Geographical Society. The continuing importance of the committee's work was recognised in 1947, when it became the recipient of a Treasury grant and permanent staff, who are accommodated at the Royal Geographical Society. *Lists* of approved geographical names in countries and regions overseas have been issued, of which a new series began in 1953, and glossaries are prepared covering foreign geographical terms; much work has also been done on conventional alphabets and transliteration systems.

957 Petermann, August Heinrich (1822-1878) was a pupil at the Geographical Art School in Potsdam, founded by Heinrich Berghaus; he then worked as cartographer for A von Humboldt and subsequently at the Johnston cartographic firm in Edinburgh. With this experience he went to London, where he compiled some remarkable population maps, including those in the 1851 census, then to Gotha in 1854, to the firm of Justus Perthes. He was responsible for the great *Stieler atlas* (*qv*) and for the *Mitteilungen* which still bears his name (see below).

958 'Petermanns Geographische Mitteilungen' from its inception in 1855 to 1938 bore the title *Dr A Petermanns Mitteilungen aus Justus Perthes Geographischer Anstalt*. It combined with *Das Ausland,* with *Aus allen Weltteilen* and with *Globus.* For some forty years it remained unequalled in scientific quality and breadth of interest in geographical studies. It contains not only original articles of great scholarship, but bibliographies and evaluative notices of new publications, and cumulative indexes add to its usefulness as a research tool.

959 The Petroleum Information Bureau, London, provides information on the oil industry on a world-wide basis. Its work includes the preparation and distribution of a wide variety of printed material, including factual memoranda about the many aspects of the oil industry and its operation in the various oil-producing areas, *eg* ' The world's oil reserves ': a series covering oil development in the USA, the Middle East, the USSR, the Commonwealth countries, Latin America, the Far East, Britain, Western Europe, Africa; ' Oil —a large scale industry '. 'A century of oil: an historical account of the oil industry from 1859 to 1959 ' and ' Notes about oil ', an illustrated outline of the origin and early history of petroleum. Librarians are welcomed on the PIB mailing list of new and revised titles. A monthly newsletter contains comment on topical matters connected with oil and statistics such as United Kingdom oil consumption and world oil production. The Bureau maintains a statistics and information department where detailed information is filed and carded and a small library is available for the consultation of journals and other publications. Other services include the provision of charts, diagrams and other visual material. A colour map is available which indicates by symbols the world's oil-producing and oil-refining countries in 1960, with relevant statistics. An associated Films Bureau has about 350 film title, loaned free of charge to schools and interested organisations.

Refer also British Petroleum: *Our industry—petroleum* (fourth edition, 1970).

960 ' Petroleum: journal of the European Oil Industry ', published monthly from 1939 by Technology Publications Limited, contains illustrated articles and news concerning the world petroleum industry; items are included towards a petroleum industrial directory.

Refer also The oil and petroleum year book (Walter Skinner, London).

961 The George Philip Group, now comprising Edward Stanford Limited, Georama Limited, the Maritime Press and Kandy, began as a Liverpool business in 1834, the London House being founded in

1836. With the passing of the 1870 Education Act, the publishing programme expanded to supply the needs of elementary education. In 1902, the London Geographical Institute came into being and, by this time, a number of educational and reference atlases, wall map series and small monographs had become established. The index should be consulted for Philip's publications mentioned in this text. Numerous textbooks on many aspects of the subject, also for History, Scripture and English, are issued, many in series; and the firm has specialised in a great variety of globes, with a wide range of use and price and in map display fittings. Among recently published reference works are *A bibliography of British geomorphology,* edited by K M Clayton, and *The Geographer's Vademecum,* compiled by J C Hancock and P F Whiteley. Reference atlases include *The atlas of the earth, The international atlas, The Observer atlas of world affairs, The library atlas, The university atlas* and the *Record atlas.* The firm is also agent for many publishers overseas and is itself publisher for national organisations, such as the Geographical Association, and plays an important part in international ventures; it is, for example, joint publisher and distributor for the *International yearbook of cartography (qv).*

962 'Philippine cartography 1320-1899', compiled by Carlos Quirino, is now in a second edition, 1963, with an introduction by R A Skelton. A historical account is given of the various manuscript and printed maps and charts of the area, from the earliest Chinese map, about 1320, to the end of the nineteenth century, when the first detailed atlas of the Philippines was published. Over 1,100 maps are listed and carefully described in chronological order.

963 'Phillip's inland navigation', written by John Phillips in 1792, has been reprinted by David and Charles of Newton Abbot, with an introduction by Charles Harefield, 1970. The reprint is of the fourth edition, 1805. The work remains a classic, covering every county.

964 The Photogrammetric Society, London, publishes the *Photogrammetric record,* with which is included the annual report with the President's address. The society aims to facilitate the exchange of information and ideas and to participate in national and international conferences, to hold meetings and to issue publications, encourage research and maintain a library and collections of photographs, plans and models.

965 'Photogrammetry and current practice in road design', by K M Keir, is a unique fifteen-page report published by Hunting Surveys and Consultants Limited, summarising the design process and attempting 'to dispel the commoner misconceptions surrounding the technique'.

966 'Physical and information models in geography: *Parts I, II and V of Models in geography'*, edited by Richard J Chorley and Peter Haggett (Methuen, University paperbacks, 1967; *see also Models in geography*), deals with 'Models, paradigms and the new geography', 'The use of models in science', 'Models in geomorphology', 'Models in meteorology and climatology', 'Hydrological models and geography', 'Maps as models', 'Hardware models in geography' and 'Models in geographical teaching'. The contributors are mainly young geographers working in British universities. There are figures and references in the text.

967 'Physical atlas of zoogeography', compiled by W E Clark and P H Grimshaw, was issued by Bartholomew in 1911 as part of the proposed *Physical atlas*. More than two hundred maps of the world show the distribution of mammals, birds, reptiles and amphibians, as well as a selection of fishes, molluscs and insects, together with ninety-two pages of explanatory text.

968 'The physical geography of China', prepared by the Institute of Geography, USSR Academy of Sciences, under the general editorship of V T Zaychikov, was originally published by the 'Thought' Publishing House, Moscow, 1964, and issued in an English translation by the United States Government Joint Publication Research Service, Washington DC, in 1965. A xerox photocopy was distributed by the United States Department of Commerce, Technical Information Service and the work was republished in two volumes by Praeger in 1969. Important as the first major work in a European language on the physical geography of the Chinese People's Republic, the text describes the overall natural conditions and the chief natural regions, based on much field work and the results of research expeditions carried out by Chinese and Russian scholars. The work contains an introduction and six articles by Soviet specialists on topography, climate, inland waters, soils, flora and vegetation, and the animals found in China. The bibliography is comprehensive, including articles, divided into Cyrillic and non-Cyrillic sources; the bibliography in the Praeger edition is arranged alphabetically by author's name.

969 'Physikalischer atlas of Heinrich Berghaus' was a unique achievement, thought to be the first thematic world atlas (Justus Perthes, 1845; second edition, 1852). It was of particular importance because of its influence on atlas production and the use of techniques. Careful thought was given to the choice of symbols; colour gradations were used to show changing relief, the colours being added by hand, the new lithographical method having been rejected on financial grounds. Maps of ethnographic interest were also included, showing population density, distribution of races, and such aspects as education and government.

Refer Gerhard Engelmann: 'Der Physikalische Atlas des Heinrich Berghaus: die kartographische Technik der ältesten thematischen Kartensammlung', *International yearbook of cartography*, 1964.

970 'Picture atlas of the Arctic', compiled by R Thorén (Elsevier, 1969), follows the plan—'The Arctic Ocean'; 'Drifting Ice Stations', 'The Arctic region of Alaska'; 'The Canadian Arctic'; 'Greenland (Denmark)'; 'Iceland'; 'Norwegian Islands in the Arctic'; 'Arctic Scandinavia' and 'The Soviet Arctic', all sections being divided as necessary. Air and ground photographs are included, also maps in the text and there are a number of useful references.

971 'The picturesque atlas of Australasia', edited by Andrew Garran (London, 1886-89), is in three volumes, consisting mainly of text, profusely illustrated with engravings and with map plates at intervals, each volume being complete with its own index. Especially interesting for geographers are sections on the early discoverers, including Captain Cook, early settlement, the topography of individual states, with sections on the chief towns and cities, such as Sydney, Melbourne and the Hunter River district, with maps showing average rainfall, railways and other significant features. It is not systematic or balanced in coverage, but contains a vast amount of information currently to hand.

972 'The Pioneer histories' series of A and C Black are intended to provide broad surveys of the great migrations of European peoples, for purposes of trade, conquest and settlement, into the non-European continents. They aim to describe a racial expansion which has created the complex world of today, so nationalistic in its instincts and so internationalised in its relationships. Each volume takes for its subject the history of an important movement and, while related to others in the series, is thus complete in itself. Outstanding in the series are Edgar Prestage: *The Portuguese pioneers;* J B Brebner: *The explorers of North America;* Eric A Walker: *The great trek;* J C Beaglehole: *The exploration of the Pacific;* F A Kirkpatrick: *The Spanish conquistadores;* James A Williamson: *The age of Drake;* W P Morrell: *The gold rushes;* Sir William Foster: *England's quest of Eastern trade;* Arthur Percival Newton: *The European nations in the West Indies.*

973 ' Plant science: an introduction to world crops ', by Jules Janick and others (Freeman, 1969), constitutes an introductory study of agronomy, horticulture and forestry throughout the world, written by four expert authors. The contents, ' Plants and men ', ' Nature of crop plants ', ' Plant environment ', ' Strategy of crop production ', ' Industry of plant agriculture ' and ' The market place ', present the essential background factors to be considered in any environmental studies. The text is extensively illustrated by small photographs and diagrams.

Refer also Ronald Good: *The geography of the flowering plant* (Longmans, 1947, 1953, 1964).

Marion I Newbigin: *Plant and animal geography* (Methuen, 1936).

974 ' Pleistocene geology, a biology, with special reference to the British Isles ', by R G West (Longmans, 1968), concerns mainly the glaciated and periglaciated parts of North-West Europe, with the two final chapters dealing superficially with the Pleistocene of the British Isles. Chapter by chapter, the text presents a general account and synthesis of the Pleistocene—' Ice and glaciers ', ' Glacial geology ', ' Non-glacial sediments and stratigraphy ', ' The periglacial zone ', ' Stratigraphical investigations ', ' Biological investigations ', ' Land/sea-level changes ', ' Chronology and dating ', ' Climatic change ', ' Pleistocene successions and their subdivision ', ' Pleistocene history of the flora and fauna of the British Isles '. There are two Appendices: ' Methods of isolating and counting fossils ' and ' Lacquer method of treating sections '; plates, figures and diagrams abound throughout the text and references have been appended to each chapter.

975 'The Polar bibliography' is produced by the United States Library of Congress, Science and Technology Division, for the Department of Defense, Washington, 1956-. Covering polar and subpolar regions, abstracts are included for unclassified reports and other documents prepared by relevant organisations since 1939; there are subject and individual author indexes.

Refer also Polar research: a survey (Washington DC, Committee on Polar Research, National Research Council, National Academy of Sciences, 1970).

976 'Polar record', a journal of Arctic and Antarctic exploration and research issued three times a year since 1931 by the Scott Polar Research Institute, is one of the most valuable sources for polar studies, giving a balanced survey of polar work. Technical articles, review articles and notes are included and a feature is 'Recent polar literature . . .', which includes indicative abstracts. An index is now issued every two years, and a cumulative index is available, 1931-1959, in several volumes.

977 'The Polar world', by Patrick D Baird (Longmans, 1964), is one of the *Geographies for advanced study* series; among the vast literature on this subject, it is probably one of the most balanced and comprehensive treatises to its date, illustrated with photographs, sketchmaps and diagrams, and including chapter references. Following a general introduction, the history of Arctic exploration is traced, the physical aspects of landforms and seas, flora, fauna and native peoples and transportation. Regional descriptions of special parts include Svalbard and Greenland, after which similar treatment is given to Antarctica and the sub-Antarctic islands.

978 Polish Scientific Publishers (PWN) include geographical studies as one of their specialities and have published many valuable texts, particularly on the geography, soils and hydrography of Poland. The *Polish geographical bibliography,* compiled by the PAN Institute of Geography, began with the year 1945 (1956-); the first of the retrospective volumes, to be issued as completed, was the volume covering 1936-1944 (1959). PWN has also published for the institute the *Polish geographical nomenclature of the world. The periglacial bulletin,* organ of the Periglacial Geomorphology Commission of the International Geographical Union, has been published since 1924. *The universal geography of PWN,* to be complete in forty-five volumes, has been issued in parts since 1963, together with the respective sheets of the *PWN atlas of the world* (*qv*).

979 'Political geography', by N J G Pounds (McGraw-Hill, 1963, 1967), the outcome of working with students, taking Hartshorne's definition: 'The study of the variation of political phenomena from place to place in intercommunication with variations in other features of the earth as the home of man. Included in these political phenomena are features produced by political forces and the political ideas which generate those forces'. Topics dealt with include 'The state and the nation', 'Area and location of the state', 'Frontiers and boundaries', 'The territorial sea', 'Population', 'Resources and power', 'Core areas and capitals', 'The geography of administrative areas', 'Geographical aspects of relations between states', 'The political geography of foreign trade'. 'The political geography of rivers', 'The political geography of international organisations', 'Colonies and colonisation', 'The undeveloped world', 'The political patterns of the world'.

Refer also Richard Hartshorne: 'Political geography in the modern world' *in Journal of conflict resolution,* 1960.

R E Kasperson and J V Minghi, *ed The structure of political geography,* 1970.

J R V Prescott: *The geography of frontiers and boundaries* (Hutchinson University Press, 1965, 1967).

H W Weigert: *Principles of political geography* (New York, 1970).

980 'Politics and geographic relationships: *readings on the nature of political geography',* by W A D Jackson (Prentice-Hall, 1964), is in many ways a pioneering work presenting concepts underlying the phenomena and motivation which dictate man's organisation of himself and of the surface of the earth. The problem of boundaries and frontiers is discussed, also core areas and capital cities, the bases of economics and population, politics and transportation, the problem of the sea, international resources and political implications and the political and geographical implications of space research.

Refer also W A D Jackson: *The geography of state policies,* in the same series, 1968.

981 The Polynesian Society, Wellington, New Zealand, exists for the study of the native peoples of the Pacific area; the quarterly *Journal*, 1892-, has provided a major forum for discussion on all aspects of the New Zealand Maori people and other Pacific Island peoples. It includes bibliographies and a substantial book review section. Other publications of the Society include Maori texts, monographs, a reprint series and other miscellaneous publications, distributed by Reed, publishers, of Wellington. Particularly interesting is *Polynesian navigation*, edited by Jack Golson.

> *Note* A W Reed is the acknowledged leading Maori scholar; his *Illustrated encyclopedia of Maori life* is the standard reference work, of which a shortened version, the *Concise Maori encyclopedia,* was published in 1964.

982 'Population studies': *a journal of demography,* founded by the Population Investigation Committee, has been published since 1947 by the London School of Economics and Political Science. The first four volumes appeared in quarterly parts, thereafter three issues a year were published. The shorter research reports of the committee appear in the journal, together with careful reviews and book lists. In addition, the results of research carried out in other countries are included, with the assistance of advisory editors in France, India, Sweden and the USA.

> *Compare Annales de démographie historique* (La Société de Démographie Historique, 1964-).

983 'Portolan charts: *their origin and characteristics with a descriptive list of those belonging to The Hispanic Society of America',* by Edward Luther Stevenson (The Hispanic Society of America, 1911), contains in addition a brief, general, annotated bibliography. The short interesting text brings together the known facts about the charts and their makers, including a number of excellently chosen reproductions. The descriptive list comprises numbered charts, arranged chronologically, mentioning known details about their authors and characteristics.

984 'The Portuguese pioneers', by Edgar Prestage (Adam and Charles Black, 1933, reprinted 1966), has become a classic and probably the best example of *The Pioneer histories* series, edited by V T Harlow and J A Williamson, which was devised to ' provide broad surveys of the great migrations of European peoples . . .' (*qv*). The expeditions here described include the discovery of Madeira, the

Azores and the Cape Verde Islands, the coasts of Africa and Brazil and the sea passages to India, Malaya, the Spice Islands, China and Japan, led by such outstanding personalities as Prince Henry the Navigator, Diogo Cão, Cadamosto, Bartholomew Dias and Vasco da Gama. There are four useful maps.

Refer also C R Beazley: *Prince Henry the Navigator: the hero of Portugal and of modern discovery 1394-1460* (reprinted, 1968).

985 ' The Pre-Cambrian along the Gulf of Suez and the Northern part of the Red Sea ', by H M E Schürmann (Leiden: Brill, 1966), is a major work, including figures and sketchmaps in the text, fifty-five plates and three folded maps in nine colours and three in black and grey. Special attention has been given to the investigation of the components of conglomerates which are used to clarify the mutual relations among the Pre-Cambrian formations. The work provides a most notable addition to knowledge of the Pre-Cambrian areas of the globe and essential reading for all bedrock geologists, representing the field and the laboratory activity of more than fifty years. Its greatest single value has perhaps been to use all the available methods in establishing the relative and absolute ages of these old formations devoid of fossils.

986 'A preliminary bibliography of the natural history of Iran ', compiled by Robert L Burgess and others (Pahlavi University College of Arts and Sciences, 1966), is in English and Farsi. More than 1,700 references have so far been entered and more information will be available when the work is more complete.

987 Presses Universitaires Français, Paris has become noted for the publication of scholarly geographical monographs; these include many regional studies prepared by French academic geographers, sometimes based on doctoral theses. The ' France de demain ' series was designed to provide a new and up to date regional geography of France.

988 ' Pressures on Britain's land resources ', by Dr G P Wibberley (University of Nottingham Department of Agricultural Economics, 1965), was the tenth Heath Memorial lecture, delivered on 5 March, 1965; though short, it provides an excellent summary of the situation and concise factual data.

989 ' **The printed maps in the atlases of Great Britain and Ireland:** *a bibliography, 1579-1870, with an introduction by F P Sprent and bibliographical notes on the map-makers, engravers and publishers* ', was compiled by Thomas Chubb, assisted by J W Skells and H Beharrell (Homeland Association, 1927). Sections deal successively with the atlases of England and Wales, Scotland and Ireland, arranged chronologically, with analyses of contents and bibliographical notes.

990 ' **The printed maps of Tasmania:** *a descriptive bibliography* ', by R V Tooley (Map Collectors' Circle, 1963), begins with the rediscovery of Tasmania towards the end of the eighteenth century and lists, with annotations, the printed maps from that time to 1900.

991 ' **Problems and trends in American geography** ', edited by Saul B Cohen (Basic Books, 1967), and sponsored by the Association of American Geographers, provides a summary of contemporary American geographic thought, contributed by nineteen authors. Key problems facing American geographers are analysed, all connected in some way with the understanding of the man-environment confrontation.

992 ' **Proceedings of the symposium on the granites of West Africa: Ivory Coast, Nigeria, Cameroon** ', March 1965, was published by Unesco in English and French, 1968. Specialists in the study of granites in Africa and other continents took part in an itinerant symposium in the field studying the granite formations, comparing and discussing their observations. Conclusions were given in the papers forming this publication.

993 ' **Progress in geography:** *international reviews of current research* ', the first English work of its kind in the subject, is edited by Christopher Board, Richard J Chorley, Peter Haggett and David R Stoddart, with a further panel of advisory editors from overseas universities, published annually by Arnold from 1969. Within the framework of five or six chapters, a wide range of individual topics is assessed and discussed, illustrated with photographs, sketch-maps and diagrams as necessary. The aim of the editors is to present regular, scholarly reviews of current developments within the field.

994 ' **Progress in oceanography,** edited by Mary Sears of Woods

Hole Oceanographic Institute, began in 1963 (Oxford: Pergamon Press, 1964-) as a medium to make the new developments in oceanography more widely known. It is a combination of original reports, reviews and bibliographies.

995 'A prologue to population geography', by Wilbur Zelinsky (Prentice-Hall International, 1970, in the *Foundations of economic geography* series), as with all the volumes in this series, concentrates on a major theme of economic geography. The series as a whole is intended to provide a broad cross-section of current research in economic geography, stemming from a concern with a variety of problems. This, the first volume of the series, acts as a bridge between economic and cultural geography, exploring ideas and methods. In three main parts—' What does the population geographer study?'; ' Distribution of the world population '; and ' Towards a typology of population regions'. An Appendix sets out ' Demographic, social and economic indicators of development level for forty-one selected countries ' and a selected annotated list of references not already cited in the text is given.

996 'A prospect of the most famous parts of the world', compiled by John Speed, 1627, may be considered the first printed atlas by an Englishman. The American maps added to the 1676 edition are reproduced in the Theatrum Orbis Terrarum reprint, 1966, with an introduction by R A Skelton.

997 ' Provinces of England: *a study of some geographical aspects of devolution'*, by C B Fawcett, 1919, was re-issued in 1960 by Hutchinson, revised and with a preface by W Gordon East and S W Wooldridge. The book, regarded by its author as ' an essay in the application of geography to a particular political problem, that of the delimitation of Provinces of England ', was in three sections: a study and criticism of the existing political divisions of England; secondly, an account of each province—North England, Lancashire, Peakdon, Yorkshire, West Midlands (or Severn), East Midland (or Trent), Devon, Wessex, Bristol, East Anglia, London, Central England—and the remaining topics to be considered, namely the Anglo-Welsh boundary; unity of the Provinces, the Provinces as educational areas, relation of the Provinces to other principle divisions. The thesis of the work was first outlined in a lecture to the Royal Geographical Society entitled 'Natural divisions of England', printed in *The geographical journal* in 1917, pp 124-41. Appendices in the book include 'Population and area of each province' in four categories. There are figures in the text and a folded map, 'Density of population per square mile', 1931. The book proved controversial when published, but must undoubtedly be regarded as a classic of original geographical thought and application.

998 Ptolemaeus, Claudius, usually known as Ptolemy, worked in Alexandria between 127 and 151 as mathematician, astronomer and geographer. He was the first to deal systematically with latitude and longitude and to attempt the scientific construction of maps; his ideas dominated geographical thinking for centuries, notably his concepts of an encircling 'unknown land' and of a series of climatic zones. His great work, the *Geography*, became a standard, in spite of its numerous errors, because in it Ptolemy aimed at the highest level of precision and completeness and also because, after the second century, no further advance in geographical exploration and knowledge was made for a very long time. The appearance of a Latin translation of the *Geography* in 1410 was particularly important; Renaissance scholars studied the work with greater interest even than Ptolemy's contemporaries had done, and it helped to stimulate a renewed zeal for geography and exploration. The work itself appeared in a great number of editions and a vast literature has grown up around it, both definitive and evaluative, in addition to the attention given to it in histories of cartography and geography. All the maps of the 1490 edition are reproduced in A E Nordenskiöld's *Facsimile atlas . . . (qv)*; the Italian editions are analysed in A M Hind: *Early Italian engraving*, 1938. Theatrum Orbis Terrarum of Amsterdam issued in 1969 a facsimile reproduction of the *Geographia*, Venice edition 1511, introduced by R A Skelton outlining the current knowledge of this edition, with a brief biography of Bernardo Silvano of Eboli, who contributed some corrections and improvements to it.

Refer G R Crone: 'Epic work of Claudius Ptolemy' *in The geographical magazine,* October 1971.

J Winsor: *Bibliography of Ptolemy's Geography,* 1888.

W H Stahl: *Ptolemy's Geography: a selected bibliography* (New York Public Library, 1953).

H N Stevens: *Ptolemy's Geography,* 1908.

L Bagrow: *The origin of Ptolemy's Geographia,* 1946.

E Lynam: *The first engraved atlas of the world* ..., 1941.

R V Tooley: *Maps and map-makers* (Batsford, second revised edition, 1952).

999 'The PWN atlas of the world', in four main parts, eight fascicules, comprising altogether five hundred original physical, demographic and economic maps on scales of from 1,250,000 to 1:10M, has been supplied with the respective parts of the *PWN universal geography* from 1963. The index includes some 150,000 graphical names. PWN (Państwowe Wydawnictwo Naukowe; Polish Scientific Publishers, Warsaw) are noted also for the publication of scientific journals in several languages and for scholarly monographs in almost every field of knowledge, geography being one of the main interests.

1000 'Quantitative geography: *technique and theories in geography'* by John P Cole and C A M King (Wiley, 1968), is in four parts: 'Introduction, mathematics and statistics'; 'Spatial distributions and relationships'; 'Dimensions of space and time'; 'Models, theories and organization'. This kind of thinking is becoming increasingly important in geography. There are numerous worked examples in this book, a glossary, figures, chapter references and a section 'Mathematical and statistical signs'.

Refer also W V Tidswell and S M Barker: *Quantitative methods,* 1971.

1001 'The quaternary era, with special reference to its glaciation', in two volumes by J K Charlesworth (Arnold, 1957), is an exhaustive work for reference on all aspects of the Pleistocene epoch. The text is illustrated and includes maps and diagrams.

1002 'The quaternary of the United States: *a review volume for the Congress of the International Association for Quaternary Research'*, prepared and edited by H E Wright jr, and D G Frey (Princeton University Press; OUP, 1965), is an essential summary of the field of study, illustrated with maps and diagrams.

1003 'A question of place: *the development of geographic thought'* (R W Beatty, Limited, 1967, 1969), by Eric Fischer, Robert D Campbell and Eldon S Miller, began as a graduate seminar on the development of geographic thought at the George Washington University; it brings together a vast amount of factual knowledge and comment, in two parts: 'Early geography' and 'Modern geography'. Greek and Roman geographers, Arab geographers, Renaissance and post-Renaissance geographers, 'Fathers of modern geography' are each examined, followed by chapters on the development of geographical studies, particularly in Germany, France, Great Britain, the USSR and the United States.

1004 'The railway encyclopaedia', compiled by Harold Starke, 1963, contains information on the different class locomotives, railways by place-name, eminent persons engaged in railway history in one capacity or another and some interesting out-of-the-ordinary entries, such as 'Amalgamations that failed'. Brief notes are given on each topic.

1005 Raisz, Erwin (1893-1968), the Hungarian geographer, died in Bangkok, while on his way to the International Geographical Congress in New Delhi. During his professional career in cartography and geography, he had drawn thousands of maps for a variety of publications and had himself written and illustrated some hundred papers and four books and had produced two atlases. His major work, *General cartography,* opened up what was virtually a new field in American geography; he began the first course in cartography at Columbia University and lectured at the Institute of Geological Exploration at Harvard. He was prominent in the first efforts to organise a cartographic section in the Association of American Geographers and was chairman of the Cartography Committee for seven years; he was also the first map editor of the *Annals* and himself designed a number of maps for the National Atlas of the United States.

1006 Rand McNally Company, Chicago, has as one of its main interests the publishing of maps, guides, atlases and globes. The atlases include world atlases at various levels, such as *Goode's world atlas* (qv), the *Collegiate world atlas,* the *Cosmopolitan world atlas, Current events world atlas, $1.00 world atlas,* the *International world atlas;* historical atlases, including the *Atlas of world history* edited by R R Palmer, the *Historical atlas of the Holy Land,* and the *Rand McNally Bible atlas* by Emil G Kraeling; regional atlases of America and the *Atlas of western Europe,* by Jean Dollfus; and a number of atlases showing communications and travel information.

A landmark in the firm's publishing history was the production of the *Rand McNally road atlas,* packed with information especially on the United States, Canada and Mexico; the atlas has gone into numerous, now annual, editions. Special features include the indication of National Parks, insets of city plans, differentiation of highways and notification of new developments. Relief is not shown, except in the form of spot heights. The *Commercial atlas and marketing guide,* first published in 1876, has world coverage, with emphasis on America; as the new editions have multiplied, the nation's expansion to the west has been traced, the development of railways and the extension of settlements and communications. The firm's series of fifteen pocket sized regional guidebooks of the United States, begun after the second world war, provided another landmark in the expansion of the firm's projects. Geo-Physical Maps Inc, an organisation specialising in producing six-foot geophysical globes for government agencies, commercial firms and educational institutions was absorbed.

1007 ' **Rare Australiana:** *catalogue of facsimiles and other publications of the Libraries Board of South Australia* ' (Adelaide: Libraries Board of South Australia, 1965) is, as the Board hopes, of historic importance and scholarly work '. Both xerographic/photographic and letterpress methods were used; the facsimile editions are exact reproductions of the original texts and they include all illustrations and charts in colour or black and white. Among the entries are many references to early books on Australia and the Pacific.

1008 ' **Rare county and other maps** ' is the title of a collection reproduced by the Royal Geographical Society and John Bartholomew and Son, Limited, of maps in colour from Blaeu's rare atlas of 1648; they include maps of Cheshire, Kent, Lancashire, Middlesex, Norfolk, Warwickshire, Worcestershire, Yorkshire and one of the British Isles.

1009 Ratzel, Friedrich (1844-1904) was a geographer of recognised scholarship, acquired in the course of his studies at a number of universities and during his wide travels; his academic influence, particularly at Berlin and Leipzig, was considerable. Of all his published works, in periodical, treatise and monograph form, the two which have achieved lasting fame were the *Anthropogeographie,* in two volumes, 1882, 1891, revised in 1903; and the *Politische Geographie,* 1897, in a new edition by E Oberhammer, in 1925. Much controversy, both informed and uninformed, has since raged around his ideas, which initiated the theory known as 'geographical determinism' and *Lebensraum.* Ratzel stimulated and wrote the philosophy of the first actual *Weltgeschichte,* carried out by his pupil, Hans Helmolt, in nine volumes 1899-1907.

Refer also Harriet Wanklyn: *Friedrich Ratzel: a biographical memoir and bibliography,* 1961, which contains extensive biographical and bibliographical notes.

E C Semple: *Influences of geographic environment (qv).*

1010 'The Reader's Digest great world atlas', 1962, 1965, demonstrates the encyclopedic trend in modern atlases and also the graphic use of relief model techniques. Planned under the direction of Professor Frank Debenham, the atlas is in three sections, 'The face of the world', 'The countries of the world', and 'The world as we know it'; there is a British Isles index and a world index. Many of the maps are pictorial and diagrammatic, accompanied by textual notes and statistics. The conventional maps are by Bartholomew. The new edition was revised and brought up to date. There is also *The Reader's Digest atlas of Australia,* which includes Papua and New Guinea, 1968.

1011 'Reader's guide to Scotland: a bibliography', published by The National Book League, 1968, covers all aspects of the country, the subjects being arranged in broad groups, subdivided as necessary. Particularly valuable to geographers are Section I, 'General', for background reading: section II, 'History', especially the subdivisions on the economy, agriculture, industry, commerce, transport and communications; III, 'Tourism', in which are to be found references to maps, as well as books; VII, 'Education', including universities; IX, 'Administration', especially 'Town and country planning'; and X, 'Agriculture, industry and commerce'. The editor of the list is D M Lloyd, Keeper of the Printed Books, National Library of Scotland. In its main outlines, the list follows the plan of that published in 1950, compiled by the late Dr Henry W Meikle and his collaborators.

1012 'Readings in economic geography', edited by H J Roepke and T J Maresh (Wiley, 1967), a collection of papers by American and Australian geographers, consists of a selection of articles, including case studies and examinations of principles and individual topics, grouped within the following framework: 'Population and resources'; 'The exploitation of biotic resources'; 'Intensive subsistence agriculture'; 'Middle-latitude mixed farming'; 'Tropical commercial agriculture'; 'Specialised farming'; 'Metallic minerals'; 'Fuel minerals and energy production'; 'Manufacturing'; 'Transportation and trade'; 'Services and urban activity'. Each main section, prefaced by a summary, contains from four to nine articles. Many maps, sketchmaps, diagrams and some small photographs illustrate the text.

1013 'Readings in the geography of North America: *a selection of articles from The geographical review* 1916 to 1950, consists of twenty-two articles re-printed; selection was in commemoration of the American Geographical Society's centennial and as a gesture to foreign geographers attending the Seventeenth Congress of the International Geographical Union (New York: The Society, 1952), illustrated with maps and diagrams and including a bibliography.

1014 'Recent history atlas: *1870 to the present day'*, edited by Martin Gilbert (Weidenfeld and Nicolson, 1966; second edition, 1967), includes 121 diagrammatic maps in black and white by John R Flower, presenting the main historical developments of the past hundred years. Each map has been specially designed to help explain some important episode, such as the founding and progress of the League of Nations 1919 to 1939, treaties, alliances or population problems. The work is prefaced by a chronological table of events during the period, mainly in Great Britain, Germany, France, Italy, the United States and the USSR.

1015 Réclus, Jacques Elisée (1830-1905) was Professor of Comparative Geography at the Université Nouvelle in Brussels, which he established as a geographical institute. His geographical work, written in the tradition of Ritter and Humboldt, is centred in the comprehensive *La terre,* in two volumes, published between 1867 and 1869, and translated into English by B B Woodward in 1871. His *Nouvelle géographie universelle* in nineteen volumes, 1876-1894 was translated and edited by E Ravenstein and A H Keane, published in London, 1878-94. Again, *L'homme et la terre*, in six volumes, 1905-1908 describes physical milieu and phenomena, the distribution of mankind and the history of human institutions and their interrelations; this work was issued in a new edition by Paul Réclus, G Goujon and others, in three volumes, 1931.

1016 Reed of Wellington has published a valuable series, each book giving an outline of its subject within a compass of thirty-two pages, illustrated: Don Wasley: *Airways of New Zealand;* A W Reed: *How the Maoris came, How the Maoris lived* and *How the white men came;* A W Reed: *Pastoral farming in New Zealand;* A N Palmer: *Railways in New Zealand;* J H Millar and A W Reed: *Roads in New Zealand;* A N Palmer: *Shipping of New Zealand.* A second series has proved equally popular and informative: S R West: *A guide to trees;* C S Woods: *Native and introduced freshwater fishes;* D C M Manson: *Native beetles;* Charles Masefield: *Native birds;* D C M Manson: *Native butterflies and moths;* R K Dell: *Native crabs;* Bruce Hamlin: *Native ferns;* David Miller: *Native insects;* J M Moreland: *Native sea fishes;* Sheila Natusch: *Native rock;* H B Fell: *Native sea-stars;* R K Dell: *Native shells;* Bruce Hamlin: *Native trees.* Reed's *Atlas of New Zealand,* compiled by A W Reed, is probably the publishers' most widely known single work (A H and A W Reed, 1952), including large scale maps of regions, followed by distribution maps of farming, vegetation, livestock and the location of air facilities. Some statistical information is included.

1017 ' Referativnyi zhurnal geografiya ' has been issued bi-monthly from 1951, then monthly from 1956, by the Akademiya Nauk SSSR, Moscow. It is a journal of the greatest scholarship, containing original studies, abstracts, bibliographies and reviews; the contents pages are in English. Some thirty-seven thousand abstracts and quotations are included annually. The *Documentatio geographica* amounts to about 4,500 abstracts yearly.

1018 ' Reference guide for travellers ' was compiled and edited by J A Neal (New York: Bowker, 1969); arrangement is regional, with an appendix section dealing with individual topics, such as ' Travel periodicals '. Annotations are included as necessary, and there are a place index, a publisher's index and an author-title index.

1019 'A reference guide to the literature of travel', *including voyages, geographical description, adventures, shipwrecks and expeditions*, by E G Cox, is an extensive work of the greatest usefulness, in three volumes (University of Washington, 1935-49). Volume 1, covering the old world, 1935, was reprinted by lithography in 1948; volume 2 deals with the new world, 1938, and volume 3 with Great Britain, 1949. Source material is covered to 1800. Volume 3 includes also chapters on maps and charts, general reference books and bibliographies; indexes of personal names are given in volumes 2 and 3. A fourth volume was planned for Ireland.

1020 'Reference pamphlets', published by the Central Office of Information, comprise short factual texts, each concerning some aspect of the Commonwealth; some of the later titles have been no 25, 'Consultation and co-operation in the Commonwealth'; no 38, 'Nigeria: the making of a nation'; no 45, 'Sierra Leone: the making of a nation'; no 48, 'Tanganyika: the making of a nation'; no 51, 'Jamaica: the making of a nation'; no 52, 'Community development: the British contribution'; no 53, 'Trinidad and Tobago: the making of a nation'; no 54 'Uganda: the making of a nation'; no 55, 'Promotion of the sciences in the Commonwealth'; no 56, 'The Federation of Malaysia'; no 58, 'The Colombo Plan' and no 59, 'Kenya'.

1021 'Regional economic analysis in Britain and the Commonwealth, a bibliographic guide', by F E Ian Hamilton, was prepared for the Commission on Methods of Economic Regionalization, International Geographical Union, established in 1960. The volume is in seven parts: 'The British Isles'; 'The British Commonwealth in general'; 'Africa'; 'Australia and New Zealand'; 'Canada'; 'South and South-east Asia'; 'The smaller territories, British Caribbean and the Atlantic Islands'. Dr Hamilton's introduction, based in part upon a paper presented to the Strasbourg Conference of the IGU Commission on Methods of Economic Regionalization, July 1967, traces the evolution of the regional concept from the works of Camden to the Reports of the Board of Agriculture, the ideas of Halford Mackinder and others through to recent quantitative studies. There is an introduction to each part and the entries in part I include abstracts.

Refer also Paul Claval et Étienne Juillard: *Région et régionalisation dans la géographie française et dans d'autres sciences sociales: bibliographie analytique* ... (Paris: Dalloz, 1967).

1022 'Regional geography of the world', by Jesse H Wheeler jr and others, was completely revised for the third edition (Holt, Rinehart and Winston, 1969). Eight regions are distinguished—Europe, the Soviet Union, the Middle East, the Orient, the Pacific world, Africa, Latin America, Anglo-America. The characteristics of each are discussed and their role in the world as a whole, based on the factual physical and cultural features. A new section was entitled 'Processes that shape the geography of areas'. There is a chapter on urban geography, using London as an example; statistics have been revised to 1968 in as many cases as possible, maps and bibliographies have been up-dated and there are many new photographs.

1023 'Regional geography, theory and practice', by Roger Minshull (Hutchinson University Library, 1967, reprinted, 1968), presents a survey of the methods, concepts and objectives of the 'regional' studies undertaken by geographers: 'The regional method of description'; 'Regions as real objects'; 'Formal and functional regions'; 'The nature of regional geography'; 'Alternatives to an inadequate concept'; 'The influence of methods of mapping'; and conclusions. There are figures in the text, notes and references at the ends of chapters and a bibliography.

1024 'A regional history of the railways of Great Britain', the overall title of a series published by David and Charles of Newton Abbot, includes 'The West Country', 'Southern England', 'Greater London', 'North East England', 'The Eastern Counties', and others in preparation. The texts, which are illustrated, trace the development of the railway systems and relate their progress to economic and social conditions.

1025 'Regional studies: *journal of the Regional Studies Association',* edited by Dr Peter Hall (Pergamon Press, May 1967-), is published twice a year and is international in scope. Papers are in English, with summaries in English, French, German and Russian. Contributions reflect the application of systematic method to the solution of problems of regional planning. Subjects covered include the machinery of regional economic planning, regional economic development, economic growth in developing countries, metropolitan regional planning and models of regional and urban development. Most articles carry references and the text is illustrated.

Refer also P Birot: *Les régions naturelles du globe* (Paris: Masson, 1970).

1026 'Régions, nations, grands espaces: *géographie générale des ensembles territoriaux'*, by Paul Claval (Paris: Editions M-Th Génin, 1968, in the series *Géographie économique et sociale*), a monumental text, illustrated with a few photographs, sketchmaps and diagrams, is presented in two main parts: 'Théorie des ensembles territoriaux et de leurs rapports' and 'Régions, nations, grands espaces'; the variations of economic geography are examined in all aspects. There is a list of the authors cited in the text, usually in the numerous footnotes.

1027 Reinbek International Forestry Documentation Centre, a unit of the German Federal Research Organization for Forestry and Forest Products, co-operates with the Commonwealth Forestry Bureau, Oxford, the Library of the Department of Agriculture in Washington and with FAO to give a completely international forestry documentation service, dealing with more than ten thousand literature items a year. The Oxford 'decimal system' is used. Work is in progress on a *World forestry atlas* (*qv*); a quarterly 'titles service' is maintained; also quarterly is the *Bibliographie des Forstlichen Schrifttums Deutschlands,* in addition to special bibliographies prepared on request.

1028 'Reinhold one-volume encyclopedias', includes a series of 'Earth sciences' volumes. *The encyclopedia of oceanography* is the first in this series, edited by Rhodes W Fairbridge (1966), a comprehensive work prepared by a large team of specialist scholars. Entries are in alphabetical order, generously illustrated throughout with photographs, sketchmaps, diagrammatic maps, graphs and diagrams, as appropriate; references follow all the main entries.

1029 Reise- Und Verkehrsverlag, Stuttgart, publishes a sales catalogue, the *RV Katalog*. This is loose-leaf, kept up to date by monthly supplements, and is probably the best guide to maps currently on sale.

1030 'Répertoire d'établissements enseignant la cartographie'

(Paris: Comité Français de Cartographie, 1968), was compiled by Brigadier D E O Thackwell, with the collaboration of delegates from member countries of the International Cartographic Association, with the aim of collecting information on the different systems in the training of cartographers now in use and of making the information available in as concise and convenient a form as possible. The project arose from discussions on 'Education in cartography' of Commission I, 1964, and was presented at the Fourth Technical Conference on Cartography, New Delhi, 1968. In the catalogue, the distinction is made between essentially cartographic activities, called 'W' and activities complementing others, marked 'V'. Entries are listed first alphabetically by country within the categories, then alphabetically by name of institution. Text is in English and French.

Refer also Brigadier R A Gardiner and H Fullard: 'National qualifications in cartography' *in The cartographic journal,* June 1971.

1031 The '**Research digest**' is a semi-annual publication, issued formerly, 1954-, by the Bureau of Community Planning, University of Illinois, now by an informal organisation, Urban Planning Research Group; it is international in scope and is most valuable in its role of bringing together information on all aspects of this growing subject, with an emphasis on methodology and the problems of urban and regional planning. Also published is the *Quarterly digest of urban and regional research.*

1032 '**Research in Japanese sources: a guide**', compiled by Herschel Webb and Marleigh Ryan (Columbia University Press for the East Asian Institute, 1965) is presented as continuous explanatory text, divided into sections. The sections on bibliography and general reference works, on geography and place names, will be of most interest to geographers.

Refer also Yoshida Tōgo: *Japanese geography: a guide to reference and research materials* (Tokyo: Dai Nihon dokushi chizu, 1939).

Atlas of Japan: physical, economic and social (Tokyo: International Society for Educational Information, 1970).

1033 '**Research index**', 1969-, is in looseleaf form, published by Business Surveys Limited; in each edition are comprehensive references to articles and news items of financial interest appearing in more than a hundred periodicals. Pink pages, I, refer to industrial and commercial news, the most useful from the geographical point of view; blue pages, II, refer to companies in alphabetical order.

1034 ' La revue Canadienne de geographie ', of which three or four issues a year have been published since 1947 jointly by the Geographical Societies of Montreal and Quebec, was the first established Canadian geographical journal of high academic merit. It is mainly in French, with English or French abstracts.

1035 ' Revue de géologie et de géographie ' has been published twice a year since 1957 by the Académie de la République Populaire Roumaine, Bucharest, in English, French, German and Russian. In addition to scholarly articles, the periodical includes reviews of new work and illustrations, maps and charts.

1036 ' Revue forestière français ', founded by Léon Schaeffer, is edited by l'Ecole Nationale des Eaux et Forêts, with the collaboration of the Société des Amis et Anciens Elèves de l'Ecole Nationale des Eaux et Forêts, Nancy. Illustrated articles are the main feature, frequently carrying references or bibliographies, followed by a correspondence section and book reviews arranged by country. Five cumulated indexes have so far appeared: covering 1862-1887 (1887); 1888-1902 (1903); 1903-1927 (1930); 1928-1948 (1950); and 1949-1960 (1962).

1037 Ritter, Carl (1779-1859) was one of the great leaders of geographical thought in Germany and holder of the first academic chair in geography at the University of Berlin. Of his numerous publications, two early works made a great impact and remained of lasting importance: the *Europa, Ein Geographisch-Historisch-Statistiches Gemaldes* . . . of which the first volume appeared in 1804 and the second in 1807, with a small atlas volume in 1806; and *Die Erdkunde im Verhältnis zur Natur und zur Geschichte des Menschen oder allgemeine vergleichende Geographie,* of which the first volume, on Africa, was published in 1817, and the second, dealing with Asia, the following year. The second edition was published in nineteen volumes between 1822 and 1859. Ritter was greatly concerned to point the connection between geography and history and to determine the influence of geography upon the human race. He was intimately connected with the direction of the Gesellschaft für Erdkunde, from its foundation in 1828.

1038 ' **The road and tourist map of Norway** ', published by Cappelen, based on the most recent material from Norges Geografiske Oppmåling and tourist information, is the result of investigations carried out by several hundred contributors throughout the country. Five double map sheets, 1-2, Southern Norway; 3-4, Central Norway; 5-6, North-Central, the Møre and Trøndelag area, all on 1:325,000; 7-8, Nordland; and 9-10, Tromsø and Finmark, both on 1:400,000, present the first modern road and tourist map on this international standard. Road networks are classified according to the latest road regulations. A key, printed on the reverse, indicates sixty symbols referring to useful information, given in English, French and German. Six colours show relative relief features. ' Southern Norway ', on 1:1M scale, shows fully-coloured, hill-shaded relief, road classifications and distances. All these maps are available in Britain through Bartholomew.

1039 ' **Road international** ', quarterly from 1950- (International Road Federation Limited, London), is a truly international periodical, illustrated, presenting new developments, ventures and projects. ' Its controlled circulation takes it to the desks of men involved in the roads and road transport field throughout the world.'

1040 ' **Roman roads in Britain** ', by I D Margary (John Baker, 1955, 1957 in two volumes; revised in one volume, 1967), is a well-told narrative, well documented and illustrated with fine aerial photographs, and with maps and diagrams. ' Only once previously has an attempt been made to give a descriptive account of all the Roman roads in Britain and that was by Thomas Codrington just over sixty years ago, a most valuable work which has remained the standard textbook till our own time . . .' (Preface).

1041 Royal Canadian Geographical Society, Ottawa, was founded in 1929 for the advancement of geographical knowledge and, in particular, for the general diffusion of information on Canadian geography. The *Canadian geographical journal,* published from 1930, is the chief publication, with text in English or French; book reviews are included, also bibliographies. A cumulative index is available covering volumes 1-59, 1930-1959.

1042 '**The Royal English atlas:** *eighteenth century county maps of England and Wales*', originally produced by Emmanuel Bowen and Thomas Kitchin in 1762, presented in one elegant folio volume the best contemporary maps of the English counties and of North and South Wales, in a standard format, based on the most accurate sources available to the authors, printed and in manuscript, including the maps in their own *Large English atlas,* published earlier. This work has now been reprinted in exact facsimile, from an uncoloured copy in the British Museum, by David and Charles of Newton Abbot. It contains forty-four finely engraved maps, distinguished by much geographical detail and beauty of craftsmanship, each map being embellished with an elaborate cartouche, portraying scenes within the county, views of cathedrals and so on. The introduction, by J B Harley and Donald Hodson, outlines the publication history of the atlas within the framework of the eighteenth century London map trade, analysing the sources used in the compilation of the maps.

1043 Royal Geographical Society, London, had its origin in the amalgamation of the Association for Promoting the Discovery of the Interior Parts of Africa, founded in 1788, and the Palestine Association, which dated from 1805, together with the Raleigh Dining Club. By 1830, the need for a more completely organised institution for the advancement of geography became apparent and the Geographical Society of London was created, which gained a Royal Charter in 1859. The development of the work of the Society through the years may be traced through the issues of *The journal,* 1830-1880, and of *The geographical journal* (*qv*). Sir C R Markham recorded the early work of the Society in *50 years work of the RGS,* 1881; the centenary volume, *The record of the Royal Geographical Society 1830-1930,* by H R Mill (RGS, 1930), was continued by G R Crone in *The Royal Geographical Society: a record, 1931-1955* (RGS, 1955). Membership includes academic geographers, explorers and interested amateurs. The Map Room is open to the general public, the library to members only. The Society's contribution to world geography has been considerable. Discovery and exploration have been actively encouraged by grants, advice, loan of instruments and the preparation of maps. Publications have been of a high standard, notably *The geographical journal* (*qv*) and *New geographical literature and maps* (*qv*) which in 1951 succeeded

Recent geographical literature, maps and photographs added to the Society's collections, 1918-1941; the original maps reproduced in *The geographical journal*, usually drawn by the Society's cartographic staff, are particularly valuable. The *Research series* and the *Library series* include many useful works, such as *Current periodicals in the library of the RGS*, 1961, *A classification for geography*, 1962 and *Publications of the Royal Geographical Society 1830-1964*, 1964. The Society has reproduced facsimiles of *The map of the world*, by J Hondius, English county maps, the Catalan world map, the Hereford world map, the Gough map and *Early maps of the British Isles, AD 1000-1579*. The library holds more than 100,000 volumes and fine collections of lantern slides and photographs, and the map room has a unique collection of maps and atlases, both historical and modern.

Refer also E W Gilbert: 'The Royal Geographical Society and geographical education in 1871' *in The geographical journal*, June 1971.

1044 Royal Geographical Society of Australasia, South Australia branch, was founded at Adelaide in 1885, and a Queensland branch at Brisbane in the same year. Publications include Annual *Proceedings* and occasional *Reports*. The library holds more than thirty thousand volumes.

1045 Royal Meteorological Society of London was founded in 1850 as The British Meteorological Society for the promotion of the science of meteorology in all aspects; a Royal Charter was granted in 1866 and the name was changed to The Meteorological Society, the present full name being acquired in 1883. In 1921, the Society amalgamated with the Scottish Meteorological Society. A Scottish centre is based on Edinburgh University, a Manchester centre at Manchester University and a Canadian branch has centres at Montreal, Toronto and Winnipeg. The Society possesses one of the greatest meteorological libraries in the world, comprising more than forty thousand books and pamphlets and some 1,500 manuscripts, besides periodicals and lantern slides. Periodical publications include the *Quarterly journal*, 1871- and *Weather: a monthly magazine for all interested in meteorology*, 1946-, which includes a 'Weather log', with map.

Note The publications and services of the Meteorological Office are listed in *Government Publications,* Sectional list no 37 (HMSO) revised at intervals.

Refer also B J Mason: 'The role of meteorology in the national economy', *in Weather,* November 1966.

T J Chandler: *Modern meteorology and climatology* (Nelson, 1972).

Michael Chisholm *et al, ed*: *Regional forecasting*: *Proceedings of the Twenty-Second Symposium of the Colston Research Society,* 1970).

1046 Royal Scottish Geographical Society was founded in 1884 with the aim of extending geographical education in Scotland, of maintaining contact with Scotsmen abroad and with others interested in the subject. A cartographic section was formed in 1960. The main publication is the *Scottish geographical magazine* (*qv*); the society has also published *The early maps of Scotland,* in a second revised edition, 1936. The library holds a valuable collection of more than forty thousand books and some sixty thousand maps, including a set of early maps of Scotland, and about two hundred periodicals. Overseas visitors holding any geographical society membership may use the society's library and rooms.

1047 The Royal Society has always taken a scientific interest in geographical matters and acts as the National Committee for Geography. A cartography sub-committee was set up in 1960 to advise on cartographical topics affecting the United Kingdom.

1048 The Royal Tropical Institute, Amsterdam, had its origins in the mid-nineteenth century and was incorporated as an Association in 1910. The aim of the institute is to collect and disseminate knowledge concerning tropical countries; the Central Library is one of the most extensive in the world in books, periodicals, maps and charts within the subject field.

1049 'Rubber developments', a central source in its subject is issued quarterly by The Natural Rubber Producers' Research Association (Malayan Rubber Fund Board), London. Articles are of technical, historical or regional application and are illustrated; the main articles are preceded by brief abstracts. Shorter notes on items of current interest are also included.

1050 '**Rumanian studies:** *an international annual of the humanities and social sciences*', edited by K Hitchens, began publication by Brill of Leiden in 1970. It had its origin in the feeling among Rumanian scholars that an opportunity was needed to inform a wider audience about the new directions in research and the changes in interpretation that are taking place in their country. Bibliographical items are to be a feature.

1051 '**Russia and the Soviet Union:** *a bibliographic guide to Western-language publications*' was edited by Paul L Horecky (University of Chicago Press, 1965), with particular emphasis on English publications. Materials are arranged under broad subject headings, divided into more specific categories. Main divisions include 'General reference aids and bibliographies'; 'General and descriptive works'; 'The land'; 'The people, ethnic and demographic features'; 'The nation: civilizations and politics'; 'History'; 'The state'; 'The economic and social structure'; 'The intellectual and cultural life'. Brief annotations are appended and there is an index.

1052 '**Russian land, Soviet people:** *a geographical approach to the USSR*', by James S Gregory (Harrap, 1968), is a study of man within his environment. The work begins with a comprehensive account of Russian geography, showing how the environment of the Great Russian Plain influenced the development of cultures, followed by studies of the regions according to climate, soil and natural vegetation, with a particular study of the progress of agriculture. The second part deals in depth with 'The regions of the USSR'. There are maps in the text and a bibliography.

1053 Sahab Geographic and Drafting Institute, Tehran, founded under the direction of Professor A Sahab, has for thirty years specialised in the printing of maps of Iran and the Middle Eastern countries for scientific and educational purposes, also atlases and globes. The current programme includes a national atlas of Iran, a geographical encyclopedia in Persian, the preparation of a complete set of maps on the Middle East at 1:1M and the completion of a regional map series of Iran, of which some sheets are already available.

1054 Saxton, Christopher (*c* 1542-1610 or 1611) attempted the first national survey of England and Wales. Under the royal patronage, he surveyed and drew maps of all the counties. The British Museum possesses one of the earliest copies of the atlas, finished in 1579, and has arranged for the reproduction of the maps in colour. They are of remarkable accuracy, considering the instruments available, showing a systematic approach to the delineation of ground features in map form; varying sizes of lettering are used to show relative importance of settlements, for example, and the maps are finished with a wealth of artistic detail.

1055 'Scandinavian lands', by Roy Millward (Macmillan, 1964) covers Norway, Sweden, Denmark and Finland, treating first the regional geography of each country, then ' Some facets of Scandinavian geography '. Brief notes are appended and a bibliography. Aerial photographs, sketchmaps and diagrams have been carefully selected; the balance between the individuality of the four countries and their essential unity is skilfully presented vis à vis the rest of the world.

1056 'The Scandinavian world', by Andrew C O'Dell (Longmans, 1957) broke new ground in considering the countries of Finland, Sweden, Denmark, Norway, the Arctic islands, Faeroes, Iceland and Greenland as an entity. The approach is direct—' Physical and historical introduction ', ' Regional geography ', ' Economic geography '. Throughout the text are numerous photographs, sketchmaps, diagrammatic maps and figures; an appendix sets out Conversion Tables, metric to British units and a selected bibliography, arranged regionally.

1057 'The scenery, antiquities and biography of South Wales', by Benjamin Heath Malkin, published by T N Longman in 1804, is probably the most important work dealing with South Wales at the beginning of the nineteenth century, based on material collected by Malkin in 1803 and embellished with many views drawn on the spot and engraved by Laporte. The work has been reproduced by SR Publications Limited, East Ardsley.

1058 School of Oriental and African Studies Library was built up on the collection of Oriental books owned by the London Institution; the University Library and the libraries of King's College transferred their Oriental books in exchange for the Western books from the London Institution Library. Acquisition policy attempts completion in all significant publications relating to Asia, Africa and Oceania in all languages, in humanities and social sciences; there is also a rapidly growing collection of lantern slides, transparencies and photographs. The catalogues were reproduced by G K Hall and Company in 1964 (dated 1963) and are available as a complete set or in sections. Analytical entries are included for periodicals, festschriften and bibliographies. Several bibliographies and special catalogues have been compiled by the staff, also the *Monthly list of periodical articles,* 1954- and *Theses on African studies,* 1964.

1059 'Schweizerischer Mittelschulatlas' (Zurich, 1898-), though not readily available in Britain, has been influential in the development of educational atlases. The chief source of the foreign topographical maps until 1910 was the *Stieler atlas;* for the revisions, 1927-32, under the direction of Professor Imhof, the principal source was the *Grande atlante del Touring Club Italiano* and, since then, the 1954 *Atlas mira, The Times mid-century edition* and the national mapping services. The atlas was redesigned in 1955 and again in a thirteenth edition, in 1962, for the Konferenz der Kantonalen Erziehungsdirektooren, an outstanding feature being the new method of relief representation evolved by Professor Imhof. The atlas begins with maps of the home country, working from it to other parts of the world.

Refer Eduard Imhof: ' The Swiss Mittelschulatlas in a new form ' *in International yearbook of cartography,* 1964.

1060 ' Science in New Zealand ', edited by F R Callaghan (Wellington: Reed, 1957), was prepared for the meeting of the Australian and New Zealand Association for the Advancement of Science at Dunedin in 1957. Individual subjects are treated by experts, such as ' Geography ' by G Jobberns. Other relevant topics include climate, meteorological progress, geophysics, sea fisheries, native and introduced birds, changed and changing vegetation, science and agriculture, history of soil science, science in the fruit industry, science in the dairy industry and post-war developments in oceanography.

1061 ' The science of geography: *report of the ad hoc Committee for Geography',* Earth Sciences Division, National Academy of Sciences, National Research Council, Washington, 1965, surveys the problems facing contemporary geographers and makes some suggestions for future development.

1062 Scientific Committee on Antarctic Research was set up by the International Council of Scientific Unions in 1958 to continue the co-operative scientific exploration of Antarctica after the completion of the IGY research, 1957-1958.

1063 Scott Polar Research Institute, Cambridge was established in 1920 to encourage and facilitate polar research and to disseminate information on polar regions. The library holds some ten thousand books and reprints, in addition to collections of maps, manuscripts, photographs, prints, films, log-books and records. The library of the British Glaciological Society is also housed here and all the collections are available for use by any genuine research worker. More than three hundred periodicals are indexed by the library staff and the feature 'Recent polar literature', a select list of books, pamphlets and articles, with indicative abstracts, is prepared for inclusion in the Institute's *Polar record* (*qv*). Practical experience in the library through eighteen years enabled Brian Roberts to adapt the Universal Decimal Classification schedules for use in polar libraries; this was published jointly by the institute, the Fédération Internationale de Documentation and the British Standards Institution (second revised edition, 1963). A series of special publications began in 1952; one of these of particular interest is the *Illustrated ice glossary,* by Terence Armstrong and Brian Roberts, 1956.

1064 'Scottish geographical magazine', the organ of the Royal Scottish Geographical Society(*qv*), published three times a year since 1885, has always contained articles of world interest, with an emphasis on Scottish affairs. A new editorial policy began in 1966, whereby a large part of most issues consists of articles on regions of Scotland and geographical studies of various aspects of Scottish life, work and natural background, on a systematic plan. References on Scottish geography are included. An index is prepared every second year. The two sections of reviews, for books and for atlases and maps, are not extensive, but sound and instructive.

1065 'The sea: *ideas and observations on progress in the study of the seas'* (Wiley International, 1962-1963, 1970), an advanced and technical work, yet readily understandable, was conceived and the first three volumes edited by M N Hill and others, illustrated by maps and diagrams. The three volumes run as follows: 'Physical oceanography', 1962, which deals with fundamentals, the interchange of properties between sea and air, the dynamics of ocean currents, the transmission of energy within the sea, waves, turbulence and the physics of sea ice; 'The composition of sea water: comparative and descriptive oceanography', 1963, concerned with the fertility of the oceans, currents, biological oceanography and oceanographical miscellanea; and 'The earth beneath the sea', including the history of geophysical exploration, topography and structure and sedimentation. After Dr Hill's death, the fourth volume, dedicated to his memory, was edited by Dr A E Maxwell; contributions were made by seventy-seven scholars. The volume shows how the study of the sea-floor has progressed through the techniques described, including seismic, magnetic, geothermal, topographic, earthquake epicentre, palaeontological, plus other geophysical and geological information on a world-wide basis.

Refer also Mary Sears, *ed*: *Oceanography* (American Association for the Advancement of Science, 1961).

Margaret Deacon: *Scientists and the sea 1650-1900: a study of marine science* (Academic Press, 1971).

Otto Kinne: *Marine ecology* (Wiley, 1970-).

Donald W Hood, *ed*: *Impingement of man on the oceans* (Wiley Interscience, 1971).

1066 '**Sea surveys:** *Britain's contribution to hydrography*', by Vice-Admiral Sir John Edgell (HMSO, 1965), is in five sections—'The early hydrographers', 'The hydrographers of the British Navy', 'Then and now', 'The work of the surveyor in war', 'Sea surveys, 1948-1965', the last named by Lt Commander P B Beazley. Included is the list of names of the Hydrographers of the British Navy and there is a selection of photographs and reproductions.

1067 '**Sedimentation:** *annotated bibliography of foreign literature*', was first edited by E Goldberg in 1965 and published for the United States Department of Agriculture and the National Science Foundation by the Israel Program for Scientific Translations, covering 1959-1964. Literature published in English from non-English countries was also included. Further volumes have covered the literature of one or two years and the index is cumulative.

1068 'The seismicity of the earth, 1953-1965' was prepared by Professor J P Rothé, in English and French (Unesco, 1969), listing all earthquakes equal to or greater than 6, recorded during the period under review. Maps of each region give the epicentres, classified according to their magnitude and depth of focus. 286 pages of tables cover fifty-one regions, accompanied by notes.

1069 'Seismicity of the earth and associated phenomena', by B Gutenberg and C F Richter (Princeton University Press, 1949, 1954) was reproduced in facsimile by Hafner Publishing Company, 1965. The 1954 edition was used; following an introduction to the materials and methods employed, a classification is given of shocks, their mapping and the frequency and energy of earthquakes. Regional discussions are concerned with the Circum-Pacific belt, the Alpine belt, non-Alpine Asia, Oceanic Arctic belts, rift zones and the seismicity of marginal to stable masses. Each section is divided into more specific chapters. Tables and references take up about a third of the book.

1070 'Seismicity of the European area', by Vít Kárník (Amsterdam: Reidel, in collaboration with Academia of Prague, 1969-), was carried out in accordance with the resolutions of the European Seismological Commission and its Subcommittee for Seismicity of the European Area, of which the author is Chairman. The history of the project is summarised in the introduction: section 2, ' Seismological information ', includes definitions, national catalogues of earthquakes, earthquake parameters entering into the catalogue, their determination and accuracy; 3, 'Uniform classification of earthquakes'; 4, ' Statistical data '. A catalogue of earthquakes 1901-1955 forms the bulk of the book. Sources of information are given. A second volume is to follow.

1071 'Seismology: *a brief historical survey and a catalogue of exhibits in the Seismological Section of the Science Museum'*, a useful and practical booklet, was edited by J Wartnaby for the Ministry of Education: Science Museum (HMSO, 1957), *Geophysics handbook,* no 1. The text provides a compact summary of aspects of the topics concerned, both historical and current: ' The causes of earthquakes ' deals with theories from myths to contemporary scientific explanations; ' The detection of earthquakes ' and ' Interpreting the records' include descriptions and figures of the instruments used.

'The distribution of earthquakes' is followed by sections on 'Microseism' and 'Seismic prospecting'. The catalogue contains forty-five items, mainly instruments, pieces of equipment, and specific items such as the 'Chang Heng seismoscope' to 'Oil exploration in Britain' and the Geophone. A bibliography of monographs and articles is appended.

1072 'Selected bibliography on Kuwait and the Arabian Gulf', compiled by Soraya M Kabeel (Kuwait University, 1969), includes some 1,300 entries of articles and books on the subject, both primary and secondary material; the first part is classified by subject order, the second is an alphabetical listing of authors, titles and subjects.

1073 'A selected list of books and articles on Japan in English, French and German', compiled by Hugh Borton and others (Harvard University Press for the Harvard-Yenching Institute, Cambridge, Mass, revised and enlarged edition, 1954), contains bibliographies, reference works, periodicals and, among other subject entries, 'Geography' and 'Economics', with sub-divisions. The index is comprehensive.

1074 'Selection of international railway documentation', monthly from January 1964, is edited jointly by the International Union of Railways, Paris, and the International Railway Congress Association, Brussels, prepared under the direction of the International Railway Documentation Bureau, with the co-operation of the documentation services of a great number of railway administrations. Three separate editions are in English, French and German. Each entry bears a running number, the UDC number, the classification number utilised by the International Railway Documentation Bureau and, if applicable, the numerical reference of a microfilm. Five broad subject headings are distinguished: 'General—economic and historic matters'; 'Railway operation'; 'Rolling stock and railway traction'; 'Fixed railway installations' and 'Technique of the other transport methods, general technique and miscellaneous'. The abstracts are initialled. Each issue includes an order form for microfilms or photocopies.

1075 Semple, Ellen Churchill (1863-1932) is remembered especially for her *Influences of geographic environment* . . . (*qv*) in which she attempted to elucidate and interpret to English-speaking readers the theories of Ratzel, to whom she refers in the preface as 'the great master who was my teacher and friend during his life, and after his death my inspiration'. Her contribution as an academic geographer in her own right was, however, considerable. She taught intermittently in several American universities and played an important part in establishing the Graduate School of Geography in Clark University, being herself trained in economics and sociology as well as in geography. Of her other publications, the best known is *American history in its geographic conditions*, 1903, which was revised by C F Jones in collaboration with the author, 1933.

1076 'Serial atlas of the marine environment' has been in process of publication by the American Geographical Society since 1962. When completed, the work will constitute a unique source of co-ordinated data; high speed electronic computers assist in sorting and the results are being plotted on loose-leaf double sheets of transparent material to facilitate comparison, in a series of separate folios.

1077 'Settlement and encounter: *geographical studies presented to Sir Grenfell Price'*, edited by F Gale and G H Lawton (OUP, 1969), comprises a series of studies linked by the central theme of the practical and political implications of recent settlement and colonisation, with special reference to Australia, a theme of particular interest to Sir Grenfell Price. Included is an appreciation of Sir Grenfell, by A Marshall, and there is a bibliography of his works.

1078 'Shell nature lovers' atlas of England, Scotland, Wales' was compiled by James Fisher (Ebury Press and Michael Joseph, 1966), with maps by John R Flower, 'inspired by the revolution in conservation since 1945'. Included in the atlas are the natural 'treasures' to which public access is possible. Full use has been made of the published documents of such conservation bodies as the Nature Conservancy, the Forestry Commission, the Council for Nature, the Society for the Promotion of Nature Reserves, the National Trust for Scotland, the network of County Naturalists' Trusts, the Royal Society for the Protection of Birds, the Scottish Wildlife Trust, the Wildfowl Trust and the Wildfowlers' Association for Great Britain and Ireland. Each category of site has been represented in text and on maps by a symbol or letter; also useful is a list of addresses of relevant national organisations. Ordnance Survey maps have been used as a base for the maps; the scale throughout is 1:760,320, except for the map of the Greater London area, which is at 1:411,840, and the work ends with a summary of the Ordnance Survey maps particularly useful in naturalist work, either in walking or motoring, with notes on the application of the National Grid system. Handy to use with the Atlas is *The Shell book of exploring Britain,* compiled by Garry Hogg (John Baker, 1970), a new kind of touring guide, presenting carefully planned itineraries and introductions to the history, geography, agriculture and way of life of each area; also the *Shell County Guides,* edited by John Betjeman and John Piper.

1079 ' Sierra Leone in maps ', edited by John Clarke, was published in 1966 by the University of London Press for Fourah Bay College, Freetown, in a second edition, 1969. The work comprises fifty-one maps, with commentaries by thirteen past and present members of the college staff, who present comprehensively the nature and resources of the country. A bibliography, divided into five categories, is arranged by author.

1080 ' Sino-Portuguese trade from 1514 to 1644: *a synthesis of Portuguese and Chinese sources',* by T'ien-Tsê Chang (Leiden: Brill, second edition, 1969), is a photomechanical reprint, with some corrections of the first edition of 1933, which brings together important source material hitherto separated or unknown. The seven sections cover 'An historical sketch of China's maritime trade down to 1513 '; ' The early Sino-Portuguese trade relations '; ' The expulsion of foreigners from China and the prohibition of foreign trade '; ' Trade or no trade '; ' The rise of Macao '; ' The arrival of other Europeans in the Far East and the position of the Portuguese '; ' Macao in days of tribulation and the decline of Sino-Portuguese trade '. A bibliography is included.

1081 ' Sir Francis Drake's voyage around the world: *its aims and achievements',* by Henry R Wagner, has been reprinted from the 1926 edition by N Israel, Amsterdam, 1969. The same firm also published *Sir Francis Drake: a pictorial biography,* by Hans P Kraus, in 1970, with an historical introduction by Lt Commander David W Waters and Richard Boulind and a detailed catalogue of the author's collection of manuscripts, printed books, maps and views, portraits and medals, mostly dating from Drake's own lifetime; included are 133 reproductions.

Refer also K R Andrews: *Drake's voyages: a re-assessment of their place in Elizabethan maritime expansion* (Weidenfeld and Nicolson, 1967).

1082 Skelton, Dr R A (1906-1970) devoted his professional life to the study of historical cartography, passing on his scholarship by means of numerous books and articles. He was Superintendent of the Map Room of the British Museum and held office on national and international bodies, such as the editorial committee of *Imago mundi,* the Hakluyt Society, the Commission of Ancient Maps, etc. Outstanding among his published works were his English edition of the *Geschichte der Kartographie, Explorers' maps, County atlases of the British Isles 1579-1850, Decorative printed maps* . . . and his contribution to the study and evaluation of *The Vinland map and the Tartar Relation;* among scores of articles, he contributed to special numbers of *The geographical magazine* devoted to maps and map-making and wrote innumerable reviews, prefaces and introductions.

1083 Smith: *J Russell Smith: geographer, educator, and conservationist,* by Virginia M Rowley (University of Pennsylvania Press, 1964) is a detailed study, including illustrations and a bibliography.

1084 Smithsonian Institution, Washington, is unique among the learned organisations of the world. *The Smithsonian Institution,* by Walter Karp, 1965, produced by the editors of *American Heritage,* tells of the bequest under which the institution was founded, its early years and its vast reputation among scholars everywhere. The *Smithsonian meteorological tables,* in a sixth revised edition, by Robert J List, is a reference work consisting of 174 tables—conversion, wind and dynamic, barometric and hypsometric, geopotential and aerological, standard atmosphere, etc. Other publications include *The large Magellanic cloud,* by Paul W Hodge and Frances W Wright, 1968; a boxed set containing a book and an identification atlas, consists of 168 separate photographic charts, with text giving a historical survey of discoveries about the nature and properties of the cloud, including a summary of current knowledge. *Opportunities in oceanography,* third edition in 1968, was prepared by the Interagency Committee on Oceanography; *Museums directory of* *the United States and Canada,* second edition, 1965, is a joint publication of the American Association of Museums and the institution. Another fine production is *Seabirds of the tropical Atlantic Ocean,* a Smithsonian identification manual, by George E Watson. *Flora of Japan,* a much revised and extended translation, by Jisaburu Ohwi, edited by F G Meyer and E H Walker, 1965, contains some 4,500 species. In addition, there are many more works on climate and meteorology, anthropology, biology, botany and forestry, exploration, regional geography and the Harriman Alaska series of reports, setting out the findings of the Harriman Alaska Expedition of 1899. *Lighthouse of the skies: the Smithsonian Astrophysical Observatory; background and history, 1846-1955* was completed by B Z Jones in 1965.

1085 'Snow structure and ice fields', by Gerald Seligman, was a key work in the subject, prepared for the International Geophysical Assembly at Edinburgh, in 1936, when the Association for the Study of Snow and Ice was formed. Soon after, the association became the British Glaciological Society and has finally progressed to become the International Glaciological Society.

1086 The Snowy Mountains Scheme, Australia, is probably the largest single venture undertaken in that continent. First envisaged as an irrigation project and later a hydro-electric operation, it is complex and far-reaching in effects which will continue to have social and technical advantages, as the following monographs point out: Lionel Wigmore: *Struggle for the Snowy: the background of the Snowy Mountains Scheme* (Melbourne: OUP, 1968); C Meeking: *Snowy Mountains conquest: harnessing the waters of Australia's highest mountains* (Hutchinson, 1968).

1087 Société de Géographie de Paris, the first specialist geographical society, has, since its foundation in 1821, exerted the greatest influence on geographical opinion and research. The *Bulletin,* with its successors, 1822-, has provided a forum for discussion of the highest scholarship and has reflected the development of geographical studies through this formative period. In 1941, the periodical joined with the *Annales de géographie* (*qv*) under the new title of *Annales de géographie et bulletin de la Société de Géographie.* The Society has published also the *Acta geographica* since 1947 and issues a monthly bibliography. The library holds a valuable collection of more than four hundred thousand volumes.

1088 Society for the Bibliography of Natural History, founded in 1936, promotes the study of natural history in the widest sense, including geology, botany and zoology. The *Journal* has included a number of facsimiles of rare natural history works and, in 1959, the first of a series known as the Sherborn Fund Facsimiles was published. The first reproduction in this series was that of Alexander von Humboldt's *Essai sur la géographie des plantes,* 1807. The Society has no central headquarters, the Secretariat being located at the British Museum (Natural History), London.

1089 Society for Environmental Education, founded in 1968, to provide opportunities for the discussion of ideas on the role of the environment in education, seeks to further the development of Environmental Studies in schools as an inter-disciplinary contribution to the curriculum, based on fieldwork. A *Bulletin* is published twice a year.

1090 The Society for the History of Discoveries was founded in 1960 for the purpose of stimulating publications in research in the history of geographical exploration. Membership, on an international basis, is interested mainly in geography, history, the relevant literature and science. The *Annals* of the society, *Terrae incognitae,* are both catholic and eclectic in publication of articles. In 1964, a book review section was added to the Society's semi-annual *Newsletter. Terrae incognitae,* an annual, was authorised at the 1966 Annual Meeting and the society also sponsors a monograph series jointly with the Newberry Library of Chicago and the University of Chicago Press.

1091 The Society for Nautical Research was founded in 1910 to encourage research into nautical antiquities, matters relating to seafaring and shipbuilding in all ages and among all nations, the language and customs of the sea and other subjects of nautical interest. Among other early activities of the society was its influence in the establishment of the National Maritime Museum at Greenwich and the *Victory* Museum at Portsmouth. *The mariner's mirror,* quarterly journal of the society, contains illustrated articles, shorter notes, careful and informative signed reviews and short notices (monthly from January 1911 to September 1914 and from July 1919 to December 1923); an index of volumes 1 to 35 has been compiled by R C Anderson. Also published are 'Occasional publications' and *A treatise on shipbuilding* . . ., edited by W Salisbury and R C Anderson.

1092 Society of University Cartographers, established in Glasgow in 1964, has now a Canadian branch, based on York University, Toronto. An annual summer school and annual meeting are held; symposia deal with such topics as the needs and interests of practising cartographers, thematic mapping of all kinds and the history of cartography. A sub-committee has been formed to investigate the training of cartographers. The *SUC bulletin,* published twice a year, contains articles and notes concerning new publications.

1093 'Socio-economic models in geography', edited by Richard J Chorley and Peter Haggett (Methuen, university paperback, 1967, 1968), deals with 'Models, paradigms and the new geography', 'Demographic models and geography', 'Sociological models in geography', 'Models of economic developments', 'Models of urban geography and settlement location', 'Models of industrial location', 'Models of agricultural activity', well illustrated with diagrams and figures. There are end-of-chapter references.

1094 'Soil biology and biochemistry', a new international journal dealing with soil organisms, their biochemical activities and influence on the soil environment and plant growth, is prepared by an editorial committee and regional editors under Professor E W Russell (Oxford: Pergamon Press). Original work is included on quantitative, analytical and experimental aspects of these studies, providing a record of the progress of such studies throughout the world. The periodical also contains short communications and notes concerning current experiments, techniques, equipment and significant observations.

1095 'Soil biology, reviews of research' (Unesco, 1969) is one of the Natural Resources Research series. Chapters are included on methodological problems in soil biology in various parts of the world, the biological fixation of atmospheric nitrogen by free-living bacteria, ecological associations among soil micro-organisms, biology and soil fertility, microbial degradation and the biological effects of pesticides in soil, with emphasis on the humid tropical and semi-arid soils and on the related problems of conservation or reclamation.

1096 Soil Survey of Great Britain: Soil Survey maps are published by the Ordnance Survey on behalf of the Soil Survey of England and Wales and the Macaulay Institute for Soil Research in Scotland. Individual *Memoirs—Bulletins of the Soil Survey of Great Britain; Memoirs of the Soil Survey of England and Wales; Memoirs of the Soil Survey of Scotland*—contribute to the full use and understanding of the soil maps. In many cases, folded copies of the maps are included in the *Memoirs*.

1097 Somerville, Mary, (1780-1872) during her long life had a great influence on academic education in several branches of the natural sciences. Her work *On the connexion of the physical sciences,* 1836, included a discussion of tides, currents, climate, plant geography and other natural phenomena. One of the best known of her works and of lasting value for its methodology was *Physical geography,* 1848, which was revised in a sixth edition by H W Bates in 1870 (last edition, 1877). In this work, a general review of the continents is followed by more detailed regional treatment; topics, such as the ocean, rivers, vegetation and the distribution of life forms from insects to humans, are discussed, incorporating regional examples throughout.

Refer: J N L Baker: 'Mary Somerville and geography in England', *Geographical journal,* 111, 1948, reprinted in *The history of geography: papers by J N L Baker,* 1963.

1098 Sorre, Maximilien (1880-1962), a pupil of Paul Vidal de la Blache and Emmanuel de Martonne, continued the inspiration of the French school of geography and was a leading figure in academic geography, especially at Lille, Clermont-Ferrand, Aix-Marseille and at the Sorbonne, where he succeeded Albert Demangeon. He wrote prolifically, especially on his main interests, the relationships of geography with biology, medicine and sociology, and the influence of environment on human development. He contributed some volumes to the *Géographie universelle* and many papers to the *Annales de géographie;* among his major works are the three-volume *Fondements de la géographie humaine* completed in 1952 and *L'homme sur la terre: traité de géographie humaine,* 1961.

1099 'Source book for geography teaching', prepared by the Commission on the Teaching of Geography of the International Geographical Union (Longmans/UNESCO, 1965), supersedes the earlier UNESCO publication, *A handbook of suggestions on the teaching of geography,* 1951. It was published 'in the belief that geography can make a meaningful contribution to the advancement of mutual understanding between peoples'. Sections include 'Importance and educational value of geography'; 'The nature and spirit of geography teaching'; 'Teaching techniques: direct observation'; 'Teaching techniques: indirect observation'; 'Teaching material'; 'The geography room'; 'The organization of geography teaching'. 'Sources of documentation' by B Brouillette, is a particularly useful section, arranged by such broad subject groupings as periodicals, bibliographies, a succinct international bibliography, teaching material and international statistics.

1100 'South African bibliography', compiled by Reuben Musiker (Crosby Lockwood, 1970) is a companion volume to the *Guide to South African reference books* (*qv*).

1101 'South and Southeast Asia: *a bibliography of bibliographies'*, compiled by G Raymond Nunn (University of Hawaii, East-West Center Library, 1966), is arranged principally by countries, subdivided by date.

1102 'South Asian government bibliographies', in three volumes, published by Mansell, London, provide the most comprehensive history of central government publications of Ceylon, India and Pakistan, prepared under the aegis of the Centre of South Asian Studies, Cambridge. Much of the material has not been systematically recorded elsewhere. The three volumes comprise *Union catalogue of the Government of Ceylon,* edited by Teresa Macdonald; *Union catalogue of the Central Government of India publications;* and *Union catalogue of the Government of Pakistan publications,* both edited by Rajeshwari Datta.

1103 'South East Asian archives', July 1968- (Kuala Lumpur: SARBICA), reflects all the major archival activities in the South East Asian region. The journal is published annually; the first issue being devoted mainly to the proceedings of the inaugural conference of the new organisation and the second mainly containing the papers on archives presented to the International Conference on Asian History held in Kuala Lumpur in August 1968.

1104 The South Sea Commission is an advisory and consultative body concerned with the economic and social development of the non-self-governing territories administered by the six Member Governments in the South Pacific region: Australia, France, Great Britain, New Zealand, the Netherlands and the United States of America. Work programmes have included research on the improvement of air and sea transport, fishing and other basic industries and agriculture. Publications include the *Proceedings,* issued twice a year; the *Quarterly bulletin,* 1951-1959, re-named the *South Pacific bulletin,* 1960-. A library was set up in 1948, with headquarters at Noumea, which collects basic reference books, official publications, relevant maps and essential standard works, as well as files of the monographs, information bulletins, technical reports and pamphlets published by the commission.

1105 'South-east Asia: *a critical bibliography'*, compiled by Kennedy G Tregonning (University of Arizona Press; University of Malaya Press, 1969), covers the countries of Burma, Thailand, Cambodia, Vietnam, North Vietnam, Malaysia, Indonesia and the Philippines. Each section is introduced by short comments and annotations are appended to the entries as necessary.

1106 'South-west Pacific: *a geography of Australia, New Zealand and their Pacific Island neighbourhoods'*, by K B Cumberland (Whitcombe and Tombs; Methuen, fourth edition, 1968), remains a basic introduction to the area, well illustrated and including maps in the text and a bibliography. The area covered includes New Guinea, New Hebrides, New Caledonia, Fiji, Samoa, Tonga and the Cook Islands. For the latest edition, all the maps have been re-drawn in a bold, clear style.

1107 'Soviet and East European abstracts' series has been edited and published quarterly at the Institute for Soviet and East European Studies, University of Glasgow, replacing the former *Information supplement to Soviet studies.* It covers Albania, Bulgaria, Czechoslovakia, the Democratic Republic of Germany, Hungary, Poland, the USSR and Yugoslavia. Each issue contains about a thousand abstracts of books, newspapers and journals published in these countries.

1108 ' **Soviet geography:** *accomplishments and tasks: a symposium of fifty chapters, contributed by fifty-six leading Soviet geographers*', was edited by a committee of the Geographic Society of the USSR, Academy of Sciences of the USSR, of which the Chairman is I P Gerasimov, and was translated from the Russian by Lawrence Ecker. The English edition was edited by C D Harris and published by the American Geographical Society in 1962, as Occasional paper no 1. The work begins with ' Geography in the Soviet Union: an introduction ' by I P Gerasimov and ' Russian geography ', by A A Grigoryev. Part I then continues with ' The history and present state of Soviet geography '; II, ' The specialised geographic sciences '; III, ' Integrated scientific problems and trends '; IV, ' The role of geographers in the transformation of nature '; V, ' Methods of geographic research '; VI ' Geographic education and popularization of scientific geographic knowledge '; VII, ' The Geographic Society of the USSR '. Chapters within these parts are by individual authors. The text is illustrated and there are references included.

Refer also G Melvyn Howe: ' Geography in the Soviet universities ' *in The geographical journal,* March 1968.

1109 ' **Soviet geography:** *review and translation* ', published by the American Geographical Society, supported by the National Science Foundation, ten times a year (September to June) from 1960, aims to make available in English reports of current Soviet research in geography. The chief editor and translator is Theodore Shabad. Most of the reports are translations from *Izvestiya Akademii Nauk SSR, seriya geograficheskaya; Izvestiya vsesoyuznogo geograficheskogo obshchestva; Vestnik Moskovskogo Universiteta, seriya geografiya; Khozyaystvo;* and *Voprosy geografii.* Brief abstracts precede the articles. Most issues also contain news notes on Soviet political and economic developments of interest to geographers. Lists of references usually follow the articles and frequent surveys of Soviet geographical literature are included. A transliteration table is given for the general information of readers and for the identification of the geographical names on the maps reproduced in *Soviet geography.*

1110 ' **Soviet trade directory** ', edited by Alec Flegon (Flegon Press, 1964), provides the first such guide to the industry of the Soviet Union to be published outside Russia; it is comprehensive, reflecting changing circumstances in trading with the non-Communist world and a changed attitude to the publication of information. The directory is based entirely on official Russian sources. A list of Soviet factories, in Russian and English, is classified by industry and included is a map of the principal centres of ferrous and non-ferrous metallurgy.

1111 'Soviet Union', a colourful monthly pictorial, covering every aspect of life in the USSR, is available in matching English and Russian editions, from Central Books, London.

1112 'Spacial analysis: *a reader in statistical geography'*, edited by B J L Berry and D F Marble (Prentice-Hall International, 1968), consists of thirty-six individual contributions on aspects of the subject.

1113 Special Committee on Oceanic Research was set up in 1957 by the International Council of Scientific Unions to frame a programme of research, especially on climate, fertility of the sea and on the disposal of radio-active waste. A special research programme was organised in the Indian Ocean during 1962-63.

1114 Special Libraries Association of New York has devoted much thought to geographical collections and maps. The Geography and Map Division was organised as a unit of the Washington Chapter in 1941. The first number of the *Bulletin* appeared in November, 1947, thereafter twice a year until 1953, when it became quarterly; it has gradually expanded to include research articles, lists of new maps and books, bibliographies, book reviews and news valuable to geographers, librarians and earth scientists. Findings of surveys and recommendations are frequently incorporated in the *Bulletin*, such as the final report on the cataloguing and classification practices of the larger American libraries in no 24, April 1956. Other papers of geographical interest appear from time to time in *Special libraries*. An important research tool published by the Association is *Map collections in the United States and Canada: a directory* (*qv*). *The cartographic research guide*, nearing completion, is to include bibliographies and selected references relating to the various aspects of maps, map-making, map research and study.

1115 'Speculum orbis terrarum', the work of Gerard de Jode

(Antwerp, 1578), was reproduced by TOT in 1965, with an introduction by R A Skelton. From a cartographic point of view it is just as important as Ortelius' great achievement.

1116 Speed, John (1552-1629) was not himself a cartographer, but he collected together a wealth of detailed information and views and so continued the work of Saxton and Norden (*qv*) in systematically making known the topographical features of the country. His most notable work was the *Theatre of the Empire of Great Britaine,* 1611. Speed's maps were the first on which were shown the territorial divisions of each county.

1117 ' The spirit and purpose of geography ', by S W Wooldridge (*qv*) and W G East (Hutchinson University Library, 1957, third edition 1966), is a classic introduction to geographical studies, dealing with the development of geography as a subject discipline and examining some of the main aspects of the contemporary subject and the concepts involved in physical geography, biogeography, mapping, historical, political, economic and regional geography.

1118 Stamp, Sir Laurence Dudley (1898-1966) was a practical as well as a theoretical geographer and the outstanding personality among academic English geographers during recent years. He worked and held appointments in many parts of the world; he was in the Chair of Geography at the London School of Economics between 1945 and 1948 and was appointed to the Chair of Social Geography when that was created in 1948. His major achievement is usually considered to have been his organisation of the Land Utilisation Survey of Great Britain (*qv*); he became the leading authority on land use and a pioneer in urging the planned use of land resources in every country. In 1942 the Minister of Agriculture appointed him to be his adviser on rural land utilisation and he thus became one of the first geographers to be used by the government in Britain. At the time of his death, he was appointed to head a committee to advise the government on the use of natural resources and land. He was director of the World Land-Use Survey of the International Geographical Union (*qv*). By his ability, academic integrity and personality, in the course of innumerable commitments, both international and national, he advanced the prestige of British geography throughout the world and through his published works he brought a fresh impetus to the understanding of geographical matters to students, to his professional colleagues and to laymen alike. In addition to contributions to periodicals, prefaces and forewords, inaugural and presidential addresses, he wrote or edited many textbooks, edited the UNESCO *History of land use in arid lands,* 1961, *A glossary of geographical terms* (*qv*) and *Longman's dictionary of geography* (*qv*). The Dudley Stamp Memorial Fund was established in 1967 for the encouragement of geographical study and research, especially for young geographers; and a Dudley Stamp Memorial cumulative index for *Geography* is being prepared.

1119 ' Standard encyclopedia of the world's mountains ' and *Standard encyclopedia of the world's oceans and islands,* both edited by Anthony Huxley, and *Standard encyclopedia of the world's rivers and lakes,* edited by R Kay Gresswell and Anthony Huxley, are all published by Weidenfeld and Nicolson, 1962, 1965, and are all constructed on approximately the same pattern. Each comprises an introductory general article, followed by alphabetical entries of varying length. Many coloured and monochrome plates and text illustrations are included, also locational maps.

1120 Stanford, Edward, Limited, was founded in 1852. In 1884, the Stationery Office transferred the entire government stock to it. The firm's policy changed in 1947 when the map printing and publishing became absorbed into George Philip & Son Ltd, and Stanford became a clearing house for sheet maps and atlases, maintaining an excellent bibliographical service. Revised catalogues are issued at frequent intervals. The Stanford ' Planfile ', designed specially to accommodate Ordnance Survey maps and plans, can accommodate up to two hundred sheets by a method of suspension filing, and an index to the contents can be mounted on the inside of the lid. *The Stanford reference catalogue of maps and atlases,* in loose-leaf format, is a unique reference guide and sales catalogue of modern cartography, compiled from information supplied by map and atlas publishers throughout the world. Individual sections include maps of each continent, region and country of the world, also the sky and planets; national official surveys, thematic maps, wall maps and atlases of the world, regions and countries, road maps and atlases, town plans, tourist maps and guidebooks are listed, by title, publisher, scale, characteristics, date, size and price in sterling. For maps consisting of a number of sheets, there is an index showing which sheets have been published and, where appropriate, the various editions of each sheet. Amendments and additions together with revised pages are issued in twice yearly bulletins. Details are given also of equipment stocked or which can be supplied, with information of other services offered; new publications are noted, price changes, corrections and deletions. There is a thumb index. At intervals, a 'catalogue' sheet presents information of catalogues and lists available, under the headings 'General', 'Tourist', 'Nautical' (including Stanford's coloured charts, Maritime Press publications and Kandy books), 'Business', 'Academic' and a 'box' left free for 'Special requirements'.

1121 'Stanford's geological atlas of Great Britain and Ireland' was first published in 1904. A revised edition, in 1964, re-written and re-drawn by T Eastwood, includes comprehensive information. Excellent geological maps of the counties are accompanied by geological descriptions of the scenery on all main railway routes and accounts of the topography, county by county.

1122 'States and trends of geography in the United States, 1957-1960', published by the Association of American Geographers in 1961, was a report prepared by the association, in collaboration with the National Academy of Sciences, National Research Council, for the Commission on Geography of the Pan American Institute of Geography and History.

1123 'Statistical analysis in geography', by Leslie J King (Prentice-Hall International, 1969), attempts a review of the applications of statistical analysis in geography to date, examining some of the achievements along these lines and the relative strengths and weaknesses of different analyses, with emphasis on the unique spatial problems of point pattern analysis, areal association and regionalisation. 'Emerging trends and future prospects' are touched upon; a glossary of terms and symbols is included, as well as a bibliography.

1124 'Statistical mapping and the presentation of statistics', by G C Dickinson (Arnold, 1963, reprinted 1964), names four special categories of reasons for compiling statistical maps, namely: to arouse greater interest in the subject matter concerned; to clarify it, simplify it or explain its more important aspects; to prove a point referred to; or to act as a statistical 'quarry' for other users. Statistical techniques are examined, as affecting the compilation of maps and diagrams, followed by sections dealing with 'Sense from statistics: the search for the significant'; 'Choosing the right method and making the most of it'; 'General aspects of map design'; 'Some worked examples' and 'Sources of statistics'. An appendix gives a 'List of Census Reports for Great Britain, 1801-1931'. Sketch-maps and diagrams illustrate the text.

1125 'Statistical methods and the geographer', by S Gregory (Longmans, 1963; 'Geographies for advanced study' series), provides an introduction to the assessment of quantitative data in geographical studies, dealing with such topics as 'The nature of the raw material'; 'The calculation and use of the mean'; 'Deviation and variability'; 'The normal frequency distribution curve and its uses'; 'Other frequency distribution curves'; 'Characteristics of samples'; 'Methods of sampling'; 'The comparison of sample values'; 'The problem of correlation'; 'Regression lines and confidence limits'; 'Fluctuating and trends' and 'Scope for the future'. There is a short bibliography.

1126 'Statistics: *African sources for market research',* by Joan M Harvey (CBD Research Limited, Beckenham, Kent, 1970), is arranged by 'Country sections'; there is an index of titles and one of organisations. For each entry, the address of the central office is given, any bibliographical material available, a statistical summary, production, external trade, a quarterly trade survey and relevant libraries.

1127 Stefansson, Vilhjalmur (1879-1962) was a vigorous personality, an original thinker and a practical pioneer. His greatest achievements lay in his exploration of the Arctic, in his development of techniques for living in Arctic conditions and in making the polar environment more widely understood. His published works on the Arctic include *The friendly Arctic* . . . , 1921; *Unsolved mysteries of the Arctic,* 1939; and *Ultima Thule* . . . , 1942. He collected an invaluable polar library, which he gave to the University of Dartmouth; the Stefansson Collection, moved to new quarters in 1959, now comprises more than seventy thousand items and tape recordings. Stefansson's second abiding interest was in exploration generally, particularly in the theory he put forward in *The northward course of empire,* 1922, a work of compelling interest. Among his other notable contributions to the literature of exploration were

Great adventures and explorations: from the earliest times to the present, as told by the explorers themselves, 1947, and *Northwest to fortune: the search of western man for a commercially practical route to the Far East,* 1960.

1128 Stembridge, J H (1889-1969) was a pioneer in geography teaching, committed early in his career to the improvement and status of geography as a major subject in the academic curriculum; he was influential in the development of the geographical equipment available and himself wrote a wide range of textbooks. His appointment as Geographical Editor of the Oxford University Press was of great significance and he supervised the production of a number of educational atlases. He is perhaps most widely known for his 'World-wide geographies', 1930-; his *Germany* went into a fourth edition, 1950; *A portrait of Canada* was published in 1943 and *Africa* in 1963.

1129 Steward, J H, Limited, London, is one of the most notable firms specialising in surveying, drawing, optical and meteorological instruments—compasses and clinometers, altimeters, telescopes, thermometers, hygrometers, stereoscopes, plane tables, planimeters, hypsometers, soil samplers, etc. Catalogues and price lists are issued at intervals, in addition to special leaflets and interim newssheets.

1130 'Stieler's atlas of modern geography' has been notable for scholarship, balance and technical excellence since the first definitive edition of 1831. Particularly interesting new editions have been the centenary edition of 1925 and the international edition of 1934, completely revised by Herman Haack and others. The chief use of the atlas is locational; the excellent index contains some three hundred thousand entries.

Refer Dr Werner Horn, in *Petermanns Geographische Mitteilungen,* 1967, part 4: an appreciation of Adolf Stieler and a list of the 144 maps produced by Stieler between 1798 and 1837. His early maps were drawn for a periodical, the *Allgemeine Geographischen Ephemeriden;* later he co-operated with Justus Perthes.

1131 'The storage and conservation of maps', a report prepared by a Committee of the Royal Geographical Society in 1954, was printed separately and also included in *The geographical journal,* June 1955.

Refer also British Standard Specification for the Storage of Documents: Topographical maps and drawings.

1132 'The story of maps', by Lloyd A Brown (Little, Brown and Company, Boston, 1949; McClelland and Stewart Limited, Canada), is now a rare classic. The text follows a chronological trend: 'The earth takes shape; the habitable world'; 'The world of Claudius Ptolemy'; 'The Middle Ages'; 'Charts and the haven-finding art'; 'The map and chart trade'; 'The latitude'; 'The longitude'. 'Survey of a country'; 'Survey of a world'. There are extensive notes and a bibliography and line drawings and map extracts in the text. Lloyd A Brown was former curator of the fine collection of maps at the William L Clements Library, University of Michigan, then Librarian, Peabody Institute of Baltimore.

1133 Strabo, Greek historian and geographer, born about 63 BC, travelled widely and with lively intelligence, setting down his observations, together with references from his equally wide reading, in a seventeen-volume *Geography,* thus providing one of the chief sources for contemporary topographic knowledge, beliefs and geographic thought. Following the introductory books, two deal with Spain and Gaul, two with Italy, one with northern and eastern Europe, three with Greece, one with Asia and the Far East, three with Asia Minor, one with Persia and India, one with the Tigris-Euphrates area, Syria and Arabia, and one with Africa. He advanced the concept that physical geography could not be separated from human geography and, in fact, confined geographical science to the inhabited world. The Loeb edition, 1917-32, by H L Jones, includes a bibliography.

1134 'The stratigraphy of the British Isles', by Dorothy H Rayner (CUP, 1967), is concerned mainly with surface and near surface processes and the rocks resulting from them, in a straightforward, not too technical style. A particular feature is the integration of Irish stratigraphy with that of the rest of the British Isles. The impact of other branches of earth sciences, such as sedimentology and geochronology, is taken into account. An appendix sets out 'Stratigraphical divisions and zonal tables'. There are a few excellent photographs, a number of diagrammatic maps and figures, and references at the ends of chapters.

1135 'Studies in cartobibliography, *British and French and in the* *bibliography of itineraries and road-books'*, a unique work by Sir H G Fordham, 1914, has been reprinted by Dawsons of Pall Mall, London, 1969. Revised papers include 'An introduction to the study of the cartography of the English and Welsh countries, with an index list of the maps of Hertfordshire, 1579-1900'; 'British and Irish itineraries and road-books'; 'Descriptive list of the maps of the Great Level of the Fens, 1604-1900'; 'John Cary, mapseller and globe-maker'; 'Descriptive catalogues of maps and methods of arrangement, with specimens of full and abridged descriptions of maps of various dates'; 'An itinerary of the sixteenth century: *La guide des chemins d'Angleterre'*; 'The cartography of the Provinces of France, 1570-1757', concluding with 'A bibliography of works of reference relating to British and French topography and cartography'. Notes and references are cited throughout.

1136 'Studies in the climatology of South Asia', compiled by V Schweinfurth and others, in 1970, consists of 'A rainfall atlas' of the Indo-Pakistan sub-continent based on rainy days. In thirty-two pages are fifteen, mostly coloured, maps at 1:7,500,000, with three text-contributions and a bibliography of rainfall conditions in India and Pakistan between 1945 and 1969.

1137 'Studies of a small democracy: *essays in honour of Willis Airey'* (Paul's Book Arcade for the University of Auckland), edited by Robert Chapman and Keith Sinclair, contains a portrait, diagrams, reference notes and commentary on the work of Willis Thomas Goodwin Airey.

1138 'Subject catalog of the Special Panama Collection of the Canal Zone Library-Museum: *the history of the Isthmus of Panama as it applies to interoceanic transportation'* was printed in one volume by G K Hall, 1964. Nearly half of this collection is related to the planning and construction of the existing Panama Canal; these items include some five thousand books, reports, engineering drawings, maps, diaries and photographs. Other holdings of the Canal Zone Library-Museum pertain to pioneer exploration and voyages in the region; early surveys for a canal and a railroad and post-1914 projects for improvement of the canal and for a sea-level waterway. Twenty-six pages of maps and photographs are included.

Refer also Ian Cameron: *The impossible dream: the building of the Panama Canal* (Hodder and Stoughton, 1971).

1139 Suess, Eduard (1831-1914) was one of the major influences in physical geography during his academic career, particularly in Vienna, where he was professor of geology. His great work, *Das Antlitz der Erde*, published in four volumes between 1883 and 1909, was translated into French by de Margerie and into English as *The face of the earth*.

1140 Suggate, Leonard S (1889-1970) was one of the leaders of the 'new' geography, which developed so rapidly after the second world war. In addition to the influence of his imaginative teaching, which helped to establish geography in schools, he produced the classic *Africa*, which went into many editions, also *Australia and New Zealand*, besides numerous notes and contributions to journals. He was one of the earliest teachers to recognise the value of field studies, illustrations and visual aids in teaching.

1141 'The surface water year book of Great Britain' (Water Resources Board and Scottish Development Department, HMSO, 1968), in its latest edition, includes hydrometric statistics for British rivers, together with related rainfalls and river water temperature for the year ended 30 September 1965. This volume is the first in a new series, designed to be in conformity with new annual publications presenting statistics of rainfall and groundwater. A supplement, published soon after the parent volume, is to be issued every five years, containing descriptions, information and explanation. Five yearbooks and one supplement will constitute one complete volume.

1142 'A survey of the mineral industry of Southern Africa', by R B Toombs, was published by the Department of Mines and Technical Surveys, Mineral Resources Division, Canada, in 1962, as a record of the Seventh Commonwealth Mining and Metallurgical Congress, 1961. A section considers the minerals of the country, followed by regional mineral production, relevant comments on the geography and geology, also on the minerals as part of the economy. Statistics are included, also illustrations and maps, in addition to maps in a pocket.

1143 Sverdrup, Harald Ulrik (1888-1957) contributed largely to research work in oceanography and polar science, especially as professor of meteorology at the Geophysical Institute in Bergen, as director of the Norwegian Polar Institute and director of the Scripps Institute of Oceanography of the University of California.

One of his major published works, in association with other scientists, was *The oceans: their physics, chemistry and general biology*, 1942 (Allen and Unwin, 1944); his *Oceanography for meteorologists* was published by Allen and Unwin in 1945.

1144 'Sweden books', a small catalogue compiled by Dillons's University Bookshop in co-operation with the Swedish Institute, London, 1970, contains a section on 'Geography' and another on 'Tourism'.

1145 SYMAP, the Synagraphic Mapping System, is capable of composing spatially distributed data of wide diversity into a map, a graph or other visual display, its main advantages being its adaptability and utilisation of widely available computer hardware, notably the line-printer, to produce the displays. The system has been developed at the Northwestern University's Technological Institute by Professor H T Fisher, Professor of City Planning at the Harvard Graduate School of Design; he was the founder and is the Director of the Laboratory for Computer Graphics and Spatial Analysis, where, with financial aid from the Ford Foundation, techniques for graphic display are being developed, using the accuracy, thoroughness, speed and low cost of computers.

1146 'A system of world soil maps', worked out by Soviet soil scientists, serves as a basis for new soil maps of the world and of separate continents; it was included in the Physico-Geographical atlas of the world. The five groups are: Polar soil formation: arctic and tundra soils; Boreal soil formation; Sub-boreal soil formation; Sub-tropical soil formation; and Tropical soil formation.

Refer also M S Simakova: *Soil mapping by color aerial photography* (Israel Program for Scientific Translations for Oldbourne Russian Translations programme).

1147 'Tabula Imperii Romani' was initiated by O G S Crawford at the International Geographical Congress of 1928—a series of maps covering the whole of the Roman Empire at 1:1M scale. Sheets were produced for the Aberdeen and Edinburgh areas, but, in fact, the British sheets were superseded by the Ordnance Survey *Map of Roman Britain.* Elsewhere, in Italy, Egypt, France and Germany, the project was carried further, according to plan; Alexandria, Cairo, Aswan and Wadi Halfa, with Lusdunum (Lyons) sheets appeared during the 1930s and Mogontiacum (Mainz) in

1940. After the war, other European sheets were completed. Each sheet is accompanied by an alphabetical catalogue of the sites shown thereon, listing under each entry the evidence for identification and other notes concerning the wealth of detail brought together. A progress report was issued by the Ordnance Survey in 1933; a further Congress concerning the project was held at the Royal Geographical Society headquarters in 1935 and the resolutions were published by the Ordnance Survey under the title ' International map of the Roman Empire, London Congress, 1935 '.

1148 ' Tanzania today: *a portrait of the United Republic'* (University Press of Africa for the Ministry of Information and Tourism, 1968), is illustrated and contains a folding map. In conjunction with this publication it would be useful to use *Tanzania in maps,* edited by L Berry (University of London Press, 1971), which consists of black and white maps, each with a facing page of text, a geographical survey including contributions by many authoritative writers. There are numerous illustrations and statistics covering every aspect of the geography of the country.

1149 ' Taxonomy and geography: *a symposium'*, edited for the Association (The Systematics Association, 1962) by David Nichols, presents papers read at the Symposium, 1959. Professor C H Lindroth, of Lund, Sweden, and Professor (later Sir) Dudley Stamp contributed a foreword and postscript respectively. There are diagrams and sketchmaps in the text and references at the ends of sections. Topics discussed included ' The taxonomic problems of local geographical variation in plant species '; ' Towards a zoogeography of the mosquitoes '; ' Pest pressure an underestimated factor in evolution '; ' Geographic variation and speciation in Africa with particular reference to *Diospyros*'; ' Some aspects of the geography, genetics and taxonomy of a butterfly '.

1150 Taylor, Eva Germaine Rimington (1879-1966) worked with A J Herbertson at Oxford, and, subsequently, as professor of geography at Birkbeck College, became a leader of geographical thought in this country, specialising in the historical geography of the Tudor and Stuart periods, in mathematical and navigational geography, and in the history of cartography, regarding these special topics as elements in the gradual development of the relationship between man and his environment. Among her prolific writings, besides innumerable articles and reviews, were a number of works on human and economic geography: *Tudor geography, 1485-1583; Late Tudor and early Stuart Geography, 1583-1650; The Haven-finding art; The mathematical practitioners of Hanoverian England 1714-1840* (CUP for the Institute of Navigation, 1966); *The mathematical practitioners of Tudor and Stuart England; Early Hanoverian mathematical practitioners;* for the Hakluyt Society, an edition of Roger Barlow's *Briefe summe of geographie; The writings and correspondence of the two Richard Hakluyts; Sketchmap geography* and *The geometrical seaman*. The annual Eva G R Taylor lecture began in 1960, at the Royal Geographical Society headquarters.

See A bibliography of her works, 1905-1966, *in Transactions of the Institute of British Geographers,* volume no 45.

1151 Taylor, Thomas Griffith (1880-1963) led a varied and adventurous life and was a pioneer in geographical thought and in the practical application of geographical concepts. His influence on academic geography was very great, especially in Australia, where he gave the first lectures in geography at Sydney in 1907, being appointed to a Chair as associate professor of geography in Sydney in 1910; between 1928 and 1935 he was at Chicago University, after which he established the Department of Geography in the University of Toronto, before returning to Australia in 1951. He made particular studies of the geography of Australia and Canada; among his chief published works are: *Australia: a study of warm environments and their effect on British settlement,* 1940; and *Canada: a study of cool continental environments and their effect on British and French settlement,* 1947. The influence of environment on human development and settlement was an abiding interest, reflected in his *Environment and nation,* and *Environment and race,* followed by *Environment, race and migration: fundamentals of human distribution, with special sections on racial classification and settlement in Canada and Australia,* and *Urban geography: a study of site, evolution pattern and classification in villages, towns and cities; Journeyman Taylor,* 1958, is an entertaining, but informative, piece of autobiography. He also edited *Geography in the twentieth century (qv).*

1152 'Teaching geography' series, published by The Geographical Association in a sensible and clear format and style, includes to date: A D Walton: *A topical list of vertical photographs in the national air-photo libraries;* P G Hookey: *Do-it-yourself weather instruments;* J A Bond: *The uses of a revolving blackboard in geography teaching,* with R A Beddis: *A technique using screen and blackboard to extract information from a photograph* and E F Trotman: *Producing a slide set with commentary for elementary fieldwork;* L J Jay: *Geography books for sixth forms;* D G Mills: *Teaching aids on Australia and New Zealand;* R J P Newman: *Fieldwork using questionnaires and population data;* E W Anderson: *Hardware models in geography teaching;* a series of map exercises using Ordnance Survey map extracts; D P Chapallez *et al*: *Hypothesis testing in field studies;* M M Baraniecki and D M Ellis: *A market survey—technique and potentialities.*

1153 'Techniques in geomorphology', by C A M King (Arnold, 1966), treats a wide range of techniques in a rapidly-developing field in a manner which no other work had yet done. Observations of form and process in the field are described, also laboratory experiments with scale models, morphometric analysis of maps and photographs, the analysis of sediments and the statistical analysis of geomorphic data.

Refer also to Dr King's *Beaches and coasts.*

1154 'Terra' has been an influential quarterly since 1888, with text in Finnish and Swedish and summaries in English and German, published by Geografiska Sällskapets i Finland Tidskrift, Helsinki. The book reviews are authoritative and bibliographies are frequently included.

1155 'Terra Australia Cognita: *or voyages to the Terra Australis or Southern hemisphere, during the sixteenth, seventeenth and eighteenth centuries'*, written by John Callander and published in three volumes 1766-1768, is a work of importance and value on the early history of Australasia, including Tierra del Fuego, Southern Patagonia, and the Falkland Islands, which was reproduced in 1967 by N Israel, Meridian and TOT.

1156 'Terrae incognitae: *the annals of the Society for the History of Discoveries'* (Amsterdam: N Israel, 1969-) covers the development of aspects of geography, history, literature and science relevant to the main theme, with neither geographic nor chronological limits. The project was authorised, on an annual basis, at the

1966 Annual Meeting of the society, to support a monograph series published jointly with the Newberry Library of Chicago, by the University of Chicago Press, and a semi-annual *Newsletter*. *Terrae incognitae* ... is an attractively produced journal; articles are welcomed from 'all who do research in the general area of the history of discoveries' and there is a useful review section.

1157 The Textile Council, Education and Information Department, Manchester, issues a number of teaching aids, wall charts, films and illustrated booklets. Wall charts include 'The production of rayon'; 'Textile processing'; 'Cotton flow chart'; 'Cotton growing countries'. Samples of cotton fabrics are available and a film, 'Britain's cotton'. Booklets include 'Growing cotton plants' and 'Introducing cotton'. The films, such as 'Spinning your future', 'Cotton—nature's wonder fibre' and 'Needle and cotton', are available on loan free of charge and there is a filmstrip on cotton growing, with an accompanying booklet. A variety of similar aids on silk are to be had on application to The Silk Education Service.

1158 'Textile history', published from 1968 by David and Charles of Newton Abbot, is aimed to bring together the work of scholars engaged in original research into all aspects of the history of textiles. The processes of spinning, bleaching and knitting, the social and economic aspects of the industry and trade, studies of Courtaulds and of James Longsdon, the fustian manufacturers, are some of the topics so far included in the journal. References to new publications in the field are included in each issue.

1159 The Textile Institute issues three periodicals of interest to geographers: the monthly *Journal of the Textile Institute,* the chief medium for the publication of original research work in textile science and technology, accounts of practical investigations, literature reviews and other surveys; *The Textile Institute and industry,* also monthly, in which are shorter papers of more practical bias, conference papers, etc; and, quarterly, *Textile progress,* each issue of which contains a critical review of recent developments in the technology of a particular sector in the industry, prepared by international authorities.

1160 'Le Théatre françois', by Maurice Bouguereau (Tours, 1594), constitutes the first national atlas of France, the sixteenth maps having been prepared by Gabriel Tavernier and others, with an introduction by F de Dainville. This rare work was reproduced in facsimile by TOT in 1966.

1161 ' **Theatrum orbis terrarum** ' was the title of the famous collection of maps by Abraham Ortelius (*qv*) and is the title given to the series of facsimile atlases and other rare works, a magnificent venture by the Theatrum Orbis Terrarum Publishing Company, Amsterdam. The late R A Skelton and Alexander O Vietor were the first advisory editors; now acting with Alexander Vietor are Dr Helen Wallis and Dr Ir C Koeman. The atlases chosen for inclusion have been those of importance in the early history of cartography and now rare in the original editions. Reproduction is in original size and in black and white as being the best way to show the delicacy and the splendour of the engraving and the work of the cartographer in its true form. They include important editions of Ptolemy's *Geography*, terrestrial atlases by Ortelius, De Jode, Mercator, Hondius, Jansson and Blaeu; nautical chartbooks by Waghenaer, Blaeu, Colom Dudley, Goos and Roggeveen; topographical works such as the *Civitates orbis terrarum* (*qv*) and special regional atlases, including *Speed's atlas of Great Britain*, and *Le théatre françois* (*qv*). One of the greatest achievements in the publication programme so far has been the reproduction of the third centenary edition of Blaeu's *Le grand atlas*, in twelve volumes, and another is *De Nieuwe Groote Ligtende Zee-fakkel*. These series make available a corpus of early atlases and other works illustrating the progress of geographical and cartographical knowledge from the time of Ptolemy to the seventeenth century. Each volume is accompanied by an English introduction, a bibliographical note, collation, references and list of literature, prepared by an acknowledged specialist in the field of the history of cartography. Sebastian Münster's *Cosmographei* . . . , 1544, has also been reproduced, from the 1550 Basle edition, with an introduction in German and English by Professor Dr Ruthardt Oehme; and *A list of geographical atlases in the Library of Congress, with bibliographical notes*, by Philip Lee Phillips (*qv*) is in process of reproduction, from the reprint of the Washington edition 1909-1920. A series begun more recently is the *Coelum Stellatum: a history of extraterrestrial cartography*, comprising facsimiles of the most important early star atlases, chosen by an international board of advisory editors. Each facsimile is accompanied by an introduction and each is reproduced in its original size and by the use of full scale photography. Some books are being written especially for this series, including Dr K H Meine: *A history of lunar cartography,* and Dr A J M Wanders: *A history of Martian cartography.* Another new series, 1970-1971, is 'Neudrucke Ausdem Geographischen jahrbuch', edited, with introduction and indexes, in twelve volumes, by Dr Werner Horn, which will provide a compact source of invaluable information. Dr Horn, who has been associated with the *Geographisches Jahrbuch* and with historical geography and cartography in particular, has selected contributions from some hundred volumes of the yearbook and the 'Geographen-Kalender'. TOT issues general catalogues usually twice a year, a monthly *Newsletter* and frequent handouts announcing individual projects.

Refer also 'Facsimile atlases', *in The cartographic journal,* June 1965.

1162 'Theoretical geography', by William Bunge (Gleerup for The Royal University of Lund, Sweden, Department of Geography, second revised edition, 1966, *Lund studies in geography* series), is 'centred on the character of geographic theory'—'A geographic methodology', 'Metacartography', 'A measure of shape', 'Descriptive mathematics', 'Towards a general theory of movement', 'Experimental and theoretical central places', 'Distance, nearness and geometry', 'The meaning of spatial relations' and 'Patterns of location' being the main topics, subdivided as required. There are a number of sketchmaps and diagrams in the text and appended is a list of *Lund studies in geography.*

1163 'Theses on Asia', compiled by B C Bloomfield, contains those accepted by the universities of the United Kingdom and Ireland, 1877-1964 (Cass, 1967). The offshore islands and Oceania are included; arrangement is by region, by country and then by a stereotyped list of subject headings.

1164 'The three voyages of Martin Frobisher in search of a passage to Cathay and India by the North-West, AD 1576-1578', was edited by Vilhjalmur Stefansson in two volumes (Argonaut Press, 1938); this fine production reproduces the original text of George Best and includes illustrations, maps, notes and a bibliography.

1165 ' 3-D junior atlas ', edited by Frank Debenham (Harrap, revised edition, 1963), was an exciting experiment in atlas design. General maps of world communications and vegetation are followed by maps of the British Isles and of the countries and regions of Europe, Asia, Africa, the Americas, Australia, New Zealand and the world oceans. The new techniques of relief presentation are emphasised by layer colouring and oblique hill shading.

Refer also ' Three-D geography: a stereoscopic aid for the study of terrain ' developed by The Geographical Association in conjunction with C F Casella and Company Limited.

1166 ' Tibet: international studies ', the quarterly journal of The Jawaharlal Nehru University School of International Studies, New Delhi, has an emphasis on current affairs and economy; bibliographies are included from time to time.

1167 ' Tijdschrift voor Economische en Sociale Geografie ' (Brill of Leiden for Nederlandsche vereeniging voor economische en sociale geografie 1910-1966; Koninklijk Nederlands aardrijkskundig genootschap, Amsterdam, 1967-), contains articles in English or Dutch, with abstracts, and a few reviews; notes and selected readings are usually included, also maps.

1168 ' The Times atlas of the moon ', edited by H A G Lewis, contains material from the Ranger, Surveyor and Orbiter photographic surveys, cartography being executed by Fairey Surveys Limited and Hunting Surveys Limited from the original United States AF charts, for John Bartholomew, Edinburgh. The work comprises 110 pages of detailed maps in full colour, numerous colour pictures and diagrams and the astronauts' maps specially prepared from lunar landing missions; there are three pages of maps of the far side of the moon. Lunar flight techniques are explained, precise physical data and a history of moon mapping and theories on the origin of moon features all make the atlas more intelligible and interesting. A key picture of the moon, with a super-imposed graticule and map page divisions enables an area to be pinpointed and there is a complete index of all lunar features, with latitude and longitude references.

1169 ' The Times atlas of the world ', compiled and published in sheets under the direction of Dr John George Bartholomew in 1922, was issued in a revised ' mid-century edition ' by Dr John

Bartholomew and his cartographic department for The Times Publishing Company between 1955 and 1960. The contents of the five volumes are arranged as follows: 1, The world, Australia, East Asia, 1958; 2, South West Asia and Russia, 1960; 3, Northern Europe, 1955; 4, Southern Europe and Africa, 1956; 5, The Americas, 1957. Geographical authorities throughout the world contributed to the accuracy of the maps, many of which are double-page spreads. The most significant areas are on the scale of 1:1M. Each map area is portrayed on the projection and by the cartographic techniques most suited to it; and a unique feature at this date was the nine-plate coverage of the USSR on the scale of 1:5M, based on the *Atlas mira.* The final maps were printed by deep-etch photo-offset, giving sharp outlines and excellent registration. The style is typical of the House of Bartholomew; the paper used is of fine texture, the entire layout is spacious, the range of colour tints judicious and aesthetically pleasing, and the lettering is legible, especially the small names in Times roman. Some historical and reference data are included and each volume carries a full gazetteer. The successor to this work was the *Comprehensive edition,* 1967, second edition, 1969, in one volume; nevertheless, this work contains greater detail, as well as considerable additional material, with no loss of scale, this being achieved by printing on both sides of the paper, using narrower margins and including a single index. Some revisions and improvements were made; endpaper keys show which parts of the world are covered by which plates; an international glossary gives the English equivalents of common name-words. Some discoveries by satellite surveys were included.

1170 'The Times index gazetteer of the world', 1966, a locational guide containing about 345,000 geographical locations, is more extensive than the index to the mid-century edition of *The Times atlas of the world* and is intended primarily as an independent work of reference, although map references to the atlas are given when applicable. Latitude and longitude co-ordinates are included.

1171 Topicards, introduced by Macmillan, were announced as 'a new approach to secondary geography, a classroom-tested card scheme, designed for group, project or individual work for C S E or revision at 'O' level. Fitting in with the increasing realisation of the utility of audio-visual aids in teaching, the cards quickly became widely used, providing a framework around which an imaginative teacher could build, and, having built up a stock, could be used in different combinations, as required. The introductory series dealt with basic concepts in geography—'Latitude and longitude'; 'Climate and weather'; 'Map interpretation'; 'Rivers'; 'Coastlines'; 'Vegetation'; 'Agriculture'; 'Manufacturing industries'; 'Communications' and 'Settlement'. Each set combined four copies of three different cards with teacher's notes. 'Map interpretation', for example, comprised 'Scale and measurement', 'Map symbols' and 'Relief', with subdivisions of each text, illustrated by photographs, sketchmaps and graphs, and including a number of exercises. Later series included sample studies of field work and more detailed topics which could be linked with the original basic ones.

1172 Touring Club Italiano is one of the great cartographic agencies, founded in Milan in 1894. The mapping department is modern and very efficient, having for many years exerted considerable influence on the standard of publication of travellers' guides, monographs, maps, atlases, yearbooks, manuals and periodicals, and has compiled and edited cartographic works of high scientific merit, notably the *Atlante internazionale della Consociazione Turistica* (*qv*) and the *Atlante fisico-economico d'Italia,* a land use map of Italy and the *Carta automobilistica d'Italia,* 1:200,000; the *Road map of Europe* 1:500,000 is also very widely used.

1173 'Town and townscape', by Thomas Sharp (Murray, 1968), presents an analysis of the elements of character and individuality in a town and the factors necessary in good building. The author then attempts to show how to examine present day towns and to appreciate the changing relationships between the buildings and their adjacent streets and open spaces. Finally, the effect of motor traffic and the trend for building high blocks are examined. There are numerous photographs, town plans and drawings.

> *Refer also* Publications of the Town and Country Planning Association, available from The Planning Bookshop, London, including *Town and country planning,* the journal of the Association; *Planning bulletin;* and *Bulletin of environmental education,* a monthly guide to sources and resources for teaching and learning about the environment.

1174 'Trade and commerce', a monthly report on Western industry, published by Sanford Evans Publishing Company, contains notes on all aspects, including photographs and statistics in the text.

1175 'Trade and industry', incorporating the *Board of Trade journal* and *New Technology,* issued weekly, contains practical news items, statistics, graphs of imports, exports, etc, and, as a main feature, ' Programme of trade promotions '.

1176 Traité de géographie physique', completed by Emmanuel de Martonne in 1909, became a standard reference work; the seventh edition, 1948, was expanded to three volumes. E D Laborde made an English translation of the work, with the co-operation of the author, published in 1927 under the title *A shorter physical geography;* in his foreword to the work, Laborde refers to de Martonne as ' the leading exponent of Physical Geography. Perhaps the first writer to realise clearly the exact amount of scientific basis required for the subject, his work is strikingly different from English texts in the absence of over much geology and physics. His method is strictly synthetic, and consists of building up geographical principles through the examination of typical regions. In this way the reader feels that he is dealing with Geography, and not with abstract science '.

1177 'Transactions of the Asiatic Society of Japan', has been prepared by the Asiatic Society of Japan, Tokyo, in a first series 1872 to 1922; second series 1924 to 1940; and a third series from 1948. It is one of the largest and most important collections of studies on all aspects of Japanese civilisation. An index to volumes 1-50 was given in the second series, 1928.

1178 'The transformation of the Chinese earth: *aspects of the evaluation of the Chinese earth from earliest times to Mao-Tse-tung'*, by Keith Buchanan (Bell, 1970), compares ' the Chinese earth to a palimpsest: today there is new writing on this ancient manuscript, writing whose characters are bold and clear and confident'. This quotation, plus the sub-title of the introduction, ' with their own strength they made the landscape . . . ', shows the approach to this study. ' The occupation of the Chinese land '; ' The unity and diversity of the Peoples of China '; ' The density of the Chinese earth '; ' Underdevelopment and development in China '; ' The rise of the people's communes '; ' Contrasts in the microgeography of selected communes '; ' The agricultural regions of China '; 'Agricultural production in China'; 'The industrial sector'; 'Transport and the integration of the Chinese living space'; 'Population: spatial patterns'; 'China: over-populated or under-populated?'; 'Towards a reappraisal of China's intellectual resources'; and 'A summing up: after the dust has settled . . .'. An appendix explores 'The basic units in the environment of China' and there is a useful bibliography. A group of excellent photographs in the centre of the text and a number of sketchmaps and cartograms demonstrates particular topics.

1179 'Transport history', published by David and Charles from 1968, three times a year, in conjunction with the University of Strathclyde, contains original articles on the history of all forms of transport, including shipping, road transport, railways and canals, particularly in the British Isles. The social and economic aspects of the subject are examined, together with relevant developments in engineering. In each issue are book reviews, notes and news items, illustrated and with figures in the text.

1180 'Transport research', which has been published quarterly by the Pergamon Press for the Institute of Transportation, Los Angeles, is prepared with the assistance of an international editorial board, for the rapid dissemination of the most significant scientific results in the field. In addition to articles, there are notes, queries and answers, a section announcing forthcoming events and abstracts in English, French, German and Russian.

> Refer also W R Siddall: *Transportation geography: a bibliography* (Kansas State University Library, 1967, Bibliography series, no 1).

1181 'Trends in geography: *an introductory survey'*, edited by Ronald U Cooke and James H Johnson (Pergamon Press, 1969), a collection of twenty-six short essays, had its origin in a conference for geography teachers organised by the University of London Institute of Education, 1968, each contribution presenting a personal view of a particular field of study. Each chapter in the four parts—physical, human, applied geography and area studies— attempts to guide the reader through the literature of the 'fifties and 'sixties, evaluating new concepts and methods, followed by a select bibliography.

1182 'Tropical agriculture', by G Wrigley (Faber 1961, revised

1969), explains the increase in agricultural production and soil fertility and how this may be maintained. Factors affecting tropical agriculture in particular are dealt with as a whole, not as they affect individual crops. Crop ecology is considered, also crop culture, crop improvement and production, with some reference to cattle-raising.

1183 'Tropical man', continuation of the *International archives of ethnography,* yearbook of the Department of the Royal Tropical Institute, Amsterdam (Leiden: Brill, 1968-), now known as The Department of Social Research, is a series of bulletins appearing at irregular intervals—'the voice of the department'. Its primary purpose is for publications of the Fellows of the department, but contributions from others may be accepted in English and French. The journal also contains book reviews; each issue includes the annual report and is illustrated.

1184 'Tropical science', the quarterly journal of The Tropical Products Institute (formerly The Colonial Products Laboratory), is distributed by HMSO. The 'News round-up' feature is most useful; each issue contains an average of two main articles in addition to reports of investigations at the institute. Careful reviews, a list of recently issued reports and an extensive bibliography of items recently added to the Library, arranged alphabetically within subject headings, serve as an excellent 'current awareness' service.

1185 'Tudor geography 1485-1583', by E G R Taylor (Methuen, 1930; Octagon Books, 1968), must surely remain a classic of scholarship amid the vast literature dealing with the history of exploration and geographical thought. Quoting from the author's preface: '. . . Elizabeth's day saw the map and the globe as the necessary furniture of the closet of scholar, merchant, noble and adventurer alike, and dreams of Empire were formulated which found expression in Drake's achievement and Humfrey Gilbert's splendid failure. The date of the latter has been chosen for the term of this study, for it was marked also by the withdrawal of John Dee, the man behind the scenes of overseas enterprise: it saw, too, the firm establishment of the younger Richard Hakluyt as the propagandist for expansion. His work as a geographer, and that of his emulator, Samuel Purchas, will form the thesis of a later volume.' Valuable appendices follow: 'Catalogue of English geographical or kindred works (printed books and *mss*) to 1583, with notes'; 'List of John Dee's geographical and related works'; 'Catalogue and bibliography of contemporary libraries' and 'Illustrative and evidential documents'. A few plates depict contemporary instruments and maps.

1186 '**Tuttitalia:** *enciclopedia dell'Italia antica e moderna*', issued by the Istituto Geografico de Agostini, Novara, in 1963, covers all aspects of the life of the country and is superbly illustrated, often in colour. Twenty-four volumes have a regional arrangement, in Italian, with a summary and index in each volume and textmaps throughout, where suitable. Beginning with Piemonte—Valle d'Aosta, the regions set out continue with Liguria, Lombardia, Le Venezie, Emilia—Romagna, Toscana, Marche, Umbria, Lazio, Abruzzo—Molise, Campania, Puglie—Basilicata, Calabria, Sicilia and Sardegna, some in two volumes, and the Venice region in three.

1187 The Uganda Geographical Association, founded in 1961, arranges lectures, meetings and excursions; the annual journal is the *East African geographical review*. Close association is maintained with the centre of academic geography in the Department of Geography at Makerere University College, and with the various government departments, also with the East African Institute of Social Research, at Makerere. A series of glaciological expeditions to the Ruwenzori Mountains made a significant contribution to the observations of the International Geophysical Year.

1188 '**Ulster and other Irish maps c1600**', edited by G A Hayes-McCoy (Irish Manuscripts Commission, 1964), contains facsimiles of twenty-three manuscript maps, fifteen of areas within Ulster, six of places in Munster, one of part of the Outer Hebrides and one a plan of Dublin Castle. The first twelve maps were the work of Richard Bartlett.

1189 'The USSR and Eastern Europe: periodicals in Western languages', compiled by Paul L Horecky and Rogert G Carlton (Library of Congress, Slavic and Central European Division, 1958; second edition, 1964), is a selective listing, by country, and including publications outside Europe, if relevant to the study of the area. Brief annotations are usually appended.

1190 '**United Kingdom glossary of cartographic terms**' had its origin in 1964, when the International Cartographic Association appointed a Commission with the object of establishing some degree of standardisation in the use of terms by map makers and users throughout the world. It was decided to publish a multi-lingual dictionary of technical terms in cartography. The Royal Society British National Committee for Geography worked on the United Kingdom contribution and the final text was completed in the autumn of 1966. This was published by the Royal Society in advance of the international work, with an introduction by W D C Wiggins. The International Cartographic Association Special Commission II agreed on a standard form and format for the dictionary; each term to be defined in full, with explanatory notes in English, French, German and Russian, possibly also in some cases, in Spanish. In addition, single-word equivalents were to be listed, but not defined, in up to eight other languages, Dutch, Hungarian, Italian, Japanese, Polish, Portuguese and Swedish.

1191 'United Kingdom publications and theses on Africa', 1963, was published by Heffer of Cambridge in 1966 for the Standing Conference on Library Materials on Africa. The work lists 1,260 books and periodical articles on Africa, published in the United Kingdom, during 1963, including also an index of references to Africa made during the sessions of both Houses of Parliament and a list of theses on Africa accepted by universities in the United Kingdom and Ireland during the academic year 1962-63. A second volume, *United Kingdom publications and theses on Africa, 1964,* was also published in 1966.

1192 United Nations Organization: Much of the work of the UN and its specialised agencies, especially of Unesco (see below), WMO (*qv*) and FAO (*qv*), is of direct or indirect interest to geographers. Publications of particular importance include *World cartography* (*qv*) and *The demographic yearbook* (*qv*). Reports and surveys of economic conditions are numerous, notably *The world economic survey* (*qv*), *The handbook of international trade and development statistics, The statistical yearbook, The commodity survey, Economic survey of Europe, Economic survey of Asia and the Far East,* etc, and such specific reports as the *Quarterly bulletin of coal statistics for Europe, Timber bulletin for Europe,* etc. Among UNESCO publications, the series of reports, *Bibliographical services throughout the world,* the first two prepared by Miss L N Malclès,

1955, and the summary 1950-1959 cumulated by R L Collison, 1961, followed by another summary, 1960-1964, by Paul Avicenne, is an invaluable series; *Bibliography, documentation, terminology,* continues the coordination of such information from all member countries. Relevant individual bibliographies include the UNESCO *international bibliography of economics.* Invaluable also are the publications which derive from specialist research, such as the 'Arid zone programme' and the 'Humid tropics research programme', or from such co-operative projects as the International Hydrological Decade (*see* 'Nature and resources'). Frequently the results of co-operative research, seminars or conferences are summarised in monograph form, such as *Arid lands: a geographical appraisal,* edited by E S Hills, 1966, and the *Source book for geography teaching* (*qv*). It has recently been decided that two United Nations publications, *Current issues: a selected bibliography on subjects of concern to the United Nations* and *New publications in the Dag Hammerskjold Library* be merged into one publication, *Current bibliographical information,* twenty-two issues to be published each year; in addition, a special feature of the new serial is the inclusion of publications of selected agencies.

Scientific maps sponsored by Unesco with the collaboration of the relevant bodies include various maps of Africa, the International geological map of Europe, the International quaternary map of Europe, the Metallogenic map of Europe, the International hydrogeological map of Europe, the Bioclimatic map of the Mediterranean region, the Soil map of the world, the Vegetation map of the Mediterranean region and the *Atlas of the international co-operative investigations of the Tropical Atlantic.* Numerous publications relate to earth sciences, such as the *Annual summary of information on natural disasters,* and on hydrology and oceanography. Periodicals include the quarterly *Impact of science on society*—reports on science as a major force for social change—and *International marine science* (temporarily suspended at the time of writing), a quarterly newsletter prepared by Unesco and FAO. The 'Natural resources research series' began in 1963; *La protection de la grande faune et des habitats naturels en Afrique centrale et orientale,* by Sir Julian Huxley, was published in 1961 (now *op* in English; French edition still available); and the 'Science policy studies and documents' series sets out factual information concerning the science policies of various Member States. The sub-section of Geography and Map

Libraries of the Special Libraries section of IFLA is, with Unesco, working on a world directory of geography and map libraries.

The most complete record of the work of the United Nations is to be found in the *Yearbook of the United Nations,* published since 1947; there is also *Everyman's United Nations . . . and Your United Nations: the official guide book.* The *UN monthly chronicle* is a central source of information. *Books in print,* the general catalogue, and monthly lists are issued, in addition to frequent brochures and advance notices.

1193 The United States National Agricultural Library, one of the focal points in the world for agricultural documentation, has introduced a new cataloguing and indexing system (CAIN) on magnetic tape. These tapes contain a store of bibliographic data encompassing the broad field of agriculture, including agricultural economics and rural sociology, agricultural products, animal industry, engineering, entomology, food and human nutrition, forestry, plant science, soils and fertilisers and other related topics previously included in the *Bibliography of agriculture,* the *National agricultural library monthly catalog* and the *Pesticides documentation bulletin.* All references cited in the *American bibliography of agricultural economics,* a new publication issued by the American Agricultural Economics Association, 1970-, are also included. Subject, author, biographical and organisational indexes can be compiled from the tapes, as well as comprehensive bibliographies. All tapes are nine track, 800 bpi, designed for use in the IBM 360 series computer.

See also 1197.

1194 United States Antarctic Research Program of the National Science Foundation issues an 'Acquisitions list of Antarctic cartographical materials', through the Office of Coordinator for Maps, Department of State; this document continued as part of the *Antarctic report,* retaining its consecutive numbering system begun in 1961. The Office of Coordinator for Maps continues to prepare the list from contributions made to the Office by participating organisations or agencies of the United States. New lists appear as the volume of material warrants. Four categories of documents are distinguished—gazetteers, geodetic data, maps and charts, and photography.

1195 US Board on Geographic Names was first set up in 1890, the present Board being established in 1947, to consider the standardisation of geographical names throughout the world and the terms and abbreviations used on maps and charts. Since 1953, the Board has issued several parts each year of the *Gazetteers* (Washington, Government Printing Office, 1955-) each issue being devoted to a separate country.

1196 United States Coast and Geodetic Survey, established as the Coast Survey in 1807, was by an Act of 1871 expanded to include responsibility for geodetic work throughout the country. Hydrographic, oceanographic and topographic surveys are undertaken and coastal surveys co-ordinated. Nautical and aeronautical charts are published and distributed. Other publications include geodetic control data, planimetric maps, coast pilots and annual tables of tide and current predictions.

1197 The United States Department of Agriculture, National Agricultural Library, in the late sixties awarded a six-month contract to Melpac Inc for the indexing of about thirty thousand agricultural books and journal articles for publication in the *Bibliography of agriculture*, a comprehensive index to world literature on agriculture and allied sciences received in the Library. Subject analysis grouped five main categories: agricultural research; agricultural economics and rural sociology; agricultural products; animal industry and forestry, plant science and soils. About 40 per cent of the literature is in foreign languages—Dutch, French, German, Italian, Japanese, Norwegian, Portuguese, Russian, Spanish and Swedish.
See also 1193.

1198 United States Geological Survey was created by Act of Congress in 1879, taking over at that time the functions and records of earlier surveys. Work is carried on through five branches: administrative, geological, topographic, conservation and water resources. Numerous publications, valuable not only to United States geologists, but to those interested in the subject throughout the world, include the *Bulletin,* 1883-: *Professional papers,* 1902-: the *Monographs series,* 1890-: *Water supply papers,* 1896-: *Annual reports,* 1880-. *Mineral resources of the United States* was published by the survey between 1882-1923, thereafter by the Bureau of Mines, to which the Mineral Resources Division was transferred in 1925. The library is considered to be the largest geological library in the world; acquisitions are as complete as possible in geology, palaeontology, mineralogy, ground and surface water, cartography and mineral resources, and are strong also in the related fields of mathematics, engineering, physics, chemistry, soil science, botany, zoology, oceanography and natural resources. The catalogue of the library was published by G K Hall in twenty-five volumes in 1964.

1199 'United States of America' is one of the series of Irish University Press Area Studies, edited by Professor P Ford and Mrs G Ford, with Professor H C Allen. The documents comprise the nineteenth century British sessional and command papers relating to the United States, covering every aspect of the relations between Britain and America during that period. They are broadly in three groups: diplomatic papers, including correspondence; trade and production reports giving economic information and statistics showing the growth and fluctuations of commerce; and, mainly after 1830, accounts and reports by English observers on aspects of American life. The papers have been arranged chronologically within the subject areas. Particularly interesting in this context are volume I, *Agriculture;* volumes 2 to 6, *Bering Sea seal fisheries;* volumes 19 to 40, *Commercial Reports;* volume 48, *Industry;* volume 50, *Samoa;* volumes 52 to 55, *Trade and tariffs.*

1200 Universities: Information about universities is to be found in *Orbis geographicus,* in geographical journals, such as *The geographical journal,* ' *University news',* and in academic journals themselves. Most geography departments now issue at least one journal, prepared usually by staff and/or students, sometimes including contributions by former students; many maintain their own presses. Some sponsor research and collaborate, if the opportunity occurs, with other relevant bodies. Most geography departments specialise in a few specific aspects of geography; nearly all are building up fine research collections and offer a range of services, bibliographical, cartographic, photographic, etc. Many have important special collections, such as the collection of aerial photographs at the University of Keele, or have specialised in periodical files, reports of symposia, area studies or have some individual acquisition policies, such as maintaining collections of early travel literature or national atlases.

1201 'The university atlas', in a new, thirteenth edition prepared by a team of international cartographers, edited for George Philip and Son by H Fullard and Professor H C Darby, is of major significance. 176 pages of maps in full colour, 24 pages of graphs concerning selected stations throughout the world and an index of more than fifty thousand place-names are a few of the features which make a first-class reference atlas.

1202 The University Tutorial Press has published three series of relevance to geographers—the *Secondary school geographies,* the *Advanced Level geographies* and, of particular interest, the *Modern geography* series, in which there are the following texts, all suitably illustrated: D M Preece: *Foundations of geography;* D M Preece and H R B Wood: *The British Isles;* D M Preece and H R B Wood: *Europe;* A W Coysh and M E Tomlinson: *North America;* W B Cornish: *Asia;* A W Coysh and M E Tomlinson: *The southern continents;* D C Money: *Australia and New Zealand.*

1203 Unstead, J F (1876-1965) was among the leaders of geographical thought in Britain in the early years of this century, revealing the influence of Halford Mackinder, who supervised his early work. He helped to establish academic geography in Britain, especially through his position at Birkbeck College, where he built up the Department of Geography and was in 1922 appointed first professor of geography. His innumerable articles and texts made notable advances in the teaching of geographical method, especially in regional geography; he advocated the study of small 'unit-areas', combining these to form regions, then world divisions.

1204 'Urban affairs', quarterly journal sponsored by the City University of New York, 1970- is world-wide in scope, with a certain bias towards America; in addition to articles, reviews are included and there is an annual index.

Refer also K J Driscoll: *Town study—a sample urban geography* (Philip, 1971).

J Haddon: *Local geography in towns* (Philip, 1971).

1205 'Urban analysis: *a study of city structure with special reference to Sunderland',* by B T Robson (CUP, 1969), attempts a 'geographical method of analysing the social structure of a single town and the use of the results of this analysis in a spatial examination of one facet of the town's sociology . . .'. The work has been published as one of the early series of books and monographs issued by The Syndics of the Cambridge University Press, *Cambridge geographical studies,* which either describe and illustrate new ideas and techniques now reshaping geographical studies, or books which embody the results of new research. They are not to be regarded as textbooks. In this study are four sections—Methodology, Analysis, Application and Conclusion—with a number of appendices, sketchmaps, diagrams and a bibliography.

Refer also Harold Carter: *The study of urban geography* (Arnold, 1971).

— and W K D Davies: *Urban essays: studies in the geography of Wales* (Longmans, 1970).

1206 ' **Urban core and inner city,** *Proceedings of the International Study week in Amsterdam, 1966*', was published by Brill of Leiden under the auspices of the Sociographical Department, 1967. The idea of making scientific researches into the problems of the inner city of Amsterdam was first conceived in the Sociographical Department of Amsterdam University in the summer of 1962. The term 'inner core' was adopted as a functional concept for that section of a city which is the centre of its life, where is a concentration of offices, stores, services, production and traffic. The study-week was planned as a means of exchanging views with workers in the same field in other European countries. The main themes under discussion included the basic concepts, definitions and theoretical approaches; the delimitation, inner tensions and shifting of the urban core; static form versus dynamic function; the application of research finding to the planning of urban cores and research methodology and technics.

Refer also Centres of art and civilisation series (Paul Elek): for example, Stewart Perowne: *Rome,* 1971.

1207 ' **Urban land: news and trends in land development** ', official publication of The Urban Land Institute, is issued monthly, with the July and August issues combined; in addition to articles and information notices, the feature ' Notes of relevant publications ' is of obvious practical value.

1208 'Urbanism and urbanization', edited by Nels Anderson (Leiden: Brill, 1964), consists of a valuable selection of articles, published in two volumes for the International Studies in Sociology and Social Anthropology.

Refer also T G McGee: *The urbanization process in the third world* (Bell, 1972).

1209 'Urbanization in newly developing countries', by Gerald Breese (Prentice-Hall International, 1966), one of the 'Modernization of traditional societies' series, provides a broad-based introduction to the characteristics of urbanisation, the various types and the implications for emerging nations compared with the urbanised western countries. Field studies of large urban areas in Europe, Africa, the Middle East, the Indian subcontinent and South-East Asia are analysed.

Refer also T G McGee: *The Southeast Asian city* (Bell, 1971).

1210 Van Nostrand Company Limited publishes a series of monographs of interest to geographers, the *Searchlight books* series, each of which considers some aspect of political geography or effect of geography on world affairs. General editors are George W Hoffman and G Etzel Pearcy. Several titles have been issued each year since 1962; most of them have been written by geographers or by specialists in allied sciences, for example: R J Harrison Church: *Environment and policies in West Africa*, 1963; W Gordon East: *The Soviet Union*, 1964; Harris B Steward: *The global sea*, 1964; V H Malmström: *Norden: crossroads of destiny*, 1965; David J M Hooson: *A new Soviet heartland*, 1964; Donald J Patton: *The United States and world resources*, 1968; Robert D Hodgson: *The changing map of Africa*, second edition, 1968 and many others.

1211 Varenius, Bernard (1622-1650) produced two works of great geographical interest: *Descriptio regni Japoniae et Siam*, Amsterdam, 1649, Cambridge, 1673; and *Geographia generalis*, Amsterdam, in four editions, 1650, 1664, 1671 and 1672, and two Cambridge editions, 1672 and 1681. A number of English translations are of varying merit. In 'The geography of Bernard Varenius', J N L Baker (Institute of British Geographers, *Transactions and papers*, 1955) gives a re-assessment of his work and a summary of other writers' comments.

Refer also S Gunther: *Varenius*, 1905, reproduced by TOT, 1970.

1212 ' **Vegetation and soils:** *a world picture* ', by S R Eyre (Arnold, 1963, second edition, 1968), provides an introduction to this complex subject, keeping the emphasis on salient features; systematic treatment of climate, for example, is limited to an appendix in which some mean monthly precipitation and temperature figures are tabulated for two or three stations within the area occupied by each vegetation type. The four main divisions are: ' Vegetation and soil development ', ' Vegetation and soils outside the tropics ', ' The British Isles ', and ' Tropical regions ', concluding with a note on ' The outlook for wild nature '. Appendices contain ' Vegetation maps of the continents ', ' Climatic correlations with vegetation ' and a ' Glossary of technical terms '. Maps and diagrams are clear and evocative; there were references and short bibliographies in the first edition, extended for the second.

Refer, edited by the same author, *World vegetation types*.

1213 ' **Vegetation map of the Mediterranean region, 1 : 5M** ' (Unesco, 1970) is one of the *Unesco scientific maps* series. The map was prepared by a panel of plant ecologists set up jointly by Unesco and FAO, led by Professor Gaussen and others; the cartography was assisted by the staff of the Institut Géographique National. Two sheets, east and west, are printed in eleven basic colours, covering an area stretching from the temperate region of Europe and Central Asia to the tropical rain forests below latitude 12 degrees north in Africa, and from western Africa to the Indus Valley. 105 main vegetation types are shown, divided into fifty-two climatic formations, forty-six edaphic formations and seven introduced or transformed vegetation; these main types are also subdivided, bringing the total number of differentiated vegetation types to 246. The general principle followed has been to show the ' potential ' vegetation as it would be unchanged by man or animals. Cultivated areas are not indicated, but vegetation produced by irrigation and afforestation is marked. Desert areas are especially carefully indicated; also the mountain areas where vegetation belts change over very short distances. An explanatory booklet, in English and French, accompanies the map. Also to be noted is the companion map, *Bioclimatic map of the Mediterranean zone* on the same scale (Unesco and FAO, 1963).

1214 Vidal de la Blache, Paul (1845-1918) studied history and geography, and throughout his life was preoccupied with the concept of the integration of historical and geographical factors, together with the influence of man on his environment; he substituted the theory of 'possibilisme' for Ratzel's environmentalism or determinism. He became the supreme influence in French geography, dominating academic geography from his position at the Sorbonne. He was one of the founders of the *Annales de géographie* and the annual *Bibliographie;* his *Etats et nations de l'Europe,* 1889, his *Atlas,* 1894, the *Tableau de la géographie de la France,* 1903 and the *France de l'Est,* 1917 are all classic works, well illustrating his methods. The *Principes de géographie humaine* was constructed after his death by de Martonne from articles and an unfinished manuscript: the work ran into five editions, the fifth being published by Colin in 1955, and a translation from the first French edition was made by M T Bingham (Constable, 1926), edited by Emmanuel de Martonne, in which the editor's preface and the translator's note analyse the plan and scope of the work. The great *Géographie universelle* (*qv*) was conceived by him before the first world war and renewed by Lucien Gallois after his death. Regional and local geographical studies owe much to his insistence on a firm understanding of the geological background, and the mutual relations of soil, climate and living organisms.

1215 'The Vinland map and the Tartar Relation', as a separate publication, was prepared by R A Skelton, Thomas E Marston and George D Painter (Yale University Press, 1965), with a foreword by Alexander O Vietor. The reproduction of this map and the manuscript caused excitement, scepticism, much discussion and published comment; in Alexander Vietor's words—' The present publication of these remarkable documents is designed to be a preliminary work; completeness or finality is not claimed for the commentaries, which are to be considered a springboard for further investigation '. One of the most balanced commentaries was that by G R Crone, ' The Vinland map cartographically considered ', in *The geographical journal,* March 1966. An interesting comparison is ' The finding of Wineland the Good: the history of the Icelandic discovery of America ' by A D Reeves, 1895.

Refer also F J Pohl: *The Viking explorers* (New York: Crowell, 1966).

J R Swanton: *The Wineland voyages* (Smithsonian Miscellaneous Collections, 107, 12, 1947).

J R L Anderson: *Vinland voyage* (Eyre and Spottiswoode, 1967).

and *compare* S E Morison: *The European discovery of America: the northern voyages AD 500-1600* (OUP, 1971).

1216 'Vocabularium geographicum', compiled by G Quencez, in co-operation with scholars representing each country, in French, German, Italian, Dutch, English and Spanish (Council for Cultural Co-operation of the Council of Europe, 1968), was based on an analysis of a number of secondary school text books of the countries involved.

Refer also H Baulig: *Vocabulaire franco-anglo-allemand de géomorphologie* (Faculté des lettres de Strasbourg, Editions Ophrys, 1970).

1217 Vogt, William (1902-1968) was a scholar devoted to the study of world population and the effects of human cultures on the natural environments. Field naturalist and lecturer for the National Association of Audubon Societies 1935-39, he then turned to studies of the climate, population and resources of Chile; his publication, *Road to survival,* 1948, made him known to a wider audience. He published a work on Scandinavia, 1950-51 and *People* in 1960, while working for the Conservation Foundation.

1218 'A voyage towards the South Pole performed in the years 1822-24 *containing an examination of the Antarctic Sea'* was the classic work of James Weddell, 1825 and 1827. David and Charles of Newton Abbot made a reprint from the second edition in 1970, with an introduction by Sir Vivian Fuchs. To the second edition, Weddell added *Observations on the probability of reaching the South Pole* and *An account of a second voyage performed by the Beaufroy to the same seas.*

1219 Waghenaer, Lucas Jansz (*d* 1593) is particularly remembered for *De spieghel der zeevaerdt,* the first atlas of sea charts of the coasts of Northern Europe, 1584. There are numerous inaccuracies on the coasts, but the indications of safe anchorages and the representations of parts of the coasts as seen from the sea provided an important aid to navigation. Admiral Lord Howard of Effingham sponsored an English version, which was published as *The* *mariner's mirrour* in 1588. *Bibliographie de l'oeuvre de Lucas Jansz Waghenaer,* describing all the editions of the atlas, page by page, and noting all known variants of the issues of the charts as well as of the text, appears in *Bibliotheca Belgica,* first series, also reprinted by N Israel of Amsterdam in the 1880-1890 volume, 1961.

1220 ' Walkabout ', 'Australia's way of life magazine', has been published monthly since 1934 by the Australian National Travel Association, Melbourne. Of interest to the layman as well as to the geographer, it is illustrated and includes informative book reviews.

1221 War Office, Geographical Section, General Staff, issued a *Catalogue of maps* in 1947, plus supplements; the edition of 1952 omitted the sheets at 1:250,000 and larger scales, which had been withdrawn from public sale. A detailed classification scheme is used by which each map sheet bears a unique location number.

1222 ' Warne's natural history atlas of Great Britain ', edited by Arnold Darlington, with illustrations by Charles King, shows the approximate location throughout Britain of the six most widely occurring land types and, in an identification section, illustrates some of the vertebrates, invertebrates and plant species associated with them.

1223 ' Water, earth and man: *a synthesis of hydrology, geomorphology and socio-economic geography',* edited by Richard J Chorley, including contributions by twenty-five authors (Methuen, 1969), is illustrated with figures and maps in the text, also with charts and references. The following is the train of thought: ' The world ', ' The Basin ', ' Precipitation ', ' Evapotranspiration ', ' Surface run-off ', ' Ground water ', ' Channel flow ', ' Snow and ice ', ' Short-term run-off patterns ', 'Annual run-off characteristics ', ' Long-term trends ' and ' Choice in water use '. Each of the chapters develops the theme by proceeding from the many aspects of water occurrence to a deeper understanding of natural environments and their fusion with the activities of man in society and the hydrological cycle as a conceptual link between the various aspects of geography.

Refer also David K Todd, *ed*: *The water encyclopedia* (New York: Water Information Center, 1971).

Water Resources Centre: *Archives*, 1890-.

H Wellish: *Water resources development, 1950-1965*: *an international bibliography* (Jerusalem: Israel Program for Scientific Translations, 1967).

Association française pour l'étude des eaux (AFEE): *Thesaurus national de l'eau.*

— *Information eaux* (formerly *Eaux et industries*). American Water Resources Association: *Water resources abstracts,* 1968-; *Water resources bulletin.* United Nations: *Water resources journal,* 1967-.

J Linton Gardner and Lloyd E Myers, *ed*: *Water supplies for arid regions* (University of Arizona, 1967).

1224 'Weather economics', based on the papers and discussions presented at the eleventh agricultural meteorology symposium held at the Welsh Plant Breeding Station near Aberystwyth on March 13, 1968, was edited by James A Taylor and published for the University College of Wales, Aberystwyth (Pergamon Press, 1970), as *Memorandum* no 11. The discussions following each paper are appended at the end of each section. The papers illustrate and measure the import of weather hazards on the budgets of weather-sensitive industries, such as farming and agriculture and forestry; the effect of snowstorms on communications in built-up areas, types of physical weather-proofing and the adjustment of day to day programmes to particular weather probabilities. The keynote of the symposium might be said to be the encouragement of more serious, long-term economic assessments of weather hazards. ' The cost of British weather ', for example, ' The effect of the weather on farm organization and farm management ', ' Weather and risk in forestry '. James Taylor includes an edited report of the discussions . . . and an ' Economic postscript ' was contributed by G N Rubra.

Refer also Alan Watts: *Weather forecasting ashore and afloat* (Adlard Coles, 1967).

1225 'Webster's geographical dictionary', first published in 1949 (second revised edition, 1960, Merriam of Springfield, Mass), is a dictionary containing names, places and geographical terms, together with geographical and historical information, pronunciations and 151 maps. The latest census figures are given for Great

Britain, Canada and the USA. The maps are useful and a thumb index helps quick reference.

1226 'Welsh landforms and scenery', by G Melvyn Howe and Peter Thomas (Macmillan, 1963), portrays the variety of Welsh landforms, tectonic, igneous and gradational. Cross-referencing is employed frequently throughout the text, which is illustrated by well-chosen photographs, block diagrams, maps and sketchmaps. There is a short bibliography.

1227 'Die Welt des Islams' is an international journal concerned with the development of contemporary Islam, edited by Otto Spies under the direction of E G Gómez, K Jahn and G Jäschke (Leiden: Brill, founded in 1913 under the title *Die Welt des Islams: Zeitschrift der Deutschen Gesellschaft für Islamkunde*, then edited by Georg Kampffmeyer). The journal ceased publication during the second world war and resumed in 1951 as an independent periodical with a new, more international character. Sections are devoted to documents, communications and reports, in addition to factual contributions by leading authorities; much space has always been given to reviews of new scholarly publications on the world of Islam.

1228 'Weltatlas: *die Staaten der Erde und ihre Wirtschaft'*, edited by Professor Edgar Lehrmann with the assistance of Professor Heinz Sanke (Leipzig: VEB *Bibliographisches Institut*, 1952), is an entirely new atlas, useful for students of economic geography. Each map faces a corresponding location map on the same scale. The relevant economic information is shown by a uniform system of colouring and conventional signs. The basis of the economic maps is land utilization, shown by solid colour, and twelve categories are distinguished, from first class arable to steppe and desert, with superimposed symbols in colour to indicate the predominant crops. Mineral resources are shown by signs and letters mostly in black, and industries similarly in red or blue. Germany is treated in great detail; in addition to general maps, the country is covered in five plates at a scale of 1 : 1·4m. For other European countries and Asia, where the information is of necessity often generalised, the scales are much smaller.

1229 'De Weltmeere: *Taschenatlas mit den wichtigsten tatsachen aus Meteorologie und Nautik'* (Gotha: Haack, 1956) is in the fifteenth edition. Twenty-three maps of oceans and seas and excellent port plans are accompanied by text covering information on weather, oceanography, navigation and navigational equipment, and shipping. A subject index increases the reference value of the work.

1230 ' **Welt-Seuchen-Atlas** ', world atlas of epidemic diseases, first edited by H Zeiss (Gotha, 1941-45), was revised in an enlarged edition by E Rosenwaldt (Hamburg, 1952, 1956). Sponsored by the Bureau of Medicine and Surgery, Navy Department, Washington, the work is an outstanding contribution to the study of the relationship of diseases, physical environment and man.

1231 ' **West African studies of Mary H Kingsley** ', with a new introduction by John E Flint, was reproduced in a third edition by Cass, 1964, including Mary Kingsley's Preface to the first edition of 1899, and the introduction to the second edition by George A Macmillan. Mr Flint, in the introduction, considers Mary Kingsley's life and achievement. The original appendices are not included.

1232 Westermann, Georg, Publishing House, Brunswick, founded in 1838, has become world-famous for the production of maps, atlases, wall maps and charts, and a variety of monographs and bibliographical works. *Westermanns Geographische Bibliographie,* inaugurated in 1954 (1955-   ), covers all geographical periodicals in German and the important ones published in other countries; in addition, all publications of German university institutes of geography and academies are included. Ten numbers are issued each year and the entries are produced on perforated slips ready for mounting on index cards. *Westermann Lexikon der Geographie,* edited by Dr Wolf Tietze in four volumes 1968-1970, also contains biographies of contributors to geography, articles on the terminology, the concepts of geographers and much material from allied sciences, such as cartography, geodesy and ethnography. In 200,000 entries and numerous maps and other illustrations, the editor aims to provide details not easily available to the general reader. Earlier theories, which have by now been amply digested, such as Ratzel's 'Anthropogeographie ', are treated exhaustively. A world oil atlas (*Erdol Weltatlas*) was published in 1966, containing admirable maps and diagrams. The *Diercke Weltatlas,* also a fine production, includes introductory pages about Germany, settlement patterns and landscape types within the country, and town plans.

1233 Whitcombe and Tombs Limited, publishers and printers, as well as book-sellers, Christchurch, have a special interest in all aspects of Australasian affairs. Catalogues are issued frequently, including special educational catalogues. The *New Zealand topical geographies* series comprises seventeen booklets on various aspects of New Zealand life and landscape, based on a systematic, rather than regional, approach. The firm is publisher for the New Zealand Council for Educational Research—the *Educational research* series and *Studies in education*. Periodicals published include *Historical news* in association with the History Department, University of Canterbury; *New Zealand geographer* (*qv*) and the *New Zealand journal of educational studies*. Many of the firm's monographs are available in Britain.

1234 'The white road: *a survey of polar exploration'*, by L P Kirwan (Hollis and Carter, 1959), was the first English language detailed and scholarly survey of Arctic and Antarctic exploration, beginning with the earliest known voyages and continuing to the transpolar voyage of the nuclear powered submarine, *Nautilus*, with an emphasis on the trends and influences at work throughout polar history. The Commonwealth Trans-Antarctic Expedition is included, also the results of research achieved during the International Geophysical Year, ending with a bibliography. There are some photographs and reproductions and a few maps, including a folding map of the North-west Passage, indicating some of the voyages undertaken in the hope of finding it.

1235 'Whyte's atlas guide', compiled by Fredrica Harriman Whyte (New York: Scarecrow Press; Bailey Brothers and Swinfen, 1962), is a helpful list, classified by region and subject, of the maps in twenty atlases most likely to be found in American libraries. Sections are devoted to groups of countries, individual countries, island groups, oceans, city plans and areas of unique interest, such as the Great Lakes and the Polar Regions. Most of the maps carry a brief annotation and there is a place-name index.

1236 'The wildscape atlas of England and Wales', planned in two volumes for publication 1971-1972, is based on selected data from the Second Land Utilization Survey of Britain. There will be approximately sixty different combinations of basic colour and overprinted habitat symbols on the maps. Text gives details of field and cartographic techniques and also quantitative, ecological and planning analyses of findings from research undertaken during European Conservation Year. A map showing the colour key, a map index to the sheets, photographs and an annotation to the first, experimental, sheet, that of Wensleydale, is given in an article by Alice Coleman, 'A wildscape atlas for England and Wales' in *The geographical magazine* for October 1970.

1237 Wiley, John, and Sons Limited: In addition to the texts of more general background interest, many of the *Interscience* series are relevant to geographical studies. Those of central interest include P J Darlington: *Zoogeography: the geographical distribution of animals;* O W Freeman: *Geography of the Pacific;* Edward Higbee: *American agriculture: geography, resources, conservation;* G H T Kimble and Dorothy Good: *Geography of the northlands;* P E Lydolph: *Geography of the USSR;* A H Robinson: *Elements of cartography;* Guy-Harold Smith, *ed: Conservation of natural resources;* L D Stamp: *Africa: a study in tropical development.*

1238 Woods Hole Oceanographic Institution, at Woods Hole, Mass, sponsors research in all aspects of the oceans, including the physics, chemistry, biology, geology, geophysics, and meteorology of the water masses; the bottom and margins of the seas and the interaction with the atmosphere. Of the many publications, the *Woods Hole Oceanographic Institution atlas* series, 1960-, is one of the most outstanding.

1239 Wooldridge, S W (1900-1963), a leading geomorphologist, especially on the London Basin and South East England, exerted also great influence in almost all aspects of geography, both practical and philosophical, particularly in his position as professor of geography at King's College, London, and through the offices he held in the British Association, section E, and the Geographical Association. He was also a founder spirit of the Institute of British Geographers and an inspiration to the Field Studies Council. He was a stylist and his writings made a considerable impact on the profession at large, especially *The spirit and purpose of geography* (*qv*), which he wrote in collaboration with W G East. *The geo-*

grapher as scientist: essays on the scope and nature of geography, 1956, is a collection of some of his most interesting work.

1240 The Wordie Collection of Polar Exploration, National Library of Scotland, was founded by Sir James Wordie, geologist and explorer (1889-1962); it covers Arctic and Antarctic exploration and research, including works in many languages, altogether some five thousand items, mainly from the eighteenth and nineteenth centuries. Included are Sir James' files as a member of the Discovery Committee of the British Colonial Office. The catalogue has been published in one volume, 1964.

1241 'The World Aeronautical Chart', at 1:1M scale, was devised by the Aeronautical Chart Service of the United States Air Force; when the International Civil Aviation Organisation was set up, the chart was put at its disposal. Features on the ground must be shown in such a way that they can be easily identified from the air; international boundaries are important, but not internal ones. The size and shape of settlements are vital features, and roads and railways are important as landmarks. Placenames are of low priority, but additional information such as radar aids and technical air traffic control data are vital, in addition to the location of airfields.

1242 'World agricultural economics and rural sociology abstracts', published quarterly from 1959, in English, French, German and Spanish, by the Commonwealth Agricultural Bureaux, Farnham Royal, in co-operation with the International Association of Agricultural Libraries and Documentalists and the International Association of Agricultural Economists, is the only abstracting journal to include world literature on agricultural economics and rural sociology (North Holland Publishing Company, Amsterdam, 1959-). Entries, covering books, reports, bulletins and articles, are classified, including a section for 'Reference material'; annual author and subject indexes are issued.

1243 'World atlas of agriculture', published by the Istituto Geografico de Agostini, Novara, under the aegis of the International Association of Agricultural Economists and prepared for the Committee for the World Atlas of Agriculture, is in four volumes, 1969-1972. The complete work will comprise sixty-two maps of land utilisation and four volumes of monographs illustrating the agricultural economy of the various states and territories. The basic scale of the maps is 1:5M; Europe, except for Russia, part of Asia Minor, the Middle East, Japan, New Zealand, parts of North and South Africa, are represented at 1:1,250,000. For a few islands of limited area the scales used are 1:1,250,000, 1:1,500,000 and 1:2,500,000. Central America is depicted at 1:3M and the less populated areas of Canada and Alaska at 1:12,500,000. The monographs, carried out on a uniform plan, and edited by Professor K C Edwards and Dr John Giggs, deal with ' Europe, USSR, Asia Minor '; ' South and East Asia, Oceania '; 'Americas ' and 'Africa '. They are illustrated by line-drawings and statistics, showing development and present-day conditions. The germ of the project was created at the ninth International Conference of Agricultural Economists and plans were discussed during the period 1956 to 1959. The General Secretariat was established at the University of Padua, the Cartographic Department at the University of Bologna and the Monograph Department at the University of Nottingham, with the initial co-operation of the Department of Geography of the London School of Economics and Political Science. Local collaborators were nominated in every country and the final phase of organisation was the adoption of the project by the International Association of Agricultural Economics on the occasion of the Eleventh International Conference of Agricultural Economists held at Cuernavaca in 1961. On the back of each map, in addition to the sheet-lines of adjoining areas, the contour lines are drawn, for they could not be shown on the maps themselves without impairing legibility; in some cases, small additional maps are given, showing selected crops of special significance to the country concerned, which could not be shown on the maps themselves, the main consideration being comparability of content. The project is the first presentation of agricultural resources on medium-scale maps on a uniform basis and a superb example of international co-operation.

1244 ' World atlas of mountaineering ', edited by Wilfrid Noyce and Ian McMorrin (Nelson, 1969), is more than an atlas; it is a finely produced review of mountaineering activity, containing an invaluable range of information on the great mountaineering areas of the world. A series of articles covers every area—the Alps and Himalaya, Japan, New Zealand, the American continent, central Africa and polar regions—each article by a writer who knows the

area from his own climbing experience. There are 260 black and white illustrations, some in colour, maps and sketchmaps.
Compare Standard encyclopaedia of the world's mountains (qv).
Les montagnes de la terre, in two volumes, by R Frison-Roche, 1964.

1245 'World cartography', issued approximately annually since 1951 by the Cartographic Office, Department of Social Affairs of the United Nations, New York, is the best single source of information on activities, progress and plans in the field of cartography throughout the world. In English and French, the fascicules include longer or shorter articles, reports, notes and valuable bibliographies. The second major source of information of world-wide scope stems from the regional cartographic conferences of the United Nations for Asia and the Far East, of which the first was held in Mussoorie, in 1955; the *Proceedings* and *Technical papers* are invaluable for informed discussion on general topics and for reviews of achievements in individual areas, not entirely confined to the Far East.

1246 'The world cities', by Peter Hall (Weidenfeld and Nicolson, *World university library,* 1966), begins with a chapter on 'The metropolitan explosion', followed by chapters devoted to the centres of population based on London, Paris, Randstad Holland, Rhine-Ruhr, Moscow, New York and Tokyo. A forecast of 'The future metropolis' completes the text, which is generously illustrated by photographs, sketchmaps and town plans; bibliographical notes are included.

1247 The World Data Center A: Glaciology, was initiated as an outgrowth of the International Geophysical Year, supported largely by a continuing grant from the National Science Foundation. *Glaciological notes* are published.

1248 'World directory of geographers', prepared in 1952 by the American Geographical Society for the International Geographical Union, on the occasion of the Washington Congress, 1952, was the first attempt at such a world list, in which the special interests of leading geographers were featured. The work has been continued as *Orbis geographicus (qv).*

1249 'The world economic survey', issued since 1948 by the United Nations Department of Economic and Social Affairs, New York, presents in English, French and Spanish an annual analysis of world trade, payments, production, industrialisation and development. Each issue since 1956 has included a special study of a major economic subject. With the 1964 edition (1965), a new publishing policy began; in a cloth edition for the first time, in two parts, part I focuses attention on development plans and provides an appraisal of targets and progress in the developing countries and part II is a review of recent developments in the world economy and a discussion of a number of topical problems.

1250 'World fisheries abstracts', prepared by FAO in English, French and Spanish since 1950, was published quarterly to 1961, subsequently bi-monthly. Abstracts of important publications and articles, prepared on cards for ready reference filing, review the world's technical literature on fisheries and related subjects. The *Handbook for world fisheries abstracts,* 1950-, available free with the subscription, contains detailed descriptions of alternative systems for filing the *Abstracts.*

1251 'World fishing' (incorporating Fish industry) is a monthly magazine covering every aspect of the commercial fishing industry in all parts of the world (Grampian Press, London). Sections include notes entitled 'Comment'; correspondence; short informative articles on new developments; 'Top boats of the month', grouped under port headings; 'European fishing vessel completions during ... (the previous year). There are many monochrome photographs and the advertisements are naturally of significance to those specialising in the industry.

1252 'World forestry atlas' has been in process of compilation for some thirty years at the Federal Research Centre for Forestry in Reinbeck, near Hamburg. Over sixty maps showing the distribution of forests throughout the world were published by 1967.

1253 'A world geography of irrigation', by Leonard M Cantor (Praeger, 1970), sets out 'to gather information from a variety of sources' and 'to provide for the student of geography a synoptic picture of world irrigated agriculture'. In the first two sections, 'Irrigation in perspective' and 'The regional geography of irrigation', the work achieves probably the most systematic and comprehensive survey to date. Forty photographs illustrate the variety of landscape and methods of agriculture referred to and sixty maps and diagrams help to show regional distributions, major developments, water resources and methods of application.

1254 'World highways', the monthy duplicated newssheet of the International Road Federation (Highway Transportation Consultant to United Nations and Co-operating Agency of the Organization of American States and Organization for European Economic Cooperation, is an invaluable document in its field. Items, arranged under the names of countries or areas concerned, give details of road projects in progress or completed, in all parts of the world.

1255 World Land Use Survey was inaugurated to show by examples what could be done on a co-ordinated plan, adapting methods where necessary, and to stimulate countries to carry out for themselves a land use survey similar to the Land Utilisation Survey of Great Britain (qv), using the scheme agreed by the Commission on a World Inventory of Land Use, brought into being for the purpose at the International Geographical Union Congress, Lisbon, 1949. Important *Reports* were presented to the Washington Congress of 1952 and to the Rio Congress of 1956. *Occasional papers* have been produced irregularly since 1956, published by Geographical Publications of Bude; regional monographs have appeared, as completed, since 1958.

1256 'World map, 1 : 2,500,000', based on the International Map of the World, 1:1M, was achieved by co-operation between the cartographic organisations of Bulgaria, Czechoslovakia, the German Democratic Republic, Hungary, Poland, Romania and the Soviet Union. It covered the world, including seas and oceans, at a uniform metric scale and, using sixty-four internationally accepted map symbols, a comprehensive survey of the physical, political and economic-geographical conditions. Three variant base maps were available for use for thematic maps, with the written agreement of the general editor. The naming of the sheets, explanation of conventional signs, etc, are in Russian and English; within the maps, only Latin lettering has been used, all names being in their official forms. Physical relief is shown by contours and a layering system at convenient intervals; twelve colours have been used.

1257 'World map of climatology' *see under Atlas of meteorology* . . . .

1258 World Meteorological Organization had its origin in the International Meteorological Organization, founded at a meeting in Utrecht in 1878, following which a system of Regional Commissions, Technical Commissions and special working groups was built up. This organisation dissolved at the meeting of directors in Washington in 1947 and the WMO gradually took shape as an intergovernment body, holding the first Congress in Paris in 1951, when the new organisation became a specialised agency of the United Nations, with a full secretariat led by professional meteorologists. The World Meteorological Congress meets at least once every four years. Regional associations, one for each continent, meet once in four years and, in the interim, working groups research into subjects of regional interest. The eight technical commissions also meet every four years. Additional working groups are convened for specific purposes, as, for example, to organise the research during the International Geophysical Year. Observations have been internationally standardised. The most important of all the WMO documents are the series known as ' Publication no 9, weather reports—stations, codes, transmissions', a complete guide on the availability of basic meteorological data to the world's meteorological services, in several volumes, kept constantly up to date, complete and accurate. In addition, a vast publications programme is maintained, including the *Technical regulations, The technical publication and technical notes series,* the *International cloud atlas* and special publications such as those connected with the International Geophysical Year. *The* WMO *bulletin* keeps all members informed. Methods have been revolutionised by the use of meteorological satellites, radio or landline teleprinter circuits and facsimile transmissions. The *Catalogue of meteorological data for research,* issued looseleaf since 1965, contains published synoptic or climatological data and information from the eighty-eight member states of WMO, grouped under subject headings—current periodical publications, former periodical publications, occasional publications, data included in other publications, future plans, address to which requests for publications should be sent. Details include a brief history of the publications and an abstract of contents; it is planned to increase the information given on unpublished data. In its first report to the United Nations, WMO recommended the creation of a ' World weather program ' (*qv*).

1259 'World mining', an excellent monthly world report, has the editorial centred in Brussels, the printing department in the Netherlands and the circulation department in San Francisco. A yearbook is also issued, in June, 'Catalog, survey and directory number', including a 'Catalog index of equipment and manufacturers'.

1260 'The world of the soil', by Sir E John Russell (Collins, 1957, second edition, 1959; in the *New naturalist* series), with its fine photographs and diagrams, figures and select bibliography, has become a classic work. The text analyses the structure of soil, how it has developed, the circulation of water and gases which give the soil an atmosphere and a succession of climates quite different from the air above, and dealing with the varied forms of life made possible or impossible by the individual conditions. Sir John Russell's other major work, *Soil conditions and plant growth,* has gone through many editions during his years as Director of the Rothamsted Experimental Station; he created a world famous institution which became the centre of the Commonwealth Soil Bureau.

> *Refer also* the pioneer work: A D Hall and E J Russell: *A report on the agriculture and soils of Kent, Surrey and Sussex* (HMSO, 1911).
>
> W L Kubiena: *The soils of Europe* (Murby, 1953).
>
> Soil Survey of Great Britain, *Reports (qv).*
>
> G R Clarke: *The study of the soil in the field* (OUP, third edition, 1941).
>
> United States Department of Agriculture: *Soil survey manual.*

1261 'World political geography', by G Etzel Pearcy and others (New York: Crowell, 1948, 1957, fifth printing, 1963), shows the world at mid-century. Thirty-eight chapters are grouped under six regional headings: 'Primary concepts', 'The western hemisphere', 'Europe', 'Africa and the Middle East', 'Eastern Asia and Australasia' and 'World political geography at mid-twentieth century'. The text is illustrated by a number of sketchmaps; there are end of chapter references and a bibliography, which is placed after the contents of each chapter. A glossary defines many of the technical terms used in the text.

1262 'World political patterns', by Lewis M Alexander (Rand McNally; Murray, 1957, 1963), is one of the *Rand McNally geo-*

graphy series. The text examines the nature and development of political geography, the structure of the state as a politico-geographic unit, the changing nature of international boundaries and the states as a viable political unit before continuing with individual studies of the United States, Canada, Latin America, the European countries, the Arab world, Africa and parts of Asia and the Pacific. More than a hundred small sketchmaps illustrate the points being made and there are end of chapter references.

Refer also H J de Blij: *Systematic political geography* (Wiley, 1967).

1263 ' World population and food supply ', by J H Lowry (Arnold, 1970), at first-year university level, provides a useful introduction to this subject and its implications. The increase in world population is analysed and potential food requirements estimated; the possibilities of extending and intensifying world argiculture and of producing unconventional and synthetic foods are then examined, including ' Unconventional and synthetic foods ' in detail. Numerous illustrations are included throughout the text, also sketchmaps and diagrams.

1264 World Population Congress, established in Rome in 1955, is concerned with population problems and with the analysis of statistical data from all parts of the world. Great stress is laid on the need for uniformity in population mapping; the official Commission of the International Geographical Union on Population Mapping, set up at the Rio Congress in 1956, has done much to achieve this, particularly on the scale of 1 : 1M.

Refer A G Ogilvie: ' The mapping of population, especially on a scale 1/M ', IGU Commission for the Study of Population Problems, *Report,* 1952.

1265 ' World railways ', edited by Henry Sampson (Sampson Low in many editions at intervals), presents 'A world-wide survey of equipment and operation of the railways of the world '. Included are illustrated reports of progress and proposed developments of major railways throughout the world—some 1,500 altogether—new features in economic development and physical characteristics affecting individual railway systems, with detailed analyses of the systems in each country, illustrated with photographs, maps, statistics and gauge diagrams. Separate sections cover underground railways, manufacturers of diesel engines for rail traction and other specialist topics.

1266 'World shipping: *an economic geography of ports and seaborne trade'*, by Gunnar Alexandersson and Göran Narström (Wiley, 1967), when published, filled a gap in the English language reference material concerning ports, the economics of transportation and cargoes. First a general survey is made of international shipping, trade, cargo and shipbuilding and repair; the rest of the study is arranged regionally, covering the ports in Atlantic Europe, the Mediterranean, Anglo-America, Latin America, Africa south of the Sahara, Asia and Oceania. Photographs, sketchmaps and diagrams abound throughout the text; there is a short glossary and a section listing statistical sources, followed by an author index, an index of persons, vessels and corporations, one of ports and places and a subject index.

1267 'World survey of climatology' is the title of a series of fifteen volumes prepared under the editorship of Professor H E Landsberg (Amsterdam: Elsevier, 1969-1973). Three introductory volumes, by H Flohn, discuss general climatology, followed by volume 4, 'Climate of the free atmosphere', by D F Rex; 5, 'Climates of northern and western Europe' and 6, 'Climates of central and southern Europe', both by C C Wallén; 7, 'Climates of the Soviet Union', by P E Lydolph; 8, 'Climates of northern and eastern Asia' and 9, 'Climates of southern and western Asia', both by H Arakawa; 10, 'Climates of Africa', by J F Griffiths; 11, 'Climates of North America', by R A Bryson; 12, 'Climates of Central and South America', by W Schwerdtfeger; 13, 'Climates of Australia and New Zealand', by J Gentilli; 14, 'Climates of the polar regions', by S Orvig; and 15, 'Climates of the oceans', by H Thomsen. Graphs and tables are included in the texts and bibliographies concerning individual areas.

1268 'World timbers', compiled and edited in three volumes by B J Rendle (Benn; University of Toronto Press, 1969), show representations of timbers in colour at every opening, with descriptions on the opposite pages. The current series of *World timbers* was preceded by the journal *Wood,* which began publication in 1936, in which a feature was the series of colour plates of timbers, accompanied by technical information on their properties and values; the first series, under the name 'Wood specimens', covered the years 1936-1960; these were followed by the work under review. A selection of one hundred wood specimens was reproduced in the first two editions, now out of print. The timbers are now selected mainly for their economic importance or interest for the world markets.

1269 'World trade annual', in four volumes (New York: Walker 1963-), is planned for publication every year, with further expansions. Imports and exports of individual countries are included, relating to over a thousand items of the United Nations Standard International Trade Classification, totals are followed by a number of sub-divisions and commodity figures are analysed, giving origin, destination and values.

1270 'World vegetation', by Denis Riley and Anthony Young (CUP, 1968), is a slim volume of fine photographs, some in colour, with explanatory text on the following themes: 'Plant communities and environment'; 'Deciduous woodlands'; 'Coniferous forests'; 'Rain forests'; 'Grass lands'; 'Savannas'; 'Vegetation of dry environments'; 'Tundra and mountain vegetation'; 'Freshwater and coastal vegetation'. Appended are an essay on soils, a map showing 'Distribution of the main world vegetation types' and an index of terms.

Refer also S R Eyre, *ed*: *World vegetation types* (Macmillan, 1971).

1271 World Weather Program, formerly the World Weather Watch, recommended by the World Meteorological Organization (*qv*), is a unique system for observing, collecting, processing and distributing weather information, using the latest developments in communication, data processing and space technology. It is designed to support the weather services of individual nations by providing them with the basic weather information that can best be handled through international co-operation. The main elements of the research plan were worked out with the US National Academy of Sciences. The Tiros Operational Satellite system, begun in 1966, marked a major advance in global weather observation; automatic picture transmission equipment permits direct readout of cloud photographs at local stations throughout the world as the satellites pass over. Three World Weather Program centres have been designated, in Washington, Moscow and Melbourne.

An invaluable report, 'World weather watch: collection, storage and retrieval of meteorological data', was published in 1969 by WMO (Planning report, no 28). The first two chapters review the problems still to be solved in this sphere; chapter 3 deals with the types of data which should be stored permanently and the main part of chapter 4 lists the data to be stored at the world, regional and national meteorological centres. The following chapters consider the methods to be used in the collection of data, emphasising the advantages of using the global telecommunication system for this purpose; the problem of quality control, with special reference to real time checking by computer; guiding principles for the storage of meteorological data; the retrieval of archive data and the necessity for a definite classification system and standard cataloguing.

1272 '**World weather records**' is the continuing title of a series of publications. The first three publications were prepared by H H Clayton and Miss F L Clayton and published by the Smithsonian Institution in 1927, 1934 and 1947. The fourth, prepared at the Blue Hill Meteorological Observatory, was published by the United States Weather Bureau, covering 1941-1950, 1959. These earlier issues, which included climatological information through 1950, had global distribution and were widely used in research. At its fourth Congress in Geneva, April, 1963, the WMO sponsored the publication of the 1951-60 *World weather records* by the United States Weather Bureau; six volumes covered North America, Canada and Mexico; Europe; South America, Central America, West Indies, Caribbean and Bermuda; Asia; Africa; Antarctica, Australia and Oceania and Ocean weather stations. Requests for data were sent to more than 150 meteorological services throughout the world and the data were transferred to punched cards for machine sorting.

1273 '**The world's landscapes**', under the editorship of J M Houston, is the title of a new series by Longmans, 1970-, beginning with *China*, by Yi-Fu Tuan; *Wales*, by F V Emery; *The Soviet Union*, by W H Parker; *Ireland*, by A R Orme; and *New Zealand*, by K B Cumberland and J S Whitelaw, all published in 1970. The purpose of the series is stated to be the explanation of man's effect on the different landscape types. Freedom of treatment has been given to the authors, as regards length of text, number of illustrations, etc, but physical format is uniform and the approach is not too technical.

1274 '**World-wide directory of mineral industries education and research**', edited by Herbert Wohlbier and others, was published by the Gulf Publishing Company, of Houston, Texas, in 1968; 512 citations have been brought together, from sixty-eight countries.

1275 Wright, John Kirtland (1891-1969), one of the outstanding scholar-geographers of modern times, Librarian and Director of the American Geographical Society, has passed on his inspiration by means of his lectures, meetings with other geographical societies throughout the world and, most important, by means of his writings. He inaugurated *Current geographical literature* . . . (*qv*); compiled, with Elizabeth T Platt, *Aids to geographical research,* second edition, 1947; wrote the classic historico-geographical work, *Geographical lore at the time of the Crusades,* compiled the history of the American Geographical Society, entitled *Geography in the making: the AGS 1851-1951.* His vintage work was *Human nature in geography* (*qv*).

1276 Wye School of Rural Economics and Related Studies, at Ashford in Kent, had its origin in 1922, with the appointment of Mr James Wyllie to the staff of the South Eastern Agricultural College; the college absorbed the former horticultural Swanley College in 1945 and the systematic study of commercial horticulture dates from 1951, when Dr R Folley joined the staff. Professor G P Wibberley, appointed in 1954, introduced the study of land economics. Research is now conducted in local studies, land economics, agrarian development, horticultural production economics, farm economics and farm management. The *Press notices* issued at intervals draw attention to the publications resulting from these surveys and research: 'The British Isles tomato survey', for example, 'Farm business statistics for South-East England', 'Optimum harvesting systems for cereals', etc. Monographs are published from time to time: I D Carruthers: *Irrigation development planning aspects of Pakistan experience,* 1969, among others. The *Farm management pocket book,* by John Nix, has now been extensively revised in a fourth edition.

1277 'The Yorkshire and Humberside Planning Region: *an atlas of population change 1951-1966',* prepared by D G Symes and E G Thomas, with R R Dean, was published by the University of Hull, Department of Geography, 1968. Sixteen maps began as an experimental exercise in mapping characteristics of population change; based on the 1961 figures, they deal with absolute change, total percentage change, natural increase, birth and death rates and fertility. Conventional mapping employs proportional spheres or 'standard score' methods, using chloropleth shading in black and red, on a fold-out sheet, giving a shading key and standard deviation graphs. All maps are on the same scale, 1:565,000. No place-names are marked, but a transparent overlay shows the administrative boundaries. Tables at the end add statistics for individual places. An introductory text explains the cartographic method used.

1278 'Yukon bibliography' was published in a preliminary edition, compiled by J R Lotz (*Yukon Research Project* series, Northern Co-ordination and Research Centre, Department of Northern Affairs and Natural Resources, Ottawa, 1964). The Yukon Research Project is a long term research programme in the social, economic and historical fields in the Yukon territory. Following a general section, items are grouped within broad subject headings, such as 'Bibliographies', 'Forests and forestry', 'Glaciology and permafrost' and so on; the *Arctic bibliography* number is given when relevant.

1279 Zeiss, Carl, VEB, Jena are acknowledged experts in the field of precision instrument making; their precision co-ordinatograph, for example, with accessories for special work, such as the fitting microscope, circle tracing device, line interrupter, dual tracing device, mirror attachment setting projector and lead sharpener. A *Precision co-ordinatograph instruction manual* is available, with illustrations in a back folder. Similar comprehensive instruction booklets describe the firm's Aerial-Photograph Converter and the Stereopantometer.

See also the *Jena review*.

1280 'Zeitschrift für Geopolitik', first issued in 1924, heralded the acceptance of the new trends of thought as an independent subject in Germany. R Kjellén, one of the promoters, called it 'a science which treats of the State as a geographical organism or a spatial phenomenon', but such an association with geography was by no means universally accepted at that time.

1281 Zinc Development Association is typical of the organisations concerned with specific resources, whose libraries and publications provide source material for the geographer. The library was formally organised in 1943 and holds a stock of nearly twenty thousand pamphlets and more than a thousand books, in addition to patents, reports and directories on the world literature of zinc and its uses. *Zinc abstracts* and its companion, *Lead abstracts,* are prepared by the Zinc Development Association and Lead Development Association Abstracting Service. Between them, the two publications review all current world literature on the uses of the metals and their products.

1282 ' Zumstein Katalog mit register ', first published in 1964 by Zumsteins Landkartenhaus, Munich, is intended as an annual publication, and is proving one of the most important sources of information on world maps, atlases and globes, although no claim is made for exhaustive coverage, except for Germany. Text is in German, with headings in English, and the ' catchword ' index is easy to use.

APPENDIX

1283 As this book is in page proof, two important revisions have become available in Britain. The first and second editions of the work by Chauncy D Harris, cited as no 53, are now out of print. An expanded and revised third edition (University of Chicago, Department of Geography, 1971, *Research paper* no 137, is entitled *Annotated world list of selected current geographical serials in English, French and German, including serials in other languages with supplementary use of English or other international languages.* 316 serials are listed from sixty-four countries, the selection having been made on the basis of the quality of geographical material, frequency, regularity and longevity of publication, citations in international bibliographies and availability in major libraries.

The second edition of the *International list of geographical serials* has been revised, expanded and updated by Chauncy D Harris and Jerome D Fellmann, with the assistance of Jack A Licate (University of Chicago, Department of Geography, 1971, *Research paper* no 138). About eight hundred periodicals are newly listed in this edition and nearly all entries have been revised. Careful notes on the purpose, scope, compilation and content of the entries are set out in the introduction.

INDEX

References are to the entry numbers

Aario, Leo: *Atlas of Finland* 120
Aberdeen: British Association volume 916
  maps 1147
  University, Department of Geography 916
  Geographical Society 922
Aberystwyth: University College of Wales 320, 460, 529
  Weather economics symposium 1224
Aborigines 155, 191
About Sweden . . . 1
Abrahams, Peter: *Jamaica* 349
Abr-Nahrain . . . 2
Abstracting services in science, technology, medicine, agriculture . . . 3
Abstracts 3, 56, 67, 68, 80, 83, 171, 182, 184, 185, 223, 238, 256, 258, 272, 287, 336, 346, 370, 468, 494, 510, 554, 563, 646, 655, 659, 666, 679, 685, 699, 729, 732, 875, 976, 1027, 1074, 1107, 1109, 1167, 1180, 1223, 1242, 1250, 1281
Abstracts of Belgian geology and physical geography 184
Academia Colombiana de Historia 90
Academy of Sciences, USA 139, 883, 975, 1122, 1271
Academy of Sciences, SSSR, Institute of Geography 37
Academy of Sciences, USSR 4, 40
'An account of the countries adjoining to Hudson's Bay', Arthur Dobbs 261
'An account of Prince Edward Island in the Gulf of St Lawrence, North America', John Stewart 261
Acta cartographica (TOT) 5
Acta geographica 6, 1087
Acta historiae Neerlandica 7
Adams, D K and H B Rodgers: *An atlas of North American affairs* 131

Adams, Robert McC and others: *The fitness of man's environment* 463
Adelaide: Royal Geographical Society of Australasia 1044
  University, Mawson Institute . . . 839
  Waite Agricultural Research Institute 339
Admiralty (UK) 842, 956
  charts 8, 240, 842
  The Admiralty chart . . . 8, 842
  'The Admiralty Hydrographic Service . . . ' 842
  Admiralty Hydrographic Department 634
  Admiralty Naval Intelligence Division, Geographical Section 504
Adolphe, H: *The bibliography of Mauritius* . . . , A Toussaint and H A 203
Advanced atlas of modern geography (Bartholomew) 413
Advanced geographies series (Methuen) 537, 823, 850
Advanced level geographies (UTP) 1202
Advanced practical geography 10
Advancement of science 234
'Aerial surveys and integrated studies' 11
Aerofilms Limited 631
Aeronautical Chart 1 : 1M 828
  and Information Centre 842
Aeronautical chart catalogue 677
Aeronomy 706
Afghanistan, bibliography 51, 184
Afghanistan 12
Afghanistan at a glance 12
Afghanistan . . . development in brief 12
Afghanistan present and past 12
Africa 17, 152, 167, 228, 249, 290, 314, 429, 499, 500, 501, 519, 526, 574, 543, 653, 665, 666, 705, 726,

429

Africa—(contd)
  727, 730, 735, 781, 796, 850, 950, 1021, 1022, 1037, 1043, 1058, 1133, 1148, 1149, 1187, 1191, 1192, 1209, 1210, 1213, 1231, 1244, 1249, 1261, 1262, 1266
  agriculture 655
  atlases 485, 673, 934
  bibliographical resources 15, 16, 17, 22, 119, 172, 179, 182, 184, 186, 187, 278, 287, 598, 665, 844, 1193
  climate 321, 1267, 1272
  discovery and exploration 448, 619, 984
  forestry 471, 1268
  granites 992
  land use 621
  maps 23, 62, 107, 141, 202, 277, 453, 609, 618, 675, 687, 853, 952, 1192, 1243
  mineral resources 1142
  North, economic atlas 934
  oil 959
  periodicals 6
  population 512
  South 138, 184, 426, 590, 1100
  theses 787, 788, 1058
  transport 518, 932
  tropical 32
Africa ..., ed Colin Legum 602
Africa ..., Sir L Dudley Stamp 1237
Africa, J H Stembridge 1128
Africa, L S Suggate 1140
Africa Collection, Stanford University 20
Africa Department, Northwestern University 20
Africa Institute, Pretoria 13
Africa: maps and statistics 13
Africa on maps dating from the 12th to the 18th century 14
Africa research bulletin 15, 24
Africa Research Limited, Exeter 15
Africa south of the Sahara 21
Africa south of the Sahara ..., comp K M Glazier 16
Africa south of the Sahara, A T Grove 16
Africa studies 24
Africa studies committee, University of Chicago 20

Africa University Department, Yale University 20
African abstracts 3, 666
African area index 278
African bibliography 17
African Book Company, Limited 735
African bulletin 20
African Documents Centre, Boston University 788
African heritage ... 18
African notes 19
African Studies Centre, University of California 20
African Studies Centre, University of Cambridge 20
African Studies in Canada Committee, University of Alberta 20
African Studies Unit, University of Leeds 20
An African survey, Lord Hailey 21
Africana 20, 22
Africana Center, International University Bookseller Inc, New York 21
Africana notes and queries 24
Afrika kartenwerk 23
Afrikaforum 24
Afrique équatoriale 184
Afshar, Iradj: *Bibliography of Iranian bibliography* 184
  Bibliography of Persia 184
  Index Iranicus 184
Afzelins, Nils: *Books in English about Sweden* 1
The age of Drake, J A Williamson 972
Agenda for survival ..., ed H W Helfrich 430
Agostini, Giovanni de 25
  see also Istituto Geografico de Agostini
An agricultural atlas of Scotland 26
Agricultural development in Nigeria ... 27
Agricultural Economics Research Institute, University of Oxford 28
Agricultural geography, L J Symons 167
Agricultural geography symposium: report 30
Agricultural index see Biological and agricultural index

Agricultural Librarians and Documentalists Association 188
Agricultural meteorology 31, 228, 418
An agricultural geography of Great Britain, J T Coppock 29
The agricultural register . . ., H Frankel 655
Agricultural research in tropical Africa, St G C Cooper 32
'Agricultural statistics', C A Halstead 246
'Agricultural typology and land use mapping' 33
Agriculture 4, 26-34, 144, 154, 156, 167, 188, 217, 220, 228, 290, 291, 320, 336, 339, 390, 440, 460, 468, 478, 505, 526, 539, 621, 631, 655, 766, 973, 1012, 1093, 1171, 1178, 1193, 1197, 1199, 1182, 1237, 1253, 1260, 1276
Board of: *Reports* 1021
Parliamentary papers 715
world atlas 1242
Agroclimatological methods, *Proceedings,* Reading symposium 34
Agronomy, bibliography 209
Aguilar Nuevo Atlas de España 35
Ahmad, K S: *A geography of Pakistan* 412
Aids to geographical research . . . J K Wright and E Platt 36, 521, 1275
Air photo packs 275
Air photographs—*Man and the land* 496
see also under Humboldt
Airways of New Zealand, Don Wasley 1016
Akademiya Nauk SSR, Moscow 37, 1017
Akram, Mohammed: bibliography of Afghanistan 184
Alaska 193, 382, 970, 1084,
maps 131, 418, 1243
Albania 184, 1107
The Albanian book 184
Albert I, Prince of Monaco 690
Alberta, atlas 105
University 20, 105
Alden, A J L: *Map book of the Benelux countries,* A J B Tussler and A J L A 807
Aldine university atlas 38

Alexander, L M: *World political patterns* 1262
Alexander von Humboldt, Hanna Beck 628
Alexander von Humboldt, L Kellner 628
Alexander von Humboldt . . . , ed J H Schultze 628
'Alexander von Humboldt and his world', Berlin Exhibition 628
see also under Humboldt
Alexandria, Library 432
maps 1147
Alexandersson, Gunnar and Göran Norström: *World shipping . . .* 1266
Algeria, bibliography 184, 788
Al-Idrisi 39
Allen, H C, on USA 1199
Allen, W J: *Longman's loops* 796
Allgemeine geographischen Ephemeriden 1130
Allgemeine geomorphologie, Herbert Louis 777
Allgemeine hydrographie, Fritz Wilhelm 777
Allgemeine klimageographie, Joachim Blüthgen 777
Allgemeine landschaftskunde, Josef Schmithüsen 777
Allgemeine siedlungsgeographie, Gabriele Schwarz 777
Allgemeine sozial- und bevölkerungsgeographie, Hans Bobek 777
Allgemeine vegetationsgeographie, Joseph Schmithüsen 777
Allgemeine wirtschafts- und verkehrsgeographie, Erich Obst 777
The allocation of economic resources . . . 460
All-Union Geographical Society of the USSR 40
Al-Maghreb, bibliography 184
Almagià, Roberto 41
Alps 65, 1069, 1244
Die alten städtebilder . . . 175
Aluminium abstracts 3, 42
The aluminium courier 42
Aluminium Federation 3, 42
Amazon 873
Amazonia 43
Amenity Lands Act (Northern Ireland) 1965 885

431

America 519, 526, 530, 574, 587, 602, 726, 991, 1013, 1022, 1244, 1249, 1266
  cartography 618, 675, 790, 892, 952
  climate 1267, 1272
  discovery and exploration 395, 448, 972, 1215
  land use 621
  timber 1268
  transport 518, 932
  see also Latin America; North America, United States American Agricultural Economics Association 1193
American agriculture . . . , Edward Higbee 1237
American Association for the Advancement of Science 72, 883, 1223
American Association of Museums 1084
American Association of Petroleum Geologists 927
American bibliography of agricultural economics 1193
The American city . . . , Raymond E Murphy 44
The American Economic Association 729
American Geographical Society of New York 45, 59, 225, 315, 333, 415, 467, 500, 508, 560, 567, 595, 637, 694, 920, 921, 944, 1013, 1108, 1248, 1275
  Aids to geographical research . . . 36
  Atlas of diseases 115
  Atlas of the historical geography of the United States 139
  Current geographical publications 358
  The geographical review 510
  Serial atlas of the marine environment 1076
  Soviet geography . . . 1109
American Geological Institute 185
American Geophysical Union 649
American heritage magazine 1084
American history in its geographic conditions, E C Semple 1075
American Library Association of Chicago 205
The American Neptune . . . 46
American Philosophical Society 927

American University of Beirut, Economic Research Institute 356
American War of Independence 240
American Water Resources Association: *Water resources abstracts* 1223
  Water resources bulletin 1223
Amherst, Field-Marshal Jeffrey 240
Amin, Abdul Amir: *British interests in the Persian Gulf* 239
Amosu, Margaret: *Nigerian theses* 184
Amsterdam University, Sociographical Department 1206
Amtmann, Bernard, Inc, Montreal 47
Analysis of census data for pigs, G T Jones 655
The analysis of geographical data, W H Theakstone and C Harrison 48
The ancient explorers, M Cary and E H Warmington 49
Anderson, E W: *Hardware models in geography teaching* 1152
Anderson, H C: 'User requirements for modern nautical charts' 8
Anderson, I G, *ed*: *Marketing and management* 835
Anderson, J R L: *The Ulysses factor* . . . 448
Anderson, John P: *The book of British topography* 224
Anderson, Nels, *ed*: *Urbanism and urbanization* 1208
Anderson, R C: *English ships* 1091
  List of English naval captains . . . 1091
  List of English men of war . . . 1091
  The Mariner's Mirror, index 1091
  A treatise on shipbuilding . . . , *ed* W Salisbury and R C A 1091
The Andes of Southern Peru, Isaiah Bowman 225
Andregg, C H: 'The scientific art of mapping and charting' 830
Andrews, K R: *Drake's voyages* . . . 1081
Anecdota cartographica, Leo Bagrow 639
Anglemeyer, Mary, *comp*: *Natural resources—a selection of bibliographies* 885

The Anglo-American cataloging rules for maps . . . 1967 284
Animal geography . . . , M I Newbigin 904
Animal health 336
Anesi, José: *Atlas of Argentina* 913
Angola 174, 184
Annales, Association of American Geographers 80
Annales de géographie 3, 50, 183, 371, 1087, 1098, 1214
Annales de spéléologie 285, 290
Annals of coal mining and the coal trade 363
Annals of the IGY 689
Annotated bibliography of Afghanistan 51
Annotated bibliography of Burma 52
An annotated world list of selected current geographical serials in English . . ., C D Harris 53, 1283
Annuaire d'Afrique du nord 290
Annuaire hydrologique (Quebec) 733
Annual summary of information on natural disasters (Unesco) 54, 1192
Antarctic 55, 354, 368, 382, 448, 464, 565, 604, 649, 689, 755, 839, 877, 976, 1062, 1194, 1218, 1234, 1235, 1240, 1267, 1272
 bibliography 47, 56, 791
The Antarctic, H G R King 61
Antarctic bibliography, John Riscoe 56
Antarctic ecology 57
Antarctic journal of the United States 58
Antarctic map folio series 45, 59
Antarctic record 55
Antarctic research 57, 60, 249
Antarctica 106, 151, 233, 249
Antarctica . . . , Frank Debenham 61
Antarctica, ed Trevor Hatherton 61
Anthropogeographie, Friedrich Ratzel 1009, 1075
Anthropology 97, 1084
Antique maps and their cartographers, Raymond Lister 62, 624
Das Antlitz der Erde, Eduard Suess 1139
Apenchénko, V S: *Atlas narodov mira,* ed S I Bruk and V S A 104

Arab world, maps 107
 bibliography 184
 see also under Egypt
Arabia 567, 1133
 bibliography 179, 184, 214
 maps 107
Arabian Gulf, bibliography 1072
Arakawa, H: 'Climates of northern and eastern Asia' 1267
'Climates of southern and western Asia' 1267
Arbeitskreis bibliographie des kartographischen schrifttums 213
Arbók (Iceland) 184
Arcano del mara 63
Archer, J E and T H Dolton: *Fieldwork in geography* 503
Arctic 57, 65, 256, 382, 565, 755, 976, 1056, 1234, 1267
 bibliography 47, 67, 196, 791, 1278
 discovery and exploration 304, 448, 619, 1127, 1240
 maps 418, 464, 970, 1235
Arctic 64
Arctic and Alpine research 65
The Arctic Basin, J E Sater 66
The Arctic bibliography 67, 1278
Arctic Institute of North America 64, 66, 67, 68, 637
Arctic Ocean 970
Area 69, 657
Argentina, atlas 913
Argentina, George Pendle 601
La Argentina: suma de geografia 70
Arid lands 71, 621, 189
Arid lands . . . , *ed:* E S Hills 1192
Arid lands in perspective . . . 72
Arid lands research institutions . . . , Patricia Paylore 73
Arid zone research 376, 621, 691, 733, 1192
Arizona University, Arid lands research 71, 73
Armstrong, Terence and Brian Roberts: *Illustrated ice glossary* 1063
Arnold, Edward (Publishers), London 74
Arnold, Thomas James I: *Bibliographie de l'oeuvre de Lucas Jansz Waghenaer* 178

The art of navigation in England in Elizabethan and early Stuart times, David W Waters 75
Artificial Satellite Sub-Committee, IGY 689
Artsmail 873
Arvill, Robert: *Man and environment* . . . 885
Ash, M: *Regions of tomorrow* . . . 547
Asher, A and Co, Amsterdam 79
Asher, G M: *A bibliographical and historical essay on the Dutch books and pamphlets relating to New-Netherlands* . . . 174
Asia 20, 76, 299, 396, 419, 480, 512, 518, 519, 526, 574, 621, 632, 730, 738, 781, 1021, 1022, 1032, 1037, 1058, 1069, 1133, 1163, 1177, 1192, 1209, 1213, 1223, 1261, 1262, 1266, 1268
  bibliographies 184, 288, 926, 1101, 1102, 1103, 1105
  climate 1136, 1267, 1272
  Colombo Plan 332
  exploration 448, 914
  maps 62, 76, 126, 136, 202, 277, 301, 330, 453, 618, 675, 934, 952
  Southeast 31, 1209
Asia, W B Cornish 1202
Asia . . . , ed: Guy Wint 602
Asia and the Far East, UN Cartographic Conferences 1245
Asia: a guide to basic books, A T Embree and others 76
The Asia bulletin . . . 76
Asia Minor, maps 1243
Asia Publishing House 76
Asian studies 496
Asian survey . . . 77, 456
Asia's lands and peoples . . . , G B Cressey 78, 804
Asiatic Society of Japan 1177
  see also School of Oriental and African Studies, University of London
Assault on the unknown . . . , Walter Sullivan 689
Associated Publishers, Amsterdam 79
Association de Géographie Français 183
Association for Promoting the Discovery of the Interior Parts of Africa 1043
Association for the Study of Snow and Ice 1085
Association français pour l'Étude des Eaux: *Thesaurus national* . . . 1223
Association of American Geographers 80, 308, 315, 886, 1122
Association of Geographical Teachers of Ireland 81, 511
Association of Japanese Geographers 719
Association of Planning and Regional Reconstruction 306
Aslib: *Index to theses* . . . 184
Astronomers, Arab 567
The astronomical and mathematical foundations of geography, Charles H Cotter 82
Aswan 1147
Atacama 225
Athens: Center of Ekistics 83
  French Institute 184
Atlante Farnesiano 567
Atlante fisico-economico d'Italia 1172
Atlante internazionale della Consociazione Turistica 84, 1172
Atlante mondiale (Agostini) 717
Atlante Veneto (Coronelli) 567
Atlantes Neerlandici . . . 85, 269
Atlantic Ocean 679, 877, 1021, 1076, 1084, 1192
Atlantic Ocean atlas 877
Atlas, figure 754
Atlas (periodical) 86
Atlas, comp Paul Vidal de la Blache 1214
Atlas aérien 87
Atlas aerofotográfico (Argentina) 70
Atlas Antarktiki 106
Atlas Belfram 100
Atlas botanic 88
Atlas československé . . . 89
Atlas de Colombia 90
Atlas de economía Colombiana 90
Atlas de France 91, 707
Atlas de la France de l'est 92
Atlas de Paris et de la région Parisienne 93
Atlas de Venezuela 94
Atlas der Schweiz 95

Atlas du Maroc 96
Atlas escolar de Colombia 90
Atlas for anthropology 97
Atlas général Larousse 90, 98
Atlas général Vidal-Lablache 99
Atlas historique (Atlas Belfram) . . . 100
Atlas historique et géographique 517
Atlas internationale Larousse 101
Atlas mira 102, 1059, 1169
Atlas nacional de España 103
Atlas narodov mira 104
Atlas novus (Blaeu) 164
Atlas of Alberta 105
Atlas of the Arab world and the Middle East 107
Atlas of Australian resources 108
An atlas of Australian soils 109
Atlas of birds in Britain 110
The atlas of Britain and Northern Ireland 111
Atlas of the British flora 110, 112, 885
Atlas of Canada 113
Atlas of Denmark 114
Atlas of diseases 115
The atlas of the earth 961
Atlas of Edinburgh 116
Atlas of the empire (China) 868
Atlas of England and Wales (Saxton) 353
Atlas of European birds 117
Atlas of European history 118
Atlas of evolution 119
Atlas of Finland 120
Atlas of Florida 121
Atlas of glaciers in South Norway 123
Atlas of the Great Barrier Reef 124
Atlas of Great Britain (Ordnance Survey) 924
Atlas of the historical geography of the United States 139
Atlas of the history of geographical discoveries and explorations 125
Atlas of the international co-operative investigations of the tropical Atlantic (Unesco) 1192
Atlas of Islamic history 126
Atlas of Israel 127
Atlas of the lithological and palaeogeographical maps of the USSR 330
Atlas of London and the London region 128
Atlas of London and suburbs ('Geographia') 491
Atlas of meteorology . . . 129, 164, 606, 750
Atlas of New Zealand geography 130
An atlas of North American affairs 131
Atlas of physical, economic and social resources of the Lower Mekong Basin 132
Atlas of planning maps 527
An atlas of population change 1951-66—the Yorkshire and Humberside Planning region 133
Atlas of the Republic of China 301
An atlas of resources of Mysore State 871
Atlas of Saskatchewan 134
Atlas of Scotland (Bartholomew) 164
Atlas of social and economic regions of Europe 135
Atlas of South-east Asia 136
Atlas of Tanzania 137
Atlas of the Union of South Africa 138
Atlas of the United States 139
Atlas of the Universe 140
Atlas of Western Europe, Jean Dollfus 141, 1006
Atlas of the world commodities, population, trade and consumption . . . 142
Atlas of world history 143, 1006
Atlas of the world's resources 144
Atlas östliches mitteleuropa 145
Atlas över Sverige 146
Atlas Porrua de la Republica Mexicana 147
Atlas sive cosmographicae meditationes . . . (Mercator) 46
Atlas Swiata 952
Atlas USSR 330
Atlas van de Europese vogels (Elsevier) 117
Atlas van Nederland 148
Atlas zur Erdkunde (Lautensach) 451
Atlases: 36, 800, 1006, 1053
  autocode programme (Camap) 26
  celestial 85, 1161

435

Atlases—(contd)
   city 491
   early 218, 363, 574, 820, 846, 996, 998, 1161
   educational 244, 309, 331, 571, 593, 939, 961, 1059, 1128
   guides 85, 176, 184, 269, 789, 1235, 1282
   historical 118, 126, 139, 143, 294, 442, 517, 612, 613, 614, 615, 816, 892, 952, 961, 1006
   national 35, 85, 89, 90, 91, 94, 96, 103, 108, 113, 114, 120, 126, 137, 139, 146, 148, 866, 867, 868, 870, 871, 1160
   pocket 164
   regional 93, 105, 106, 107, 111, 116, 121, 124, 128, 131, 132, 133, 134, 135, 136, 141, 145, 147, 164, 274, 301, 330, 374, 375, 377, 378, 405, 418, 724, 851, 895, 909, 913, 949, 970, 971
   road 138, 164
   sea 830, 1219, 1229
   special 26, 88, 97, 104, 109, 112, 117, 119, 125, 129, 142, 144, 164, 177, 321, 322, 323, 324, 330, 464, 527, 606, 681, 725, 752, 845, 850, 862, 869, 885, 934, 967, 969, 1006, 1014, 1076, 1078, 1121, 1136, 1222, 1236, 1243, 1252, 1258, 1277
   transparency 755
   world 84, 98, 99, 102, 164, 170, 232, 277, 309, 331, 361, 368, 413, 418, 451, 464, 492, 571, 573, 575, 578, 579, 672, 717, 725, 780, 852, 874, 933, 936, 952, 961, 978, 1006, 1010, 1130, 1165, 1169, 1172, 1201, 1214, 1228
Atmosphere, weather and climate, R G Barry and R J Chorley 149, 850
Atmospheric tides, thermal and gravitational, Sydney Chapman and R S Lindzen 150
Atolls 921
Auckland, maps 130
  University, New Zealand Geographical Society 903
Aurada, Fritz: 'Entwicklung und Methodik der Freytag-Berndt Schulwandkarten' 475
Aus allen weltteilen 958

Das ausland 958
Australasia 454, 526, 1077, 1233, 1261
  maps 892, 971
Australia 155, 163, 332, 339, 379, 420, 443, 448, 512, 602, 621, 656, 847, 858, 901, 932, 1007, 1021, 1104, 1106, 1152, 1220, 1268
  bibliographical sources 2, 153, 154, 184, 190, 191, 212, 345, 708
  climate 1267, 1272
  maps 62, 108, 109, 124, 160, 184, 331, 345, 375, 618, 675, 934, 939, 1010
Australia, *Parliamentary papers* 715
Australia and New Zealand, D C Money 1202
Australia and New Zealand, L S Suggate 1140
Australia and the Pacific Islands ..., Allen Keast 156
Australia in its physiographic and economic aspects, T G Taylor 1151
Australia journal of applied science 339
Australia, New Zealand and the South Pacific..., Charles Osborne 152
Australia: a study of warm environments ..., T G Taylor 151, 1151
  see also under Oceania
Australian and New Zealand Association for the Advancement of Science 1060
Australian bibliography ..., D H Borchardt 153
An Australian bibliography of agricultural economics ... 154
Australian Development Research Foundation 379
The Australian encyclopedia 155
The Australian environment (CSIRO) 156
The Australian environment, T G Taylor 1151
Australian geographer 157
Australian geographical studies 656
The Australian Geography Teachers' Association 158
Australian Institute of Cartographers 159, 272

Australian journal of agricultural research 339
Australian maps 160
Australian national Antarctic Research Expeditions (ANARE) 61
Australian National Travel Association 1220
Australian National University: Research School of Pacific Studies 736
Australiana 1007
Austria, bibliography 182, 184
Autocode programme (Camap) 26
Automatic cartography and planning 445
Automatic Documentation Centre, Frankfurt 658
Automation, in cartography 940
'Automation: hydrographic charting', A R Boyle 634
Automobile Association of South Africa: *Road atlas* 138
Avicenne, Paul, ed: *Bibliographic services throughout the world* 1192
Aylward, John: 'The retail distribution of Ordnance Survey maps ...' 924
Azores, discovery 984

Bachmann, Frederich Von: *A bibliographical guide and index to the principal collections of town plans* ... 175
Background notes on the countries of the world 161
Background to geography, G R Crone 162
Background to political geography, G R Crone 163
Baedeker's Russia ... 363
Baer, Dr, ed: INSPEL 684
Baffin 304
Bagnall, A G: on New Zealand reference material 588
Supplement to *Guide to New Zealand reference material* ... 204
Bagrow, Leo 164, 639
Anecdota cartographica 639
Geschichte der kartographie 164, 562
Imago mundi ... 164
The origin of Ptolemy's Geographia 998

Baija, F J Rio: *Bibliography of economic geography,* Galicia 184
Bailey, Sir Edward: *Charles Lyell* 801
Geological Survey of Great Britain 551
Bailey, J H *et al, comp*: *A guide to paperbacks on Asia* 76
Baird, Patrick D: *The Polar world* 976
Bakaev, U G, ed: *Atlas Antarktiki* 106
Baker, Alan and J B Harley, ed: *Studies in historical geography* 363
et al: Geographical interpretations of historical sources 505
Baker, D L ed: *The Daily Telegraph world atlas* 361
Baker, J N L: 'The geography of Bernard Varenius' 1211
The history of geography ... 620
'Mary Somerville and geography in England' 1097
Balchin, W G V, ed: *Geography and man* ... 526
Geography: an outline for the intending student 534
Baldock, E D: *Manual of map reproduction techniques* 818
Balkans 904
Banerjee, B, ed: *Essays on agricultural geography* ... 460
Banks, Arthur, cartographer 723
Banks, Joseph: *Endeavour journal* 427
Baraniecki, M M and D M Ellis: *A market survey* 1152
Baranov, A N *et al*: *Atlas mira* 102
Baratier, E *et al*: *Atlas historique* ... 100
Barker, S M: *Quantitative methods,* W V Tidswell and S M B 1000
Barlow, Roger: *Briefe summe of geographie* 596, 1150
Barnes, Michael, translator 864
Barr, John, ed: *The environmental handbook: action guide for the United Kingdom* 430
Barrow, John: *A chronological history of voyages into the Arctic Region* 304

437

Barry, R G 796
  on mechanical map production 823
  and R J Chorley: *Atmosphere, weather and climate* 149, 850
Bartholomew, George 164
Bartholomew, John 309
Bartholomew, John George 164, 229, 1169
  and A J Herbertson: *Atlas of meteorology* . . . 129, 750
Bartholomew, John and Son, Limited 164, 218, 262, 269, 413, 1008, 1038, 1168, 1169
  The citizen's atlas of the world 309
  Early map series 164
  Physical atlas 129, 606, 967
 'Bartholomews—on the map' 164
Bartlett, Richard, cartographer 1188
Baschin, Otto, ed: *Bibliotheca geographica* . . . 563
Bascom, P, ed: English language *Concise encyclopedia of explorers* 341
The bases of a world commonwealth, C B Fawcett 458
Bassett, D K 942
Bastié, Jean: *Atlas de Paris* . . . 93
Bataillon, Lionel 506
Bates, H W 1097
Bates, Marston 817
  and Philip S Humphreys, ed: *The Darwin reader* 921
Baulig, Professor Henri 92
  The changing sea-level 657
  Le plateau central de la France 329
  Vocabulaire franco-anglo-allemand de géomorphologie 1216
Bautier, R H 100
Bavaria, maps 378
Baynton-Williams, Roger: *Investing in maps* 624
Beaglehole, J C, on Captain Cook 596
  The Endeavour journal of Joseph Banks 427
  The exploration of the Pacific 972
Beasley, W G et al: *China and Japan* 301
Beaufort, Sir Francis 8
Beaujeu-Garnier, Jacqueline: *Géographie de la population* 512
  Paris (atlas) 949

Beaujeu-Garnier, Jacqueline—(*contd*)
  and Jean Bastié, ed: *Atlas de Paris* . . . 93
  et al, ed: *La géographie française* . . . 514
The beauties of scenery . . ., V Cornish 348
Beaver, S H 512, 712
  'The Le Play Society and field work' 775
Beazley, Sir Charles Raymond 582
  The dawn of modern geography 165, 365
  Prince Henry the Navigator 984
Beazley, P B, on sea surveys 1066
Beck, Hanna: *Alexander von Humbolt* 628
  and Wilhelm Bonacker: *Mexican atlas* 851
Beddis, R A: *A technique using screen and blackboard to extract information from a photograph* 1152
Bederman, Sanford H: *A bibliographical aid to the study of the geography of Africa* 172
Beginning practical geography, Arthur Bray 807
Behaim, Martin, globe-maker 166, 567, 867
Beharrell, H 989
Bekker-Nielsen, Hans and Thorkil D Olsen: *Bibliography of Old Norse-Icelandic studies* 184
Belfast 809
  Queen's University, Department of Geography 714
Belgium 371, 807, 841
  bibliography 182, 184
  wool trade 708
Belgrade, maps 240
Bell's advanced economic geographies 167
Bengtson, Hels A: *Fundamentals of economic geography*, William Van Royen and N A B 478
Benham, Frederic: *The Colombo Plan* . . . 332
Benito, Juan: *Bibliography of agricultural geography* (Spain) 184
Benjamins, John: Scientific Periodical Trade, Amsterdam 79

Beresford, Maurice: *The lost villages of England* 798
New towns of the Middle Ages 898
Berge der welt, ed, H R Muller 168
Bergen Geophysical Institute 1143
Berger, H: *Die geographischen fragmente des Hipparch* 610
Berghaus, Heinrich Karl 169, 628, 957
  Physikalischer atlas . . . 168, 725, 969
Berichte zur deutschen landeskunde 184
Berio Library, Genoa 334
Berlin, atlas 949
Berlin, Geographical Institute 951
  Geographical Society 563, 628
  maps 378
  University, Chair of Geography 1037
Berlinghieri, Francesco: *Geographia* 492
Bermuda, weather records 1272
Bernleithner, Ernst and Rudolf Kienauer: *Bibliographiekirchengeschichtlicher* 177
Berr, Henri 506
Berry, B J L: *Geography of market centres* 542
  and D F Marble, ed: *Spacial analysis* . . . 1112
Bertarelli, Luigi V 84
Bertelsmann Verlag 709, 745
  Cartographic Institute 170, 578
Bertelsmann atlas international 170, 578
Bertelsmann hausatlas 170
Berton, Peter and Eugene Wu: *Contemporary China* 344
Best, George, voyages of Martin Frobisher 1164
Bevan, W L and H W Philliott: *Mediaeval geography* . . . 607
Biafra 950
Bialik Institute of the Jewish Agency 127
Bible, atlas 1006
Bibliografia aborigen de Costa Rica, Jorge A Lines 828
Bibliographia geodaetia 171
A bibliographic aid to the study of the geography of Africa 172

A bibliographic guide to population geography 173
A bibliographical and historical essay on the Dutch books and pamphlets relating to New-Netherland . . . 174
Bibliographical bulletin of American oceanography and geophysics 948
A bibliographical guide and index to the principal collections of town-plans and rivers published in the sixteenth and seventeenth century . . . 175
Bibliographical services throughout the world (Unesco) 1192
Bibliographical Society of America 211
Bibliographie cartographique internationale 176, 257, 269
Bibliographie de cartographie ecclésiastique 177
Bibliographie de la France 184
Bibliographie de l'oeuvre de Lucas Jansz Waghenaer 178, 1219
Bibliographie des forstlichen schrifttums Deutschlands 1027
Bibliographie des oeuvres relatifs à l'Afrique et à l'Arabie 179
Bibliographie du Laos 184
Bibliographie du Népal 181
Bibliographie de Tahiti et de la Polynésie française 180
Bibliographie géodésique internationale 171, 182
Bibliographie géographique de Belgique 184
Bibliographie géographique internationale 50, 183, 290, 315, 329
Bibliographie hydrologique 669
Bibliographie sommaire de l'oeuvre mercatoriennes 846
Bibliographiekirchengeschichtlicher Karten Deutschlands 177
Bibliographiekirchengeschichtlicher Karten Österreichs 177
Bibliographies 36, 45, 51,
  Africa 184, 186, 590, 665, 788, 844, 1100
  agriculture 154, 188
  Alaska 193
  America 211
  Antarctic 56

Bibliographies—(contd)
  Arabia 214
  Arabian Gulf 1072
  arid lands 71, 72, 189
  Asia 76, 926, 1101, 1102, 1103, 1105
  atlases 789
  Australia 153, 190, 345, 1071
  Australian aborigines 191
  Australiana 212
  British Columbia 194
  Bulgaria 184, 245
  Canada 196, 298
  Caribbean 263
  cartography 213, 789
  charts 822
  China 344
  Commonwealth 873
  Costa Rica, maps 828
  Denmark 184
  Dutch-American 280
  Dutch West Indies 174
  forestry 943, 1027
  geodesy 182
  geography 623
  geology 185, 237
  geomorphology 238
  Ghana 198
  Hungary 630
  hydrology 187
  International Map 1:1M 695
  Iran 986
  Iraq 585
  Ireland 584
  Japan 1032, 1073
  Kerala State, India 199
  Kuwait 1072
  Latin America 184, 235, 584, 770
  J H G Lebon 776
  Mackenzie Delta 805
  Malaya 201
  ' Man and environment ' 873
  Maori 200, 206
  maps 160, 202, 269, 415, 989
  Mauritius 203
  Middle East 356
  national 184
  National Book League 873
  natural history 1088
  natural resources 885
  New Zealand 200, 204, 588, 878
  Nigeria 906
  Pacific Islands 197, 314, 345, 946

Bibliographies—(contd)
  Pakistan 412
  place-name literature 205
  Poland 978
  Polar regions 791, 975
  population 173
  Portugal 235
  railways 195
  Friedrich Ratzel 648
  rivers 175, 873
  Scotland 873, 1011
  seismology 208
  Singapore 184
  soil science 209
  Spain 235
  Sweden 1
  Tasmania, maps 990
  town plans 175
  USSR 207, 246, 589, 1051
  vegetation maps 675
  Wales 252
  water resources 1223
  West Indies 235
  Yukon 1278
  Bibliography and index of geology exclusive of North America 185
  Bibliography, documentation, terminology (Unesco) 1192
  Bibliography of Africa 186
  Bibliography of African bibliographies 665
  Bibliography of African hydrology 187
  Bibliography of agriculture 188, 1193, 1197
  A bibliography of arid lands bibliographies 189
  Bibliography of Australia 184, 190
  Bibliography of Australian Aborigines . . . 191
  Bibliography of bibliographies of East Asian studies in Japan 288
  Bibliography of books on Alaska . . . 193
  A bibliography of British Columbia 194
  A bibliography of British geomorphology 238, 961
  A bibliography of British railway history 195
  A bibliography of Canadiana . . . 196

Bibliography of cartography (Congress) 783
A bibliography of Fiji, Tonga ... 197
The bibliography of the first letter of Christopher Columbus ... 334
A bibliography of Ghana ... 198
A bibliography of the Gold Coast 184, 198
Bibliography of Iranian bibliographies 184
Bibliography of Kerala State, India 199
Bibliography of the literature relating to New Zealand 200
Bibliography of Malaya ... 201
Bibliography of maps and charts published in America before 1800 202
The bibliography of Mauritius ... 203
A bibliography of New Zealand bibliographies 204
A bibliography of non-periodical literature on Sierra Leone ... 184
Bibliography of Persia 184
Bibliography of place-name literature ... 205
Bibliography of printed Maori to 1900 206
Bibliography of Ptolemy's geography 998
Bibliography of the Punjab 184
Bibliography of reference materials for Russian area studies ... 207
Bibliography of seismology 208
Bibliography of soil science ... 209
The bibliography of West Africa 184
Bibliography of the West Indies ... 184
A bibliography of works relating to Scotland 184
Biblio-mer ... 210
Biblioteca Danica 184
Bibliotheca Americana ... 211
Bibliotheca Australiana 212
Bibliotheca Belgica ... 184
Bibliotheca cartographica 213, 269, 695
Bibliotheca celtica 184
Bibliotheca geographica ... 563, 847

Bibliotheca geographorum Arabicorum 214, 228
Bibliotheca Hibernicana 715
Bibliothèque nationale, Paris 176, 215, 269, 277, 284, 315
Bickmore, D P 940
Bickmore, E *ed*: *Automatic cartography* ... 445
Bingham, M T, translator 1214
Bioclimatic map of the Mediterranean zone 1192, 1213
Biogeography 219, 1117
Biogeography and ecology in South America 216
Biological and agricultural index 217
Biology 228, 1084
  marine 279
  pleistocene 974
Birch, J W: *The Isle of Man* 716
Bird, James: *The geography of the Port of London* 633
The major seaports of the United Kingdom 809
Birds, atlas maps 110, 117
Birdwood, L 480
Birmingham atlas 297
Birmingham University, Centre of West African Studies 20
  Department of Geography 491, 749
Birkbeck College, Department of Geography 1203
Blache *see* Vidal de la Blache
Black Country, computer atlas 297
Black Sea, maps 164, 277
Blackut, T J, cartographer 446
Blaeu, Jan Willenz 218
Blaeu, Willem Jansz 85, 94, 218, 1161
  Atlas novus 164, 218
  biography 218
  catalogue 218
  globes 567
  Le grand atlas ... 218, 574, 1008, 1161
Blakeney, T S: 'A R Hinks ...' 609
Black, A and C: *The pioneer histories* 972
Bloomfield, B C, *comp*: *Theses on Asia* 1163
Blue Hill Meteorological Observatory 1272

Blumea 219
Bluthgen, Joachim: *Allgemeine klimageographie* 777
Board, Christopher *et al, ed*: *Progress in geography* 993
Board of Agriculture, London 220
The review and abstract of the County Reports ... 219
Board of Trade journal 221
Board of Trade Library, London 221
Bobek, Hans: *Allgemeine sozial- und bevölkerungsgeographie* 777
Boccio, H O, cartographer 70
Bocker, T W *et al*: *The flora of Greenland* 466
Bodleian Library, Oxford: Map Room 222, 269, 315, 572, 765
Boggs, S W, classification 315
and D C Lewis: *The classification and cataloging of maps and atlases* 284, 315, 316
Bogs 230
Boletin de información cientifica Cubana 223
Bolivia, bibliography 184
Bologna University, Cartographic Department 1243
Bonacher, Wilhelm: *Mexican atlas*, Hanno Beck and W B 851
Das schrifttum zur globenkunde 567
Bonapace, Umberto: 'La production cartographique de l'Institut Géographique de Agostini ...' 717
Bond, Major C C J, cartographer 612
Bond, J A: *The uses of a revolving blackboard in geography teaching* 1152
Bonn University, Institute for Cartography and Topography 213
Bonne, Rigobert, globes 567
Bonset, E J, publishers 79
The book news (Heffer) 605
The book of British topography 224
The book of wool 708
Books 873
Books about Nigeria ..., W John Harris 184
Books about Singapore and Malaysia ... 184

Books for grammar and high schools (Edward Arnold) 74
Books in English on Sweden 1
Books in print (UN) 1192
Borchardt, D H: *Australian bibliography* ... 153
Bordeaux, Institut de Géographie de la Faculté des Lettres 251
Borjeson, Hj: *Swedish ships* 1091
Bormann, Werner 803
on the Bertelsmann Cartographic Institute 170
Der grosse Bertelsmann weltatlas 578
Borneo 282, 349
Borton, Hugh *et al, comp*: *A selected list of books and articles on Japan* ... 1073
Boston University: African Studies Centre 20, 788
Library: *Catalog of African government documents* ... 278
List of French doctoral dissertations ... 788
Botanical Society of the British Isles: *Atlas of the British flora* 112, 885
Botany 88, 1084, 1088
Bouguereau, Maurice: *Le théatre français* 1160, 1161
Boulind, Richard, on Sir Francis Drake 1081
Boulnois, L and H Millot: *Bibliographie du Népal* 181, 184
Boundaries 530, 979, 980
Boundary layer meteorology 226
Bowen, E G, on J A Fleure 465
Wales 850
et al, ed: *Geography at Aberystwyth* 529
Bowen, Emmanuel and Thomas Kitchin: *The Royal English atlas* ... 1042
Bowman. Isaiah 225. 476, 515, 860
The Andes of Southern Peru 225
Desert trails of the Atacama 225
Forest physiography ... 225
Geography in relation to the social sciences 225
Limits of land settlement ... 225
The new world 225, 899
Pioneer fringe 225
Bowman, J 2

Boyle, A R 940
'Automation in hydrographic charting' 634
Boyle, G M and Marjorie Colbeck: Supplement to *A bibliography of Canadiana*... 196
Braconnier, Raymond and Jacques Glandard: *Larousse agricole*... 766
Bradshaw, R and M M Owen, *trans*: *Larousse encyclopedia of the earth* 767
Bradshaw's canals and navigable rivers of England and Wales... 227
Brass research 561
Braun, Georg and Franz Hogenberg: *Civitates orbis terrarum* 313
Bray, Arthur: *Beginning practical geography* 807
Brazil 782, 984
  bibliography 174, 184
Brebner, J B: *The explorers of North America* 972
Breese, Gerald: *Urbanization in newly developing countries* 1209
Briault, E W H and D W Shave: *Geography in the secondary school*... 496, 536
Bicker, Charles: *A history of cartography*..., R V Tooley and C B 618
Briefe summe of geographie, Roger Barlow 596, 1150
Brill, E J, Leiden, publishers 5, 39, 177, 214, 228, 409, 692, 727, 730, 748, 905, 950, 985, 1050, 1080, 1183, 1206, 1208, 1227
Brill's news 228
Brill's weekly 228
Brinkman's catalogus van boeken 184
Bristol 809
  University *The Isle of Man* 716
  Spelaeological Society 285
Britannia, William Camden 231, 363, 907
Britannia..., John Ogilby 919
The Britannica world atlas 232
British books in print 269
Britain and the British seas, Halford J Mackinder 229, 806
Britain's green mantle..., A G Tansley 230

'Britain's new forests' 472
Britain's structure and scenery, Sir Dudley Stamp 331
British Antarctic Expedition 1907-09 604
British Antarctic Survey 233, 565
British Association for the Advancement of Science 234, 342, 348, 569, 806, 920
British Association for Russian Studies 587
The British bulletin of publications on Latin America 235
British Cartographic Society 236, 239, 271
British Colonial Office, Discovery Committee 1240
British Columbia, bibliography 194
  University 660
'British Columbia and Vancouver Island', W C Hazlitt 261
British Copper Institute 561
British Council 761
British daily weather reports 242
British geological literature 237
British Geomorphological Research Group 238, 555, 786
British Glaciological Society Library 1063, 1085
British Guiana 674
British Guiana, Michael Swan 349
British interests in the Persian Gulf, A A Amin 239
British Isles 371, 526, 841, 1212
  aerial photographs 757
  climatological atlas 850, 1045
  field studies 462
  gazetteer 164
  maps 274, 277, 353, 451, 551, 765, 1008
  National Agricultural Advisory Council 322
  National Atlas Committee 866
  pleistocene 974
  stratigraphy 1134
  tomato survey 1276
  transport 1179
  vegetation 460
The British Isles, Albert Demangeon 371
The British Isles, D M Preece and H R B Wood 1202
British landscapes through maps 496

443

British Medical Association 869
The British Meteorological Society (now the Royal Meteorological Society of London) 1045
British Museum 240, 1042, 1054
Catalogue of the manuscript maps ... 283
cataloguing of maps 284
classification 315
Department of Oriental Printed Books and Manuscripts 240
Map Room 269, 820, 1082
topographical collection 224
British Museum quarterly 240
British national bibliography 184
British National Committee for Geography 241
British parliamentary papers 715
British railways monthly review of technical literature 1074
British South Africa Company 184
British Speleological Association 285
British Standards Institution 1063, 1131
British topography, Richard Gough 572
British Trust for Ornithology: *Atlas of British birds* 110
British Waterways Board 352
British weather in maps, J A Taylor and R A Yates 242
Britton, G P: *Marine meteorology and oceanography* 830
Broadway travellers series 635
Brockhaus, world atlas 579
Brooke, M Z: *Le Play, engineer and social scientist* ... 774
Brouillette, B: 'Sources of documentation' 1099
Brouwer, G J 280
Bruk, S I *Numbers and distribution of the peoples of the world* 104
and V S Apenchénko: *Atlas narodov mira* 104
Brun, Christian: *Maps and charts published in America before 1800* ..., J C Wheat and C B 202, 822
Brunei 299
bibliography 282
Brunet: *Manuel du libraire et de l'amateur de livres* 387

Brunhes, Jean 243
La géographie humaine 243, 515
Géographie humaine de la France 243
and C Vallaux: *La géographie de l'histoire* 243
Brunswick International School Book Institute 244
Brussels: Institut belge des sciences administratives 184
University, Institut de Sociologie 738
Bryant, Paul T, ed: *From geography to geotechnics* 476
Bryson, R A: 'Climates of North America' 1267
Bucharest, Meteorological Institute 323
Buchan, Alexander, ed: *Atlas of meteorology* ... 129
Buchanan, Keith: *The transformation of the Chinese earth* ... 1178
Buchanan, R O, ed: *Bell's advanced economic geographies* 167
Industrial activity and economic geography ..., R C Estall and R O B 643
Buchon, J A C and J Tastu: *Notice d'un atlas en langue catalane* 277
Buedeler, Werner: *The International Geophysical Year* 689
Bulgaria 1107, 1256
bibliography 182, 184
Centre for Agricultural Scientific Information ... 184
Bulgaria: a bibliographic guide 184, 245
Bulletin, School of Oriental and African Studies 20
Bulletin de l'Afrique noire 24
Bulletin d'IFAN 653
Bulletin of material on the geography of the USSR 246
Bulletin of quantitative data for geographers 247
Le bulletin signalétique 290
Bunbury, E H 582
A history of ancient geography among the Greeks and Romans 617
Bundesanstalt für Bodenforschung 688
Bundesanstalt für Landskunde und Raumforschung 213

Die Bundesrepublik Deutschland in karten 248
Bunge, William: *Theoretical geography* 1162
Bureau of Commercial Fisheries 881
Bureau of Community Planning 1031
Bureau de Recherches géographiques et minières, Paris 848
Burgess, Robert L et al: *A preliminary bibliography of the natural history of Iran* 986
Burma 299, 419, 1105
bibliography 184
maps 136
Burma Research Project, New York University 52
Burns, Sir Alan: *Fiji* 349
Burns, W A: *The McGraw-Hill illustrated world geography,* ed: Frank Debenham and W A B 804
Bushnell, V C, ed: *Antarctic map folio* series 59
and R H Ragle, ed: *The Icefield Ranges Projects, Scientific results* 637
Bushong, Allen D, compilation of American and Mexican theses 184
Business Surveys Limited: *Research index* 1032
Butcher, A L: *Cave survey* 285
Butterworth Group, publishers 249
Buxton, E: *List of modern thematic maps in the Oxford libraries* 942

Cabot, Sebastian 567, 867
Cadamosto 984
Cahen, Claude and W F Leemans, ed: *Journal of the economic and social history of the Orient* 730
Cahiers de géographie de Québec 250
Cahiers Népalais series 181
Cahiers d'outre-mer . . . 251
Cairngorms, map 10
Cairo, maps 1147
California University: African Studies Centre 20
Institute of International Studies 77
Latin American Center 584

California University—*(contd)*
Scripps Institution 279, 1143
Water Resources Center 1223
Callaghan, F R ed: *Science in New Zealand* 1060
Callander, John: *Terra Australia cognita* . . . 1155
Calmann, John, ed: *Western Europe* . . . 602
Camap, programme for distribution maps 26
Cambodia 132, 299, 419, 1105
The Cambrian bibliography 252, 847
Cambridge: Block order scheme 253
Centre of South Asian studies 1102
Scott Polar Research Institute 566
University 525
Africa Studies Centre 20
Department of Geography 254
Explorers' and travellers' club 253
Farm economics branch 28
Geographical handbook series 504
Geographical Society 254
Cambridge expeditions journal 253
Cambridge geographical series 619
Camden, William: *Britannia* 231, 353, 363, 907, 1021
Cameroons 906, 992
Camp Fortune, Gatineau Park 446
Campbell, E M J, ed: *Imago mundi* 639
Campbell, J D: revision, *The Oxford atlas,* Sir Clinton Lewis and J D C 933
Campbell, R D et al: *A question of place* . . . 1003
Canada 152, 332, 363, 446, 500, 512, 625, 841, 1021, 1128, 1262
Africa collections 20
Arctic regions 418, 970
bibliography 47, 184, 196, 298, 708, 805
census figures 1225
Committee on African Studies, University of Alberta 20
Department of Mines and Technical Surveys 3, 113, 208, 256, 818, 843
Geological Survey 401, 550
Humanities Research Council 261, 298
Institute of Surveying, Ottawa 259
maps 38, 105, 113, 131, 269, 612,

Canada maps—(contd)
792, 816, 819, 843, 867, 918, 1006, 1243
Museums directory 1084
National Library 260
Oxford school atlas 939
Public Archives 257
Revue ... de géographie 3
Royal Meteorological Society, branches 1045
Social Science Research Council 261
weather 747, 1272
wool trade 708
Canada in maps 257
Canada: a geographical interpretation 255
Canada: a study of cool continental environments, T G Taylor 1151
Canadian Association of Geographers 255, 258
Canadian cartography 259
Canadian Federation 255, 261
Canadian geographer 258
Canadian geographic journal 1041
Canadian government publications ... 184
Canadian Historical Association 612
Canadian theses 184
Canadiana 260
Canal Zone (Panama) Library-Museum 1138
Canals 227, 274, 1179
Cannenburg, V: *comp*: *Library catalogue of Netherlands Historical Maritime Museum* 888
Canterbury Library, New Zealand 1233
Cantor, Leonard M: *A world geography of irrigation* 1253
Cão, Diogo 984
Cape Verde Islands 184, 984
Cappelens, J W: maps 164, 262, 1038
Norge 909
Cardiganshire, county map 274
Cardinall, A W: *A bibliography of the Gold Coast* 184, 198
Cardy, W F G: ' Report on the geology of the Lower Swansea Valley ' 799

Caribbean 602, 1021, 1272
bibliography 184, 263
Council 263
Organization 184
Caribbean review 264
Caribbean studies 265
Caribbeana 1900-1965 ... 266
Carillo, Alejandro: *Atlas Porrua de la Mexicana,* Jorge H Millares and A C 147
Carl Zeiss Jena, V E B 721
Carlsberg Foundation, Denmark 114
Carlsund, Bo: *Atlas of the world commodities ...,* Olaf Jonasson and B C 142
Carlton, R G: *The USSR and Eastern Europe ...,* P L Horecky and R G C 1189
Carnegie Institution: *Atlas of the historical geography of the United States* 139
Carnegie Trust 26
Carpenter, Nathaniel 620
Carrington, Richard: *A million years of man* 855
Carruthers, I D: *Irrigation development planning aspects of Pakistan* ... 1276
Carta automobilistica d'Italia 1172
' *Cartactual* ' 267
' Carte internationale du monde au millionième ', E Meynen 695
Carter, C C: *Landforms and life* ... 762
Carter, Charles F, *ed*: *Manchester and its region* 234
Carter, E F, *ed*: *The railway encyclopedia* 1004
Carter, G F: *Man and the land* ... 812
Carter, Harold: *The study of urban geography* 1205
and W K D Davies: *Urban essays: studies in the geography of Wales* 1205
et al: *Geography at Aberystwyth* 529
Cartographers 85, 436, 624, 800, 820, 846, 907, 919, 928, 957, 1092
training 241, 314, 415, 1070
Cartographia, Budapest 267, 269, 270, 319, 672, 870

The cartographic journal 236, 271, 676, 695, 820
Cartographic Research Guide 284
'Cartographic survey' (in *The geographical journal*) 507
Cartography 3, 85, 95, 99, 102, 106, 125, 148, 222, 235, 259, 269, 279, 366, 369, 386, 390, 394, 395, 414, 417, 436, 445, 508, 509, 522, 562, 567, 628, 676, 709, 717, 743, 744, 777, 790, 804, 822, 827, 841, 843, 862, 948, 1005, 1030, 1059, 1082, 1114, 1115, 1190, 1195, 1213, 1237, 1243, 1245, 1256
   aerial 911
   automatic 785, 940
   bibliography 176, 177, 213, 269, 494, 783, 789, 1135
   conferences 1192
   extra-terrestrial 1161
   historical 62, 164, 365, 452, 562, 582, 618, 639, 824, 825, 1132, 1150, 1161
   institutes 168, 169, 170
   marine 830
   three-dimensional 368
Cartography 159, 269, 272, 962 (Philippines)
   see also International Cartographic Association; Kartographie; Maps
Cartography of the northwest coast of America to the year 1800 273
Cary, G and J, London, mapmakers 274
   New British atlas 274
   . . . new itinerary 274
   . . . traveller's companion 274
Cary, John 274, 1135
   . . . new and correct English atlas 274
Cary, M and E H Warmington: *The ancient explorers* 49
Cary, William and John: globes 567
Casella, C F and Co, Limited, London 275
Cassini family 276, 567, 924
Castiglioni, Manlio 84
Catalan atlas 277, 1043
Catalogue des cartes nautiques sur vélim . . . 215
Catalog of African government documents . . . 278

Catalogue of books, maps, plates on America . . . (Muller) 280
A catalogue of early globes . . . 587
'A catalogue by Joan Blaeu . . .' 218
A catalogue of Latin American flat maps . . . 281
Catalogue of the Malaysia/Singapore Collection 282
Catalogue of the manuscript maps, charts, and plans . . . in the British Museum 283
Catalogue of maps, atlases, globes . . . G and J Cary 274
Catalogue of meteorological data for research (WMO) 1258
Catalogues: geographical material 45, 47, 240, 358, 381, 382, 468, 496, 781, 783, 844, 876, 888, 900, 925, 943, 1029, 1058, 1120, 1135, 1138, 1198, 1223
   computer 1193
   instruments 275
   maps 222, 269, 582, 654, 1221
   publishers 74, 79, 164, 228, 451, 605, 802, 807
   Scripps Institution of Oceanography Library 279
   Yale University Library, South East Asia Collection 291
Cathay, North West passage 1164
Cathay and the way thither, Henry Yules 596
Cave, A C and B S Trinder, *ed*: *Pergamon general historical atlas* 952
Cave survey, A L Butcher 285
Cave fauna, E A Glennie and M Hazleton 285
Cave Research Group of Great Britain 285
Cave science 285
   see also Speleology
CBD Research Limited, Beckenham 357, 439
Census data 297, 308, 1124
Central America 184, 1243
Central Mortgage and Housing Corporation: *Habitat* 594
Central Office of Information, London 286
   Reference pamphlets 1020
Centre d'analyse documentaire pour l'Afrique noire (CADAN) 287

447

Centre de Documentation Cartographique et Géographique 269, 841
Centre for East Asian Central Studies, Tokyo 288
Centre for Environmental Studies, London 69, 289
Centre International de Développement de l'Aluminium 42
Centre National de la Recherche Scientifique 93, 176, 181, 183, 184, 285, 287, 290, 683, 841
Centre of African Studies, University of Edinburgh 20
Centre of African Studies, University of Warsaw 20
Centre of West African Studies, University of Birmingham 20
Centres of art and civilisation series (Paul Elek) 1206
Cereals 1276
Ceres 291
Ceylon 184, 332, 501, 1102
Chabot, Georges: *Géographie régional de la France* 516
et al, ed: *La géographie française* . . . 514
Chad, bibliography 788
Chalifour, J E: *The national atlas of Canada,* revised edition 867
The Challenger Expedition 292
The Challenger reports 292
The Challenger Society 293
Chambers, J D: *Laxton* . . . 773
Chambers's historical atlas of the world 294
Chambers's world gazetteer and geographical dictionary 295
Chandler, T J: *Modern meteorology and climatology* 1045
Chang Heng Seismoscope 1070
Chang T'ien-Tsê: *Sino-Portuguese trade from 1514 to 1644* . . . 1080
The changing map of Africa, R D Hodgson 1210
The changing nature of geography, Roger Minshull 296
The changing sea-level, Henri Baulig 657
Channel Islands, gazetteer 164
Chapallaz, D P et al: *Hypothesis testing in field studies* 1152

Chapman, Robert and Keith Sinclair, ed: *Studies of a small democracy* . . . 1137
Chapman, Sydney and R S Lindzen: *Atmospheric tides* . . . 150
Character of a conurbation . . . 297
Character of races, Ellsworth Huntington 632
Charles Lyell, Sir Edward Bailey 801
Charlesworth, J K: *The quaternary era* . . . 1001
Chartbooks, nautical 1161
Charts 8, 283, 369, 398, 507, 603, 624, 634, 677, 678, 690, 822, 830, 842, 845, 850, 983, 1045, 1132, 1196, 1219, 1241
Admiralty 240, 398
Aeronautical 1:1M 828
bibliography 269
celestial 274
maritime 553
portolan 215, 277
Charts and maps relating to Australia 345
Chase, Stuart 476
Chatterjee, S P 865
National atlas of India 871
Chavanne, J et al: *Die literature über die Polar-Regionen der erde* . . . 791
A checklist of Canadian literature . . . 298
Checklist of Southeast Asian serials 299
Cheeseman, H R, ed: *Bibliography of Malaya* . . . 201
Cheltenham classification 315
Cheshire, maps 1008
Chicago University: Committee for the Comparative Study of New Nations 20
Committee on African Studies 20
Department of Geography 53, 173, 300, 694, 1151
Chile 1217
Chilver, Henry 289
China 46, 78, 163, 249, 344, 408, 530, 603, 715, 738, 758, 804, 968, 984, 1080, 1178, 1249
bibliography 184, 873
Institute of Chinese Culture 301
maps 277, 301, 613, 868, 934

China, Yi-Fu Tuan 1273
China and Japan 301
China proper 504
China reconstructs 302
China Welfare Institute, Peking 302
China's geographic foundations, G B Cressey 758
Chinese Geographical Institute 868
Chisholm, A H *et al, ed*: *The Australian encyclopaedia* 155
Chisholm, Michael: *Geography and economists* 167
  Research in human geography . . . 515
  et al, ed: *Regional forecasting* . . . 1045
Chi-Yun, Chang, *ed*: *National atlas of China* 868
Cholera, maps 115
Chorley, R J: *Atmosphere, weather and climate,* R G Barry and R J C 149, 850
  Network analysis in geography, Peter Haggett and R J C 891
  Water, earth and man . . . 1223
  and Peter Haggett: *Frontiers in geographical teaching* 477, 850
  Integrated models in geography 856
  Models in geography 850, 856, 966
  Physical and information models in geography 856, 966
  Socio-economic models in geography 1093
  et al: *The history of the study of landforms* . . . 622
  Progress in geography . . . 993
Chriss, M and G R Hayes: *An introduction to charts and their use* 842
Christian, Garth: *Tomorrow's countryside* . . . 885
Christianity, historical maps 127
Christopher Columbus: documents and proofs of his Genoese origin 334
A chronological history of voyages into the Arctic Region . . . 304
Chubb, Thomas 305, 820
  A descriptive list of the printed maps of Norfolk . . . 305
  A descriptive list of the printed maps of Somersetshire . . . 305

Chubb, Thomas—*(contd)*
  The printed maps in the atlases of Great Britain and Ireland . . . 305, 353
Chulalongkorn University: *Bibliography of material about Thailand* . . . 184
Church, R J Harrison 904
  Environment and policies in West Africa 1210
  West Africa 796
The citizen's atlas of the world 309
Cities 44, 234, 297, 306, 307, 308, 312, 355, 592, 594, 611, 772, 795, 949, 1161, 1246
  see also Town plans
Cities in evolution, ed Patrick Geddes 306
Cities of destiny, ed Arnold Toynbee 307
Cities of the Soviet Union . . ., C D Harris 308
City 310
City and region . . ., R E Dickinson 693, 312
'The city as center of change in modern Venezuela', D J Robinson 772
The city in history, Lewis Mumford 311, 355
City, region and regionalism . . ., R E Dickinson 312, 693
City University of New York 1204
Civilization and climate, Ellsworth Huntington 632
Civitates orbis terrarum 313, 1161
Clapham, A R 461
Clapham, H G *et al*: 'The soil mechanics and foundation engineering survey of the Lower Swansea Valley Project area' 799
Clapp, T A, illustrator 1084
The Clarendon Press, Oxford: Cartographic Department 111, 314, 939, 940
Clark, Audrey N 929
Clark, Colin: *The economics of irrigation in dry climates* 28
Clark, K R *Poultry and eggs in Britain* . . ., K E Hunt and K R C 28, 655
The state of British agriculture . . ., K E Hunt and K R C 28, 655

449

Clark, Peter K: 'Engraved os one-inch maps'... 924
Clark, W E and P H Grimshaw: *Physical atlas of zoogeography* 967
Clark University Graduate School of Geography 407, 1075
Clarke, G R: *The study of the soil in the field* 1260
Clarke, H J and W J Allen: *Longman's loops in geography* 796
Clarke, J I: *Sierra Leone in maps* 1079
Clarke, R V: 'The use of watermarks in dating old series one-inch os maps' 924
Classification 45, 109, 284, 315, 496, 1063, 1221
The classification and cataloging of maps and atlases, S W Boggs and D C Lewis 284, 315, 316
A classification for geography (Royal Geographical Society) 1043
Classification of geographical books and maps in libraries (IGU) 315
Claval, Paul: *Régions, nations, grands espaces*... 1025
Clawson, Marion and C L Stewart: *Land use information*... 760
Clayton, Miss F L 1272
Clayton, H H and Miss F L: *World weather records* 1272
Clayton, K M 185, 961
  Bibliography of British geomorphology 238
  Geographical abstracts 494
  Guide to London excursions 390, 586
Clear, T: 'A review of twenty-one years of Irish forestry' 713
'Clements Robert Markham and the geographical department of the India Office' 836
Climate 31, 34, 149, 242, 390, 478, 527, 529, 530, 538, 555, 632, 710, 733, 750, 747, 776, 777, 804, 850, 966, 1084, 1097, 1113, 1136, 1212, 1214, 1258, 1267
  maps 10, 129, 164, 321, 445, 1045
Climate, W G Kendrew 747
Climate and the British scene, Gordon Manley 331
The climate near the ground, Rudolf Geiger 317

'Climates of Africa', J F Griffiths 1267
'Climates of Australia and New Zealand', J Gentilli 1267
'Climates of central and South America', W Schwerdtfeger 1267
'Climates of central and Southern Europe', C C Wallén 1267
The climate of central Canada, W G Kendrew and B W Currie 747
The climates of the continents, W G Kendrew 318, 747
'Climates of north America', R A Bryson 1267
'Climates of northern and eastern Asia', H Arakawa 1267
'Climates of northern and western Europe', C C Wallén 1267
'Climates of the oceans', H Thomsen 1267
'Climates of the Polar regions', S Orvig 1267
'Climates of southern and western Asia', H Arakawa 1267
'Climates of the Soviet Union', P E Lydolph 1267
Climatic atlas of Europe 319
Climatic factors and agricultural productivity, ed J A Taylor 320
Climatological atlas of Africa 321
Climatological atlas of the British Isles 322, 850, 1045
Climatological atlas of Rumania 323
Climatological atlas of the world 129, 324
Close, Sir Charles: *The early years of the Ordnance Survey* 924
Cloud studies in colour, Richard Scorer and Harry Wexler 325
Cloud types for observers 1045
Clouds 275, 325, 1045
Clozier, R et al, ed: *La géographie française*... 514
Clydeside 491, 809
Coal 363, 1192
Coalfields in Great Britain, map 551
The coast of England and Wales... 326
Coasts 230, 364, 462, 503, 529, 531, 576, 657, 686, 762, 1171
  USA survey 1196

Coates, Austin: *Western Pacific Islands* 349
Cocks, J V Somers, *comp*: *The Dartmoor bibliography* 362
Cocoa statistics (FAO) 468
Codrington, Thomas, on Roman roads in Britain 1040
Coelum stellatum . . . 327, 1161
Coffee Publicity Association Limited 328
The Coffee Information Bureau, London 328
Cohen, Barry: *Monsoon Asia* . . . 738
Cohen, S B: *Problems and trends in American geography* 991
Colbeck, Marjorie: *Supplement to A bibliography of* . . . *Canadiana* 196
Cold climate environments and processes, Symposium 954
Cold Regions Bibliography Section, Library of Congress 56
Cole, J P: *Bulletin of quantitative data for geographers* 247
' Industrial statistics ' 246
Latin America 249
' Nationalities of the USSR in 1959 ' 246
' New economic regions of the USSR ' 246
' Population statistics ' 246
' Selected data from the Soviet statistical yearbook for 1958 ' 246
' Soviet foreign trade ' 246
' The Soviet iron and steel industry ', R E H Mellor and J P C 246
and F C German: *A geography of the USSR* . . . 249, 548
and C A M King: *Quantitative geography* . . . 1000
Cole, Monica: *South Africa* 850
Coleman, Alice: ' The Second land use survey: progress and prospect ' 761
'A wildscape atlas for England and Wales ' 1236
and K R A Maggs: *Land use survey handbook* 761
Colin, Armond, publisher 176, 183, 329, 371, 473, 517, 709, 1214
Collège de France, Department of Human Geography 243

Collegiate world atlas 1006
Colless, B E, on the pearl trade 2
Collet's Holdings Limited, Wellingborough 330
Collinder, Per: *History of marine navigation* 842
Collins, Commodore K St B 603
Collins, William, Sons and Company 331, 901, 1260
educational atlases 331
New naturalist series 882
Collison, R L, *ed*: *Bibliographical services throughout the world* 1192
Collocott, T C and J O Thorne: *Chambers's world gazetteer* . . . 295
Colombia: atlases 90
bibliography 184
Instituto geográfico Augustín Codazzi, Bogota 90
The Colombo Plan 332, 1020
The Colombo plan (journal) 332
The Colombo Plan and other essays, Frederic Benham 332
Colonisation 1077
Colorado University, Alpine and arctic research 65
The colour encyclopaedia of clouds 325
Columbia 674
Columbia University, Geographical research staff 333
Columbia-Lippincott gazetteer of the world 333
Columbus, Christopher 350, 394, 728, 869
Collection, at Genoa 334
Journal 334
Letters 334
Columbus . . ., Björn Landström 334
Columbus Memorial Library, Pan-American Union 184
Comitas, Lambros: *Caribbeana* . . . 266
Comité français de Stratigraphie 683
Comité National français de Géographie 176
Comité National de Géographie (France) 91
Commerce 221, 478, 1174

451

Commercial atlas and marketing guide (Rand McNally) 1006
Commercial course atlas 961
Commercial gazetteer of Great Britain 491
Commission for Agricultural Typology, IGU 33
Commission internationale d'Histoire Ecclésiastique Comparée 177
Commission of Ancient Maps 1082
Commodities 142, 539
The commodity survey (UN) 1192
The common lands of England and Wales 331
Common Market 141, 438, 540, 603
Commons and village greens . . . 335
Commonwealth 20, 182, 1020, 1021
Commonwealth Agricultural Bureau 209, 336, 655, 1242
Commonwealth Bureau of Agricultural Economics 336
Commonwealth Bureau of Animal Breeding and Genetics 336
Commonwealth Bureau of Animal Health 336
Commonwealth Bureau of Animal Nutrition 336
Commonwealth Bureau of Dairy Science and Technology 336
Commonwealth Bureau of Helminthology 336
Commonwealth Bureau of Horticulture and Plantation Crops 336
Commonwealth Bureau of Pastures and Field Crops 336
Commonwealth Bureau of Plant Breeding and Genetics 336
Commonwealth Bureau of Soils 336, 1260
 Bibliography of soil science . . . 209
Commonwealth Forestry Bureau 336, 1027
Commonwealth Geographical Bureau 337
The Commonwealth Institute 973
Commonwealth Institute of Biological Controls 336
Commonwealth Institute of Entomology 336
Commonwealth Mining and Metallurgical Congress 1142
Commonwealth Mycological Institute 336
CSIRO, Soils Division 109, 339
 The Australian environment 156
 Symposium, 'Land evaluation' 756
Commonwealth survey 340
Commonwealth Trans-Antarctic Expedition 1955-58 354, 1234
Communications 1171
see also under individual headings
Comores Islands 16
A compact geography of the Netherlands, Utrecht State University 890
Comparative study of new nations committee, Chicago University 20
'A comparison of map cataloging systems', Mary Ellis Fink 284
A comprehensive checklist of serials of geographical value, C D Harris and J D Fellmann 694
Comprehensive Edition, The Times atlas of the world 164
Computation Research and Development, Ministry of Housing and Local Government 785
Computers 69
Comtat (maps) 100
Concepts in geography series (Longmans) 796
Concise encyclopedia of explorations 341
Concise encyclopaedia of world timbers 471
A concise glossary of geographical terms 342, 961
Concise Maori encyclopedia 981
Condliffe, J B: *The development of Australia* . . . 379
Condry, W M: *The Snowdonia National Park* 882
Congo-Brazzaville 917
Congress *see* Library of Congress
Conkling, E C *et al*: *The geography of economic activity* 539
Connaissance du monde 343
Conover, Helen F: *Nigerian official publications* . . . 184
Cons, G J, *ed*: *Handbook for geography teachers* 598
Conservation 57, 431, 440, 451, 472, 763, 813, 884, 885, 886, 1078, 1095, 1236, 1237

Conservation, Joyce Joffe 885
Conservation Foundation 1217
Conservation of natural resources, ed Guy-Harold Smith 1237
'Construction of a map of the world on a scale of 1:1 Million', Albrecht Penck 695
Contemporary China . . ., Peter Berton and Eugene Wu 344
Continental drift, ed G D Garland 930
Continental drift, S K Runcorn 927
Continental drift, D H and M P Tarling 929, 930
'Continental drift', J T Wilson 927
see also Kontinentaldrift
'Contribution of geographical congresses . . . to the development of cartography', K A Salichtchev 269
Contribution to Asian studies, ed K Ishwaran 738
Contribution to a bibliography of Australia and the South Sea Islands 345
Cook, Captain James 398, 427, 603, 718
Voyages of discovery 596
Cooke, H Lester jr, cartographer 126
Cooke, R U and J H Johnson: *Trends in geography* . . . 1181
Cooper, St G C: *Agricultural research in tropical Africa* 32
Copper, research 561
Copper 346
Copper abstracts 3, 346
Copper Development Association, 3, 346
Coppock, J T: 'An agricultural atlas of Scotland' 26
An agricultural geography of Great Britain 29
Copyright 827
Coradi, G, Limited, Zurich 347
The coral reef problem, W M Davis 921
Coral reefs and atolls . . ., J S Gardiner 921
Corals and atolls . . ., F Wood-Jones 921
Corbellini, Pietro 84
Cork Islands 1106
Cornish, Vaughan 348
The beauties of scenery . . . 348

Cornish, Vaughan—*(contd)*
The great capitals . . . 348
The poetic impression of natural scenery 348
The preservation of our scenery 348
Scenery and the sense of sight 348
The scenery of England 348
The scenery of Sidmouth 348
Cornish, W B: *Asia* 1202
Cornwall 363, 551
The Corona Library 349
Coronelli, P Vincenzo Maria 567
Atlante Veneto 567
Coronelli-World League of Friends of the Globe 567
The corridors of time, H J Fleure and H J E Peake 465
Cosmographei oder beschreibung aller länder stetten 350
Cosmography 508
Cosmopolitan world atlas 1006
Costa Rica, maps 828
Cotter, Charles H: *The astronomical and mathematical foundations of geography* 82
Cotton 505, 1157
Cotton Collection (BM) 240
Council for Nature 1078
Council of Cultural Co-operation 351
Council of Europe 141, 244, 437, 440, 886, 1216
Atlas of social and economic regions of Europe 135
Education in Europe series 351
Countries of Europe as seen by their geographers 351
Countries of the world information series (Pergamon) 651
Countryside Act 1968 352
Countryside Commission 69, 352, 440, 882
The Countryside Commission for Scotland 440
The Countryside in 1970 Committee for Wales 440
The Countryside in 1970 Conference 440, 885
County atlases of the British Isles . . . 353, 820, 1082
Country Naturalists' Trusts 1078
Courtaulds 1158
Courtenay, P P: *Plantation agriculture* 167

Cox, Edward G: *A reference guide to the literature of travel* ... 1018
Coysh, A W and M E Tomlinson: *North America* 1202
The southern continents 1202
Crace Collection 240
Crackington Haven, map 10
Craddock, Campbell: 'Geologic maps of Antarctica' 59
Craig, A S: *Dictionary of rubber technology* 389
Crates 567
Crawford, O G S: *Tabula Imperii Romani 1 : 1M* 1147
Crawford R M: *Evolution of Australia* 443
Crespigny, R R C de, on Asian studies in Australia 2
Cresques, Abraham: *Catalan atlas* 277
Cresques, Yehuda: *Catalan atlas* 277
Cressey, George B: *Asia's lands and peoples* ... 78, 804
China's geographic foundations 758
Crossroads ... 78
Land of the 500 million ... 758, 804
Crick, B R et al, ed: *A guide to manuscripts relating to America* ... 587
The Crompton Bequest 830
Crone, G R 618
Background to geography 162
Background to political geography 163
classification scheme 315
The discovery of America 394
'The epic work of Claudius Ptolemy' 994
The explorers 448
'The future of the International million map of the world' 695
'Geography in Great Britain, 1956-60', K C Edwards and G R C 535
'Jewells of antiquitie...' 596
Maps and their makers 633, 825
'The Mariner's Mirror, 1588' 833
Modern geographers 860
'New light on the Hereford map' 607
The Royal Geographical Society ... 1043

Crone, G R—(*contd*)
'Seventeenth century Dutch charts of the East Indies' 85
'The Vinland map cartographically considered' 1215
Cronulla, Fisheries and Oceanography Research Units 339
Crops 336, 505, 973, 1182
Cross, M F and P A Daniel: *Fieldwork for geographical classes* 503
The crossing of Antarctica ... 354
Crossley, J M D: 'Notes on Africana in the Yale University Library' 665
Crossroads: land and life in South-west Asia 78
Crowell Collier and Macmillan Inc 188
Crusades, maps 126
Cuba 184, 223, 411
Culican, W, on Phoenician settlement 2
The culture of cities, Lewis Mumford 311, 355
Cumberland, K B: *South-west Pacific* 1106
and J S Whitelaw: *New Zealand* 1273
A cumulation of a selected and annotated bibliography of economic literature on the Arabic-speaking countries of the Middle East ... 356
Cundall, Frank: *Bibliography of the West Indies* (excluding Jamaica) 184
Current bibliographical information (UN) 1192
Current Caribbean bibliography 184, 263
Current European directories 357
Current events world atlas 1006
Current geographical publications 45, 315, 358
Current issues... (UN) 1192
Current national bibliographies of Latin America 184
Current periodicals in the library of the Royal Geographical Society 1043
Currie, B W: *The climate of central Canada*, W G Kendrew and B W C 747

Cvijić, Johan 489
Cyclopedia of New Zealand . . . 359
Czechoslovakia 1107, 1256
  bibliography 182, 184
  national atlas 89
Czechoslovakia: a bibliographic guide 360

De Capo Press, reprints 212
de Gama, Vasco 984
Dahomes 184
Daily Telegraph world atlas 361
Dairy Research Unit, Melbourne 339
Dairy science abstracts 336
Dakar University, Organisation of African Unity 673
Dalrymple, Alexander 8
Daniel, P A: *Fieldwork for geography classes*, M F Cross and P A D 503
Danish-Norwegian ships, P Holck 1091
'Danish topographic mapping' W M Gertsen 114
Dapples, E C et al, ed: *Lithofacies maps* 792
Darby, H C: 'Domesday Book—the first land utilization survey' 761
  Domesday geography of England 397
  Geographical handbook series 504
  The library atlas, ed Harold Fullard and H C D 780
  The university atlas, ed Harold Fullard and H C D 1201
Darlington, Arnold, ed: *Warne's natural history atlas of Great Britain* 1222
Darlington, P J: *Zoogeography* . . . 1237
Dartmoor 335
Dartmoor, L A Harvey and D St Leger-Gordon 331
Dartmoor, R H Worth 363
The Dartmoor bibliography . . . 362
Dartmoor Preservation Association 362
Dartmouth College Library, Stefansson Collection 382

Darwin, Charles: *On the structure and distribution of coral reefs* . . . 921
The Darwin reader, ed, Marston Bates and Philip S Humphreys 921
Data, quantitative, for geographers 246
Datta, Rajeshwari, ed, *Union catalogue of the Central Government of India publications* 1102
  Union catalogue of the Government of Pakistan publications 1102
David and Charles, Newton Abbot, publishers 220, 227, 231, 304, 363, 463, 472, 505, 644, 645, 827, 834, 924, 963, 1024, 1042, 1158, 1179, 1218
Davies, Arthur, festschrift 460
Davies, W K D: *Urban geography* . . ., Harold Carter and W K D D 1205
Davis, D J, translator 554
Davis, William Morris 364, 476, 622
  The coral reef problem 921
  Geographical essays 364, 498
  The dawn of modern geography 165, 365
Dawson, Commander L S: *Memoirs of hydrography* 842
Day, Vice-Admiral Sir Archibald 842
Guide to IGY World Data Centres 689
Daysh, G H J, festschrift 910
  ed: *Studies in regional planning* 961
  and A C O'Dell: 'Geography and planning' 527
De Aparicio, Francisco et al: *La Argentina* . . . 70
de Beer, Sir Gavin, comp: *Atlas of evolution* 119
de Bell, Garrett, ed: *The environmental handbook* 430
de Blij, H J: *Systematic political geography* 1262
de Chavanne *see* Saueur de Chavanne
de Dainville, F 1160
de Geer, Sten 489

455

de Goeje, M J, *ed*: *Bibliotheca geographorum Arabicorum* 214
Description de l'Afrique et de l'Espagne, texte Arabe, R Dozy and M J de G 39
De Graaf, publisher, Nieuwkoop 79
de Gruyter, Walter and Co, Berlin 366, 433, 744
Lehrbuch der allgemeinen geographie 777
de Jode, Gerard 1161
Speculum Orbis Terrarum 1115
De la Beche 551
De Langhe, Henri A, cartographer 874
de Launay, L: *Géologie de la France* 329
de Margerie, French translation, *Antlitz der erde* 1139
de Martonne, Emmanuel 99, 367, 1098, 1214
Europe central 367
Traité de géographie physique 367
de Nuce, M S: *Madagascar bibliography* 184
De Salis, H R, *comp*: *Bradshaw's canals* ... 227
De Smet, L: *Bibliographie géographique de le Belgique*, *comp* M E Dumont and L De S 184
d'Urville, Dumont 398
de Visscher, Paul and Jaques Putzeys: *Répertoire bibliographique du Conseil d'Etat* (Belgium) 184
de Vries, S *et al*: *Atlas of world history* 143
de Weerdt, Denise: retrospective bibliography of Belgium 184
Deacon, G E R: 'Matthew Fontaine Maury' 838
Deacon, Margaret: *Scientists and the sea* ... 1065
Dead Sea Scrolls 614
Dean, J R: 'The International Hydrographic Bureau' 690
Dean, R R 133
et al: *The Yorkshire and Humberside Planning Region: an atlas* ... 1277
Dean, W G, *ed*: *Economic atlas of Ontario* 405
'An experiment in atlas structure ...' 405

Debenham, Frank 368
Antarctica ... 61
Discovery and exploration ... 368, 393
The McGraw-Hill illustrated world geography 368, 804
Map making ... 368
Nyasaland 349
The Reader's Digest great world atlas 368, 1010
3-D junior atlas 368, 1165
Le déclin de l'Europe, Albert Demangeon 371
Decorative printed maps ..., R A Skelton 369, 1082
Dee, John 567, 1185
Deep-sea research and oceanographic abstracts, ed Mary Sears and Mary Swallow 370
Deffontaines, Pierre 515
Géographie universelle Larousse 768
Defoe, Daniel 620
Deighton, H S: *Atlas of European history*, ed E W Fox and H S D 118
Delamarre, Jean-Brunhes 515
Delft, Topographic Service 148
Dell, R K: *Native crabs* (New Zealand) 1016
Native shells (New Zealand) 1016
della Vida and Francesco Gabrieli: new edition, work of al-Idrisi 39
Demangeon, Albert 371
ed, *Annales de géographie* 371
Le déclin de l'Europe 371
Problèmes de géographie humaine 371
The democratic yearbook (UN) 372, 1192
Demography 173, 445, 682, 982, 995
Denman, D R *et al*: *Commons and village greens* 335
Denmark 544, 1055, 1056
agriculture 320
bibliography 182, 184,
and the European Community 438
national atlas 114
National Committee for the International Hydrological Decade 908
University Geography Laboratory 114
wool trade 708

Denmark: collected papers . . . 373
Denmark: literature, language, history, society, education, arts: a select bibliography 184
Denoyer-Geppert, Chicago: *Maps, globes, charts* . . . 269
Denudation 364
Department of Environment: *Atlas of planning maps* 527
Department of Health, Scotland 866
Department of Middle Eastern studies, University of Melbourne 2
'Deposits of useful minerals of the USSR' 330
'Deposits of useful minerals of the the world' 330
Derbyshire lead mining glossary, N Kirkham 285
Deschamps, Pierre: *Dictionnaire de géographie ancienne et moderne* . . . 387
Descriptio regni Japoniae et Siam, Bernard Varenius 1211
A descriptive atlas of New Zealand 374
A descriptive atlas of the Pacific Islands 375
A descriptive list of the printed maps of Norfolk 305
A descriptive list of the printed maps of Somersetshire . . . 305
Desert trails of the Atacama, Isaiah Bowman 225
Deserts 72, 73, 189, 364, 914
Deserts of the world . . ., ed William G McGinnies *et al* 376
'Determinism' 1009, 1214
Deutsche Bibliographie . . . 184
Deutsche Eisenbahn Technik 1074
Deutsche Forschungsgemeinschaft 135
Deutscher Generalatlas 377
Deutsche Generalkarte 808
Deutsche Gesellschaft für Kartographie 213, 269, 745
Deutsche planungsatlas 378, 949
Deutsche Staatsbibliothek, Berlin 14
Deutsches Hydrographisches Institut, Hamburg 556
The development of Australia . . ., J B Condliffe 379

The development of country towns in the South-West Midlands . . . 655
'The development of earth science' 811
'Developments in British hydrography . . .' 9
Developments in sedimentology series 418
Devon 335, 363, 551
Dewdney, J C, ed: *Durham County and City with Teeside* 234
Diamonds 3, 616
Dias, Bartholomew 984
Diccionario geografico-estadistico-historico de España . . . 380
Dickinson, G C: *Maps and air photographs* 821
Statistical mapping and the presentation of statistics 1124
Dickinson, R E: *City and region* . . . 312, 693
City, region and regionalism . . . 312, 693
The makers of modern geography 810
Dicks, D R, ed: *The geographical fragments of Hipparchus* 610
Dictionaries 295, 380, 384, 385, 386, 388, 389, 480, 683, 797, 1225
Dictionary catalog of the Department Library, US Department of the Interior 381
A dictionary of geography, F J Monkhouse 385
Dictionary catalogue of the Hawaiian Collection, University of Hawaii 383
Dictionary catalog of the Stefansson Collection . . . 382
Dictionary of discoveries, I A Langnas 384
Dictionary of geography . . ., W G Moore 385
A dictionary of mapmakers, R V Tooley 386, 820
Dictionary of national biography 797
A dictionary of natural resources and their principal uses 388
Dictionary of rubber technology 389
Dictionnaire des explorations 384
Dictionnaire de géographie ancienne et moderne . . . 387

457

Diercke weltatlas 1232
Difrieri, H A et al: *La Argentina* ... 70
Digest of agricultural economics 28
Digest of agricultural economics and marketing 655
Dillon, J L and G C McFarlane: *An Australian bibliography of agricultural economics* ... 154
Dillon's University Bookshop 1144
Dinstel, Marion, comp: *List of French doctoral dissertations on Africa* ... 788
Directorate of Overseas Surveys 390
Directories 357
Directors' guide to Europe 391
Directory of Indian geographers 641
Directory of institutions engaged in arid zone research (Unesco) 73
Directory of meteorite collections and meteorite research 392
Disasters, natural 1192
Discovery 125, 140, 254, 304, 384, 448, 449, 553, 596, 604. 619, 984, 1043, 1090, 1156
Discovery and exploration . . ., Frank Debenham 368, 393
Discovery Committee 877
The discovery of America, G R Crone 394
The discovery of North America ..., H Harrisse 273, 395
' Discuss the economic geography of New Zealand ', David Morgan 902
Disease, mapping 115, 869, 1230
Djambata, Amsterdam, cartographer 107
Dobbie McInnes (Electronics) Ltd, Glasgow 940
Dobbs, Arthur: 'An account of the countries adjoining to Hudson's Bay ' 261
Dr A Petermanns mitteilungen aus Justus Perthes geographischer anstalt 958
Documentatio geographica 269, 1017
Documentation 171, 658
Documentation on Asia 396
Dodge, R E, translator 515
Do-it-yourself weather instruments 1152
Dokumentazja Geograficzna 269
$1.00 world atlas 1006

Dollfus, Jean: *Atlas of Western Europe* 141, 1006
France: its geography and growth 474
Dolton, T H: *Fieldwork in geography*, J E Archer and T H D 503
Domesday, woodlands 505
' Domesday Book—the first land utilization survey ', H C Darby 761
Domesday geography of England 397
Dominican Republic, bibliography 184
Don River 873
Dorries, Hans: *Siedlung- und Bevölkerungsgeographie* ... 173
Dorset 416, 551
Douglas, R K, comp: *Catalogue of the printed maps, plans and charts in the British Museum* 283
Doumani, George, ed: *Antarctic bibliography* 56
Dovedale, map 10
Dover 809
Doxiadis, Dr C A: *Ekistics* ... 83
Dozy, R and M J de Goeje: *Description de l'Afrique et de l'Espagne* ... 39
Drainage 755
Drake, Sir Francis 972, 1081, 1185
The world encompassed 596
Drake's voyages . . ., K R Andrews 1081
Drawings, topographical, in British Museum 283
Dresden: Geodatisches Institut 182
Sächsische Landbibliotek 350
Technical University, International Documentation Centre for Geodesy 171
Drinkwater, T A et al, ed: *Atlas of Alberta* 105
Driscoll, K J: *Town study* ... 1204
Du Toit, A L: *Our wandering continents* ... 927, 930
Dublin University Geographical Society 86
Dubois, Marcel 50
Duckham, A N and G B Masefield: *Farming systems of the world* 457
Dudley, Colom 1161
Dudley, Sir Robert: *Arcano del mare* 63

Dudley Stamp Memorial, cumulative index for *Geography* 1118
The Dudley Stamp Memorial Fund 1118
Duigman, Peter, comp: *Handbook of American resources for African studies* 598
Dumont, M E and L D Smet: *Bibliographie géographique de la Belgique* 184
Dunkle, John R: text of *Atlas of Florida* 121
Durham County and City with Teesside, ed J C Dewdney 234
Durham University Department of Geography 487
Dury, G and J A Morris: *Land from the air* 757

Eager, Alan R: *A guide to Irish bibliographical material*... 583
Eames, Wilberforce: *Bibliotheca Americana*...211
Early charts of New Zealand... 398
Early Hanoverian mathematical practitioners, E G R Taylor 1150
Early Italian engraving, A M Hind 998
Early maps of the British Isles... 1043
The early maps of Scotland 1046
'Early steam navigation in China' 46
The early years of the Ordnance Survey... 924
The earth and its resources 804
Earth Science Editors 399
Earth science reviews 400
Earth Science Symposium on Hudson Bay 401
Earth sciences 189, 290, 486, 558, 549-557
Earth sculpture and the origin of land forms 482
Earthquakes 54, 1068, 1069, 1070, 1071
Earthquakes—atlas of world seismicity 402
East, W Gordon 458
The geography behind history 530
The Soviet Union 1210

East, W Gordon—*(contd)*
The spirit and purpose of geography, S W Wooldridge and W G E 633, 1117, 1239
and S W Wooldridge: revised edition, Fawcett: *Provinces of England* 997
East Anglian bibliography 184
East Anglian Economic Planning Council 445
The East African economic review 403
The East African geographical review 403, 1187
The East African Literature Bureau 32
East African Studies Programme, Syracuse University 20
East Asian Institute, USA 1032
East India Company 239
East Indies, Dutch charts 85
East Indies, parliamentary papers 715
East Midland geographer 405, 499, 502
Eastern Mongolia, V A Obruchev 914
Eastwood, T: *Geological maps* 1121
Echeit, Georg 244
Echo sounding 565
Echer, Lawrence, translator 1108
l'Ecole nationale des Eaux et Forêts 1036
The ecologist 817
Ecology 216, 312, 528, 691, 884
Economic abstracts 3
Economic and social history of the Orient 730
Economic aspects of agricultural development in Africa 655
Economic aspects of pigmeat marketing 655
Economic atlas of Ontario 405
Economic bulletin for Latin America 406
Economic Commission for Latin America 406
Economic development in the Tropics 850
Economic geography 407
'Economic geography in the USSR, 1960', F C German 246

An economic geography of China, T R Tregear 249, 408
An economic geography of oil, Peter R Odell 167
An economic geography of West Africa 167
Economic geology series (Geological Survey of Canada) 550
Economic implications of the size of nations ... 409
Economic, social and political studies of the Middle East 410
Economic survey of Asia and the Far East (UN) 1192
Economic survey of Europe (UN) 1192
Economic survey of Latin America (UN) 411
Economics 228, 662, 729, 1249
The economics of irrigation in dry climates 28
The economy of Pakistan ... 412
Ecosystems 34
Edge, R C A: 'Ordnance Survey at home' 924
Edgell, Sir John: *Sea surveys* ... 1066
Edinburgh: atlas 116
  British Honduras-Yucatan Expedition 69
  Geographical Institute (Bartholomew) 164
  Institute of Geological Sciences 54
  maps 116, 1147
  Public Library, BCS collection 236
  University, Centre of African Studies 20
  Department of Geography 164, 236, 920
  Royal Meteorological Society, Scottish Centre 1045
The Edinburgh world atlas (Bartholomew) 164, 413
Editions Rodopi N V, publishers 79
Edlin, H L: 'The Forestry Commission in Scotland' 472
Eduard Imhof ... 414, 460
Education, Ministry of: *Geography and education* 525
Education and training in the mapping sciences ... 415
'Education in cartography in the United Kingdom' 241
Education in Europe series 351
Educational field studies 416

Edwards, K C 404, 499
  Nottingham and its region ... 234
  The Peak District 331
  and G R Crone: 'Geography in Great Britain, 1956-60' 535
Efrat, Elisha: *Géographie d'Israel,* Efraim Orni and E E 513
Eggs 655
Egypt 857, 1147
  agriculture 621
  bibliography 184, 788
  maps 107
Eidgenössische Landestopographie, Berne 95
Ekistic index 83
Ekistics, Athens Centre 83
Ekistics: an introduction to the science of human settlements, C A Doxiadis 83
El Salvador, bibliography 184
Eldridge, H T: *The materials of demography* 173
Elementary atlas (Philip) 961
Elements of cartography, A H Robinson 1237
Elements of cartography, A H Robinson and R D Sale 417
Elements of geography, V C Finch et al 804
Elements of geology, Sir Charles Lyell 801
Elkington, T T and M C Lewis: *The flora of Greenland* (translation) 465
Ellesmere Island 565
Elliott, F E: *A German and English glossary of geographical terms,* E Fischer and F E E 560
Ellis, D M: *A market survey* ..., M M Baraniecki and D M E 1152
Elsevier, Abraham 418
Elsevier, Louis 418
Elsevier Publishing Company Limited 127, 418, 483, 970
  Atlas of the Great Barrier Reef 124
  Atlas van Europese vogels 117
  Earth science reviews 400
  Glossaria interpretum 568
  Oceanography series 418
Elsevier's weekly 418
Embleton, Clifford and C A M King: *Glacial and periglacial geomorphology* 564

Embree, A T *et al, comp*: *Asia: a guide to basic books* . . . 76
Emden *see* Gierloff-Emden
Emerging Southeast Asia . . ., D W F 419
Emery F V: *Wales* 1273
Emmering, S, publishers 79
L'Empire Britannique, Albert Demangeon 371
Encyclopedia Nigeriana 735
Encyclopaedia of American woods 471
Encyclopaedia of Australia 420
Encyclopaedia of Ireland 421
An encyclopaedia of the iron and steel industry 422
Encyclopaedia of Latin-American history 423
An encyclopaedia of London, ed William Kent 424
An encyclopaedia of New Zealand 425
The encyclopedia of oceanography (Reinhold) 1028
Encyclopaedia of South Africa 426
Encyclopaedias 155, 325, 341, 359, 420, 421, 422, 423, 424, 425, 426, 471, 802
The Endeavour journal of Joseph Banks . . . 427
Energy 539
Engelmann, Gerhard: 'Der physikalische atlas des Heinrich Berghaus . . .' 969
Engelmann, W: *Bibliotheca geographica* 847
England 75, 227, 326, 331, 335, 348, 397, 458, 465, 472, 505, 545, 587, 657, 708, 726, 798, 898, 997, 1054, 1097, 1276
  cartography 369
  geographical knowledge 620
  geographical literature 1485-1583 769
  globe-making 567
  Laxton village 773
  maps 164, 353, 363, 907, 989, 1042, 1043, 1078, 1236
England's quest for Eastern trade 972
English Channel 877
English county maps . . . 284
English ships, R C Anderson 1091
'Engraved os one-inch maps . . .' 924

Engravers 624, 989
Environment 226, 281, 306, 331, 440, 506, 526, 537, 539, 555, 564, 648, 774, 812, 815, 855, 991, 993, 1009, 1075, 1088, 1094, 1098, 1151, 1214, 1217, 1223, 1230
  desert 376
  marine 1076
Environment and nation 1151
Environment and policies in West Africa 1210
Environment and race 1151
The environmental handbook 430
The environmental revolution . . . 431, 885
Environmental studies 289, 592
'The epic work of Claudius Ptolemy' 994
Eratosthenes 432, 520
Die erde . . . 3, 366, 433, 563
Erdei, Ferenc, *ed*: *Information Hungary* 651
Erkunde . . . 434
Die erdkunde im verhältnis zur Natur und zur geschichte des menschen 1037
Erosion 364, 669
'Error and revision in early os maps' 924
Espenshade, E B, jr, *ed*: *Goode's world atlas* 571
Essai sur la géographie des plantes 1088
Essays in geography for Austin Miller 460
Essays in political geography, ed: C A Fisher 435
Essays on agricultural geography . . . 460
Esselte Corporation, Stockholm 436, 672, 676, 800, 909
The Daily Telegraph world atlas 361
  map catalogues 269
  map service 709
Essentials of geography, O W Freeman and H F Raup 804
Essex, maps 907
Estall, R C: *New England, a study in industrial adjustment* 167
  and R O Buchannan: *Industrial activity and economic geography* . . . 643

Etats et nations de l'Europe 1214
Ethiopia, bibliography 184
The ethnographic survey of Africa 666
Euphrates 873
Europe 33, 118, 152, 163, 351, 357, 371, 391, 394, 399, 465, 508, 512, 526, 530, 518, 519, 540, 574, 602, 749, 796, 850, 904, 932, 959, 960, 972, 974, 984, 1006, 1022, 1045, 1070, 1107, 1133, 1189, 1192, 1209, 1213, 1214, 1249, 1251, 1260, 1261, 1262, 1266, 1267, 1268, 1272
  bibliographies 184
  maps 117, 118, 134, 141, 145, 164, 202, 270, 319, 330, 451, 618, 675, 687, 796, 688, 808, 848, 934, 952, 1172, 1243
Tabula Imperii Romani 1147
Europe, D M Preece and H R B Wood 1202
Europe centrale, Emmanuel de Martonne 367
Europe Economic Community 391
Europe, ein geographisch-historisch-statistisches gemäldes . . . 1037
Europe in maps (Longman) 796
The European bibliography 184, 437
European companies . . . 439
European communities . . . 438
European Community 141, 438
The European Community in maps 438
European Conservation Year 440, 1236
European Cultural Centre, Geneva 184, 437
The European discovery of America . . . 1215
European Free Trade Association 391
European Information Centre for Nature Conservation 886
The European nations in the West Indies 972
The European peasantry . . . 441
European research resources 437
European Seismological Commission 1070
Evans, D Silvan, ed: *Cambrian bibliography* 252
Evans, E Estyn 714
Evans, Sir Frederick J O 8

Evans, S, on echo sounding 565
Everest 1921 expedition 609
Everyman's atlas of ancient and classical geography 442
Everyman's encyclopaedia 413
Everyman's United Nations 1192
Evolution 855
Evolution of Australia, R M Crawford 443
The evolution of Scotland's scenery 444
Exercises on OS maps series 496
Exeter essays in honour of Arthur Davies 460
Expeditions 253
'An experiment in atlas structure: *The economic atlas of Ontario*' 405
Experimental cartography . . . 445
Experimental orthophotomap of Camp Fortune . . . 445
Explanation in geography 447
Exploration 49, 66, 118, 125, 165, 273, 309, 334, 341, 345, 365, 368, 384, 393, 507, 553, 604, 681, 847, 875, 941, 944, 947, 1043, 1084, 1090, 1127, 1215, 1240
  polar 66, 382, 977, 1234
The exploration of the Pacific 972
The explorers . . ., G R Crone 448
The explorers of North America 972
Explorers' maps . . ., R A Skelton 449, 1082
Explorers' maps of the Nile sources . . . 450
The Export Council of Norway 912
Exposé des travaux de l'IGU 654
Eyre, S R: *Vegetation and soils* . . . 1212
  World vegetation types 1212, 1270
  and G R J Jones, ed: *Geography as human ecology* . . . 528

Faber, Karl-George et al: *Bibliographiekirchengeschichtlicher karten Deutschlands* 177
The Faber atlas 451
The face of the earth, English translation, Eduard Suess: *Antlitz der erde* 1139
'Faces of regional metamorphism' 330

Facsimile atlas to the early history of cartography . . ., A E Nordenskiöld 452, 998
Facsimiles *see* Reprints
Fact sheet (Directorate of Overseas Surveys) 390
Fact sheets on the Commonwealth 338
Faeroes 1056
Fairbridge, Rhodes W, ed: *The encyclopedia of oceanography* 1028
Fairbrother, Nan: *New lives, new landscapes* 885
Fairey Surveys Limited 453, 1168
Plotterscope 453
Falke, Horst: *Die geologische karte* . . . 366
Falkland Islands 233, 1155
Far East 39, 228, 499, 720, 730, 738 932, 959, 1080, 1127, 1133, 1192, 1223
bibliography 184
maps 126, 892
UN cartographic conference for Asia and the Far East 1245
The Far East and Australasia 454
Far East trade and development 455
Far Eastern survey 456
Farm Economics Branch, University of Cambridge 28
The farm economist 28
Farm management pocket book 1276
Farm studies 1276
Farming 33, 451, 488, 655, 1224
Farming systems of the world 457
Favence, E: *The history of Australian exploration* . . . 847
Fawcett, Charles Bungay 458, 775
The bases of a world commonwealth 458
Frontiers . . . 458
Provinces of England . . . 458, 997
Febvre, Lucien: *A geographical introduction to history* 506
Feddon, Robin: *The continuing purpose* . . . 885
Fédération Aéronautique Internationale 678
Féderation Internationale de Documentation (FID) 3, 1063
Fell, H B: *Native sea-stars* (New Zealand) 1016

Fellmann, J D: *A comprehensive checklist of serials of geographical value,* C D Harris and J D F 694
International list of geographical serials, ed C D Harris and J D F 694
The fenland in Roman times, ed C W Phillips 459
Fens, maps 112, 1135
Ferguson, Sir John: *Bibliography of Australia* . . . 184, 190
Fertilizers, bibliography 209
Festschrift für Hans Kinzl 460
Festschriften 414, 460, 499, 500, 620, 714, 793, 840, 910, 1077
Fiches, faune et flore de la Méditerranée 680
Field geology, V A Obruchev 914
Field Group Studies 461
'The field of geography' series (Methuen) 447
Field studies 461
Field Studies Council 461, 1078, 1239
Field studies in the British Isles 462
Field work 416, 461, 488, 495, 502, 503, 533, 552, 764, 774, 775, 778, 1140, 1152, 1171
Fieldwork for geography classes 503
Fieldwork in geography 503
Fieldwork using questionnaires and and population data 1152
Fieldworker: the environmental studies magazine 430
50 years work of the Royal Geographical Society, Sir C R Markham 1043
Fiji 1106
bibliography 197
Fiji, Sir Alan Burns 349
Films 275
Filmstrips 275, 511
Final report on the classification of geographical books and maps 315
Finch, V C *et al*: *The earth and its resources* 804
Elements of geography 804
Physical elements of geography 804
The finding of Wineland the Good . . . 1215
Fink, Mary Ellis: 'A comparison of map cataloging systems' 284

463

Finland 120, 544, 1055, 1056
  bibliography 182, 184
  national atlas 120
Finnish Geodetic Institute 182
Finsterwalders, Richard: *Photogrammetrie* 366
'The first English globe ...', Helen M Wallis 567
The first engraved atlas of the world, E Lynam 998
The first hundred years of the Geological Survey of Great Britain 551
First lessons in human geography (Longmans) 796
The first venture atlas (Philip) 961
Fish industry 1251
Fischer, E and F E Elliott, *comp*: *A German and English glossary of geographical terms* 560
Fischer, Eric *et al*: *A question of place* ... 1003
Fisher, C A: *Essays in political geography* 435
Geographical essays on British tropical lands, ed R W Steel and C A F 501
Modern Asian studies 76
Fisher, H T: The Synagraphic Mapping System 1145
Fisher, James, *comp*: *Shell nature lovers' atlas of England, Scotland Wales* 1078
Fisher, W B: *The Middle East* ... 854
Fisheries 144, 468, 526, 679, 1250 1251
Fisheries Research Unit, Cronulla 339
Fite, E D and A Freeman, on the early cartography of America 273
The fitness of man's environment, ed Robert McC Adams *et al* 463
Fittkau, E J *et al,* ed: *Biogeography and ecology* 216
Fiziko-geografičeskij atlas mira 464
Flegon, Alec, *ed*: *Soviet trade directory* 1110
Flett, Sir J S: *The first hundred years of the Geological Survey of Great Britain* (now Institute for Geological Sciences) 551

Fleure, Herbert John 465, 529
French life and its problems 465
'The geographical distribution of the major religions' 545
Guernsey 465
Human geography in Western Europe 465
A natural history of man in Britain 331, 465
Peoples of Europe 465
The races of England and Wales 465
The races of mankind 465
and H J E Peake: *The corridors of time* 465
The Fleure Library 465
Flint, J E 1231
Flohn, H, on climatology 1267
Flora, atlas 885
The flora of Greenland 466
Flora of Japan 1084
Florida, atlas 121
Florinsky, M T *et al,* ed: *McGraw-Hill encyclopedia of Russia and the Soviet Union* 802
Florio, John, translator 595
Flower, J R, cartographer 1014, 1078
Focus 45, 467
Foncin, Myriem *et al*: *Catalogues des cartes nautiques sur vélim* ... 215
Fondements de la géographie humaine 1098
Food 28, 163, 336, 1263
Food and Agriculture Organization (FAO) 27, 28, 291, 468, 696, 702, 1027, 1192, 1213, 1250
Cocoa statistics 468
Commodity series 468
Food and agricultural legislation 468
Monthly bulletin of agricultural economics ... 468
Plant protection bulletin 468
The state of food and agriculture 468
Unasylva 468
World fisheries abstracts 468
FAO *Documentation Center* 468
FAO Documentation—*Current index* 468
The FAO review 468

Food, clothing and shelter, L W Stevens 807

Forbes, Alexander *et al*: *Northernmost Labrador mapped from the air* 911

Ford, G *et al*: *China and Japan* 301

Ford, P *et al*: *China and Japan* 301 and Mrs Ford: *United States of America* 1199

Ford Foundation 289, 943

Fordham, Sir H G: *John Cary . . .* 274

Studies in cartobibliography . . . 1135

Foreign maps and landscapes, Margaret Wood 469

Forest environments in tropical life zones . . . 470

Forest physiography . . ., Isaiah Bowman 225

Forest Products Research publications 471

Forest resources of the world 470

Forest service . . . 472

Forestry 144, 336, 390, 440, 451, 468, 478, 500, 631, 943, 1027, 1036, 1084, 1224, 1252, 1278

Forestry abstracts 3, 336

Forestry and timber technology catalogue 471

Forestry Commission 352, 472, 882, 1078
  publications 471
'The Forestry Commission and conservation' 472
'Forestry in England' 472
'Forestry in Scotland' 472
'Forestry in Wales' 472

Formosa 249

Forrest, Thomas: *A voyage to New Guinea* . . . 942

Forsaith, D M, *ed*: *Handbook for geography teachers* 598

Forschungsgesellschaft für das strassenwesen 699

Forsdyke, Lionel, cartographer 375

Foster, Sir William 596
  England's quest of Eastern trade 972
  Guide to the India Office Records . . . 641

Foundations of economic geography (Prentice-Hall series) 542

Foundations of geography, D M Preece 1202

Founders of geology 482

Fourah Bay College, Freetown 1079

Fox, David J: 'Urbanization and economic development in Mexico' 772

Fox, E W and H S Deighton, *ed*: *Atlas of European history* 118

France 243, 249, 329, 367, 440, 465, 514, 516, 517, 611, 654, 708, 766, 859, 1003, 1104, 1147, 1214
  bibliography 182, 184
  maps 62, 87, 92, 93, 268, 276, 369, 469, 918, 924, 1014, 1135, 1160

France . . ., Philippe Pinchemel 473

France de demain series (PUF) 987

France de l'Est, Paul Vidal de la Blache 1214

France: its geography and growth, Jean Dollfus 474

Frankel, H: *The agricultural register* . . . 655

Frankfurt am Main: Soziographisches Institut 135
  University: *Atlas of social and economic regions of Europe* 135

Frazer, Col Sir Augustus 240

Frazer, R M: *Atlas of New Zealand geography*, G J R Linge and R M F 130

Frederick Soddy Trust 128

Frederiksen, Ingeborg, illustrator 466

Freeman, A, on the early cartography of America 273

Freeman, E L: 'Loose print mosaicing' 453

Freeman, O W: *Geography of the Pacific* 1237
  and H F Raup: *Essentials of geography* 804

Freeman, T W: *The geographer's craft* 489
  Geography and planning 527
  A hundred years of geography 629
  Ireland . . . 850
  The writing of geography 534

French life and its problems, H J Fleure 465

French ships, Pierre le Conte 1091

French-speaking West Africa . . . 184

465

Frenzel, Konrad, *ed*: *Atlas of Australian resources* 108
Frey, D G: *The Quaternary of the United States* 1002
Freytag-Berndt und Artaria 475, 709
Friedrich Ratzel . . ., Harriet Wanklyn 648, 1009
The friendly Arctic . . ., V Stefansson 1127
Friends of the earth series (Ballantine) 430
Friis, Herman R, *ed*: *The Pacific Basin* 944
Frisius, Gemma: globes 567, 846
Frison-Roche, R: *Les montagnes de la terre* 1244
Frobisher, Martin 304, 1164
From geography to geotechnics 476
Frontier Dzungaria, V A Obruchev 914
Frontiers 163, 530, 979, 980
Frontiers . . ., C B Fawcett 458
Frontiers in geographical teaching 477, 850
Fryer, Donald W: *Emerging Southeast Asia* . . . 419
Fuchs, Sir Vivian 341
  and Sir Edmund Hillary: *The crossing of Antarctica* . . . 61, 354
Fullard, Harold 961
  'Atlas production of the 1970' 269
  Geographical digest 497
  and H C Darby, *ed*: *The library atlas* 780
  The New Cambridge modern history atlas 892
  The university atlas 1201
Fundamentals of economic geography . . . 478
Fung, K I, cartographer 134
Furnivall, J S: *Netherlands India* . . . 889
'Further light on the Molyneux globes', Helen M Wallis 567
'The future of the International million map of the world', G R Crone 695

Gabrieli, Francesco: new edition of the work of al-Idrisi 39
Gadgil, P D: 'Soil biology of the Lower Swansea Valley' 799

Gale, F and G H Lawton, *ed*: *Settlement and encounter* . . . 1077
Gallois, Lucien 479, 517, 1214
  Les géographes allemands de la Renaissance 479, 847
  Régions naturelles et noms de pays 479
Galloway, R L: *Annals of coal mining and the coal trade* 363
Galton, Francis 489
Gambia 286
Gandjei, M: *Geographical bibliography* (Iran) 184
Ganges 873
Gannett, Henry 476
Gardiner, J S: *Coral reefs and atolls* . . . 921
Gardiner, R A: 'A re-appraisal of the International Map of the World (IMW) on the millionth scale' 695
Gardner, J L and Lloyd E Myers, *ed*: *Water supplies for arid regions* 1223
Garland, G D, *ed*: *Continental drift* 930
Garnett, Alice, on H J Fleure 465
Garnier *see* Beaujeu-Garnier
Garran, Andrew: *The picturesque atlas of Australasia* 971
Gas, natural 478
Gascony, towns 898
Gatineau Park, Camp Fortune 446
Gaul (Strabo) 1133
Gaussen, Henri: vegetation maps 707, 1213
Gay, J: *Bibliographie des ouvres relatifs à l'Afrique et à l'Arabie* 179
Gay, John D: *The geography of religion in England* 545
Gay, Peter, *comp*: *Bibliography of reference materials for Russian area studies* . . . 207
Gaza Plain, maps 127
Gazetteer of the British Isles 164
Gazetteer of the Persian Gulf 480
Gazetteers 36, 164, 295, 333, 491, 1170
Gazetteers, US Board on Geographic Names 1195
Geddes, Arthur 306, 481

Geddes, Sir Patrick 606, 774, 476, 481, 775
Cities in evolution 306
'The Geddes diagrams' 306
Geiger, Rudolf: *The climate near the ground* 317
Geikie, Sir Archibald 482
Earth sculpture and the origin of land forms 482
Founders of geology 482
The scenery of Scotland 444
Structure and field geology . . . 482
Textbook of geology 482
General bathymetric chart of the oceans 690
General cartography, Erwin Raisz 804, 1005
'General introduction to the statistical account of Upper Canada' 261
General sketch of the history of Persia 836
Generalstabens Litografiska Anstalt 800
Geneva, African Institute 20
Genoa, City: *Christopher Columbus* . . . 334
Gentilli, J: 'Climates of Australia and New Zealand' 1267
Geocom bulletin 558
Geoderma 483
Geodesy 171, 182, 269, 706
Geodesy and Cartography, East German Department 270
Geo-ecology of the mountainous regions of the tropical Americas 484
Geo-forum 486
Geografia dell'Africa 485
Geograficheskii atlas 102
Geografisch tijdschrift 3
Geografiska Sallskapets i Finland Tidskrift 1154
The geographer and urban studies, David Thorpe 487
The geographer as scientist 488, 1239
Geographers 214, 228, 351, 367, 368, 371, 373, 432, 451, 458, 460, 465, 473, 479, 481, 489, 499, 500, 517, 522, 576, 606, 608, 609, 620, 627, 628, 629, 632, 685, 726, 747, 749, 753, 759, 774, 776, 786, 806, 810,

Geographers—(*contd*) 820, 836, 847, 860, 914, 916, 920, 923, 951, 998, 1003, 1009, 1015, 1037, 1075, 1082, 1083, 1097, 1098, 1108, 1118, 1122, 1128, 1133, 1140, 1143, 1150, 1151, 1176, 1203, 1214, 1217, 1239, 1248, 1275
The geographer's craft, T W Freeman 489
A geographer's reference book 490
The geographer's vademecum of sources and materials (Philip) 961
Les géographes allemands de la Renaissance 479, 847
'Geographia', London 491
Geographia, Francesco Berlinghieri 492
Geographia generalis, Bernard Varenius 1211
Geographic school bulletin 875
Geographia Polonica 33, 493
Geographical abstracts 3, 494, 238
Geographical analysis: an international journal of theoretical geography 495
Geographical aspects of Balkan problems . . ., Marion I Newbigin 904
Geographical Association 116, 275 315, 348, 465, 496, 524, 532, 536, 606, 806, 961, 1152, 1239
Air photographs—man and the land 496
Asian studies 496
British landscapes through maps 496
Exercises on OS maps 496
Geography in primary schools 496
The geography room 496
Library catalogue 496
Sample studies 496
Secondary schools section committee 496
Standing committee for field studies 495
Standing committee for sixth form/ university geography 496
The Geographical Association 1893-1953 . . . 496
Geographical bibliography, M Gandjei 184
Geographical bulletin, Canada, Department of Mines and Technical Surveys 3

Geographical digest (Philip) 497
'The geographical distribution of the major religions', H J Fleure 545
Geographical education 158
Geographical essays, W M Davis 364, 498
Geographical essays in honour of K C Edwards 460, 499
Geographical essays in memory of Alan G Ogilvie 500, 920
Geographical essays on British tropical lands 501
Geographical field classes in Benelux and the Eifel 416
Geographical Field Group 502, 774
Geographical fieldwork ... 503, 533
Geographical fragments—Hipparchus 610
'Geographical globes' 567
Geographical Handbook Committee 490
Geographical handbook series 504
Geographical Institute, Edinburgh (Bartholomew) 164
Geographical interpretations of historical sources ... 505
A geographical introduction to history 506
The geographical journal (RGS) 507, 609, 894, 1043, 1131
Geographical journal of the Hungarian Academy of Sciences 3
The geographical lore at the time of the Crusades 508, 1275
The geographical magazine 440, 449, 509, 860, 1082
'The geographical pivot of history' 806
The geographical review 45, 115, 225, 269, 510, 1013
Geographical Society of Finland 120
The Geographical Society of Ireland 511
Geographical Society of New South Wales 157
Geographical viewpoint 81
Die Geographie ..., Alfred Hettner 608
Géographie de la France, Philippe Pinchemel 329
Le géographie de l'histoire 243
Géographie d'Israel 513

Géographie de la population 512
Geographie der Staaten, Martin Schwind 777
La géographie du Talmud, A Neubauer 847
Géographie économique et sociale series 1026
La géographie française au milieu du XXe siècle 514
La géographie humaine 243, 515
Géographie humaine de la France 243
Géographie régionale de la France 516
Géographie universelle 367, 371, 479, 517, 1098, 1214
Géographie universelle Larousse 519, 767, 768
Géographie universelle des transports 518
Geographies for advanced studies series 796, 976
Die geographischen fragmente des Eratosthenes neu gesammelt ... 520
Geographisches jahrbuch 521, 1161
Geographisches taschenbuch ... 522, 923
Geographisches Zeitschrift 523, 608
Geography 4, 36, 115, 162, 183, 243, 244, 296, 315, 320, 331, 363, 365, 432, 458, 460, 494, 495, 525, 529, 545, 623, 740, 768, 769, 777, 806, 860, 916, 995, 1000, 1035, 1043, 1112, 1123, 1125, 1149, 1150, 1162, 1167, 1195, 1204
applied 759, 840, 1181, 1244
economic 167, 303, 407, 408, 478, 494, 531, 539, 542, 643, 716, 711, 752, 769, 840, 904, 995, 1012, 1021, 1093, 1117, 1167, 1223, 1228, 1249, 1266
educational 477, 487, 490, 496, 498, 503, 524, 525, 536, 598, 731, 753, 796, 850, 1099, 1128, 1140, 1152, 1171, 1192
historical 363, 432, 442, 505, 508, 529, 620, 904, 1150
human 371, 465, 500, 515, 526, 531, 606, 626, 627, 693, 710, 748, 749, 753, 769, 777, 794, 812, 813, 837, 840, 904, 950, 1009, 1015, 1098, 1151, 1181, 1214

Geography—(contd)
  physical 464, 494, 498, 576, 628, 776, 804, 813, 837, 840, 904, 961, 1003, 1097, 1117, 1133, 1139, 1176, 1181, 1185, 1211
  plant and animal 184, 219, 904, 973, 1088, 1097, 1222
  political 434, 458, 529, 539, 542, 777, 961, 979, 980, 997, 1009, 1117, 1210, 1261, 1262, 1280
  regional 516, 529, 741, 920, 1022, 1023, 1025, 1084, 1117, 1203, 1214
  urban 481, 547, 763, 1022, 1151, 1204-1209
Geography, Ptolemy 846, 998
Geography, Strabo 1133
Geography (The Geographical Association) 496, 524, 786, 1118
Geography and economics 167
Geography and education 525
Geography and growth series (Murray) 474
Geography and man . . ., ed W G V Balchin 526
Geography: an outline for the intending student, ed W G V Balchin 534
'Geography and planning', Daysh and O'Dell 527
Geography and planning, T W Freeman 527
Geography and technical education . . . 532
Geography as human ecology . . . 528
The geography behind history 530
Geography books for sixth forms 1152
Geography from the air, F Walker 531
Geography in Aberystwyth . . . 460, 529
Geography in education . . . (RGS) 532
Geography in the field, K S Wheeler 533
Geography in the making . . ., J K Wright 45
Geography in primary schools 496
Geography in relation to the social sciences, Isaiah Bowman 225
Geography in secondary schools . . . 496, 536

Geography in the Soviet Union . . . 1108
'Geography in the Soviet universities', G Melvyn Howe 1108
Geography in the twentieth century . . . 537, 1151
Geography in world society . . . 538
'The geography of Bernard Varenius' 1211
The geography of economic activity 539
A geography of Europe 540
The geography of the flowering plants 973
The geography of frontiers and boundaries 979
A geography of geographical terms 796
Geography of Great Britain, 1956-60 535
The geography of greater London . . . 541
Geography of international trade 542
The geography of iron and steel 633
Geography of London River 726
Geography of market centres . . . 542
The geography of the Mediterranean region, E C Semple 1075
Geography of the Middle East 776
The geography of modern Africa 543
The geography of Norden . . . 544
Geography of the northlands 1237
Geography of the Pacific 1237
A geography of Pakistan, K S Ahmad 412
The geography of the Port of London 633
The geography of religion in England 545
The geography of Romania 546
The geography of state policies 163, 980
The geography of towns 547
A geography of the USSR . . ., J P Cole and F C German 548
Geography of the USSR, P E Lydolph 1237
A geography of urban places . . 547
The geography room 496
Geology for geographers 576
Geological Society of America 557

Geological Survey of Great Britain 482, 551
The Geologists' Association 552
Geology 4, 54, 59, 184, 329, 330, 366, 401, 482, 549, 550, 801, 848, 863, 884, 914, 974, 1035, 1088, 1198
  bibliography 185, 237
  maps 445, 687, 848, 914, 931, 1121, 1192
Geo-magnetism 706
The geometrical seaman . . . 553, 1150
Geomorphological abstracts 238, 494, 554
Geomorphologie, F Machatschek 554
Geomorphology 238, 330, 364, 390, 488, 564, 622, 764, 786, 811, 966, 978, 1153, 1223, 1239
Geomorphology, B W Sparks 796
Geomorphology in a tropical environment 238, 555
Geomorphology of cold environments 555
Geophysical abstracts 3
Geo-physical memoirs 1045
Geophysical Society of Finland 908
Geophysikalische Bibliographie von Nord- und Ostsee 556
Georama 567, 961
George III, King, book collection 240
George Washington University 1003
Georgia State University, Bureau of Business and Economic Research 172
Geo Science abstracts 3
Geoscience documentation 558
The Geoscience Information Society, USA 557
Geo Services, London and Alberta 558
Geo-Stat system 559
Geo titles weekly 558
Gepert *see* Denoyer-Geppert
Gerasimov, I P 1108
Gérard Mercator . . ., J van Raemdonck 846
'Gerard Mercator . . .', E G R Taylor 846
Gerhard Mercator und die Geographen unter seinen Nachkommen 846

Gerlach, A C: 'The national atlas of the United States ' 139, 872
German, F E: *A geography of the USSR* . . . 249, 548
'Economic geography in the USSR, 1960 ' 246
A German and English glossary of geographical terms 560
German Cartographic Society 676
The German Copper Institute, Berlin 561
German Democratic Republic, National Committee for Geodesy and Geophysics 171
German Federal Research Organization for Forestry and Forest Products 1027
German Research Association 662
German Research Community 244
German ships 1091
Germany, J H Stembridge 1128
Germany 175, 248, 441, 522, 708, 1003, 1107, 1147, 1232, 1256
  bibliography 182, 184
  maps 248, 369, 377, 378, 469, 808, 918, 1014
Gertsen, W M: 'Danish topographic mapping ' 114
Geschichte der kartographie, Leo Bagrow 164, 562, 1082
Gesellschaft für erdkunde zu Berlin 563, 1037
Ghana: bibliography 184, 198
  Library Board 198
  University 184
Ghana agriculture . . . *bibliography* 184
Ghent Geological Institute 184
Gibb, Hamilton 635
Gibson, Bishop Edmund 231, 353
Gibson, Mary Jane, *comp*: *Portuguese Africa: a guide to official publications* 184
Gidwani, N N and K Navalani: *Indian periodicals* . . . 184
Gierloff-Emden, Hans-Günter: *Mexico* . . . 366
Gilbert, E W 806
'Andrew John Herbertson . . .' 606
'The Right Honourable Sir Halford J Mackinder ' 806
'The RGS and geographical education in 1871 ' 1043

Gilbert, E W—(contd)
'Vaughan Cornish 1862-1948 . . .' 348
and W H Parker: 'Mackinder's democratic ideals . . .' 806
Gilbert, Humfrey 1185
Gilbert, Martin: *Jewish history atlas* 723
Recent history atlas . . . 1014
Gilmour, Carter: revision, *Chisholm's handbook of commercial geography* 303
Ginsburg, Norton S, *An historical atlas of China* 613
Girtin, Tom: 'Mr Hakluyt, scholar at Oxford' 595
Girwic, Pelagia, on Soviet bibliography 184
Glacial and periglacial geomorphology 564
Glaciation 576, 755
Glacier sounding in the Polar regions symposium 565
Glaciers, maps 123
Glaciological notes 1247
Glaciological Society 566, 636, 732, 1085
Glaciology 45, 637, 669, 732, 1085, 1187, 1247, 1278
Glacken, C J 508
Glamorganshire, map 274
Glandard, Jacques: *Larousse agricole* . . . 766
Glasgow 809
  marketing maps 491
  University, Department of Geography 236, 491
  Institute of Soviet and East European Studies 1107
Glasscock, R E: *Irish geographical studies*, ed N Stephens and R E G 714
Glazier, K M, *comp*: *Africa south of the Sahara* . . . 16
Gleave, M B: *An economic geography of West Africa* 167
Glennie, E A and M Hazleton: *Cave fauna* 285
The global sea, Harris B Stewart 1210
The globe and its uses, George Goodall 567

Globes 166, 218, 269, 274, 553, 567, 624, 717, 846, 867, 888, 961, 1006, 1053, 1282
Globus 958
Der globus im wandel der zeiten . . . 567
Glossaries 227, 285, 309, 325, 335, 342, 441, 457, 468, 537, 542, 560, 568, 569, 1118, 1123, 1190
Glossary of cartographic terms 1190
Glossary of geographical names in six languages 568
Glossary of geographical terms 342, 569, 1118
Glossary of meteorology 559
Goblet, Y M: *Political geography and the world map* 961
Godwin, Harry 460
Gold Coast *see* Ghana
The gold rushes 972
Gold seekers in the desert 914
Goldberg, R ed: *Sedimentation* . . . 1067
Goldberg, R D: 'Circumpolar characteristics of Antarctic waters 59
The golden encyclopaedia of geography 570
Goldman, B J *et al*, ed: *Deserts of the world* . . . 376
Goldman, Bram J: *Arid lands research in perspective* . . . 72
Goldring, Frederick: *The weald* 331
Golson, Jack, ed: *Polynesian navigation* 981
Gómez, E G 1227
Gondwana 930
Gondwanaland revisited . . . 927
Good, Dorothy: *Geography of the northlands* 1237
Good, Ronald: *The geography of the flowering plants* 973
Goodall, D M: *The seventh continent* 61
Goodall, George: *The globe and its uses* 567
Goodman, Marie C, ed: *Map collections in the United States and Canada* . . . 819
Goodson, J B and J A Morris, *ed*: *The new contour dictionary* 893
Goode's world atlas 571, 1006
Goos, nautical chartbooks 1161

471

Gordon, A L and R D Goldberg: 'Circumpolar characteristics of Antarctic waters . . .' 59
Gordon *see* Leger-Gordon
Gores 567
Gosling, Peter L A: *Maps, atlases and gazetteers for Asian studies* . . . 76
Göteborg Akademiforlaget/Gumperts 142
Gotha, Cartographic Institute Justus Perthes 168
Gottheil, Richard: *Modern Egypt* . . . 857
Gottmann, Jean: *Geography of Europe* 540
Gotz, Antonin, ed: *Atlas Československé Socialistiché Republiky* 89
Goude, plan (Blaeu) 618
Gough, Richard: *British topography* 572
Gough Collection 222
The Gough Map 222, 572, 1043
Goujon, G et al: new ed J E Reclus: *L'homme et la terre* 1015
Goulding, W G 1487
Gourlay, R F: 'General introduction to the statistical account of Upper Canada' 261
'Statistical account of Upper Canada' 261
Government information and the research worker 866
Government publications 184, 278
Graham, M: *Understanding ecology* 528
Gran atlas Aguilar 573
Le grand atlas (Blaeu) 218, 574
Grande atlante del Touring Club Italiano 1059
Grande atlante geografico 25, 575, 717
Granite 992
Grant, maps 450
Grasslands 230
Graves, N J: *Geographical Association handbook* 496
Grazzini, Athos D: 'Problèmes que présente la préparation de l'Atlas mondial de la National Geographical Society' 874
Great adventures and explorations . . . 1127

Great Barrier Reef, atlas 124
Great Britain 28, 29, 60, 167, 184, 185, 195, 224, 229, 230, 242, 295, 301, 331, 332, 465, 472, 488, 491, 499, 529, 535, 576, 626, 655, 715, 726, 761, 762, 885, 924, 959, 988, 1003, 1021, 1024, 1040, 1104, 1124, 1141, 1199, 1225
 conservation 431, 885
 Gazetteer 924
 Geodetic Office 182
 Geological Survey 482, 551
 Land use surveys 761, 1118, 1236, 1255
 maps 62, 104, 110, 112, 231, 238, 269, 305, 507, 559, 572, 866, 885, 918, 924, 952, 989, 1014, 1116, 1121, 1135, 1147, 1222
 National Committee for Geography 676, 920
 sea surveys 1066
 Soil Survey 1096
 Trigonometrical Survey, 1791 924
 Great Britain . . ., ed A G Ogilvie 920
 see also under British Isles; United Kingdom; and under individual headings, *eg* Ordnance Survey
The Great capitals . . ., Vaughan Cornish 348
Great Lakes, maps 1255
The great trek E A Walker 972
Greater London . . ., J E Martin 167
Greece 1133
 bibliography 182, 184, 873
Greeks, geographical knowledge 617
Green, G H 235
Greenaway, John: *Bibliography of the Australian aborigines* 191
Greenhill, Basil: *James Cook . . .* 718
Greenland 418, 466, 565, 970, 977, 1056
Greenwich, Royal Observatory 586
Gregg M Sinclair Library, University of Hawaii 946
Gregor, Howard F: *Environment and economic life* 428
Gregory, K J and W L Ravenhill, ed: *Exeter essays* . . . 460
Gregory, James S: *Russian land, Soviet people* 1052

Gregory, S: *Statistical methods and the geographer* 1125
Gresswell, Ronald Kay 576
Geology for geographers 576
Physical geography 576
The physical geography of glaciers ... 576
The physical geography of rivers ... 576
Sandy shores in South Lancashire 576
and Anthony Huxley, eds: *Standard encyclopedia of the world's rivers and lakes* 1119
The Griffin 577
Griffiths, J F: 'Climates of Africa' 1267
Grigoryev, A A: 'Russian geography' 1108
Grimshaw, P H: *Physical atlas of zoogeography* 967
Der grosse Bertelsmann weltatlas 170, 578, 803
Der grosse Brockhaus atlas 579
Grosse Elsevier atlas 418
Grosse Shell-atlas of Germany and Europe 808
Grosvenor, G H: 'The National Geographic Society and its magazine' 875
Ground water year book 580
Grove, A T: *Africa south of the Sahara* 16
Grünwald *see* Iványi-Grünwald
Guatemala, bibliography 184
Gudmandsen, P: on echo sounding 565
Guernsey, H J Fleure 465
Guest, Arthur: *Advanced practical geography* 10
Guggisberg, C A W: *Man and wildlife* 815
Guide bibliographique d'hydrogéologie 581
A guide for the study of the British Caribbean ... 184
Guide to British periodicals 184
Guide to climatological practices (WMO) 319
A Guide to historical cartography ... 582
Guide to the India Office Records ... 641
Guide to Indian periodical literature ... 184
Guide to IGY World Data Centres 689
A guide to Irish bibliographical material ... 583
A guide to Latin American studies 584
A guide to literature on Iraq's natural resources 585
Guide to London excursions 586
A guide to manuscripts relating to America ... 587
Guide to New Zealand reference material ... 204, 588
Guide to reference material 521
Guide to Russian reference books 589
Guide to South African reference books 590, 1100
A guide to trees (New Zealand) 1016
Guides rouges (Michelin) 853
Guides verts (Michelin) 853
Guides 274, 363, 591, 1006
Guilbert, Aristide *et al*: *Histoire des villes de France* ... 611
Guinea, bibliography 184
Günther, S: *Varenius* 1211
Gussisberg, C A W: *Man and wildlife* 885
Gustav, King of Sweden 639
Gutenberg, B and C F Richter: *Seismicity of the earth and associated phenomena* 1069
Gutkind, E A: *Our world from the air* 592, 817
Gutkind, Erwin E 592
Gutkind, Peter C W: *The passing of tribal man in Africa* 950
Guttentag, Werner: *Bibliografía Boliviana* 184
Guyana 184, 286
Gwam, L C, comp: *A handlist of Nigerian official publications* ... 184
Glydendal, Copenhagen: atlases 84, 593

Haack, Hermann 1130
VEB Geographisch-Kartographisch Anstalt 521
Habitat 594

Haddon, J: *Local geography in towns* 1204
Hadera paper mills 127
Haffner, Vilhelm, bibliography of Norway 184
Haggett, Peter: *Frontiers in geographical teaching* 477, 850
Integrated models in geography 856
Locational analysis in human geography 794
Models in geography 850, 856, 966
Network analysis in geography 891
Physical and information models in geography 856, 966
Progress in geography . . . 993
Socio-economic models in geography 1093
Hailey, Lord: *An African survey* 21
Hake, G: *Kartographie* . . . 743
Hakluyt, Richard (Elder and Younger) 595, 769, 1150, 1185
Divers voyages . . . 596
Principall navigations . . . 595, 769
Hakluyt Society 595, 596, 635, 1082, 1150
Haldingham, Richard: *Hereford mappa mundi* 607
Haley, Bernard Francis 460
'H J Mackinder and the new geography', J F Unstead 806
Hall, A D and E J Russell: *A report on the agriculture and soils of Kent, Surrey and Sussex* 1260
Hall, D G E 136
Hall, G K, publishers 45, 279, 282, 299, 356, 381, 382, 383, 1058, 1198, 1223
Hall, Peter: *London 2000* 795
Regional studies 1025
The world cities 1246
Hall, Robert B and Toshio Noh: *Japanese geography* . . . 719
Hallwag maps 597
Halstead, C A: 'Agricultural statistics' 246
Hamburg, maps 378
Hamburger geographische studien . . . 366
Hamburger geophysikalische einzelschriften 366
Hamilton, F E I: *Regional economic analysis in Britain* . . . 1021

Hamilton, F E I—(*contd*)
Yugoslavia: pattern of economic activity 167
Hamlin, Bruce: *Native ferns* (New Zealand) 1016
Native trees (New Zealand) 1016
Hampshire, maps 907
Hamshere, John D *et al*: *Geographical interpretations of historical sources* . . . 505
Hance, W A: *The geography of modern Africa* 543
Hancock, J C and P F Whiteley, *ed*: *The geographer's vademecum* 961
Hancock, P D: *A bibliography of works relating to Scotland* 184
Handbook for geography teachers 598
The handbook of African languages 666
Handbook of American resources for African studies, Peter Duigman 599
Handbook of aviation meteorology 1045
Handbook of Hispanic source materials . . . 600
The handbook of international trade and development statistics (UN) 1192
Handbook of Jamaica 184
Handbook of Latin American Studies 601
Handbook of meteorological instruments 1045
A handbook of suggestions on the teaching of geography (Unesco) 1099
Handbooks for world fisheries abstracts 1250
Handbooks to the modern world series (Blond) 152, 602
A handlist of Nigerian official publications in the National Archives . . . 184
Handbuch der klimatologie (Köppen) 750
Hanover, Academy for Area Research and Land Use Planning 378
Hanover Institute for Space Research 248
Harding, M: *Geographical fieldwork* . . . 503, 533

Hardware models in geography teaching 1152
Harleian Collection 240
Harley, J B 1042
'Error and revision in early os maps' 924
First Edition, One Inch os maps, reprint 924
Studies in historical geography 363
Harlow, V T 984
Harriman Alaska Expedition, 1899 1084
Harris, G D: *An annotated world list of selected current geographical serials in English, French and German* 53, 1283
Cities of the Soviet Union ... 308
Soviet geography ..., English edition 1108
and J D Fellmann: *A comprehensive checklist of serials of geographical value* 694
International list of geographical serials 300, 694, 1283
A union list of geographical serials 694
Harris, John, comp: *Guide to New Zealand reference material* ... 214, 588
Harris, William John: *Books about Nigeria*... 184
Harrison, C: *The analysis of geographical data* 48
Harrison, S G et al: *The Oxford book of food plants* 942
Harrison Church, R J see under Church
Harrisse, Henry: *The discovery of North America* ... 273, 395
Hart, Henry H: *Marco Polo* ... 829
Hartling, Bertram et al: *Bibliographiekirchengeschichtlicher karten Deutschlands* 177
Hartshorne, Richard: *The nature of geography* ... 887
Persepective on the nature of geography 887
'Political geography in the modern world' 979
Harvard, Houghton Library 164
Institute of Geological Exploration 1005
Harvard-Yenching Institute 613, 1073

Harvey, Anthony: *British geological literature* 237
Geomorphology in a tropical environment 238, 555
Harvey, David W: *Explanation in geography* 447
Harvey, Joan M: *Statistics: African sources for market research* 1126
Harvey, L A and D St Leger-German: *Dartmoor* 331
Harwich 809
Hatherton, Trevor: *Antarctica* 61
Hattersley-Smith, G, on radio echo sounding 565
Hauri, Hans et al: *Edward Imhof* ... 414
The haven-finding art ..., E G R Taylor 603, 1150
Hawaii University, East-West Center Library 1101
Hawaiian Collection 383
Pacific Collection 946
Hayes, G R: *An introducton to charts and their use* 842
Hayter, A: *The year of the quiet sun* 689
'Hayter' Asian Centres 76
Hazard, Harry W: *Atlas of Islamic history* 126
Hazleton, M: *Cave fauna* 285
Hazlitt, W C: 'British Columbia and Vancouver Island' 261
The heart of the Antarctic ..., Sir E H Shackleton 604
Heawood, Edward: 'A history of geographical discovery in the 17th and 18th century' 619
Hebrew University, Jerusalem 127, 722
Heffer, W and Sons, Limited, Cambridge 605
Heimpel, Hermann 177
Heissler, Viktor: *Kartographie* 366
and G Hake: *Kartographie* 743
Helfrick, H W ed: *Agenda for survival* ... 430
Helmolt, Hans 1009
Helsinki University: Finnish national bibliography 184
Institute of Geography 120
Library 184
List of theses published from Finnish universities 184

475

Henderson, G P, *comp*: *Current European directories* 357
European companies 439
Henry, Prince (The Navigator) 165, 240, 984
Herbage abstracts 336
Herbert, Jean 568
Herbertson, A J 458, 606, 774, 806, 1150
  Atlas of meteorology 129, 750
  Man and his work . . . 606
Herbertson, F D 606
Hereford mappa mundi 607, 1043
Herring atlas 681
Herrmann, Albert, *comp*: *An historical atlas of China* 613
Hertfordshire, maps 907, 1135
Hespéris 184
Hesse, maps 378
Hettner, Alfred 523, 608, 810
  Die Geographie . . . 608
Higbee, Edward: *American agriculture* . . . 1237
The Highlands and islands of Scotland 916
Hill, M N *et al, ed*: *The sea* . . . 1065
Hillaby, John: *Nature and man* 885
Hillary, Sir Edmund: *The crossing of Antarctica* . . . 61, 354
Hills, E S: *Arid lands* . . . 71, 1192
Hilton, K J *ed*: *The Lower Swansea Valley project* 799
Hilton, Ronald, *ed*: *Handbook of Hispanic source materials* . . . 600
Himalaya 1244
Hind, A M: *Early Italian engraving* 998
Hinks, A R 609
  Hints to travellers series 609
  Maps and survey 826
  Map projections 712
  Technical series 609
  A traveller's guide to health 609
Hipparchus: *Geographical fragments* 610
HMS Beagle 921
HM Surveying Service 842
Hispanic America 1:1M map 920
Hispanic and Luso-Brazilian Council 235, 770
Hispanic Society of America 983

Histoire de la cartographie, A Libault 562
Histoire des villes de France . . . 611
Histoire universelle des explorations 384
The historian's guides to Ordnance Survey maps 924
Historical and commercial atlas of China 613
The Historical and Scientific Society of Manitoba 816
Historical abstracts 3
A historical atlas of Canada 612
An historical atlas of China 613
The historical atlas of the Holy Land 614, 1006
Historical atlas of Latin America . . . 615
The historical geography of the Shetland Islands 916
An historical introduction to the economic geography of Great Britain 167
Historical Sciences, International Congress 177
History—and geography 506, 530
  atlases 1014
  mapping of data 445
The history and use of diamond 616
A history of ancient geography . . . 617
The history of Australian exploration . . . 847
A history of cartography . . ., R V Tooley and Charles Bricker 618
History of Cartography, International Conferences 567
A history of geographical discovery in the seventeenth and eighteenth centuries, Edward Heawood 619
The history of geography . . ., J N Baker 620
The history of the geological exploration of Siberia 914
History of the Hudson's Bay Company 625
History of the Isle of Man 749
A history of land use in arid regions 621, 1118
History of lunar cartography 1161
History of marine navigation 842
A history of Martian cartography 1161

History of Science, International Congress 567
The history of the study of landforms... 622
Hitchens, K, *ed*: *Rumanian studies*... 1050
Hocken, T M: *Bibliography of the literature relating to New Zealand* 200
Hodder, B W: *Economic development in the Tropics* 850
Hodge, Paul W and Frances W Wright: *The large Magellanic cloud* 1084
Hodgkiss, A G: *Maps for books and theses* 827
Hodson, Donald 353, 1042
Hodgson, R D: *The changing map of Africa* 1210
Hodgson, W C. *ed*: *Herring atlas* 681
Hofmann, Gustav, on microclimatology 317
Hoffman, George W 1210
Hofman, Walther *et al*: *Photogrammetrie* 366
Hogben, N and F E Lumb: *Ocean wave statistics*... 915
Hogenberg, Franz: *Civitates orbis terrarum* 313
Hogg, Garry, *comp*: *The Shell book of exploring Britain* 1078
Hogg, Helen S 811
Holbein: 'The ambassadors' 567
Holck, P: *Danish-Norwegian ships* 1091
Holdgate, M W: *Antarctic ecology* 57
Holdridge, L R: *Forest environments in tropical life zones*... 470
Hole, William, map engraver 231
Holmbäck, Bure *et al*: *About Sweden*... 1
Holman, Kjeld *et al*: *The flora of Greenland* 465
Holt, G: 'Tips and tip working in the Swansea Valley' 799
Holy Land, maps 127, 614, 1006
Holyhead 809
Hondius, Jodocus 567, 1043, 1161
Honduras, bibliography 184
Honduras-Yucatan Expedition 69

Honegger, D, cartographer 446
Honeybone, R C, *ed*: *Handbook for geography teachers* 598
Hood, P J *et al*: *Earth science symposium on Hudson Bay* 401
Hookey, P G: *Do-it-yourself weather instruments* 1152
Hooson, D J M: *A new Soviet heartland* 1210
Hoover Institution on War, Revolution and Peace 598
Africa Collection 20
Guide to Russian reference books 589
Hop industry 505
Horecky, Paul L, *ed*: *Russia and the Soviet Union*... 1051
and R G Carlton: *The USSR and Eastern Europe*... 1189
Horn, Werner 1130, 1161
Horticulture 336, 1276
Hosking, Eric 331
Hoskins, W G: *The common lands of England and Wales* 331
Hough: *Encyclopaedia of American woods* 471
Houston, J M, *ed*: *The world's landscapes* 1273
Houghton Library, Harvard College 164
How the Maoris came, A W Reed 1016
How the Maoris lived, A W Reed 1016
How the white men came, A W Reed 1016
How to find out in geography... 623
How to identify old maps and globes 567, 624,
Howard, Admiral Lord, of Effingham 1219
Howe, G Melvyn: 'Geography in the Soviet universities' 1108
National atlas of disease mortality in the United Kingdom 869
and Peter Thomas: *Welsh landforms and scenery* 1226
Hsieh, Chiao-Min: *Taiwan-Ilha Formosa*... 249
Hudson 304
Hudson's Bay 261, 401

477

Hudson's Bay Company Record Society 596, 625
Huggett, F E: *The modern Netherlands* 890
Hugh Robert Mill . . ., Sir Dudley Stamp 496
Hughes, L Robert 423
Hull, Oswald: *London* 807
  Topics in geography 807
  Transport 807
Hull 809
Hull University, Department of Geography 1277
  An atlas of population change . . . 133
Human geography 515
Human geography, A V Perpillou 753
Human geography from the air 626
Human geography in Western Europe 465
The human habitat, Ellsworth Huntington 632
Human nature in geography . . . 627, 648, 1275
Humanistiska Fonden 639
Humanities Research Council of Canada 261
Humberside, maps 133, 1277
Humboldt, Alexander von 168, 810, 957, 1015
  Essai sur la géographie des plantes 1088
  Foundation 628
  Kosmos . . . 628
  map of Mexico 851
'Humboldt's map of isothermal lines . . .', A H Robinson and H M Wallis 628
Humid tropics research programme (Unesco) 1192
Humlum, J: *Kulturgeografisk atlas* 752
Humphreys, A L: *Old decorative maps and charts* 369
Humphreys, Philip S: *The Darwin reader* 921
A hundred years of geography 629
Hungary 33, 651, 1107, 1256
  bibliography 182, 184, 630
  maps 870
Hungarica 630
Hunt, A J, translator 473

Hunt, K E and K R Clark: *Poultry and eggs in Britain* 28, 655
  The state of British agriculture . . . 28, 655
Hunting Group review 631
Hunting Survey Companies 631
Hunting Surveys Limited 965, 1168
Huntington, Ellsworth 489, 632
  Character of races 632
  Civilization and climate 632
  The human habitat 632
  Mainsprings of civilization 632
  Palestine and its transformation 632
  The pulse of Asia 632
Hurd, Thomas 8
Huschke, R E, ed: *Glossary of meteorology* 569
Hutchinson University Library series 163, 296, 443, 458, 527, 547, 633, 643, 809, 825, 979, 997, 1023, 1117
Huxley, Anthony: *Standard encyclopedia of the world's rivers . . .* 1119
  Standard encyclopedia of the world's mountains 1119
  Standard encyclopedia of the world's oceans and islands 1119
Huxley, Sir Julian: *La protection de la grande faune et des habitats naturels en Afrique . . .* 1192
Huxley, Michael 509
Hydro-biologische anstalt der Max-Planck 43
Hydrographic Department, Admiralty 634
'Hydrographic surveying and data processing', A Thunberg 830
Hydrography 8, 187, 270, 445, 634, 681, 690, 777, 830, 842, 1066, 1196
Hydrology 390, 494, 529, 669, 706, 733, 884, 908, 966, 1192, 1223
Hydrometeorology 908
Hypothesis testing in field studies 1152

Iasbez, Liliana *et al:* *Glossary of geographical names in six languages* 568
Ibadan University, Institute of African Studies 19
Ibn Battuta 635
  Travels 635
Ice 564, 637, 908, 1063, 1085

Ice 566, 635
Iceland 544, 602, 970, 1056
  bibliography 184
  maps 418
Idrisi *see* Al-Idrisi
Ifo Institute for Economic Research, Munich 24
Illinois Unversity: *Research digest* 1031
Illustrated encyclopedia of Maori life 981
Illustrated ice glossary 1063
Imago mundi ... 164, 269, 639, 1082
Imhof, Eduard 95, 414, 709, 1059
  Karte und luftbild ... 777
  Kartographische geländedarstellung 744
  'The Swiss Mittelschulatlas in a new form' 1059
Impact of science on society (Unesco) 1192
Imperial Bureau of Soil Science (later Commonwealth Bureau of Soils) 209
Index bibliographicus 3
Index India 640
Index Iranicus 184
Index to maps in books and periodicals 45
India 33, 78, 239, 332, 621, 640, 708, 738, 865, 984, 1102, 1133, 1136, 1209
  bibliography 184, 199
  maps 126, 641, 707, 871, 934
  National Committee for Geography 641
  Northwest passage 1164
India Council of World Affairs 396
India Office Library and Records 641, 836
India: regional studies, ed R L Singh 642
Indian Ocean 16, 896, 1113
Indian periodicals ... 184
Indian railways technical bulletin 1074
Indian School of International Studies 703
Indochina, maps 136
Indonesia 299, 419, 738, 1105
  bibliography 184
  Colombo Plan 332
  maps 126, 136

Industrial activity and economic geography ... 643
Industrial archaeologists' guide 644
Industrial archaeology 363, 645
Industrial diamond abstracts 3, 646
Industrial Diamond Information Bureau 3, 646
Industrial diamond review 646
Industrial diamond trade name index 646
Industrial minerals 647
Industrial Revolution, factories, Parliamentary papers 715
'Industrial statistics', J P Cole 246
Industrialisation 289, 748
Industry 246, 363, 422, 440, 445, 478, 526, 527, 1175, 1199
  research 339
Influences of geographic environment 648, 1009, 1075
Information bulletin, Soviet Antarctic Expedition 55, 649
L'Information géographique 3, 650
Information Hungary 651
Information supplement to Soviet Studies 1107
Information USSR ... 652
Ingrams, Harold: *Uganda* 349
Innsbruck University Geography Department 460
Institute for Environmental Studies 592
Institut française d'Afrique Noire 653
Institut für Angewandte Geodäsie, Frankfurt 182
Institut für Landeskunde, Bad Godesberg 177, 213, 269
Institut Géographique National, Paris 654, 1213
Institut National de la Statistique et d'Etudes Économiques 93
Institute of African Studies, University of Ibadan 19
Institute of Agricultural Economics, University of Oxford 655
Institute of Agricultural History 69
Institute of Arctic and Alpine Research 65
Institute of Arid Lands Research, University of Arizona 73
Institute of Australian Geographers 656

479

Institute of British Geographers 69, 238, 458, 460, 576, 759, 776, 920, 1239
Institute of Commonwealth Studies, University of London 20
Institute of Directors, London 391
Institute of Documentation Studies, Frankfurt 658
Institute of Geodesy, Photogrammetry and Cartography, Ohio 269
Institute of Geological Sciences, Edinburgh 54
Institute of International Studies, University of California 77
Institute of Marine Engineers, London 659
Institute of Marketing 835
Institute of Navigation 603, 1150
Institute of Pacific Relations, American Council 456
Institute of Pacific Relations 660
Institute of Transport, London 661
Institute of Transportation, Los Angeles 1180
Institute of Urban Studies 1209
Institute of World Economics, Kiel 662
Institution of the Rubber Industry, London 663
Instituto Fototopográfico Argentino 70
Istituto Geografico de Agostini 25
Instituto Geográfico Militar, Argentina 70
Instituto Nacional de Pesquixes de Amazonía, Manáus 43
Instructions for the preparation of weather maps 242
Instruments 274, 275, 347, 370, 453, 553, 603, 631, 846, 1043, 1045, 1054, 1129, 1152, 1279
Integrated models in geography 856
Inter Nationes: film on Alexander von Humboldt 628
Interagency Committee on Oceanography 1084
Inter-American Geodetic Survey 182
Intergovernmental Oceanic Commission, *Technical* series 664
International African bibliography 665
International African books 17

International African Institute 17, 665, 666
International Association of Agricultural Economists 1242
International Association of Agricultural Librarians and Documentalists 188, 1242
International Association of Geochemistry and Cosmochemistry 849
International Association of Geodesy 171, 182, 269, 706
International Association of Geomagnetism and Aeronomy 706
International Association of Meteorology and Atmospheric Physics 706
International Association of Oceanography 706
International Association of Physical Oceanography, Götenborg 667
International Association for Quaternary Research 668, 1002
International Association of Scientific Hydrology 669 ,706
International Association of Sedimentologists 418
International Association of Sedimentology 670
International Association of Seismology and Physics of the Earth's Interior 671, 706
International Association of Volcanology 706
International Association on Water Pollution Research 1223
The international atlas 672, 961
International atlas of West Africa 673
International Atomic Energy Agency 849
International Bank for Reconstruction and Development 674
International bibliography of vegetation maps 675
International Biological Programme, World Conservation Committee 431
International Botanical Congress 315
International Cartographical Association 23, 269, 436, 676, 709, 1030, 1190
International Civil Aviation Organisation 677, 1241
International cloud atlas 1258

International Commission for Aeronautical Charts 678
International Commission for the North-west Atlantic Fisheries 679
International Commission for the Scientific Exploration of the Mediterranean sea 680
International Conference of Agricultural Economists 1243
International Conference on African Bibliography 186
International Council for the Exploration of the Sea 681
International Council of Scientific Unions 370, 706, 1113
  Joint Commission on Oceanography 370
International Councils of Museums 392
International Demographic Symposium, Zakopane: *Proceedings* 682
International dictionary of stratigraphy 683
International Federation of Documentation (FID) 315
International Federation of Library Associations, Sub-Section for Geography and Map Libraries 684
International Federation of Surveyors 171
International Geodetic Documentation 182
  see also Bibliographia geodaetic
International Geographical Union 33, 113, 115, 176, 183, 225, 241, 367, 484, 522, 686, 867, 870, 883, 923, 1021, 1099, 1118, 1147, 1248, 1264
  Congresses 30, 33, 238, 315, 337, 373, 390, 462, 535, 544, 586, 639, 642, 676, 677, 686, 806, 865, 923, 1013, 1255
  Newsletter 685, 686
International Geological Congress 687, 688
International geological map of Europe 687, 688, 1192
International geological map of the world 687
International Geophysical Assembly 1085
International Geophysical Year (IGY) 649, 689, 803, 877, 1062, 1187, 1234, 1247, 1258

International Geophysical Year (IGY) —(*contd*)
  Artificial Satellite Sub-Committee 689
  The International Geophysical Year, Werner Buedeler 689
' International History of City Development' project, University of Philadelphia 592
International Hydrological Decade 376, 908
International hydrogeological map of Europe 1192
International hydrographic bulletin 690
International Hydrographic Bureau 690
The international hydrographic review 690
The International Hydrological Decade 123, 691, 884, 1192
International journal of comparative sociology 692
International Library of Sociology and Social Reconstruction 312, 693
International list of geographical serials 300, 694, 1283
*International map of the world, 1:1*M 677, 695, 951, 1147
International marine science 696, 1192
International Meteorological Organization 1258
International Nickel 697
International Railway Congress Association, Brussels 1074
International Railway Documentation Bureau 1074
International Road Federation, 698, 1039, 1254
International Road Research Documentation Scheme 699
International rubber digest 663
International Rubber Study Group, London 663
International seismological summary 671
International Society for Photogrammetry 269, 700
International Society of Soil Science 701, 702
International Soil Museum 702
International studies 703

481

International studies in sociology and social anthropology 208
International Study Week, Amsterdam 1206
International Symposium on Photomaps and ortho-photomaps, Ottawa 446
International Tin Research Council 704
International Union for Conservation of Nature and Natural Resources 705
International Union of Geodesy and Geophysics 669, 689, 706
Oceanographic Section 667
International Union of Geological Sciences 399, 849
International Union of Railways, Paris 1074
International Union for the Scientific Study of Population 173
International University Bookseller, Inc, New York, Africana Center 21
*International vegetation map 1:1*M 707
International Wool Secretariat, London 708
International world atlas (Rand McNally) 1006
International yearbook of cartography 269, 436, 676, 709, 961
International yearbook for the teaching of history and geography 244
Internationales Afrikaforum see Afrikaforum
Interscience series (Wiley) 1237
An introduction to the Canadian Arctic 256
An introduction to charts and their use 842
An introduction to climate, G T Trewartha 804
An introduction to human geography, J H G Lebon 710, 776
An introduction to physical geography, M I Newbigin 904
An introduction to quantitative analysis in economic geography 711
An introduction to the study of map projections 712
An inventory of geographical research on world desert environments 189, 376

Investing in maps 624
Iran 239, 814
 bibliography 184, 986
 Colombo Plan 332
 maps 107, 126, 1053
Iraq 107, 184, 585
Ireland 81, 86, 224, 320, 421, 438, 462, 587, 885, 1045, 1135
 bibliography 182, 184, 269, 583, 715
 maps 111, 164, 305, 453, 924, 989, 1121, 1188
Ireland . . ., T W Freeman 850
Ireland, A R Orme 1273
Irrigation 28, 390, 585, 621, 631, 1086, 1253
Irrigation development planning aspects of Pakistan experience 1276
Irish forestry 713
Irish geographical studies 714
Irish geography 511
Irish historical studies 184
Irish Manuscripts Commission 1188
Irish studies in honour of E Estyn Evans 460
Irish University Press (International) Limited, London 480, 715
 Area studies 301, 1199
Iron 633
Iron and steel industry 422, 478
Iron ore memoirs 551
Isaiah Bowman School of Geography 225
Isakou, I S, *ed: Morskoy atlas* 862
Ishwaran, K, *ed*: Contributions to Asian studies, *International journal of comparative sociology* 692
 Journal of Asian and African studies 727
Islam 107, 126, 1227
Islands 363, 1119
Isle of Man 749, 761
The Isle of Man . . . J W Birch 716
Isle of Thanet Geographical Association 761
Israël, B M, publishers 79
Israel, N, reprints publisher, Amsterdam 174, 178, 211, 212, 218, 224, 252, 280, 520, 574, 639, 791, 846, 888, 1081, 1155, 1156, 1219
Israel 127, 184, 513, 854, 895
Israel Program for Scientific Translations Limited 513, 1067, 1146, 1223

Istituto Geografico de Agostini, Novara 141, 368, 567, 575, 709, 717, 1186, 1243
L'Italia, Robert Almagià 41
Italian Institute for the Middle and Far East 39
Italy 3, 182, 184, 441, 708, 1133, 1147, 1186
maps 118, 369, 1014, 1172
National Committee for Geography 41
Itenberg, I M 102
Iványi-Grünwald, Béla: *Hungarica* 630
Ivory Coast 184, 992
Izvestiya, Geological series 914
seriya geograficheskaya 37
Izvestiya Akademii Nauk SSR ... 1109
Izvestiya vseroyuznogo geograficheskogo obshchestva 40, 1109

J Russell Smith ..., Virginia M Rowley 1083
Jackson, Nora and P Penn: *A dictionary of natural resources* ... 388
Jackson, S P, ed: *Climatological atlas of Africa* 321
Jackson, William Vernon: *Library guide for Brazilian studies* 782
Jackson, W A D: *Politics and geographic relationships* ... 980
Jacobsen, N K and R H Jensen, ed: *Denmark* 373
Jagellonian University Museum, Cracow: globes collection 567
Jahn, K 1227
Jakobsen, Knud et al: *The flora of Greenland* 465
Jamaica 184, 286, 501, 1020
Jamaica, Peter Abrahams 349
James Cook ..., Basil Greenhill 718
Jane, Cecil, translator 334, 728
Janick, Jules et al: *Plant science* ... 973
Jansson 1161
Japan 54, 78, 288, 801, 708, 715, 738, 984, 1084, 1177, 1211, 1243, 1244, 1249
Association of Japanese Geographers 719

Japan—(*contd*)
bibliography 182, 184, 1032, 1073
Oxford economic atlas 934
see also under Nippon
Japanese geography, 1966 ... 719
Japanese geography ..., R B Hall and Toshio Noh 719
Japanese geography: a guide ..., Yoshida Tōgo 1032
Japanese researchers in Asian studies 288
Japanese studies on Japan and the Far East ... 720
Das Japanische inselreich ... 366
Jäschke, G 1227
Jawaharlal Nehru University School of International Studies 1166
Jay, L J: *Geography books for sixth forms* 1152
Jellinek, L, Austrian bibliography 184
Jena review 721
Jensen, R H ed: *Denmark* 373
Jerusalem: Hebrew University 127
maps 107
Jerusalem studies in geography 722
'Jewells of antiquitie ...', G R Crone 596
Jewish history atlas 723
Joan Blaeu and his Grand Atlas, C Koeman 218
Joel, Issachar: *Index of articles on Jewish studies* 184
Joffe, Joyce: *Conservation* 885
John Cary ..., Sir H G Fordham 274
The John Harvard Library series 813
John Speed's Atlas of Wales 724
Johns Hopkins University, Isaiah Bowman School of Geography 225
Johnson, A F: *A bibliography of Ghana* ... 184, 198
Johnson, D W, ed: *Geographical essays*, W M Davis 498
Johnson, Edon: *Atlas for anthropology* 97
Johnson, J H: *Trends in geography* ... 1181
Johnson Reprint Corporation 18
Johnston, Alexander Keith 725
Johnston, W and A K, Edinburgh 725, 957
Physical atlas 725
Royal atlas 725

Jonasson, Olof and Bo Carlsund: *Atlas of the world commodities* ... 142
Jones, B Z: *Lighthouse of the skies* ... 784, 1084
Jones, C F 1075
Jones, Emrys and D J Sinclair: *Atlas of London* ... 128
Jones, G R J: *Geography as human ecology* 528
Jones, G T: *Analysis of census data for pigs* 655
Jones, H L: Strabo's *Geography*, Loeb edition 1133
Jones, Sir Harry David: European military maps collection 240
Jones, J Winter, ed: *Divers voyages* ... 596
Jones, Llewellyn Rodwell 726
London river 726
memorial volume 726
Jones, Mary Jane, ed: *Mackenzie Delta bibliography* 805
Jones, Ruth: *Bibliography of Africa* 186
Jones *see also* Wood Jones
Jordan 107
Jorré, Georges: *The Soviet Union* ... 753, 796
Journal of Asian and African studies 727
Journal of Christopher Columbus 334, 728
The journal of economic abstracts 729
Journal of the economic and social history of the Orient 730
The journal of geography 731
Journal of glaciology 566, 636, 732, 1085
Journal of hydrology 733
Journal of Latin American studies 734
Journal of Nigerian studies 735
Journal of Pacific history 736
Journal of regional science 737
Journal of the Società Geografica Italiana 3
Journal of South-East Asia and the Far East 738
Journal of the Textile Institute 1159
The journal of transport history 739
Journal of tropical geography 740
Journal of West Midlands regional studies 741
Journals *see also* Periodicals
Journeyman Taylor 1151
Judea, maps 127
Jugoslavia 504
Junker, C, Austrian bibliography 184
Jusatz, H: *World map of climatology* 129
Justus Perthes geographischer anstalt 958
Juynboll, T G J: *Lexicon geographicum* 214

Kabeel, S M, *comp*: *Selected bibliography on Kuwait and the Arabian Gulf* 1072
Kamal, Yousuf: *Monumenta cartographica Africae et Aegypti* 789
Kampffmeyer, George 1227
Kandy, publishers 961
Kansas Council of Geographic Education 742
Kansas Geographer 742
Kansas State Library, Division of geography 742
Kansas University: *International bibliography of vegetation maps* 675
Kapp, K S: *The printed maps of Jamaica to 1825* 184
Kara Kum Sands 914
Kárník, Vít: *Seismicity of the European area* 1070
Karp, Walter: *The Smithsonian Institution* 1084
Karpf, A *et al*: *Die literatur über die Polar-Regionen der erde* ... 791
Karte und luftbild . . ., Eduard Imhof *et al* 777
Kartografiska Institutet, General-Stabens Litografiska Anstallt, Stockholm 146
Kartographie . . ., V Heissler and G Hake 366, 743
Kartographische geländedarstellung 366, 744
Kartographische nachrichten 269, 745
Kasperson, R E and J V Minghi, *ed*: *The structure of political geography* 979

Keast, Allen: *Australia and the Pacific islands* . . . 156
Keele University: air photographs collection 1200
Keir, K M: *Photogrammetry and current practice in road design* 965
Keltie, Sir John Scott 746
Kellner, L: *Alexander von Humboldt* 628
Kendrew, W G 949
Climate 747
The climate of central Canada 747
The climates of the continents 318, 747
Climatology 747
Kendrick, Thomas 779
Kéne, Thao: *Bibliographie de Laos* 184
Kennedy, T F: *A descriptive atlas of the Pacific Islands* . . . 375
The Kennedy Round 391
Kent, William, ed: *An encyclopaedia of London* 424
Kent 505, 1260
 maps, 924, 1008
Kenya 286, 429, 1020
 atlas 331
 bibliography 788
Kerala State, India: bibliography 199
Kerr, D G G, ed: *A historical atlas of Canada* 612
Kerr, J F 778
Khalaf, Nadim G: *Economic implications of the size of nations* . . . 409
Khozyaystvo 1109
Kienauer, Rudolf: *Bibliographiekirchengeschichtlicher karten Österreichs* 177
Kimble, G H T and Dorothy Good: *Geography of the Northlands* 1237
King, L J 495
King, C A M: *Glacial and periglacial geomorphology* 564
Quantitative geography . . . 1000
Techniques in geomorphology 1153
King, Charles, illustrator 1222
King, H G R: *The Antarctic* 61
King's College/LSE, Joint School of Geography 726
Kingsley, Mary H: *West African studies* 1231

Kinship and geographical mobility 748
Kinvig, R H 749
History of the Isle of Man 749
Kinvig Geographical Society 749
Kinzl, Hans 460
Kip, William, map engraver,231
Kirkham, N: *Derbyshire lead mining glossary* 285
Kirkpatrick, F A: *The Spanish conquistadores* 972
Kirwan, L P: *The white road* 1234
Kitchin, Thomas: *The Royal English atlas* . . . 1042
Kjelleén, R 1280
Kleiner Bertelsmann weltatlas 170
Klemp, Egon, ed: *Africa on maps dating from the 12th to the 18th century* 14
Das klima der bodennahen luftschrift 317
Die klimat der geologischen vorzeit 750
Knuf, Fritz, publisher 79
Københavns Universitets Geografiske Institut 373
Koch, Howard, jr, ed: *Contemporary China* 344
Koeman, Dr Ir C 85, 148, 574, 896
 'A catalogue by Joan Blaeu . . .' 218
 Joan Blaeu and his Grand Atlas 218, 574
Kontinentaldrift und geologie des Südatlantischen Ozeans 366
Kopal, Zdeněk: *The moon* 861
Köppen, Vladmir Peter 750
 climatic regions 750
 Handbuch der klimatologie 750
 and Alfred Wegener: *Die klimat der geologischen vorzeit* 750
Korea 78, 738
 bibliography 184
 Colombo Plan 332
Ksach, Hans-Peter et al: *Bibliographiekirchengeschichtlicher karten Deutschlands* 177
Kosmos . . ., Alexander von Humboldt 628
Kozenn, Blasius 451
Kozenn-atlas 451
Kraeling, E G: *The historical atlas of the Holy Land* 614
 Rand McNally Bible atlas 1006

485

Krasnoyarsk Institute of Physics 4
Kraus, reprints 77
Kraus, Hans P: *Sir Francis Drake* ... 1081
Kraus, Th *et al*: *Atlas Östliches Mitteleuropa* 145
Krumbein, W C *et al, ed*: *Lithofacies maps* ... 792
Kubiena, W L: *The soils of Europe* 1260
Küchler, A W, on classification of world vegetation 315
  International bibliography of vegetation maps 675
Kultura see Cartographia of Budapest
Kulturgeografisk atlas, comp J Humlum 752
Kumar, G and V Machwe: *Documentation on Asia* 396
Kupfer-mitteilungen 561
Kurzauszüge aus dem schrifttum für das eisenbahnwesen 1074
Kuwait, bibliography 1072

Laboratoire des Ponts et Chaussées 699
Laborde, E D, translator 371, 753, 1176
Labrador 625, 911
Laclavère, G 182
Lada-Mocarski, Valerian: *Bibliography of books on Alaska* ... 193
Laffitte, Pierre, *ed*: *Metallogenic map of Europe* 848
Lafreri, Antonio 754
Laing, Muriel: *A bibliography of British Columbia* ... 194
Lakes 1119
Lalande, Joseph Jérome le Français: globes 567
Lambert, A M: *The making of the Dutch landscape* 890
Lana, Gabriella *et al*: *Glossary of geographical names in six languages* 568
Lancashire 505, 552, 576, 1008
The land and people of Peru 601
Land and water, Patrick Thornhill 755
Land classification 756

Land economics 1276
Land evaluation, ed G A Stewart 756
Land from the air ... 757
The land of Britain: its use and misuse 761
The land of Britain (First Land Use Survey) 761
  see also Second Land Use Survey
The land of Britain and how it is used 761
Land of the 500 million ... 758, 804
Land research series (CSIRO) 339
Land use and resources ... 460, 657, 759
Land Use Survey handbook 761
Land Utilisation Survey of Great Britain 761, 1118, 1255
Land Utilisation Survey of Zetland 916
Land utilisations 33, 335, 378, 390, 429, 451, 501, 505, 527, 621, 631, 749, 760, 776, 777, 1243, 1255
Landsberg, H E *et al*: *World map of climatology* 129, 1267
Landforms 444, 482, 488, 554
Landforms and life ..., C C Carter 762
Landscape 763
Landscape studies ... 764
Landscapes 43, 462, 469, 472, 476, 496, 503, 529, 564, 744, 762, 778, 885, 1273
'Landscapes of Britain' series 762, 807
Landström, Björn: *Columbus* ... 334
Langnas, I A: *Dictionary of discoveries* 384
Langrens, Van (or Florentius) 567
Laos 299, 419
  bibliography 184
  maps 132
The large-scale county maps of the British Isles ... 765
Large English atlas 1042
The large Magellanic cloud 1084
Larousse, publishers 519
Larousse agricole ... 766
Larousse encyclopedia of the earth 767

Larousse encyclopedia of world geography 767, 768
Atlas générale ... 98
Atlas internationale ... 101
Late Tudor and early Stuart geography 769, 1150
Latin America 152, 163, 216, 244, 411, 423, 448, 512, 715, 734, 749, 959, 1022, 1249, 1262, 1266, 1266, 1272
  bibliography 182, 184, 235, 584, 600, 601
  maps 202, 281, 615, 618, 892, 918, 934
Latin America ..., J P Cole 249
Latin America ... (bibliography) 770
Latin American: economic bulletin 406
Latin American and the Caribbean ... 602
The Latin American markets ... 771
Latin American Publications Fund 772
Latin American Regional Conference on tropical mountains 484
Latin American research review 184
Latin American statistical bulletin 406
Latorre, Edwardo Acevedo, *ed*: *Atlas de Colombia* 90
Latter, Dr, on volcanoes 54
Laughton, John *et al*: *Geographical interpretation of historical sources* 505
Lauwerys, J A: *Man's impact on nature* 885
Laval Université, Quebec, Institut de Géographie 250
Law, B C, *ed*: *Mountains and rivers of India* 865
Lawrence, G R P: *Cartographic methods* 785
Lawton, G H: *Settlement and encounter* ... 1077
Lawton, Richard: *Liverpool essays in geography* 793
Problems in modern geography 363
La Barre, Weston, *ed*: *African heritage* ... 18
Laxton: the last England open field village 773

Le Compte. E C, translator 515
le Comte, Pierre: *French ships* 1091
Le Monnier, F *et al*: *Die literatur über die Polar-Regionen de erde* ... 791
Le Play, Frédéric 481, 774, 775, 810
Le Play Society 481, 502, 774, 775
Le Play, engineer and social scientist ... 774
'The Le Play Society and field work' 775
Lead abstracts 1281
Lead Development Association 1281
Lead mining 285
Learmonth, A T A and A M: *Encyclopaedia of Australia* 420
Lebanon 107, 184, 409
Lebensraum 1009
Lebon, J H G 776
  Geography of the Middle East 776
  An introduction to human geography 710, 776
  Land use in the Sudan 776
Ledger, D C: 'Report on the hydrology of the Lower Swansea Valley' 799
Lee, R J: *English county maps* ... 284
Lee, Samuel, translation, *Travels of Ibn Battuta* 635
Leeds African studies bulletin 20
Leeds University, African studies unit 20
Leemans, W F: *Journal of the economic and social history of the Orient* 730
LeGear, Clara E: *A guide to historical cartography* ... 582
List of geographical atlases in the Library of Congress 789
Leger-Gordon, D St: *Dartmoor* 331
Legum, Colin, *ed*: *Africa* ... 602
Lehrbuch der allgemeinen geographie 366, 777
Lehrmann, Edgar and Heinze Sanke, *ed*: *Weltatlas* ... 1228
Leicester University, Department of Geography 236
The Leicestershire Association 778
Leicestershire landscapes ... 778
Leighly, John 838
Leland's itinerary in England and Wales 779

Lelarge, M: *Biblio-mer* ... 210
Lenger, Marie-Thérèse: *Bibliotheca Belgica* ... 184
Leningrad branch, All-Union Geographical Society of the USSR 40
Leprosy, maps 115
Letts, Malcolm: 'The pictures in the Hereford mappa mundi' 607
Leverhulme Trust 26
Lewis, Sir Clinton and J D Campbell: revision, *The Oxford Atlas* 933
Lewis, D C: *The classification and cataloging of maps and atlases* 284, 315, 316
Lewis, H A G: *The Times atlas of the moon* 1168
Lewis, M C: *The flora of Greenland*, English translation 466
Lewis, M G: 'Some cartographical works in the National Library' (Wales) 880
Lewis, Roy: *Sierra Leone* 349
Lexicon geographicum 214
Lexique anglaise-français des termes appartenant aux techniques en usage à l'Institut Géographique National 654
Leyden Rijksherbarium 219
Libault, A: *Histoire de la cartographie* 562
Liberia, bibliography 788
Libraries 40, 45, 56, 68, 184, 196, 217, 221, 222, 224, 236, 240, 245, 254, 269, 279, 284, 285, 299, 315, 334, 338, 344, 358, 381, 382, 383, 390, 432, 465, 496, 549, 550, 561, 708, 783, 888, 1044, 1045, 1046, 1058, 1063, 1152, 1192, 1193, 1197, 1240, 1281
The library atlas (Philip) 38, 780, 961
Library catalogue of the School of Oriental and African Studies 781
Library classification and cataloging of geographical materials 315
'The library classification of geography', R T Porter 315
Library guide for Brazilian studies 782
Library of Congress 55, 56, 184, 217 245, 260, 269, 360, 582, 694, 783, 789, 790, 906, 975, 1189

Libro dei globi 567
Libya, maps 107
Lieth, H: *Climatological atlas of the world* 129
Life of Sir Clements R Markham 836
Lighthouse of the skies ... 784, 1084
Lim, Beda: *Malaya, a background bibliography* 184
Lim Wong Pui Huen: 'Current Malay serials' 184
Limits of land settlement ... 225
Lincolnshire: Board of Agriculture report 220
Lindroth, C H 1149
Lindzen, R S: *Atmospheric tides* ... 150
Lines, J A: *Bibliografia aborigen de Costa Rica* 828
Linge, G J R and R M Frazer: *Atlas of New Zealand geography* 130
Linmap system 785
Linton, David Leslie 786
  distribution maps 933
  Sheffield and its region ... 234
  Structure, surface and drainage in South-east England 657, 786
Lippincott geography series 538
  see also Columbia-Lippincott
List, Robert J: *Smithsonian meteorological tables* 1084
A list of American doctoral dissertations on Africa 787
List of English men-of-war ... 1091
List of English naval captains ... 1091
List of French doctoral dissertations on Africa ... 788
List of geographical atlases in the Library of Congress 789
A list of maps of North America in the Library of Congress 790
Lista de mapas parciales o totales de Costa Rica 828
Lister, Raymond: *Antique maps and their cartographers* 62, 624
  How to identify old maps and globes ... 567, 624
Die literatur über die Polar Regionen der erde ... 791
Lithofacies maps ... 792
The little Oxford atlas 939

Liverpool 363, 505, 809
  marketing map 491
  University, Department of Geography 30, 793
  Tidal Institute and Observatory 54
  Liverpool essays in geography . . . 460, 793
'Livestock husbandry on Dartmoor' 335
Livingstone, maps 450
Lloyd, D M, *ed*: *Reader's guide to Scotland* 1011
Lloyd's Register of Shipping 659
Local geography in towns 1204
Locational analysis in human geography 794
Lock Muriel: *Modern maps and atlases* . . . 269
Lockwood, S B, *comp*: *Nigeria* . . . 906
Lomnitz, C: *Earthquakes atlas of world seismicity* 402
London 167, 353, 423, 541, 586, 633, 809, 1022, 1246
  Geological Society 549
  Institute of Directors 391
  Institute of Education 598
  maps 128, 164, 240, 270, 274, 491
  Northwestern Polytechnic, Geography Department 577
London, Oswald Hull 807
London Geographical Institute (Philip) 961
London Institution 1058
London map directory ('Geographia') 491
London Press Exchange 491
London School of Economics and Political Science 128, 494, 554, 726, 981, 1243
London Traffic Survey 128
London 2000 795
London University, Institute of Commonwealth Studies 20
  Institute of Education 541, 1181
  School of Oriental and African Studies 20, 76, 356
London University, LSE and PS, Anthropological and Geographical Research Division 128
London essays in geography . . . 726
Long, M, *ed*: *Handbook for geography teachers* 598

Longmans, Green and Company, Limited 761, 774, 793, 796, 799, 839, 885, 973, 974, 976, 1056, 1099, 1273
  Geographies for advanced studies 512
  A glossary of geographical terms 569
  Longman's dictionary of geography 796, 797, 1118
  Longman's geography paperbacks
  Longman's loops: geography 796
Longsdon, James 1158
Lorimer, John Gordon: *Gazetteer of the Persian Gulf* . . . 480
The lost villages of England 798
Lotz, J R: *Yukon bibliography* 1278
Louis, Herbert: *Allgemeine geomorphologie* 777
Louisiana State University, Coastal Studies Institute 686
Lovell, Sir Bernard 140
Lowenthal, David 813
The Lower Swansea Valley project 799
Lowry, J H: *World population and food supply* 1263
Lowther, B J and Muriel Laing: *A bibliography of British Columbia* . . . 194
Lucini, Antonio Francesco 63
Lumb, F E: *Ocean wave statistics* . . . 915
Lund studies in geography 1162
Lundqvist, Gösta 800
Lüttig, Gerd 905
Luxembourg 371, 807
  bibliography 182, 184
Lydolph, P E: 'Climates of the Soviet Union' 1267
  Geography of the USSR 1237
Lyell, Sir Charles 801
  Elements of geology 801
  Principles of geology 801
  The student's elements of geology 801
Lyell Club 801
Lynam, Edward 596, 638
  The first engraved atlas of the world 998
  Richard Hakluyt and his successors 596
Lyons (Lusdunum), map 1147

489

Maack, R: *Kontinentaldrift und geologie des Südatlantischen Ozeans* 366
Macao 1080
Macaulay Institute for Soil Research 924, 1096
McBain, F C A, ed: *The Oxford atlas for Nigeria* 933
McCormick, Jack, bibliography of vegetation maps of America 675
Macdonald, Teresa, ed: *Union catalogue of the Government of Ceylon* 1102
McFarlane, G C: *An Australian bibliography of agricultural economics*... 154
McGee, T G: *The Southeast Asian city* 1209
 The urbanization process in the third world 1208
McGill University, Library of Arctic Institute of North America 68
McGinnies, William G and Bran J Goldman: *Arid lands in perspective*... 72
 et al, ed: *Deserts of the world*... 376
McGraw-Hill Publishing Company Limited 539, 711, 802, 803, 804
 Series in geography 758, 804
McGraw-Hill Encyclopedia of Russia and the Soviet Union 802
The McGraw-Hill illustrated world geography 804
McGraw-Hill international atlas 578, 803
McGregor, D R: 'Geographical globes' 567
Machatschek, F: *Geomorphologie* 554
Machwe, V: *Documentation on Asia* 396
MacKaye, Benton: *From geography to geotechnics* 476
Mackenzie Delta bibliography 805
Mackinder, Sir Halford John 458, 606, 726, 746, 806, 943, 1021, 1203
 Britain and the British seas 229, 806
 'Modern geography: German and English' 806
 'The geographical pivot of history' 806

Mackinder, Sir Halford John—(*contd*)
 'The scope and methods of geography' 806
 'Mackinder's democratic ideals...' 806
McLintock, A H, ed: *An encyclopaedia of New Zealand* 425
Macmillan, George A 1231
Macmillan and Company, Limited, London 555, 626, 756, 807, 1226
 Educational series 807
 Topicards 1171
McMorrin, Ian: *World atlas of mountaineering* 1244
Madagascar 543
 bibliography 184, 788
 see also Malagasy Republic
Madingley lectures 477, 856
Madeira Islands, discovery 984
Madrid, Instituto Geográfico y Catastre 103
 maps 270
Magazines *see* Periodicals
Magellan 212
Maggs, K R A: *Land use survey handbook* 761
Mainsprings of civilisation 632
Mair, Volkmar: 'Strassenkarten aus Mairs Geographischen Verlag' 808
Mair Geographical Publishing House, Stuttgart 164, 377, 808
La Maison des Sciences de l'Homme, Paris 261
Major, R H: *The bibliography of the first letter of Christopher Columbus*... 334
The major seaports of the United Kingdom 809
Makerere University College, Department of Geography 1187
The makers of modern geography 810
The making of the Dutch landscape 890
The making of a nation booklets 286
The making of New Zealand 1060
Malagasy Republic 16
 see also Madagascar
Malawi 286, 429
 atlas 331
Malaya 501, 983
 bibliography 201, 282

Malaya—(*contd*)
  maps 136
  University, Kuala Lumpur, Department of Geography 740
Malaya, a background bibliography 184
Malayan Rubber Export Registration Board 663
Malayan Rubber Fund Board 1049
Malaya Rubber Goods Manufacturers' Association 663
Malaysia 299, 419, 1020, 1105
  bibliography 184
Malclès, L N, *ed*: *Bibliographical services throughout the world* 1192
Maldive Islands, Colombo Plan 332
Mali 184
Maling, Peter Bromley: *Early charts of New Zealand* . . . 398
Malkin, B H: *The scenery, antiquities and biography of South Wales* 1057
Malmström, V H: *Norden* . . . 1210
Malta 286
Man: Research Institute of the Study of Man, Washington 266
Man, in Britain 331
Man, and environment 592
'Man and environment', bibliography 873
Man and environment (NBL) 885
Man and his work . . . 606
Man and his world 811
Man and the land . . ., G F Carter 812
Man and the land, Sir Dudley Stamp 331
Man and nature . . ., George Perkins Marsh 813, 837
Man and Society in Iran 814
Man and wildlife, C A W Guggisberg 815, 885
Man is not lost . . . 876
Man, nature and history 885
Manchester 809
Manchester and its region . . . 234
Manchester University, Royal Meteorological Society Centre 1045
Manitoba historical atlas . . . 816
Mankind, racial distribution 97, 104
Manley, Gordon: *Climate and the British scene* 331

Man's impact on the global environment . . . 430
Man's impact on nature, J A Lauwerys 885
Man's role in changing the face of the earth 817
Manson, D C M: *Native beetles* (New Zealand) 1016
  Native butterflies and moths (New Zealand) 1016
Manual of map classification and cataloguing prepared for use in the Directorate of Military Survey 315
Manual of map reproduction techniques 818
Manufacture 539, 1171
Maoris 398, 981, 1016
  bibliography 200, 206
Map book of the Benelux countries 807
'Map cataloging', B M Woods 284
Map collections in the United States and Canada . . . 269, 819, 1114
Map Collectors' Circle 160, 353, 386, 820, 990
Map collectors series 820
Map Curators' Group, BCA 230
Map making . . ., Frank Debenham 368
Map of Hispanic America 600, 920
Map projections, A R Hinks 712
Mapping 26, 106, 274, 386, 445, 446, 453, 559, 631, 785, 989, 1114, 1117, 1145
'The mapping of population' 1264
Maps 10, 33, 85, 106, 109, 110, 112, 113, 117, 133, 160, 231, 240, 242, 248, 262, 267, 268, 274, 276, 281, 283, 297, 315, 319, 322, 323, 330, 345, 374, 378, 438, 449, 450, 459, 469, 491, 496, 507, 529, 559, 597, 609, 624, 654, 675, 676, 707, 709, 762, 765, 790, 800, 808, 816, 818, 843, 846, 848, 909, 911, 918, 919, 934, 939, 944, 957, 962, 969, 1006, 1008, 1043, 1053, 1054, 1079, 1114, 1124, 1132, 1146, 1171, 1188, 1196, 1213, 1230, 1264, 1277
  bibliography 176, 184, 269, 271, 272, 352
  catalogues 222, 269, 274, 284, 315, 390, 654, 1029, 1135, 1282

491

Maps—(contd)
  collections 40, 164, 215, 222, 236, 240, 257, 269, 369, 684
  early 62, 164, 202, 273, 277, 305, 363, 369, 452, 530, 562, 567, 572, 607, 624, 639, 820, 928, 1043, 1046, 1116
  educational 244, 475
  engravers 820
  geological 549, 550, 551, 914
  historical 98, 99, 100, 118, 119, 952
  plastic relief 938
  projections 82, 610, 712
  storage and conservation 1131
  tourist 164, 262, 475, 808, 853, 1038, 1172
  trade 270, 353, 369, 989, 1120
  transparency 755
Maps: Commission of Ancient Maps 1082
Maps and air photographs 821
Maps and charts published in America before 1800 . . . 822
Maps and diagrams . . . 823
Maps and map-makers 824, 998
Maps and survey 826
Maps and their makers 633, 825
Maps, atlases and gazetteers for Asian studies . . . 76
Maps for books and theses 827
Maps of Costa Rica 828
  see also under Cartography; International Cartographic Association; Ordnance Survey; other individual names and names of countries
Marble, D F: *Spacial analysis* . . . 1112
Marchant, E C, *comp*: *Countries of Europe as seen by their geographers* 351
Marco Polo . . ., Henry H Hart 829
Maresh, T J: *Readings in economic geography* 1012
Margery, I D: *Roman roads in Britain* 1040
Margat, Jean, *comp.*: *Guide bibliographique d'hydrogéologie* 581
Marine and shipbuilding abstracts 659
'Marine charting today', V T Miscoski 830
Marine climatic atlas of the world 129

Marine engineering 659
Marine engineers review 659
Marine cartography 830
Marine cartography in Britain . . . 830
Marine geology 418
Marine geophysical researches 832
Marine meteorology and oceanography 830
Marine observer 850, 1045
Marine science 696
Marinelli, Olinto 84
The mariner's mirrour 833, 1091, 1219
'The Mariner's mirrour, 1588', G R Crone 833
Maritime history 363, 834
Maritime Press 961
A market survey . . . 1152
Marketing 655
Marketing and management . . . 835
Markets 542
Markham, A H: *Life of Sir Clements R Markham* 836
Markham, Sir Clements R 569, 836
  50 years work of the R G S 1043
  General sketch of the history of Persia 836
  Report on the Geographical Department of the India Office 836
Mars, cartography 1161
Marsh, George Perkins: *Man and nature* . . . 813, 837
Marshall, A, on Sir Grenfell Price 1077
Marshall, William: *The review and abstract of the County Reports to the Board of Agriculture* 220
Marshall's union list of higher degree theses in Australian University libraries 184
Marston, T E 1215
Martellus, Henricus 567
Martin, D C: 'The International Geophysical Year' 689
Martin, E L and Anthony Harvey: *British geological literature* 237
Martin, J E: *Greater London* . . . 167
Martin, Michael Rheta *et al*: *Encyclopaedia of Latin-American history* 423

Martin Behaim: his life and his globe 166
Martova, K B *et al*: *Atlas of the history of geographical discoveries and explorations* 125
'Mary Somerville and geography in England' 1097
Maryland University, Department of Geography 144
Masefield, Charles: *Native birds* (New Zealand) 1016
Masefield, G B: *Farming systems of the world* 457
Mason, B J: 'The role of meteorology in the national economy' 1045
Mason, Lt General K: *Geographical handbook* series 504
Mason, W S, *comp*: *Bibliotheca Hibernica* 715
The materials of demography . . . 173
The mathematical practitioners of Tudor and Stuart England 1150
'Matthew Fontaine Maury', G E R Deacon 838
Matthew Fontaine Maury . . ., F L Williams 838
Mauritania 184
Mauritius 16
  bibliography 184, 203
Maury, Matthew Fontaine 838
  Physical geography of the sea . . . 838
  Whale charts 838
  Wind and current charts 838
Mawson, Sir Douglas 839
Mawson, Paquita: *Mawson of the Antarctic* . . . 839
Mawson Institute of Antarctic Research 839
Max Planck Institute 43, 177
Max Planck Society 658
Maxwell, Robert, *ed*: *Information USSR* . . . 652
Maxwell, W G H: *Atlas of the Great Barrier Reef* 124
May, Jacques M 621
  map of diseases 115
  Studies in medical geography 115
Meak, Lidia *et al*: *Glossary of geographical names in six languages* 568

Meally, Victor *et al*: *Encyclopaedia of Ireland* 421
The Medeba map 127
Mediaeval geography . . . 607
Mediterranean 2, 39, 329, 540, 680, 812, 896, 904, 1075, 1192, 1213
Mediterranean lands . . . 904
'Megalopolis 131
Mehra, P L: *The Younghusband Expedition* 76
Meikle, H W *et al*: *Bibliography of Scotland* 1011
Meine, K H: *A history of lunar cartography* 1161
Mekong Basin, atlas 132
Melanesia 375, 945, 946
Mélanges de géographie . . . 840
Melbourne: Bureau of Meteorology 31
  Dairy Research Unit 339
  Sugar Research Unit 339
  University 2, 109
  World weather program centre 1271
Mellor, R E H: 'Transport statistics and notes' 246
  and J P Cole: 'The Soviet iron and steel industry' 246
Mémoires et documents . . . 290, 841
Memoirs, geological 551
Memoirs of hydrography . . . 842
Men and meridians . . . 843
Menard, H W 921
Mendelssohn's South African bibliography 844
Mendips 363
Mercantine marine atlas (Philip) 845
Mercator, Gerard 85, 567, 846, 928, 1161
  Atlas . . . 846
  Geographia 846
  projection 63, 845, 846
Mercator, Rumold 754, 846
Mercator: a monograph on the lettering of maps . . . 846
Meridian Publishing Company 79, 174, 178, 218, 224, 252, 520, 574, 791, 846, 847, 1155
Merseyside 809
  marketing map 491
Mesopotamia 723
Metal Bulletin Limited, London 647
Metallogenic map of Europe (Unesco) 848, 1192

493

Metals 478, 1281
Meteorite collections 392
Meteorite research ... 849
Meteorite research 392, 849
Meteoritical Society 849
Meteorological abstracts and bibliography 3
Meteorological glossary 1045
Meteorological magazine 850, 1045
Meteorological Office 242, 322, 586, 850, 1045
Meteorology 31, 34, 129, 164, 226, 317, 319, 321, 325, 418, 494, 569, 606, 706, 733, 747, 750, 830, 838, 966, 1045, 1084, 1143, 1229, 1258, 1267, 1271, 1272
  instruments 275
Meteorology for mariners 1045
Methuen and Company Limited, London 131, 149, 151, 429, 435, 447, 477, 537, 616, 622, 755, 768, 823, 850, 854, 856, 927, 966, 1093, 1106, 1185, 1223
Mexico 772, 1272
  bibliography 184
  maps 147, 1006
  University 184
Mexico . . ., Hans-Günter Gierloff-Emden 366
Mexican atlas ... 851
Meyer, Alfred H and John H Strietelmeier: *Geography in world society* 538
Meyer, F G and E H Walker, ed: *Flora of Japan* 1084
Meyers neuer geographischer handatlas 852
Meynen, E 315, 522
  'Carte internationale du monde au millionième' 695
  Orbis geographicus 923
  et al: *Atlas Ostliches Mitteleuropa* 145
Michaelis, A R, on space 361
Michelin, André 853
Michelin Tyre Company 853
Michigan University, William L Clements Library 1132
Micro-climatology 317
Micronesia 375, 945, 946
Midlands, England 491, 499, 552, 655, 741, 749

Middle East 2, 39, 152, 410, 467, 512, 602, 705, 776, 959, 1022, 1209, 1261
  bibliography 184, 356
  maps 107, 126, 277, 907, 1053, 1243
The Middle East . . ., W B Fisher 854
The Middle East and North Africa (Oxford economic atlas) 934
Middlesex, England 274, 1008
Mikesell, Marvin W 672
Mikesell, Martin W 672
Mill, H R 496, 604
The record of the Royal Geographical Society 1043
Millar, J H and A W Reed: *Roads in New Zealand* 1016
Millares, J H and A Carillo: *Atlas Porrua de la Republica Mexicana* 147
Miller, Austin 460
Miller, David: *Native insects* (New Zealand) 1016
Miller, E S et al: *A question of place* ... 1003
Miller, F M: *The physical basis of geography* 961
Miller, R and J W Watson, ed: *Geographical essays in memory of Alan G Ogilvie* 500, 920
A million years of man 855
Millman, P M, ed: *Meteorite research* ... 849
Millot, H, bibliography of Nepal 181, 184
Mills, D G: *Teaching aids on Australia and New Zealand* 1152
Millward, Ray: *Scandinavian lands* 1055
  and Adrian Robinson, ed: 'Landscapes of Britain' series 762
Mineral statistics 551
Mineralogy 400, 1142
Minerals 144, 330, 478, 647, 931, 1012, 1274
  United States 1198
The mini pocket atlas (Bartholomew) 164
Mining districts, Parliamentary papers 715
Ministerio de Marina, Argentina 70
Ministry of Housing and Local Government 866
  Urban Planning Directorate 785

Ministry of Local Government and Planning 866
Ministry of Housing and Local Government, Computation Research and Development 785
Minshull, Roger: *The changing nature of geography* 296
Human geography from the air 626
Regional geography, theory and practice 1023
Minto, C S: *How to find out in geography* . . . 623
Mirror of the world 350
Miscoski, V T: 'Marine charting today' 830
Mississippi 873
Mocarski *see* Lada-Mocarski
Model, Fritz: *Geophysikalische bibliographie von Nord- und Ostsee* 556
Models 269, 495, 794, 966, 1093, 1152
Models in geography, R J Chorley and Peter Haggett 850, 856, 966
Modern Asian studies 76
Modern Egypt . . . 857
The modern encyclopaedia of Australia and New Zealand 858
Modern France . . . 249, 859
Modern geographers . . . 860
'Modern geography: German and English' 806
Modern geography series (UTP) 1202
'The modern index to Richard Hakluyt's *Principall navigations*' 595
Modern maps and atlases . . . 269
Modern meteorology and climatology 1045
The modern Netherlands 890
Modernization of traditional societies series 1209
Moir, A L: 'The world map in Hereford Cathedral' 607
Moll, H, maps 620
Moluccas 942
Monaco, maps 100
Monatwissenschaftliche literaturberichte 171
Mond nickel 697
Il mondo attuale 41
Money, D C: *Australia and New Zealand patterns of settlement* . . . 627

Mongolia 868, 914
Monkhouse, F J: *A dictionary of geography* 385
Western Europe 796
and H R Wilkinson: *Maps and diagrams* . . . 823
Monsoon Asia 738
Les montagnes de la terre 1244
Monteiro, Palmyra V M: *A catalogue of Latin American flat maps* 281
Monthly bulletin of agricultural economics and statistics (FAO) 468
Montreal Geographical Society 1034
Monumenta cartographica Africae et Aegypti 789
Monuments, industrial 644
The Moon 140, 150, 1161, 1168
The Moon, Zdenek Kopal 861
The Moon: an international journal of lunar studies 861
Moore, Patrick: *Atlas of the universe* 140
Moore, W G: *Dictionary of geography* . . . 385
Moraru, Tiberiu *et al*: *The geography of Romania* 546
Morden, Robert, map engraver 231
Moreland, J N: *Native sea fishes* (New Zealand) 1016
Moretus *see* Plantin-Moretus
Morgan, David: 'Discuss the economic geography of New Zealand' 902
Morgan, F W: *Ports and harbours* 633
Morison, S E: *The European discovery of America* . . . 1215
Morocco 96, 126, 184, 788
Morrell, W P: *The gold rushes* 972
Morris, J A: *Land from the air* 757
The new contour dictionary 893
Morskoy atlas 862
Mortensen, H *et al*: *Atlas Östliches Mitteleuropa* 145
Moruzi, Constanta: *Atlas botanic* 88
Moscow 1246
Branch of the All-Union Geographical Society of the USSR 40
Chief Directorate of Geodesy and Cartography 102

World Weather Program centre 1271
Moslems: calendar 126
  contributions to geographical thought 508
The mountain geologist 863
The mountain world 864
Mountains 168, 230, 484, 503, 1119, 1244
Mountains and rivers of India 856
Mountford, E G, translator 506
Mozambique 23, 184
'Mr Hakluyt, scholar at Oxford' 595
Muir, historical atlases 961
Mukerji, B N 460
Muller, Frederik: *Catalogue of books, maps, plates on America* . . . 280
Muller, H R, ed: *Berge der welt* 168
Mumford, Ian and Peter K Clark: 'Engraved OS one-inch maps . . .' 924
Mumford, Lewis 476, 813, 817
The city in history 311, 355
The culture of cities 311, 355
The story of Utopias . . . 355
Technics and civilization 355
Munn, R E, ed: *Boundary layer meteorology* 226
Münster, Sebastian 350
Muris, Oswald and Gert Saarmann: *Der globus im wandel der zeiten* . . . 567
Murphy, R E: *The American city* . . . 44
Murray, Sir John 292
  Report on the scientific results of the voyage of HMS Challenger . . . 292
Murray's handbook for Devon and Cornwall 363
Murray's handbook for Scotland 363
Musée de l'Homme, Paris 180
Museums directory of the United States and Canada 1084
Museums of Natural History Committee 392
Musiker, Reuben: 'The bibliographical scene in South Africa' 665
  Guide to South African reference books 590

Musiker, Reuben—(*contd*)
  South African bibliography 590, 1100
Myers, Lloyd E: *Water supplies for arid regions* 1223

Naples: Royal Museum 567
  University, Oriental Institute 39
Natick: Earth Sciences Laboratory 73
National Agricultural Library 188
National atlas of Britain 866
The national atlas of Canada 867
National atlas of China 868
National atlas of disease mortality in the United Kingdom 869
National atlas of Hungary 870
National atlas of India 871
National atlas of the USA 872, 1005
'The national atlas of the United States of America', A C Gerlach 139
National Book League, London 770, 797, 873, 1011
National Centre for Scientific Research, Toulouse 11
National Council for Geographic Education 731
National Environmental Teach-in 430
National Geographic atlas of the world 874, 875
National geographic magazine 874, 875
'The National Geographic Society and its magazine', G H Grosvenor 875
The National Geographical Society 875
National grid system (UK) 1078
National Institute of Oceanography 293, 370, 877
National Library of New Zealand 878
National Library of Scotland, Map Room 879, 1240
National Library of Wales 880
National Maritime Museum Library 842, 876, 1091
National Oceanographic Council 877
National Oceanographic Data Center, Washington 881
National Parks 815, 1078

National parks, Vaughan Cornish 348
National Parks Commission 352, 882
   see also Countryside Commission
National Record of Industrial Monuments 644
National Research Council, Canada: Photogrammetric Research Section 446
National Research Council, Associate Committee on Geodesy and Geophysics 401
National Research Council of the National Academy of Sciences 883
National Sciences Foundation, Washington 649, 675
Office of Antarctic Programs 56, 58
The National Trust 885
National Trust for Scotland 1078
The National Urban Coalition, Washington 310
National Wildlife Exhibition 440
'Nationalities of the USSR in 1959' 246
Native and introduced freshwater fishes (New Zealand) 1016)
Native beetles (New Zealand) 1016
Native birds (New Zealand) 1016
Native butterflies and moths (New Zealand) 1016
Native crabs (New Zealand) 1016
Native ferns (New Zealand) 1016
Native insects (New Zealand) 1016
Native rock (New Zealand) 1016
Native sea fishes (New Zealand) 1016
Native sea-stars (New Zealand) 1016
Native shells (New Zealand) 1016
Native trees (New Zealand) 1016
Natural history, bibliography 1088
A natural history of man in Britain 331, 465
Natural resources, research 691, 1095, 1192
Natural resources research series 1095, 1192
Natural resources—a selection of bibliographies 885
The Natural Rubber Producers' Research Association, London 1049
Nature, conservation 705
   see also under Conservation
Nature and man 885

Nature and resources 691, 884
Nature Conservancy 230, 352, 431, 576, 873, 882, 885, 1078
Nature conservation in Britain 331, 885
Nature in focus 886
The nature of geography . . . 887
Nature Reserves Committee 440
Nature trails in Northern Ireland 885
Nature's network 885
Natusch, Sheila: *Native rock* (New Zealand) 1016
Nautical almanac 876
Nautilus 1234
Navalani, K: *Indian periodicals* . . . 184
'The navigating manual of Columbus' 334
Navigation 394, 603, 634, 769, 876, 981, 1150, 1229
Neal, J A: *comp: Reference guide for travellers* 1018
Near East 107, 126, 228, 277
Neath, River: map 10
Needham, Joseph 603
Nel, Professor A 6
Nelson and Sons Limited, London 112, 462, 488, 500, 530, 612, 924
Atlas of European birds 117
Atlas of evolution 119
Atlas of world history 143
Nepal, bibliography 181, 184
Netherlands 5, 62, 148, 371, 708, 807, 890, 1091, 1104
   bibliography 174, 182, 184, 269
   globe-making 567
   maps 85, 148, 369, 469
Nederlands Historisch Scheepvaartmuseum, Amsterdam 888
Netherlands India . . . 889
Netherlands journal of economic and social geography 890, 1167
Network analysis in geography 891
Neubauer, A: *La géographie du Talmud* 847
Neudrucke aus dem geographischen jahrbuch 1161
Der Neue Brockhaus 579
Neundörfer, Ludwig, *ed: Atlas of social and economic regions of Europe* 135
Neva, River 873

497

Neville-Rolfe, E: *Economic aspects of agricultural development in Africa* 655
New and current English atlas 274
New British atlas 274
New Caledonia 180, 1106
The New Cambridge modern history atlas 892
The new contour dictionary 893
'New economic regions of the USSR' 246
New England 167
New England: a study in industrial adjustment 167
New geographical literature and maps 269, 894, 1043
New Guinea 155, 299, 942, 1010, 1106
New Hebrides 180, 1106
The new Israel atlas . . . 895
The new large shining sea-torch 896
New lives, new landscapes 885
The new naturalist series . . . (Collins) 331, 882, 1260
The New Product Centre, London 897
New publications in the Dag Hammerskjold Library 1192
New school atlas 961
A new Soviet heartland 1210
New towns of the Middle Ages . . . 898
New United Nations publications 1192
The new world . . . 225, 899
New York 1246
  Copper and Brass Research Association 561
  Public Library 857, 900
  University 52, 76, 207
New Zealand 152, 155, 332, 359, 425, 454, 499, 512, 602, 708, 715, 858, 932, 937, 981, 1016, 1021, 1060, 1104, 1106, 1140, 1152, 1202, 1243, 1244, 1267, 1268
  bibliography 184, 200, 204, 588, 878
  maps 130, 374, 375, 398, 934, 937, 939
  See also under Oceania
New Zealand, K B Cumberland and J S Whitelaw 1273
New Zealand Antarctic Society 55, 61
New Zealand contemporary dictionary 901
New Zealand geographer 902, 1233
New Zealand Geographical Society 902, 903
New Zealand Hydrological Society 733
New Zealand journal of educational studies 1233
New Zealand Library Association 184, 204
New Zealand tropical geographies series 1233
Newberry Library, Chicago 1090, 1156
Newbigin, Marion Isabel 904
  Plant and animal geography 973
  Southern Europe . . . 850
Newbigin Prize 904
Newbury, C H, on the Pacific Islands 152
Newcastle 809
Newman, R J P: *Fieldwork using questionnaires and population data* 1152
Newsletter, American Geographical Society of New York 45
Newsletter, Athens Center of Ekistics 83
Newsletters on stratigraphy 905
Newson, D W: *The general bathymetric chart of the oceans* . . . 690
Newton, A P: *The European nations in the West Indies* 972
Nice, maps 100
Nichols, David, ed: *Taxonomy and geography* 1149
Nicholson, Max: *The environmental evolution* 431, 885
Nickel bulletin 697
Nielsen, Niels, ed: *Atlas of Denmark* 114
Nielsen *See* Bekker-Nielsen
Nigeria 27, 184, 289, 429, 501, 735, 950, 992, 1020
  bibliography 184, 906
  maps 23, 933
Nigeria: a guide to official publications 906
Nigerian library resources in science and technology 184
Nigerian geographical journal 3

498

Nigerian official publications . . . 184
Nigerian theses 184
Nile 107, 450, 873
De Nieuwe Groote Ligtende Zeefakkel 1161
Nippon 719
See also Japan
Nix, John: *Farm management pocket handbook* 1276
Noble, William A: *Bibliography of Kerala State, India* 199
Noh, Toshio: *Japanese geography . . .* 719
Nomenclature, geographical 956, 978
Noranda lectures 811
Norden, John 907, 919, 1116
Speculum Britanniae 907
Norden: crossroads of destiny 1210
Nordenskiöld, A R: *Facsimile atlas to the early history of cartography . . .* 452, 998
Nordic Council 391
Nordic hydrology 908
Norfolk, maps 305, 1008
Norge 909
Norges Geografiche Oppmåling 262, 1038
Nörström, Göran: *World shipping . . .* 1266
North, F J: *Sir Charles Lyell . . .* 801
North America 131, 244
North America, A W Coysh and M E Tomlinson 1202
NATO 141
North Borneo 349
North Holland Publishing Company 733
North Sea 877
Northcote, K H: handbooks accompanying *An atlas of Australian soils* 109
Soil classification 109
Northern geographical essays . . . 460, 910
Northernmost Labrador mapped from the air 911
The northward course of empire 1127
Northwest Passage 304, 1234
Northwest to fortune . . . 1127
Northwestern University 665, 792, 1145

Norway 438, 544, 708, 970, 1055, 1056
bibliography 182, 184
maps 123, 164, 262, 469, 1038
National Committee of the International Hydrological Decade 908
Norway exports 912
Norwegian Geographical Institute 909
Norwegian National Committee of Geography 544
Norwegian Polar Institute 1143
See also *Norge*
Notes africains 653
Nottingham 773
University, Department of Geography 124, 246, 247, 404, 461, 499, 502
Nottingham and its region . . . 234
Nouvel atlas général (Bordas) 451
Nouvelle carte de France 268
Nouvelle géographie universelle 1015
Noyce, Wilfrid and Ian McMorrin: *World atlas of mountaineering* 1244
Nuevo atlas geográfico de la Argentina 913
Nuffield Foundation 799
Nunn, G Raymond: *South and southeast Asia . . .* 1101
Nuovo atlante geografico moderno 717
Nyasaland 184
Nyasaland, Frank Debenham 349
Nystrom, A J, ed: *The journal of geography* 731

Oberhammer, E: new edition, *Politische geographie* (Ratzel) 1009
Obruchev, V A 914
Eastern Mongolia 914
Field geology 914
Frontier Dzungaria 914
The geological map of the Lena gold-bearing region 914
Geologie von Sibirien 914
Gold seekers in the desert 914
The history of the geological exploration of Siberia 914
Observations on the probability of reaching the South Pole 1218
The observer atlas of world affairs 961

499

Observer's handbook 1045
Obst, Eric: *Allgemeine wirtschafts- und verkehrsgeographie* 777
Ocean wave statistics . . . 915
Oceania 518, 519, 619, 781, 946, 1163, 1249, 1266, 1272
Oceanography 59, 292, 293, 370, 418, 445, 494, 664, 667, 706, 830, 838, 842, 862, 866, 877, 881, 948, 994, 1028, 1065, 1076, 1084, 1113, 1143, 1192, 1196, 1229, 1238
  Scripps Institution 279
Oceanography, ed: Mary Sears 1065
Oceanography for meteorologists 1143
Oceanography Research Unit, Cronulla 339
Oceans 1119, 1267
The oceans . . ., H U Sverdrup 1143
O'Dell, Andrew Charles 916
  ' Geography and planning ' 527
  The historical geography of the Shetland Islands 916
  Railways and geography 916
  The Scandinavian world 796, 916, 1056
  and Kenneth Walton: *The Highlands and islands of Scotland* 916
Odell, Peter R: *An economic geography of oil* 167
Odysseus 603
Oehme, Ruthardt 350
Oehser, Paul H, *ed*: *Abstracts,* National Geographical Society 875
Office de la Recherche Scientifique et Technique Outre-Mer 917
OSTI 185
Official map publications . . . 269, 918
Official publications of Sierra Leone and Gambia 184
Ogilby, John: *Britannia* . . . 919
Ogilvie, A G 489, 500, 920
  Great Britain . . . 920
  ' The mapping of population . . . ' 1264
Ogunsheye, F A: *Nigerian library resources in science and technology* 184
Ohio State University: Department of Geography 495
  Institute of Geodesy, Photogrammetry and Cartography 269

Ohwi, Jisaburo: *Flora of Japan* 1084
Oil 107, 167, 631, 959, 1232
The oil and petroleum year book 960
Ojakangas, Richard 120
Old decorative maps and charts 369
Olsen, T D: *Bibliography of Old Norse-Icelandic studies* 184
On the connexion of the physical sciences 1097
On the structure and distribution of coral reefs . . . 921
One hundred foreign maps of Australia 1773-1887 160
1 : 1M Map of the world 609, 1256
O'Neill, T P: *British Parliamentary Papers* 715
Ontario economic atlas 405
Ontario Institute of Chartered Cartographers 269
Ontario University, Department of Geography 405
Opportunities in oceanography 1084
Orb 922
Orbis geographicus 522, 685, 686, 923, 1200, 1248
Orbiter surveys 1168
Ordnance Survey maps 240, 276, 496, 586, 712, 764, 821, 893, 924, 956, 1078, 1120
  map extracts 1152
  Map of Roman Britain 1147
  One-inch reprint 363
  Publication reports 269
  ' Ordnance Survey at home ' 924
  Ordnance Survey Gazetteer of Great Britain 924
  ' The Ordnance Survey of Northern Ireland . . .' 924
O'Reilly, Patrick and Edouard Reitman: *Bibliographie de Tahiti* . . . 180
Organisation for Economic Co-operation and Development 925
Organization of African Unity . . ., University of Dakar 673
Orient *See* Asia; Far East
Oriental and Asian bibliography . . . 926
Orientalists, Congress of 2
The origin of continents and oceans 927

The origin of Ptolemy's Geographia 998
The original writings and correspondence of the two Richard Hakluyts 595
'The origins of continents and oceans...' 927
Orme, A R: *Ireland* 1273
Orni, Efraim and Elisha Efrat: *Géographie d'Israel* 513
Ortelius, Abraham 846, 928, 1115, 1161
Theatrum orbis terrarum 164, 618, 928
Orthophotomaps 446
Orvig, S: 'Climates of the Polar regions' 1267
Osborne, A K and M J Wolstenholme, *comp: An encyclopaedia of the iron and steel industry* 422
Osborne, Charles, *ed: Australia, New Zealand and the South Pacific* 152
Osborne, R H *et al, ed: Geographical essays in honour of K C Edwards* 499
Osley, A S, on Mercator 846
Oslo: Norwegian Water Resources and Electricity Board 123
Österreichische bibliographie 184
Österreichischer mittelschulatlas 451
Ostrem, G and T Ziegler: *Atlas of glaciers in South Norway* 123
Ottley, George, *comp: A bibliography of British railway history* 195
Ottoman Empire, maps 126
Our developing world 929
Our evolving civilization 1151
Our wandering continents... 927, 930
Our world from the air... 592, 817
Outlook Tower Association, Edinburgh 306
OECD, road documentation scheme 699
Overseas Geological Surveys 931
Overseas geology and mineral resources 931
Overseas railways 932
Owen, M M: *Larousse encyclopedia of the earth* 767
Owen, Peter: revised edition, *Dictionary of discoveries* 384

Oxford 620, 747
The Oxford atlas 933
The Oxford atlas for Nigeria 933
The Oxford book of food plants 942
The Oxford books 942
The Oxford card catalogue of world forestry literature 943
Oxford: Clarendon Press 314
The Oxford economic atlas of the world 934, 939
Oxford economic papers 935
Oxford forestry classification 1027
The Oxford home atlas of the world 939
The Oxford junior atlas 935
Oxford junior encyclopaedia 937
Oxford New Zealand encyclopaedia 937
Oxford plastic relief maps 938
Oxford Preservation Trust 348
The Oxford school atlas 939
Oxford studies in African affairs series 314
Oxford symposium on 'Experimental cartography' 445
Oxford system of automatic cartography 940
Oxford: sub-centre for work on *Geographical handbook* series 504
Oxford University 525
  Agricultural Economics Research Institute 28
  Exploration Club 941
  Institute of Agricultural Economics 655
  Institute of Commonwealth Studies 20
  Press 942
  School of Geography 315, 746, 806, 843
Oxfordshire, county map 274

Pacific 31, 47, 152, 212, 349, 454, 456, 660, 718, 736, 972, 1007, 1022, 1069, 1104, 1106, 1237, 1262
  bibliography 47, 211, 314, 345, 945, 946
  maps 124, 944
The Pacific Basin... 944
A Pacific bibliography... 314, 945
Pacific island bibliography 946
Pacific islands 156, 197, 375, 981

501

Pacific Science Congress 944
See also Oceania
'Packs' 275
Padmore Research Library, Ghana 184
Padua University 41, 1243
Pageant of New Zealand 1016
Painter, G D 1215
Pakistan 78, 332, 412, 621, 708, 738, 1102, 1209
  bibliography 184, 412
  maps 939, 1136
Pakistan Institute of Development Economics 412
Palermo University, Institute of Oriental Studies 39
Palestine 127, 184, 723
Palestine and its transformation 632
Palestine Association 1043
Palestine Exploration Fund 127, 947
Palestine exploration quarterly 947
Palmer, A N: *Railways in New Zealand* 1016
  Shipping of New Zealand 1016
Palmer, R R, ed: *Atlas of world history* 1006
Palmerlee, Albert E: *Maps of Costa Rica* ... 828
Pan American Institute of Geography and History 948, 1122
Pan American Union, Conservation Section 1217
  Columbus Memorial Library 184
Panama Canal Zone, geodetic bibliography 182
Panama Collection 1138
P A N Institute of Geography (Poland) 978
Panofsky, H E: 'The African studies library of Northwestern University' 665
Panorama 761
Papua, aborigines 155
  maps 1010
Paraguay ..., George Pendle 601
Parias, L H, ed: *Histoire universelle des explorations* 384
Paris 1246
  Atlas 93, 949
  Bureau de Recherches Géologiques et Minières 581
  Comité français de Cartographie 1030

Paris—(*contd*)
  Geographical Society 1087
  La Maison des Sciences de l'Homme 261
  University, Institute de Géographie 93
Parker, W H: 'Mackinder's democratic ideals ...' 806
  The Soviet Union 1273
Parks, G B: *Richard Hakluyt and the English voyages* 595
Parks, national 882
Parry, Sir William E 8
Parsons, G J S: *Manual of map classification and cataloguing* ... 315
Pascoe, John: *Oxford New Zealand encyclopaedia* 937
The passing of tribal man in Africa 950
Pastoral farming in New Zealand 1016
Pastures 336, 339
Patagonia 1155
Patrick Geddes in India 481
Patterns of settlement ... 627
Patton, D J: *The United States and world resources* 1210
Paullin, C O: *Atlas of the historical geography of the United States* 139
Paxton, P: *A bibliography of arid lands bibliographies* 189
Paylore, Patricia: *Arid lands research institutions* ... 73
  Seventy-five years of arid-lands research at the University of Arizona 71
  on sources for arid-lands research 72
  et al: *Deserts of the world* ... 376
Payne, W A 866
Peabody Museum of Salem, Massachusetts 46
The Peak District 331
Peake, H J E: *The corridors of time* 465
Pearcy, G Etzel 1210
  World political geography 1261
Pearl trade 2
Pearson, J D: *Oriental and Asian bibliography* ... 926

Pearson, J D—(contd)
and Ruth Jones, eds: *Bibliography of Africa* 186
Peary, Robert E 1127
Pehrsson, H and H Wulf, eds: *The European bibliography* 184, 437
Peking 363
Peking review 302
Pendle, George: *Argentina* 601
The land and people of Peru 601
Paraguay . . . 601
South America, a reader's guide 770
Penck, Albrecht 695, 810, 951
'Construction of a map of the world on a scale of 1:1 million' 695
Penn, Philip: A *dictionary of natural resources* . . . 388
Pennsylvania University, Department of Regional Science of the Wharton School 737
People, W Vogt 1217
Peoples of Europe, H J Fleure 465
Perambulation of Dartmoor 362
Perez, J M, bibliography of agricultural geography, Spain 184
Pergamon general historical atlas 952
The Pergamon world atlas 952
The periglacial bulletin 953, 978
The periglacial environment 954
Periglacial Geomorphology Commission, IGU 953, 978
Periodicals 2, 4, 17, 36, 37, 41, 43, 45, 50, 58, 64, 65, 69, 76, 77, 80, 81, 86, 157, 158, 159, 188, 219, 221, 223, 226, 228, 234, 236, 240, 247, 250, 251, 253, 258, 259, 269, 271, 285, 290, 291, 302, 310, 332, 339, 346, 358, 363, 366, 370, 400, 403, 404, 406, 407, 418, 430, 433, 438, 455, 456, 461, 467, 468, 483, 486, 493, 495, 496, 507, 509, 510, 523, 524, 558, 561, 563, 566, 577, 594, 636, 639, 640, 645, 646, 647, 653, 656, 657, 659, 666, 686, 690, 692, 694, 696, 697, 698, 703, 704, 708, 713, 719, 721, 722, 727, 729, 730, 731, 732, 733, 734-739, 740, 741, 742, 745, 763, 832, 834, 850, 861, 863, 864, 873, 875, 876, 884, 886, 890, 902, 904, 905, 908, 912, 922, 932, 947, 948, 953, 955, 957, 958, 960, 976, 978, 982, 1017,

Periodicals—(contd)
1031, 1034, 1035, 1036, 1039, 1041, 1045, 1046, 1049, 1050, 1063, 1064, 1074, 1085, 1090, 1091, 1094, 1104, 1109, 1111, 1118, 1154, 1156, 1158, 1159, 1166, 1167, 1174, 1175, 1177, 1179, 1183, 1184, 1187, 1189, 1192, 1204, 1207, 1220, 1223, 1227, 1232, 1251, 1254, 1259, 1280
abstracting 3
bibliography 53, 184
reprints 5
Permafrost 564, 1278
Permanent Committee on Geographical Names for British Official Use 609, 780, 956
Perpillou, A V: *Human geography* 753
Perrier, G 182
Perring, F H and S M Walters, eds: *Atlas of the British flora* 112
Persia (Strabo) 1133
See also Iran
Persian Gulf 239, 480
Perspective on the nature of geography 886
Perspectives in oceanography 664
Perthes, Justus, Gotha 521, 628, 957, 969, 1130
See also under Justus
Peru 184, 225, 601
Pesticides documentation bulletin 1193
Petermann, A H 957, 958
Petermanns geographische mitteilungen 164, 957, 958
Petro, Julius, cartographer 375
Petroleum 478
Petroleum Information Bureau, London 959
Petroleum, journal of the European oil industry 960
Pettersen, Hjalmar: *Bibliotheca Norvegica* 184
Péwé, T L, ed: *The periglacial environment* 954
Philadelphia University: 'International history of city development' project 592
Philip, George, and Son Limited, London 218, 238, 342, 501, 567, 672, 709, 780, 845, 961, 1120, 1201

503

Philip, George, and Son Limited, London—(contd)
Aldine university atlas 38
Philip's modern school atlas 961
Research Department: *Geographical digest* 497
The George Philip Group 961
Philippines 299, 419, 738, 1105
bibliography 184
maps 136, 375
Philippine cartography . . . 962
Philliott, H W: *Mediaeval geography* . . . 607
Phillips, C W: *The Fenland in Roman times* 459
Phillips, Philip Lee: *A list of maps of America in the Library of Congress* 790
and Clara Egli LeGear: *List of geographical atlases in the Library of Congress* 789
Philip's inland navigation 963
Photogrammetria 700
Photogrammetric record 964
Photogrammetric Research Section, National Research Council, Canada 446
The Photogrammetric Society, London 964
Photogrammetrie 366
Photogrammetry 269, 275, 415, 529, 654, 700
Photogrammetry: current practice in road design 965
Photographs, aerial: University of Keele collection 1200
Photography, aerial 11, 114, 132, 390, 417, 496, 531, 544, 626, 631, 757, 821, 1146, 1152
Photomaps 446
Physical and information models in geography 856, 966
Physical atlas (Bartholomew) 129, 164, 967
Physical atlas (Johnston) 725
Physical atlas of zoogeography 164, 967
The physical basis of geography 961
Physical elements of geography 804
Physical geography, R K Gresswell 576
Physical geography, Mary Somerville 1097

The physical geography of China 968
The physical geography of glaciers and glaciation 576
The physical geography of rivers and valleys 576
Physical geography of the sea and its meteorology 838
Physikalischer atlas . . . 169, 725, 969, 1146
Picardy 371
Picture atlas of the Arctic 418, 970
The picturesque atlas of Australasia 971
Piddington, Ralph, *ed*: *Kinship and geographical mobility* 748
Pielow, F A: 'Special purpose navigation charts' 634
Piggott, Mary: *Government information and the research worker* 866
Piggott, Stuart 231
Pigs 655
The Pilgrim Trust 830
Pillsbury, A F, *comp*: *Annual report,* Water Resources Center 1223
Pinchemel, Philippe: *France* . . . 473
Géographie de la France 329
Pioneer fringe 225
The pioneer histories series (Black) 972, 984
Pitcher, G M: *Bibliography of Ghana* 184
Pittsburgh University Book Centre 782
Placenames, bibliography 205
Planck, Max *see* Max Planck Institute
Planning 306, 527
Plans, in British Museum 283
Plant and animal geography 973
Plant science . . . 973
Plantation agriculture 167
Plants 88, 112, 219, 468
Le plateau central de la France . . . 329
Platt, Elizabeth 358
Aids to geographical research . . . 36, 1275
Pleistocene 564
Pleistocene geology and biology . . . 974

Pocketbook map of London (Bartholomew) 164
The poetic impression of natural scenery 348
Pohl, F J: *The Viking explorers* 1215
Pointe-Noire et la façade maritime du Congo-Brazzaville 917
Poland 493, 953, 978, 1107, 1256
 bibliography 182, 184
The Polar bibliography 975
Polar record 3, 566, 1063
Polar regions: Stefansson Collection 382
Polar research: a survey 975, 976
The Polar world 977
Polderlands 850
Polish Academy of Sciences, Institute of Geography 269, 493
Polish geographical bibliography 978
Polish geographical nomenclature of the world 978
Polish Geographical Society 493
Polish Scientific Publishers (PWN) 978, 999
Political geography, N J G Pounds 979
Political geography and the world map 961
'Political geography in the modern world' 979
Politics and geographic relationships ... 980
Politische geographie 1009
Pollock, N C: *Studies in emerging Africa* 249
Polo, Marco 277, 350
Polynesia 180, 375, 945, 946
Polynesian navigation 981
The Polynesian Society, Wellington 981
Pompidou, President 100
Popovici, Lucia and Constanta Moruzi: *Atlas botanic* 88
Population 83, 104, 173, 308, 372, 501, 505, 512, 539, 682, 995, 1217, 1263, 1264, 1277
 maps 10, 133, 869
Population Investigation Committee 981
'Population statistics', J P Cole 246
Population studies ... 982

Porter, R T: 'The library classification of geography' 315
Portolan charts ... 983
A portrait of Canada 1128
Port Arthur 363
Ports 363, 834, 1266
 plans 845, 1229
Ports and harbours 633
Portugal 184, 504, 1080
 bibliography 182, 184, 235, 269
Portuguese Africa ... 184
The Portuguese pioneers 972, 984
'Possibilisme' 1214
Posthumus, N W: *Journal of the economic and social life of the Orient* 730
Poultry and eggs in Britain ... 28 655
Pounds, N J G: *The geography of iron and steel* 633
Political geography 979
Powell, John Wesley 872
Prablu Book Service 184
Pratt, I A, comp: *Modern Egypt* ... 857
The Pre-Cambrian along the Gulf of Suez ... 985
Preece, D M: *Foundations of geography* 1202
 and H R B Wood: *Europe* 1202
A preliminary bibliography of the natural history of Iran 986
Prescott, J R V: *The geography of frontiers and boundaries* 979
The geography of state policies 163
The present state of the Empire of Morocco 18
The presentation of our scenery 348
Presses Universitaires Français, Paris 897
Pressures on Britain's land resources 988
Prestage, Edgar: *The Portuguese pioneers* 972, 984
Preston 505
Preussische Geologische Landesanstalt, Berlin 688
Price, Sir Grenfell 1077
Priestley, Sir Raymond et al: *Antarctic research* ... 60, 249
I primi esploratori dell' America 41
Prince Henry the Navigator ... 984
Principal voyages ... 769

Prince Edward Island 261
Principall navigations . . ., Richard Hakluyt 595
Principes de géographie humaine 1214
Principles of cartography 804
Principles of geology 801
Principles of political geography 979
The printed maps in the atlases of Great Britain and Ireland . . . 269 305, 989
The printed maps of Jamaica . . . 184
The printed maps of Tasmania . . . 990
Printers, of maps 624
Problèmes Africains 24
Problèmes de géographie humaine 371
Problems and trends in American geography 991
Problems in modern geography 363
Proceedings of the symposium on the granites of West Africa . . . 992
Proctor, M C F: revision, *Britain's green mantle* . . . 230
Producing a slide set with commentary for elementary fieldwork 1152
'La production cartographique de l'Institut Géographique de Agostini . . .' 717
Products: The New Products Centre 897
Professional geographer 80
Progress in geography . . . 993
Progress in oceanography 994
A prologue to population geography 995
A prospect of the most famous parts of the world 996
La protection de la grande fauna et des habitats naturels en Afrique . . . 1192
Provence, maps 100
Provinces of England . . . 458, 997
Ptolomaeus, Claudius *see* Ptolemy
Ptolemy 369, 452, 492, 1132, *Geography* 846, 998, 1161
Ptolemy's Geography . . ., W H Stahl 998
Ptolemy's Geography, H N Stevens 998

Publications of the Royal Geographical Society . . . 1043
Publishers 164, 249, 329, 331, 363, 366, 386, 409, 436, 451, 491, 534, 579, 593, 605, 624, 802, 807, 808, 820, 847, 850, 853, 961, 972, 978, 987, 1006, 1016, 1029, 1120, 1232, 1233, 1237
see also under individual names, e.g. Bartholomew
Puerto Rico University, Institute of Caribbean Studies 265
The pulse of Asia 632
Pundeff, M V, *comp*: *Bulgaria: a bibliographic guide* 184, 245
Purchas, Samuel 769, 1185
Putnam, R G *et al*: *A geography of urban places* . . . 547
Putzeys, Jaques: *Répertoire bibliographique* . . . (Belgium) 184
PWN *atlas of the world* 978, 999
PWN *universal geography* 999

Qualitative geography . . . 1000
Quantitative methods 1000
Quarries 503
Quarterly bulletin of coal statistics for Europe (UN) 1192
Quarterly bulletin of the International Association of Agricultural Librarians and Documentalists 188
Quarterly digest of urban and regional research 1031
Quaternary 637, 668
The Quaternary era . . . 1001
The Quaternary of the United States . . . 1002
Quebec 625, 1034
Queensland, maps 124
Royal Geographical Society of Australasia 1044
Quencez, G, *comp*: *Vocabularium geographicum* 1216
A question of place . . . 1003
Quinn, Alison M: 'The modern index to Richard Hakluyt's *Principall navigations*' 595
Quinn, David B: *Richard Hakluyt, Editor* . . . 595
Quirino, Carlo: *Philippine cartography* 962

Races 632, 812, 1151
The races of England and Wales 465
The races of mankind 465
Radcliffe Meteorological Station 747
Radó, Sándor 267
  National atlas of Hungary 870
Ragatz, Lowell J: *A guide for the study of the British Caribbean* 184
Ragle, E H, *ed*: The Icefield Ranges Research Project, *Scientific Results* 637
The railway encyclopaedia 1004
Railway engineering abstracts 1074
Railway research and engineering news 1074
Railways 195, 363, 932, 1024, 1074, 1179, 1265
Railways and geography 916
Railways in New Zealand 1016
Rainfall, Ceylon 501
Rainfall atlas of the Indo-Pakistan sub-continent 1136
Raisz, Erwin 1005
  et al: *Atlas of Florida* 121
  General cartography 804, 1005
  Principles of cartography 804
Raleigh, Professor Sir Walter: 'Essay on the life and work of Hakluyt' 595
Raleigh Dining Club 1043
Ramusio, Giovanni Battista, on Marco Polo 829
Rand McNally and Co, Chicago 269, 614, 672, 676, 709, 1006, 1262
Rand McNally-Hobbs *Guides* 1006
Randstad Holland 1246
Ranger surveys 1168
Rare Australiana . . . 1007
Rare county and other maps 1008
Ratsimandrava, J: *Madagascar bibliography* 184
Ratzel, Friedrich 648, 810, 1009, 1075, 1214, 1232
  Anthropogeographie 1009, 1075, 1232
  Politische geographie 1009
Raup, H F: *Essentials of geography* 804
Raveneau, L 183
Ravenhill, W L: *Exeter essays in geography* . . . 460
Ravenstein, E G: *Martin Behaim: his life and his globe* 166

Ravenstein, E G—(*contd*)
  trans. *Nouvelle géographie universelle* 1015
Raynor, Dorothy H: *The stratigraphy of the British Isles* 1134
Rayon 1157
The Readers' Digest atlas of Australia 1010
The Readers' Digest great world atlas 368, 1010
'Readers' guide to the Commonwealth' 873
'Reader's guide to Scotland' 873, 1011
Readers Union, book club 363
Reading: symposium on agroclimatological methods, 1966 34
  University, Institute of Agricultural History 69
Readings in economic geography 1012
Readings in the geography of North America . . . 1013
'A re-appraisal of the International Map of the World (1:1M) on the millionth scale' 695
Recent geographical literature . . . (now New geographical literature and maps) 1043
Recent history atlas . . . 1014
Recherches Africaines 184
Reclamation 1095
Réclus, Jacques Elisée 1015
  La terre 1015
  L'homme et la terre 1015
  Nouvelle géographie universelle 1015
Réclus, Paul *et al*: new ed *L'homme et la terre* 1015
The record atlas (Philip) 961
A record of agricultural policy 28
The record of the Royal Geographical Society . . . 1043
Recreation news 352
Red Sea 985
Reed, A W, publishers, Wellington 130, 1016, 1060
  Concise Maori encyclopedia 981
  How the Maoris came 1016
  How the Maoris lived 1016
  How the white men came 1016
  Illustrated encyclopedia of Maori life 981

Reed, A W, Publishers—*(contd)*
Nature in New Zealand series 1016
Pageant of New Zealand series 1016
Pastoral farming in New Zealand 1016
Roads in New Zealand 1016
Reed's atlas of New Zealand 1016
Reeves, A D: *The finding of Wineland the Good* . . . 1215
Referativnyi zhurnal geografiya 3, 37, 1017
Railway transport section 1074
Reference atlas—Greater London 164
Reference guide for travellers 1018
A reference guide to the literature of travel . . . 1019
Reference pamphlets series (COI) 286, 1020
Regional economic analysis in Britain . . . 1021
Regional forecasting . . . 1045
Regional geography of the world 1022
Regional geography, theory and practice 1023
A regional history of the railways of Great Britain 1024
Regional Science Research Institute 737
Regional studies 1025
Regional Studies Association 1025
Regionalisation 1021
Regionalism 1023
Régions naturelles et noms de pays 479
Regions of tomorrow . . . 547
Régions, nations, grands espaces . . . 1026
The regions of the world series 229
Register of current research (geomorphology) 238
Reid, Keith: *Nature's network* 885
Reigate, maps 491
Reinbeck, Federal Research Centre of Forestry 1252
  International Forestry Documentation Centre 1027
Reinhold one-volume encyclopedias series 1028
RV katalog 1029
Reise- und Verkehrsverlag, Stuttgart 1029

Reitman, Edouard: *Bibliographie de Tahiti* . . . 180
Religion, historical geography 177, 545
  maps 177
Rendle, B J, *ed*: *World timbers* 1268
Rennell, Major James 620
Répertoire d'établissements enseignant le cartographie 1030
A report on the agriculture and soils of Kent, Surrey and Sussex 1260
Reprints 79, 174, 178, 179, 211, 212, 218, 224, 227, 231, 252, 261, 273, 280, 301, 304, 313, 350, 363, 456, 480, 485, 492, 498, 520, 567, 574, 639, 715, 724, 779, 790, 791, 833, 846, 847, 896, 924, 942, 963, 996, 998, 1007, 1042, 1057, 1080, 1081, 1088, 1115, 1155, 1160, 1161, 1164, 1211, 1218, 1219,
Research 4, 15, 21, 28, 30, 37, 40, 42, 45, 52, 65, 69, 77, 83, 168, 172, 189, 194, 219, 238, 244, 249, 261, 266, 285, 288, 289, 290, 298, 336, 339, 344, 352, 368, 370, 376, 379, 385, 392, 437, 438, 447, 483, 495, 545, 555, 558, 561, 566, 621, 637, 639, 641, 649, 662, 668, 676, 689, 690, 704, 706, 714, 722, 727, 736, 737, 782, 832, 839, 849, 864, 875, 877, 883, 884, 885, 953, 975, 981, 993, 1002, 1027, 1031, 1033, 1050, 1062, 1063, 1090, 1091, 1095, 1104, 1113, 1158, 1180, 1194, 1206, 1223, 1252, 1258, 1275
  Antarctic 60, 61
  arid lands 71, 72, 73
Research digest 1031
Research in human geography . . . 515
Research in Japanese sources . . . 1032
Research index 1033
Research Institute for Asian Studies in Japan 288
Research Institute for the Study of Man, Washington 266
Research Institutes and researchers of Asian studies in Thailand 288
Research papers, Chicago University, Department of Geography 300

Research reports series, Aluminium Federation, London 42
Research series, American Geographical Society of New York 45
Resources, natural 11, 144, 388, 460, 478, 488, 490, 705, 759, 804, 813, 1012, 1237
Réunion 16
Revista cartografica 948
Revista geographica 948
Revue Canadienne de géographie 3, 1034
Révue de géographie de Lyon 3
Révue de géologie et de géographie 1035
Révue forestière français 1036
Rex, D F: *Climates of the free atmosphere* 1267
Reyes, J M: *Bibliografía geográfica de Bolivia* 184
Rhine 873
Rhine-Ruhr 1246
Rhine-Westphalia, maps 378
Rhineland-Palatinate, maps 378
Rhodesia 950
The Rhodesias and Nyasaland . . . 184
Rice 32
Rich, E: *History of the Hudson's Bay Company* 625
Richard Hakluyt and the English voyages 595
Richard Hakluyt and his successors 596
Richard Hakluyt, Editor . . . 595
Richards, Sir George H 8
Richards, J Howard, *ed: Atlas of Saskatchewan* 134
Richey, M W: *The geometrical seaman* . . . 553
Richter, C F: *Seismicity of the earth and associated phenomena* 1069
Riley, Denis and Anthony Young: *World vegetation* 1270
Riscoe, John, *comp: Antarctic bibliography* 56
'The rise and growth of Edinburgh' 116
Ristow, Walter R 684
— and Clara E LeGear: *A guide to historical cartography* . . . 582
Ritchie, G S: *The Admiralty chart* . . . 8, 842

Ritchie, G S—*(contd)*
'Developments in British hydrography since the days of Captain Cook' 9
'Surveyors of the oceans' 9
Ritter, Carl 168, 810, 1015, 1037
Die erdkunde im verhältnis zu natur und zur geschichte des menschen . . . 1037
Europa . . . 1037
Riverain, Jean: *Concise encyclopedia of exploration* 341
Dictionnaire des explorations 384
Rivers 503, 576, 865, 873, 1119, 1141, 1171
Rivista geografica Italiana 41
The road and tourist maps of Norway 1038
Road atlas and touring guide of Southern Africa 138
The road atlas of Europe (Bartholomew) 164
Road books 274, 919
Road international 698, 1039
Road map of Europe 1172
Road Research Laboratory 699
Road to survival 1217
Roadmaster motoring atlas of Great Britain (Bartholomew) 164
Roads 274, 518, 698, 699, 965, 1006, 1040, 1179, 1254
Roads in New Zealand 1016
Robert, W C H, *comp: Contribution to a bibliography of Australia* . . . 345
Roberts, Brian: *Illustrated ice glossary* 1063
UDC, adaptation for polar libraries 1063
Robin, G de Q: on long-range echo flights 565
Robinson, A H W: *Elements of cartography* 1237
Marine cartography in Britain . . . 417, 830
— and Helen M Wallis: 'Humboldt's map of isothermal lines . . .' 628
Robinson, Adrian, *ed:* 'Landscapes of Britain' series 762
Robinson, D J: 'The city as centre of change in modern Venezuela' 772

Robson, B T: *Urban analysis* . . . 1205
Rochdale Committee Report 809
Roche *see* Frison-Roche
Rocky Mountain Association of Geologists 863
Rodenwaldt, E, new edition, *Welt-Suechen atlas* 1230
and H Jusatz, *World map of climatology* 129
Rodger, Elizabeth M: *The large-scale county maps of the British Isles* 765
Rodgers, H B: *An atlas of North American affairs* 131
Rodier, J: *Bibliography of African hydrology* 187
Roepke, H J and T J Maresh: *Readings in economic geography* 1012
Roggeveen, nautical chartbooks 1161
'The role of meteorology in the national economy' 1045
Rolfe *see* Neville-Rolfe
Roman roads in Britain 1040
Romans, geographical knowledge 617
settlement 459
Rome, maps 270
University, Department of Geography 41
Rose, J, *ed*: *Technological inquiry* . . . 885
Rosenthal, Eric *et al*: *Encyclopaedia of Southern Africa* 426
Rosing, Kenneth A and Peter A Wood: *Character of a conurbation* . . . 297
Rothamsted Experimental Station 1260
Rothé, Professor J P 54
The seismicity of the earth, 1953-1965 1068
Row, E F, translator 515
Rowe: *Perambulation of Dartmoor* 362
Rowlands, W: *Cambrian bibliography* 252
Rowley, Virginia, M: *J Russell Smith* . . . 1083
Roxby, P M 489
Roy, General William 276, 924
Royal Afghan Embassy, London, Information Bureau 12
Royal atlas (Johnston) 725
Royal Canadian Geographical Society 1041
Royal College of Art Experimental Cartography Unit 445
Royal Danish Geographical Society 114
Royal Dutch Geographical Society 890
The Royal English atlas . . . 363, 1042
Royal Geographical Society, London 129, 164, 218, 315, 342, 450, 459, 507, 525, 532, 565, 596, 606, 609, 657, 694, 695, 746, 806, 860, 869, 956, 1008, 1043, 1131, 1147
New geographical literature . . . 894
Royal Geographical Society of Australia 1044
Royal Greenwich Observatory 586, 876
Royal Institute of International Affairs 21, 332, 601, 660
Royal Meteorological Society of London 1045
Royal Netherlands Geographical Society 148
Royal Scottish Geographical Society 164, 904, 920, 956, 1046, 1064
The Royal Society 830, 927, 1047
British National Committee for Geography 241, 1190
Royal Society for the Protection of Birds 1078
Royal Tropical Institute, Amsterdam 701, 1048, 1183
Royal United Service Institution 240
Roze, R and M Lelarge: *Biblio-mer* . . . 210
Rubber 389, 470, 663
Rubber abstracts 3
Rubber Assocation of Singapore 663
Rubber developments 1049
Rubra, G N 1224
Ruggles, Richard: *Manitoba historical atlas* . . . 816
Ruhr 1246
Rumania 546, 1256
bibliography 182, 184
maps 323
Romanian scientific abstracts 184
Rumanian studies . . . 1050
Runcorn, S K *ed*: *Continental drift* 927

Russell, Sir E J: *A report on the agriculture and soils of Kent* ... 1260
Soil conditions and plant growth 1260
The world of the soil 1260
Russell, E W, ed: *Soil biology and biochemistry* 1094
Russell, W M S: *Man, nature and history* 885
Russia and the Soviet Union ... 1051
Russian Arctic and Antarctic Scientific Research Institute 649
Russian land, Soviet people ... 1052
Ruwenzori Mountains 1187
Ryan, Marleigh: *Research in Japanese sources* ... 1032
Rydings, H A: *The bibliography of West Africa* ... 184
Ryle, George: *Forest service* ... 472

Saarmann, Gert: *Der globus im wandel der zeiten* ... 567
Sabin, John: *Bibliotheca Americana* ... 211
Sable, M H: *A guide to Latin American studies* 584
Sahab, A 1053
Sahab Geographic and Drafting Institute, Tehran 1053
St Albans, map 491
St Lawrence 261
Saito, Shiro: *Pacific Islands bibliography* 946
Sakhalin Island, Scientific Research Institute 4
Sale, R D: *Elements of cartography* 416
Salichtchev, K A 125
'Contribution of geographical congresses ... to the development of cartography' 269
Salisbury, W and R C Anderson, ed: *A treatise on shipbuilding* ... 1091
Salt, Laura E: *Oxford New Zealand encyclopaedia* 937
Salter, B R: 'Afforestation of the Lower Swansea Valley' 799
Samaria, maps 127
Samoa 1106, 1199

Sampson, Henry, ed: *World railways* 1265
Sands, classification 914
Sandy shores in South Lancashire 576
Sanke, Hans: *Weltatlas* ... 1228
Sanuto, Livio: *Geografia dell'Africa* 485
Saraceni, G C 485
Sater, J E: *The Arctic Basin* 66
Sauer, C O 817
Saueur de Chavanne, Louis: *The present state of the Empire of Morocco* 18
Sawyer, K E: *Landscape studies* ... 764
Saxelby, C H, ed: *A geographer's reference book* 490
Saxony, maps 378
Saxton, Christopher 820, 1054, 1116
Atlas of England and Wales 353
Scandinavia 576, 916, 1217
maps 62, 418, 824
Scandinavian Institute of African Studies, University of Uppsala 20
Scandinavian lands 1055
The Scandinavian world 796, 916, 1056
Scandinavian University Books 142
Scarborough, field studies 416
Scenery 348
'The scenery, antiquities and biography of South Wales' 1057
Scenery and the sense of sight 348
The scenery of England 348
The scenery of Scotland 444
The scenery of Sidmouth 348
Schaeffer, Léon 1036
Schippers, P, publishers 79
Schlenger, H et al: *Atlas Ostliches Mitteleuropa* 145
Schleswig-Holstein, maps 378
Schluter, Otto 810
Schmithusen, Josef: *Allgemeine vegetationsgeographie* 777
School of African and Asian Studies, University of Sussex 20
School of Oriental and African Studies, University of London 20, 76, 356, 781, 1058
Das schrifttum zur globenkunde 567

Schultze, J H: *Alexander von Humboldt* ... 628
Schurmann, H M E: *The Pre-Cambrian along the Gulf of Suez* ... 985
Schweinfurth, V et al: *Studies in the climatology of South Asia* 1136
Schweizerischer mittelschulatlas 1059
Schwerdtfeger, W: ' Climates of Central and South America ' 1267
Schwerz, Gabriele: *Allgemeine siedlungsgeographie* 777
Schwind, Martin: *Geographie der staaten* 777
Das Japanische inselreich ... 366
Science Museum, London 1070
The science of geography ... 1061
Science in New Zealand 1060
Science policy studies and documents series (Unesco) 1192
' The scientific art of mapping and charting ' 830
Scientific Committee on Antarctic Research 1062
Scientists and the sea ... 1065
SCOLMA: *Theses on Africa* ... 787
' The scope and methods of geography ' 806
Scorer, Richard and Harry Wexler: *Cloud studies in colour* 325
The colour encyclopaedia of clouds 325
Scotland 363, 444, 462, 472, 500, 529, 587, 879, 916, 1045, 1046
bibliography 184, 873, 1011, 1240
maps 26, 164, 761, 879, 989, 1046, 1078
Scotland's countryside 1970 440
Scott, Captain 368
Scott Polar Research Institute, Cambridge 315, 368, 566, 976, 1063
Scottish geographical magazine 904, 1046, 1064
Scottish Meteorological Society 1045
Scottish Wildlife Trust 1078
Scripps Institute of Oceanography, University of California 1143
Sea 210, 478, 518, 529, 681, 696, 832, 838
The sea ... 1065
The sea coast 331
Sea surveys ... 1066

Seabirds of the tropical Atlantic Ocean 1084
Sealock, R B and P A Seely: *Bibliography of place-name literature* ... 205
Searchlight books series (Van Nostrand) 1210
Sears, Mary, ed: *Deep-sea research and oceanographic abstracts* 370
Oceanography 1065
ed: *Progress in oceanography* 994
Second land use survey of Britain 761, 1236
' The second land use survey . . .' 761
Secondary school atlas (Philip) 961
Secondary school geographies (UTP) 1202
Sedimentation 686, 1067
Sedimentology 418, 670
Seely, P A: *Bibliography of place-name literature* ... 205
Segreda, L D: *Lista de mapas parciales o totales de Costa Rica* 828
Seismology 208, 402, 671, 706
Seismology ..., ed, J Wartnaby 1071
Seismicity of the earth and associated phenomena 1069
The seismicity of the earth, 1953-1965 1068
Seismicity of the European area 1070
Select bibliography of Mauritius 203
Selected bibliography on Kuwait ... 1072
' Selected data from the Soviet Statistical Yearbook of 1958 ' 246
A selected list of books and articles on Japan ... 1073
Selection of international railway documentation 1074
Seligman, Gerald: *Snow structure and ice fields* 1085
Seltzer, Leon E, ed: *Columbia-Lippincott gazetteer of the world* 333
Semple, E C 1075
American history in its geographic conditions 1075
The geography of the Mediterranean region 1075
Influences of geographic environment ... 648, 1009, 1075
Senegal 184
Senex, John: globes 567

Serial atlas of the marine environment 45, 1076
Serials *see* Periodicals
Serials for African studies 184
Service Géographique de l'Armée, Paris 654
Service Hydrographique de la Marine 215
Services de Tourisme du Pneu Michelin 853
Settlements 10, 66, 83, 105, 151, 225, 335, 459, 488, 505, 531, 592, 629, 1151, 1171
  bibliography 173
Settlement and encounter . . ., eds: F Gale, G H Lawton 1077
'Seventeenth century Dutch charts of the East Indies' 85
The seventh continent 61
Seventy-five years of arid-lands research at the University of Arizona 71
Seychelles Islands 16
Shabad, Theodore, *ed*: *Soviet geography* . . . 1109
  and P M Stern, *ed*: *The golden encyclopaedia of geography* 570
Shackleton, Sir E H: *The heart of the Antarctic* . . . 604
Shaler memorial series 921
Sharp, Thomas: *Town and townscape* 1173
Shave, D W: *Geography in secondary schools* . . . 496, 536
Sheffield and its region . . ., ed: David Linton 234
Shell, maps 164
The Shell book of exploring Britain 1078
Shell nature lovers' atlas of England, Scotland, Wales 1078
Shetland Islands 916
Shipbuilding 659
Shipping 850, 1045, 1179, 1229, 1266
Shipping, Parliamentary Papers 715
Shipping of New Zealand 1016
Ships of the United Netherlands 1091
The shorter Oxford economic atlas 934
The shorter Oxford school atlas 939
A shorter physical geography 1176
Shurova, S I *et al*: *Atlas mira* 102
Siam 1211

Siberia 164, 914
Siddiqui, A H, *comp*: *The economy of Pakistan* . . . 412
Sidmouth 348
Sierra Leone 184, 286, 1020
Sierre Leone, Roy Lewis 349
Sierra Leone in maps 1079
Silvano, Bernardo 998
Simakova, M S: *Soil mapping by color aerial photography* 1146
Sinclair, D J: *Atlas of London* . . . 128
  The Faber atlas, new edition 451
Sinclair, Keith: *Studies of a small democracy* . . . 1137
Sinclair Library, University of Hawaii 383
Singapore 299, 419
  bibliography 184, 201, 282
  maps 136
Singapore Chamber of Commerce Rubber Association 663
Singapore library journal 184
Singapore national bibliography 184
Singapore University, Department of Geography 740
  Library 282
Singh, Ganda: *Bibliography of the Punjab* 184
Singh, R L, *ed*: *India: regional studies* 641
Sinkiang 868
Sinnhuber, Mrs: bibliography of works of H J Fleure 465
Sino-Portuguese trade from 1514 to 1644 . . . 1080
Sir Charles Lyell . . . 801
Sir Charles Lyell's scientific journals . . . 801
Sir Francis Drake . . . 1081
Sir Francis Drake's voyage around the world . . . 1081
Sissons, J B: *The evolution of Scotland's scenery* 444
Skells, J W 989
Skelton, R A 485, 492, 596, 728, 833, 962, 996, 998, 1082, 1115, 1161, 1215
  County atlases of the British Isles . . . 353, 820, 1082
  Decorative printed maps . . . 369, 1082
  Explorers' maps . . . 449, 1082

513

Skelton, R A—(contd)
Geschichte der kartographie, English edition 562, 1082
The Vinland map ... 1082
Skerl, J G A, translator 927
Sketchmap geographies 796, 1150
Slides 509, 1152
Sloane collection 240
Sloss, L L et al, ed: *Lithofacies maps* ... 792
Slovakia, bibliography 184
Smailes, A E: *The geography of towns* 547
Smiley, J McA, cartographer 126
Smith, Guy-Harold, ed: *Conservation of natural resources* 1237
Smith, J Russell 1083
Smith, L T: edition of Leland's *Itinerary* ... 779
Smith, Wilfred: *An historical introduction to the economic geography of Great Britain* 167
Smith *see* Hattersley-Smith
Smith and Son, globemakers 961
Smithsonian Institution 363, 463, 784, 1084, 1215, 1272
Snow, Philip A: *A bibliography of Fiji, Tonga* ... 197
Snow 669, 908
Snow structure and ice fields 1085
Snowdonia, field studies 416
The Snowdonia National Park 882
The Snowy Mountains Scheme, Australia 1086
Social Science Research Council 515
Sociedad Geográfica de Colombia 90
Società di Studi Geografici, Florence 41
Société d'Edition Géographique et Touristique, Paris 343
Société de Géographie de Paris 50, 215, 829, 1087
Société Geografica Italiana 3
Société de l'Histoire de Maurice 203
Société des Océanistes 180
Societies 55, 61, 566, 875, 1043, 1044, 1046
 cartographic 271, 272, 923, 1092
 geographical 40, 41, 50, 129, 148, 157, 164, 176, 215, 254, 255, 258, 275, 403, 493, 502, 507, 510, 511,

Societies—(contd)
 522, 532, 560, 563, 598, 628, 761, 890, 902, 903, 923, 1041,1087
 geological 549, 552, 557
Society for Environmental Education 1089
The Society for the History of Discoveries 1156, 1090
The Society for Nautical Research 1091
Society of Irish Foresters 713
Society of Professional Geographers 80
Society for the Promotion of Nature Reserves 1078
Society of University Cartographers 1092
Socio-economic models in geography 1093
Soil biology and biochemistry 1094
Soil biology, reviews of research (Unesco) 1095
Soil conditions and plant growth 1260
'The soil landscapes of Australia' 109
Soil mapping by color aerial photography 1146
Soil Science, International congresses 109
Soil Survey of England and Wales 924, 1096, 1260
Soil survey manual (US) 1260
Soil surveys 631, 924, 1096, 1260
The soils of Europe 1260
Soils 109, 209, 320, 336, 339, 390, 478, 483, 488, 501, 503, 529, 531, 585, 691, 701, 702, 884, 1182, 1193, 1212, 1214, 1260
 maps 109, 148, 1146
'Some cartographical works in the National Library' (Wales) 880
'Some distinctive features of marine cartography' 830
Somerset, maps 305, 551
Somerville, Mary 620, 904, 1097
On the connexion of the physical sciences 1097
Physical geography 1097
Sømme, Axel: *The geography of Norden* ... 544
Somov, M M, translator 649
Sorbonne, Institute de Géographie 367

Somitasagama, E: 'Guide to Singapore government departments and serials' 184
Sorre, Maximilien 1098
　Fondements de le géographie humaine 1098
　L'homme sur la terre ... 1098
　Source book for geography teaching 1099, 1192
South Africa, Ministry of Information and Tourism 1148
South African bibliography 590, 1100
South Africa, Monica Cole 850
South America see *Latin America*
South and Southeast Asia ... 1101
South Asian Government bibliographies 1102
South East Asian Archives 1103
South Magnetic Pole 604
South Pacific bulletin 1104
The South Sea Commission 1104
Southampton 809
　University, Department of Geography 315
South-east Asia ... 1105
The Southeast Asian city 1209
South Eastern Agricultural College 1276
Southeastern geographer 80
The southern continents 1202
Southern Europe ... 850, 904
South-west Pacific ... 1106
Soviet and East European abstracts series 1107
Soviet Antarctic Expedition 649
　Information bulletin 55
　Reports 418
'Soviet foreign trade' 246
Soviet geography 3, 40, 45, 464, 1109
Soviet geography: accomplishments and tasks ... 1108
'The Soviet iron and steel industry' 246
Soviet statistical yearbooks 246
Soviet trade directory 1110
Soviet Union 1111
The Soviet Union, W Gordon East 1210
The Soviet Union ..., Georges Jorré 753, 796
The Soviet Union, W H Parker 1273
　see also USSR
Spaak, Paul-Henri 141

Space 140, 361
Space Research Institute, Hanover 248
Spacial analysis ... 1112
Spain 380, 500, 621, 1133
　bibliography 182, 184, 235, 873
　maps 35, 103
Spain and Portugal 504
The Spanish conquistadores 972
Sparhawk, W N: *Forest resources of the world* 470
Sparks, B W: *Geomorphology* 796
Special Committee on Oceanic Research 1113
Special libraries 1114
Special Libraries Association, Geography and Map Division 284, 819, 1114
Special publications series (American Geographical Society) 45
'Special purpose navigation charts' 634
Speculum Britanniae 907
Speculum orbis terrarum 1115
Speed, John 724, 907, 996, 1116, 1161
Speke, maps 450
Speleology 285, 290
　see also Caves
Spence, S H: 'The prospects for industrial use of the Lower Swansea Valley' 799
Spencer, Robert F and Eldon Johnson: *Atlas for anthropology* 97
Spheres, armillary 567, 624
Spice Islands 984
De spieghel der zeevaerdt 178, 1219
Spies, Otto 1227
The spirit and purpose of geography 633, 1117, 1239
Sporck, José A, ed: *Mélanges de géographie* ... 840
Sprent, F P 989
Srivastava, A P: *A guide to Iraq's natural resources* 585
Ssu Yü, Teng: *Japanese studies on Japan and the Far East* 720
Stacey, Margaret: 'The Lower Swansea Valley: housing report' 799
'Lower Swansea Valley: open space report' 799

515

Staffordshire, Geologists' Association 552
Stahl, W H: *Ptolemy's geography* ... 998
Stamp, Sir L Dudley 331, 460, 657, 759, 797, 1118, 1149
  Africa ... 1237
  Britain's structure and scenery 331
  Chisholm's *Handbook of commercial geography* 303
  A glossary of geographical terms 569, 1118
  A history of land use in arid regions 621, 1118
  Hugh Robert Mill ... 496
  The land of Britain ... 761
  Land Use Survey of Great Britain 761
  Longman's dictionary of geography 1118
  Man and the land 331
  Nature conservation in Britain 885
  Our developing world 929
Standard encyclopedia of the world's mountains 1119
Standard encyclopedia of the world's oceans and islands 1119
Standard encyclopedia of the world's rivers and lakes 1119
Standing Conference on Library Materials on Africa 1191
Stanford, Edward, Limited 961, 1120
  'planfile' 1120
  reference catalogue 269, 1120
Stanford University, Africa collection 20
  Hoover Institute 344, 589, 598
Stanford's geological atlas of Great Britain and Ireland 1121
The state of British agriculture ... 28, 655
The state of food and agriculture (FAO) 468
States and trends of geography in the United States ... 1122
'Statistical account of Upper Canada' 261
Statistical analysis in geography 1123
Statistical bulletin for Latin America 406
Statistical mapping and the presentation of statistics 1124

Statistical methods and the geographer 1125
Statistical news ... 184
A statistical summary of the world mineral industry 931
The statistical yearbook (UN) 1192
Statistics 10, 28, 48, 135, 154, 184, 221, 246, 261, 468, 470, 495, 497, 522, 559, 1112, 1124, 1125, 1192
Statistics: African sources for market research 1126
Staton, F M and Marie Tremaine; *A bibliography of Canadiana* ... 196
Steel, R W and C A Fisher, *eds*: *Geographical essays on British tropical lands* 501
  and Richard Lawton: *Liverpool essays in geography* ... 793
Steel 422, 478, 633
Steers, J A 241
  on coasts 326
  Field studies in the British Isles 462
  An introduction to the study of map projections 712
  The sea coast 331
Stefansson, Vilhjalmur 1127
  Collection 382, 1127
  The friendly Arctic 1127
  Great adventures and explorations ... 1127
  The northward course of empire 1127
  Northwest to fortune ... 1127
  The three voyages of Martin Frobisher ... 1164
  Unsolved mysteries of the Arctic 1127
  Ultima Thule 1127
Steiner, Franz, Verlag 522, 523
Stellenbosch University, Department of Geography 6
Stembridge, J H 1128
  Africa 1128
  Germany 1128
  A portrait of Canada 1128
Stephens, C G: 'The soil landscapes of Australia' 109
Stephens, N and R E Glasscock, *eds*: *Irish geographical studies* 714
Stern, P M: *The golden encyclopaedia of geography* 570

Stevens, H N: *Ptolemy's geography* 998
Stevens, L W: *Food, clothing and shelter* 807
Stevenson, E L: *Portolan charts*... 983
 Terrestrial and celestial globes... 567
 Willem Janszoon Blaeu... 218
Stieler, Adolf 1130
Stieler atlas 957, 1130
Stewart, C L: *Land use information*... 760
Stewart, G A, ed: *Land evaluation* 756
Stewart, Harris B: *The global sea* 1210
Stewart, John: 'An account of Prince Edward Island in the Gulf of St Lawrence...' 261
Stobart and Son, Limited, London: *Forestry and timber technology catalogue* 471
Stockdale, John, map-maker 274
Stockholm: Svenska Sällskapet för Antropologi och Geografi 144
Stoddart, David R et al, ed: *Progress in geography*... 993
The storage and conservation of maps 1131
The story of maps 1132
The story of Utopias... 355
Strabo 810
 Geography 1133
' Strassenkarte aus Mairs Geographischen Verlag ' 808
Stratigraphy 683, 905
The stratigraphy of the British Isles 1134
Streyffert, T: *World timber*... 471
Strietelneier, John H: *Geography in world society*... 538
Structure and field geology... 482
Structure and surface drainage in South-East England 786
The structure of political geography 979
Structure, surface and drainage in South-east England 657
Struggle for the Snowy... 1086
The student's elements of geology 801
Studies in cartobibliography... 1135

Studies in the climatology of South Asia 1136
Studies in emerging Africa 249
Studies in historical geography 363
Studies in medical geography 115
Studies in regional planning 961
Studies in urban geography series 45
Studies in the vegetation history of the British Isles... 460
Studies of a small democracy 460, 1137
Study atlas (Collins) 331
Study of critical environmental problems 430
The study of the soil in the field 1260
The study of urban geography 1205
Sturm, Rudolf: *Czechoslovakia*... 360
Subject catalog of the Special Panama Collection 1138
Sub-Saharan Africa: a guide to serials 184
Sudan 107, 184, 429, 501, 776
Suess, Eduard: *Das antlitz der erde* 1139
Suez, Gulf 985
Sugar Research Unit, Melbourne 339
Suggate, L S 1140
 Africa 1140
 Australia and New Zealand 1140
Sullivan, Walter: *Assault on the unknown*... 689
Sunderland 1205
Suomen Maantieteen Käsikirj 120
The surface water year book of Great Britain... 1141
Surrey, maps 907, 1260
 Fieldwork Society 533
Surridge, Mrs Thérèse, translator 341
A survey of bibliographies in Western languages concerning East and South East Asian studies 288
Survey of France 276
A survey of Japanese bibliographies concerning Asian studies 288
A survey of the mineral industries of Southern Africa 1142
Surveying 82, 259, 368, 415, 453, 830
Surveyor surveys 1168
' Surveyors of the oceans ' 9,

517

Surveys 9, 11, 127, 240, 390, 550, 551, 631, 769, 826, 914, 919, 923, 1196
Sussex 10, 1260
  Board of Agriculture *Report* 220
  maps 907
  University, African and Asian studies 20
Svalbard 977
Sverdrup, H U 1143
  The oceans . . . 1143
  Oceanography for meteorologists 1143
Swallow, Mary: *Deep sea research* . . . 370
Swan, Michael: *British Guiana* 349
Swanley College 1276
Swansea 809
Swansea Valley 799
Swanton, J R: *The Wineland voyages* 1215
Swayne, J C: *A concise glossary of geographical terms* 342, 961
Swaziland 429
Sweden 544, 708, 1055, 1056
  bibliography 1, 182, 184
  maps 146
  National Committee for the International Hydrological Decade 908
Sweden books 1144
The Swedish Institute 1
Swedish ships 1091
Sweeting, G S: *The Geologists' Association* . . . 552
Swiss Foundation for Mountain Research 168
Swiss Foundation of Alpine Research 864
' The Swiss Mittelschulatlas in a new form ' 1059
Swithinbank, Charles, on airborne radio echo sounding 565
Switzerland 708
  bibliography 182, 184
  maps 95, 468
Sydney University, Department of Geography 1151
Sydney, Wheat Research 339
SYMAP 297
Symes, D G and E G Thomas: *An atlas of population change* . . . 133

Symes, D G and E G Thomas—(*contd*)
  et al: The Yorkshire and Humberside Planning Region: an atlas . . . 1277
Symons, L J: *Agricultural geography* 167
Symonson, Philip: *Map of Kent* 924
Symons's meteorological magazine 850, 1045
The Synagraphic Mapping System (SYMAP) 1145
SYNTOL 287
Syracuse University: East African studies 20
Syria 107, 184, 1133
A system of world soil maps 1146
Systematic political geography 1262
The Systematic Association 1149
Szymowski, S: *German ships* 1091

Tableau de la géographie de la France 1214
Tabula Imperii Romani 1 : 1M 1147
Taff, River, map 10
Tahiti, bibliography 180
Taiwan 344
  maps 301, 868
Taiwan—Ilha Formosa . . . 249
Taiwan National University 184
Talbot, A M and W J: *Atlas of the Union of South Africa* 138
The Talmud, geography 847
Tanganyika 286, 1020
Tanghe, Raymond, *comp: Bibliographie des bibliographies* (Canada) 184
Tansley, A G: *Britain's green mantle* . . . 230
  The British Islands and their vegetation 230
Tanzania, atlas 137
  Bureau of Resource Assessment and Land Use Planning 137
  Five Year Development Plan, *Reports* 137
  Ministry of Lands, Settlement and Water Development, Surveys and Map Division 137
Tanzania today . . . 1148
Tradi, P 182
Tarling, D H and M P: *Continental drift* 929, 930

Tartar Relation
  see under Vinland Map
Tasman 398
Tasmania, maps 109, 990
Tastu, J: *Notice d'un atlas en Langue Catalane* 277
Tatton, Miss M 775
Tavernier, Gabriel, cartographer 1160
Taylor, C J: *Tropical forestry* . . . 471
Taylor, C R H, *comp: A Pacific bibliography* 945
Taylor, E G R 595, 596, 1150
  bibliography of works 1150
  Early Hanoverian mathematical practitioners 1150
  The geometrical seaman 553, 1150
  'Gerard Mercator . . .' 846
  The haven-finding art 603, 1150
  Late Tudor and early Stuart geography . . . 768, 1150
  The mathematical practitioners of Tudor and Stuart England 1150
  'The navigating manual of Columbus' 334
  'The origins of continents and oceans' . . . 927
  Sketchmap geography 1150
  Tudor geography 1150, 1185
Taylor, James A, *ed: Climatic factors and agricultural productivity* 320
  Weather economics 1224
  and R A Yates: *British weather in maps* 242
  et al: Geography at Aberysthwyth 529
Taylor, Th Griffith 1151
  Australia . . . 151, 1151
  Canada . . . 1151
  Environment and race 1151
  Environment, race and migration . . . 1151
  Geography in the twentieth century . . . 537, 1151
  Journeyman Taylor 1151
  Our evolving civilization 1151
  Urban geography . . . 1151
Taylor, Lt Col W R: 'The Ordnance Survey of Northern Ireland . . .' 924
Taxomomy and geography . . . 1149

Teachers, geographical associations 81
Teaching, use of models 966
Teaching aids on Australia and New Zealand 1152
'Teaching geography' (Geographical Association) 496
Teaching geography series (Geographical Association) 1152
Technical and commercial dictionary of wood 471
Technics and civilization 355
A technique using screen and blackboard to extract information from a photograph 1152
Techniques in geomorphology 1153
Techniques in human geography 627
Technological inquiry . . . 885
'Tectonic map of Eurasia' 330
'Tectonic map of Europe . . .' 330
'Tectonic map of the USSR' 330
Teheran 363
  University: *Bibliography of Iranian bibliographies* 184
  Index Iranicus 184
Teikoku-Shoin Company Limited, Tokyo 672
Tennessee Valley Authority 476
Terms, geographical 342, 380, 560, 567, 569, 797, 961
Terra 1154
Terra Australia cognita . . . 1155
Terrae incognitae . . . 1090, 1156
La terre (Réclus) 1015
La terre, nôtre planète 767
Terrestrial and celestial globes . . . 567
Texas University, Institute of Latin American Studies 281
Textbook of geology 482
The Textile Council, Manchester 1157
Textile history 1158
The Textile Institute 1159
The Textile Institute and industry 1159
Textile progress 1159
Thackwell, Brigadier D E O 1030
Thailand 299, 419, 1105
  Asian studies 288
  bibliography 184
  maps 132, 136, 331
Thames 809, 873

519

Theakstone, W H and C Harrison: *The analysis of geographical data* 48
Le Théatre françois 1160, 1161
The Theatre of Great Britaine 724, 907, 1116
Theatrum, Abraham Ortelius 164, 928
Theatrum Orbis Terrarum Limited: *Acta geographica* 5
reprints 79, 174, 178, 179, 218, 224, 313, 350, 485, 492, 520, 567, 574, 595, 607, 790, 791, 833, 846, 896, 996, 998, 1115, 1155, 1160, 1161, 1211
Theoretical geography 1162
The theory of continental drift, symposia 927
Thesaurus national de l'eau 1223
Thesaurus des termes géographique 385
Theses 184, 329, 827, 894, 1191
Theses on Africa ... 787
Theses on African studies 1058
Theses on Asia 1163
Thiele, Walter: *Official map publications* ... 918
'This changing world' in *Geography* 524
Thoman, Richard S and Edgar C Conkling: *Geography of international trade* 542
et al: *The geography of economic activity* 539
Thomas, E G: *An atlas of population change* ... 133
et al: *The Yorkshire and Humberside Planning Region: an atlas* 1277
Thomas, M F and G W Whittington, ed: *Environment and land use in Africa* 429
Thomas, Peter: *Welsh landforms and scenery* 1226
Thomas, William L, jr, et al: *Man's role in changing the face of the earth* 817
Thompson, Godfrey, rev ed *An encyclopedia of London* 424
Thompson, I B: *Modern France* ... 249, 859
Thompson, J Walter, Company: *The Latin American markets* ... 771

Thomsen, H: 'Climates of the oceans' 1267
Thomson, Charles Wyville (later Sir) 292
Thomson, Don W: *Men and meridians* ... 843
Thomson, J Arthur 904
Thomson, J Oliver, on early geographical knowledge 442
Thorén, R: *Picture atlas of the Arctic* 418, 970
Thorne, J O: *Chambers's world gazetteer* ... 295
Thornhill, Patrick: *Land and water* 755
Thorpe, David: *The geographer and urban studies* 487
3-D junior atlas (Harrap) 368, 1165
The three voyages of Martin Frobisher ... 1164
Tibet 868
Tibet: international studies 1166
Tidal Institute, University of Liverpool 54
Tides 150
Tidswell, W V and S M Barker: *Quantitative methods* 1000
Tierra del Fuego 1155
Tietze, Wolf: *Westermann lexikon der geographie* 1232
Tigris 873
Tigris-Euphrates (Strabo) 1133
Tijdschrift voor economische en social geografie 1167
Timber 471, 1268
Timber bulletin of Europe (UN) 1192
The Times atlas of the moon 1168
The Times atlas of the world 164, 1069, 1169, 1170
The Times index gazetteer of the world 1170
Timor 299
Tin and its uses 704
Tin Research Institute, Greeford 704
Tiros, Operational Satellite System 1271
Titmuss, F H: *Concise encyclopaedia of world timbers* 471
Toase, Mary: *Guide to British periodicals* 184
Tobago 184, 286, 1020

Todd, D K, ed: *The water encyclopedia* 1223
Tōgo, Yoshida: *Japanese geography* . . . 1032
Togo 184
Tokyo 55, 1246
Tolansky, S: *The history and use of diamond* 616
Tomlinson, M E: *North America* 1202
The southern continents 1202
Tomorrow's countryside . . . 885
Tonga 197, 1106
Toniolo, Sandro 84
Tooley, R V 313
A dictionary of mapmakers 386, 820
Map collectors' series 820
Maps and map-makers 824, 998
One hundred foreign maps of Australia 160
The printed maps of Tasmania . . . 990
and Charles Bricker: *A history of cartography* . . . 618
Toombs, R B: *A survey of the mineral industries of Southern Africa* 1142
A topical list of vertical photographs in the air-photo libraries 1152
Topicards 1171
Topics in geography 807
Toronto Public Library 261
Canadiana 196
Toronto University, Department of Geography 1151
Torres Strait, bibliography 191
Toulouse, Conferences on aerial surveys 11
Toulouse University 11
Touring Club Italiano 84, 1059, 1172
Atlante fisico-economico d'Italia 1172
Atlante internazionale . . . 1172
Road map of Europe 1172
The Tourmaster maps of Britain (Bartholomew) 164
Toussaint, A and H: bibliography of Mauritius 203
Town and townscape 1173
Town Planning Institute 866
Town plans 98, 100, 101, 102, 107, 136, 147, 313, 453, 491, 547, 575, 1235

Town study . . . 1204
Towns, M Turner 807
Towns 350, 377, 487, 529, 530, 547, 611, 655, 795, 898, 1204, 1205, 1206, 1207, 1208, 1209
Toynbee, Arnold, ed: *Cities of destiny* 307
Toyne, Peter: *Techniques in human geography* 627
Trade 221, 455, 468, 542, 1012, 1266, 1269
Trade and commerce 1174
Trade and industry . . . 1175
Trager, Frank N et al: *Annotated bibliography of Burma* 52
Traité de géographie physique 367, 1176
Traité de géomorphologie 555
Transactions of the Asiatic Society of Japan 1177
The transformation of the Chinese earth . . . 1178
Transparencies, bibliography 269
Transport 445, 501, 503, 518, 530, 661, 739, 1254
Transport, Oswald Hull 807
Transport history 1179
Transport research 1180
'Transport statistics and notes' 246
Transportation 478, 1012
Transvaal, bibliography 788
maps 23
Travel 254, 595, 596, 1017, 1018
A traveller's guide to health 609
Travels, Ibn Battuta 635
A treatise on shipbuilding . . . 1091
Tregear, T R: *Economic geography of China* 249, 408
Tregonning, K G: *North Borneo* 349
South-east Asia . . . 1105
Tremaine, Marie, ed: *The Arctic bibliography* 67
A bibliography of Canadiana . . . 196
Trends in geography . . . 1181
Trewartha, G T: *An introduction to climate* 804
Tricart, Jean: *European research resources* 437
Geomorphology of cold environments 555
Trinder, B S: *Pergamon general historical atlas* 952

521

Trinidad 184, 286, 1020
Trinity College, Dublin: Geography Department 511
Trippodo, G: *Technical and commercial dictionary of wood* 471
Troll, Carl: *Geo-ecology of the mountainous region of the tropical Americas* 484
Trollope, Christine, translator 473
Tropical abstracts 3
Tropical agriculture 184, 1182
Tropical forestry ... 471
Tropical man ... 1183
Tropical Products Institute 1184
Tropical science 1184
Tropics 470, 501, 1048, 1212, 1213
Trotman, E F: *Producing a slide set with commentary* ... 1152
Trust for National Parks and Areas of Outstanding Natural Beauty 1078
Tsunami Warning Centre, Honolulu 54
Tsunamis 54
Tsuneta Yano Memorial Society 719
Tuan, Yi-Fu: *China* 1273
Tudor geography ... 1150, 1185
Tulippe, M O: festschrift 840
Tunisia 23, 184, 707
Turkey 107, 184, 602
Turner, M: *Towns* 807
Turning points in history series 394
Tussler, A J B and A J L Alden: *Map book of the Benelux countries* 807
Tuttitalia ... 1186
Typology, agricultural 33
Tyrwhitt, Jacqueline, ed: *Patrick Geddes in India* 481

Uganda 286, 1020
  bibliography 184
  Geographical Association 403, 1187
Uganda, Harold Ingrams 349
Ulmer, Paul, cartographer 874
Ulrich's periodicals directory 3
Ulster and other Irish maps ... 1188
Ultima Thule ... 1127
The Ulysses factor ... 448
Understanding ecology 528
Union Internationale de Secours 54

A union list of geographical serials 694
USSR 40, 78, 125, 152, 182, 246, 249, 308, 330, 363, 512, 518, 540, 548, 589, 602, 619, 621, 649, 652, 715, 753, 796, 802, 959, 970, 1003, 1022, 1052, 1107, 1108, 1109, 1110, 1210, 1237, 1249, 1256, 1267, 1273
  Academy of Sciences 4, 37, 464, 968, 1017, 1108
  bibliography 182, 184, 207, 246, 873, 1051
  Department of Geodesy and Cartography 464
  maps 62, 102, 104, 106, 125, 164, 330, 675, 934, 1014
The USSR and Eastern Europe: periodicals ... 1189
  see also Russia; and under Soviet 'Unit-areas' 1203
United Arab Republic, WMO seminar 31
United Kingdom 20, 221, 430, 438, 497, 708, 809, 850
  bibliography 184
  disease mortality maps 869
  National Committee for Geography 1047
The United Kingdom contribution to the International Geophysical Year 689
United Kingdom glossary of cartographic terms 1190
United Kingdom publications and theses on Africa 1191
  see also under British Isles; Great Britain
United Nations 132, 163, 221, 372, 406, 411, 695, 848, 1192, 1223, 1245, 1248, 1249, 1258, 1269
  see also FAO
Unesco 11, 34, 54, 71, 72, 73, 176, 184, 187, 315, 319, 392, 484, 585, 621, 664, 688, 689, 691, 696, 702, 756, 796, 848, 849, 884, 992, 1095, 1099, 1118, 1192, 1213
United States of America 20, 44, 56, 58, 66, 73, 152, 163, 184, 189, 225, 333, 334, 376, 381, 394, 431, 470, 471, 497, 512, 539, 557, 621, 686, 708, 715, 760, 782, 838, 842, 881, 959, 1002, 1003, 1006, 1032, 1084,

United States of America—(*contd*)
1104 1122, 1194, 1198, 1225, 1262, 1272
  bibliography 182, 184, 205, 280, 600
  Board on Geographical Names 780, 1195
  Department of Agriculture 188, 217, 655, 1027, 1067, 1193, 1197, 1260
  Geological Survey 45, 872, 1198
  maps 38, 62, 129, 131, 132, 139, 182, 202, 269, 273, 315, 469, 792, 819, 822, 872, 874, 881, 909, 918, 1006, 1014, 1196, 1241
  National Academy of Sciences 139, 883, 975, 1061, 1122, 1271
  National Science Foundation 59, 649, 675, 817, 881, 944, 1067, 1109, 1194, 1247
The United States and world resources 1210
United States of America 1199
United States research reports 184
  see also under America; American; North America
Universal Decimal Classification 315, 1063
The universal geography of P W N 978
Universities 5, 20, 41, 1200
  See also under individual names
The university atlas (Philip) 780, 1201
Universe, atlas 140
Unsolved mysteries of the Arctic 1127
Unstead, J F 1203
  'H J Mackinder and the new geography' 806
Upper Volta 184
Uppsala University, Scandinavian Institute of African Studies 20
Urban affairs 1204
Urban analysis . . . 1205
Urban core and inner city 1206
Urban essays . . . 1205
Urban geography . . . 1151
Urban land . . . 1207
The Urban Land Institute 1207
Urban planning 592
Urban Planning Development, Ministry of Housing and Local Government 785
Urban Planning Research Group, Illinois 1031
Urban studies 44, 487

Urbanisation 291, 350, 440, 503, 529, 547, 748, 772, 1208
'Urbanization and economic development in Mexico' 772
Urbanization in newly developing countries 1209
The urbanization process in the third world 1208
Urbanism and urbanisation 1208
Urbanistica 1209
Uruguay 184
'The use of watermarks in dating old series one-inch Ordnance Survey . . . maps' 924
'User requirements for modern nautical charts' 8
The uses of a revolving blackboard in geography teaching 1152
Utrecht State University 890
Vening Meinesz Laboratorium 832
Uttar Pradesh, agriculture 33

Vail, R W G: *Bibliotheca America* . . . 211
Valck, Gerhard, globes 567
Vallaux, C: *La géographie de l'histoire* 243
van Baren, F A 702
Van Heusden, Gerard Th, publishers 79
Van Keulen, Johannes and Gerard: *The new large shining sea-torch* 896
Van Nieuwenhuijze, C A O, *ed*: *Economic, social and political studies of the Middle East* 410
Van Nostrand Company, Limited 1210
Van Ortroy, F: *Bibliographie sommaire de l'oeuvre mercatorienne* 846
Van Raemdonck, J: *Gérard Mercator* . . . 846
Van Richthofen, Ferdinand 810
Van Royen, *ed*: *Atlas of the world's resources* 144
  and Nels A Bengtson: *Fundamentals of economic geography* . . . 478
Vancouver Island 261
Varenius, Bernard 620, 1211
  Descriptio regni Japoniae et Siam 1211

Varenius, Bernard—(*contd*)
  Geographia generalis 1211
Varenius, S Gunther 1211
Vegetation 230, 315, 529, 750, 1270
  maps 445, 675, 707
Vegetation and soils . . . 1212
Vegetation maps of the Mediterranean region (Unesco) 1213
Véliz, Claudio, *ed*: *Latin America and the Caribbean* . . . 602
Venetian adventurer, H H Hart 829
Venezuela 31, 94, 674, 772
Vennetier, Pierre: *Pointe-Noire et la façade maritime du Congo-Brazzaville* 917
Vernet, A: Tunisia, vegetation map 707
Vespucci, Amerigo 350, 394, 867
Vestnik Moskovskogo Univesiteta . . . 1109
Victoria, Lake 501
Victoria University, Social Sciences Research Centre 194
Vidal de la Blache, Paul 50, 243, 367, 371, 473, 479, 489, 506, 517, 810, 1098, 1214
  Atlas général . . . 99
  Etats et nations de l'Europe 1214
  France de l'Est 1214
  Géographie universelle 1214
  Principes de géographie humaine 1214
  Tableau de la géographie de la France 1214
Viet-Nam 132, 184, 299, 419, 1105
Vietor, Alexander O 1161, 1215
The Viking explorers 1215
Villecrosse, J 182
Vilnay, Z: *The new Israel atlas* . . . 895
The Vinland Map . . . 394, 1082, 1215
'The Vinland Map cartographically considered' 1215
Vince, S W E: 'Towards a national atlas' 866
Visa 253
Visible regions atlas (Collins) 331
Visintin, L 717
  Grande atlante geografico 575
Vocabulaire franco-anglo-allemand de géomorphologie 1216
Vocabularium geographicum 1216

Vogel, W and S Szymowski: *German ships* 1091
Vogt, William 1217
  on Chile 1217
  People 1217
  Road to survival 1217
  on Scandinavia 1217
Voisin, Russell 672
Volcanoes 54
Volcanology 54, 400
  see also under Vulcanological
Volga 873
Voous, K H, *comp*: *Atlas of European birds* 117
Voprosy geografii 40, 1109
A voyage to New Guinea . . . 942
A voyage towards the South Pole . . . 1218
Voyages and travel 876
Voyages of discovery (Captain Cook) 596
Vrenghenhil, A: *Ships of the United Netherlands* 1091
Vulcanological Society of Japan 54

Wadi Halfa, maps 1147
Wageningen, Geological Institute 670
Waghenaer, Lucas Jansz 178, 1161
  The mariner's mirrour 833
  De spieghel der Zeevaerdt 1219
Wagner, H R: *Cartography of the northwest coast of America* . . . 273
  Sir Francis Drake's voyage around the world . . . 1081
Wagret, Paul: *Polderlands* 850
Waite Agricultural Research Institute, University of Adelaide 339
Waldseemuller, Martin, gores 567
Wales 227, 326, 331, 335, 462, 465, 472, 499, 505, 529, 552, 587, 757, 761, 779, 809, 880, 898, 1045, 1057, 1205
  bibliography 1226
  maps 164, 274, 353, 363, 724, 880, 989, 1042, 1054, 1078, 1135, 1236
Wales, E G Bowen 850
Wales, F V Emery 1273
Wales in European Conservation Year 440

Walford, A J 235
Guide to reference materials 521
Walkabout . . . 1220
Walker, A: *Official publications of Sierra Leone and Gambia* 184
The Rhodesias and Nyasaland . . . 184
Walker, D and R Gilbert, *eds*: *Studies in the vegetational history of the British Isles* . . . 460
Walker, E A: *The great trek* 972
Walker, E H: *Flora of Japan* 1084
Walker, F: *Geography from the air* 531
Wallén, C C: 'Climates of northern and western Europe' 1267
'Climates of central and southern Europe' 1267
Wallis, Helen 684
The first English globe . . . 567
'Further light on the Molyneux globes' 567
'Humboldt's map of isothermal lines . . .' 628
Libro dei globi, bibliographical note 567
Walschot, L, *ed*: *Abstracts of Belgian geology and physical geography* 184
Walsh, E M, cartographer 612
Walters, Gwyn 231
Walters, S M: *Atlas of the British flora* 112
Walton, A D: *A topical list of vertical photographs in the national air photo libraries* 1152
Walton, Kenneth: *The highlands and islands of Scotland* 916
Wanders, A J M: *A history of Martian cartography* 1161
Wanklyn, Harriet: *Friedrich Ratzel* . . . 648, 1109
War Office: *Catalogue of maps* 284, 1221
classification of maps 315
Warkentin, John, *ed*: *Canada* . . . 255
and Richard Ruggles, *ed*: *Manitoba historical atlas* . . . 816
Warmington, E H: *The ancient explorers* 49
Warne's natural history atlas of Great Britain 1222

Warren, Andrew, on desert dunes, sources 72
Warren, D E: 'Surveys for development' 390
Warren Reports 947
Warsaw University, Centre of African Studies 20
Wartnaby, J, *ed*: *Seismology* . . . 1070
Warwickshire 505, 1008
Washington: Naval Department, Bureau of Medicine and Surgery 1230
Public Affairs, Bureau 161
World Weather Program Centre 1271
Wasley, Don: *Airways of New Zealand* 1016
Water 669, 755
Water, earth and man . . . 1223
The water encyclopedia 1223
Water Information Centre, Port Washington, New York 1223
'*Water importation into arid lands*' 72
Water research. . . . 1223
Water resources abstracts 494, 1223
Water Resources Act 1963 580
Water Resources Board 352
Water resources bulletin 1223
Water Resources Centre, *Archives* 1223
Water resources development . . . 1223
Water resources journal (UN) 1223
Water supplies for arid regions 1223
Water supply memoirs 551
Waters, Lt Cdr David W: *The art of navigation in England* . . . 75
on Sir Francis Drake 1081
Waterways, inland 227, 363
Watson, Edward, translator 555
Watson, G E: *Sea-birds of the tropical Atlantic Ocean* 1084
Watson, J Wreford: *Geographical essays in memory of Alan G Ogilvie* 500, 920
Watters, R E, *comp*: *A checklist of Canadian literature* . . . 298
Watts, Alan: *Weather forecasting ashore and afloat* 1224
The Weald 10, 462

The Weald, S W Woodridge and F Goldring 331
Weather 10, 149, 242, 275, 527, 850, 1152, 1229, 1271, 1272
Weather . . ., W G Kendrew 747
Weather 1045
Weather economics 1224
Weather forecasting ashore and afloat 1224
Weather map 1045
Webb, Herschel and Marleigh Ryan: *Research in Japanese sources . . .* 1032
Webster's geographical dictionary 1225
Weddell, James, on South Polar exploration 1218
Wegener, Alfred 930
  continental drift theory 927
  Die klimat der geologischen vorzeit 750
  The origin of continents and oceans 927
Weigert, H W: *Principles of political geography* 979
Wellish, H: *Water resources development . . .* 1223
Welsh landforms and scenery 1226
Die welt des Islams 1227
Weltatlas . . . 1228
Weltforum Verlag, Munich 24
Weltgeschichte 1009
Die weltmeere . . . 1229
Welt-seuchen atlas 1230
Wenner-Green Foundation for Anthropological Research 817
Wensleydale, map 1236
West, R G: *Studies in the vegetational history of the British Isles . . .* 460
West, S R: *A guide to trees* (New Zealand) 1016
West Africa 796
West African studies 1231
West India Company 174
West Indies 972, 1249, 1268, 12772
  bibliography 184, 235
  maps 202
  University, Faculty of Agriculture 184
Westermann, Georg 269, 1232
Westermann bildkarten welt-lexikon 570

Westermann lexikon der geographie 1232
Westermanns geographische bibliographie 1232
Western Europe . . ., ed: John Calmann 602
Western Europe, F J Monkhouse 796
Western Pacific Islands 349
Westminster, plan 274
Westphalia, maps 378
Wexler, Harry: *Cloud studies in colour* 325
  The colour encyclopaedia of clouds 325
Weymouth, field courses 416
Whale charts 838
Wharton, Sir William J L 8
Wheat, J C and C F Brun: *Bibliography of maps and charts published in America before 1800* 202, 822
Wheat Research Unit, Sydney 339
Wheatley, Paul, on China 613
Wheeler, Jesse H, *jr, et al*: *Regional geography of the world* 1022
Wheeler, K S and M Harding, *eds*: *Geographical fieldwork . . .* 503, 533
  Geography in the field 533
Wheeler, P T: 'The development and role of the Geographical Field Group' 502
Whitaker's cumulative booklist 269
Whitcombe and Tombs, Limited, Christchurch 1233
White, H P and M B Gleave: *An economic geography of West Africa* 167
The white road . . . 1234
The Whitehall atlas (Philip) 961
Whitelaw, J S: *New Zealand* 1273
Whiteley, P F, *ed*: *The geographer's vade mecum* 961
Whittington, G W: *Environment and land use in Africa* 429
Whittow, J B and P D Woods: *Essays in geography for Austin Miller* 460
Whyte's atlas guide 1235
Wibberley, G P 1276
  Pressure on Britain's land resources 1276

Wiggin, Henry and Company, Limited 697
Wiggins, W D C 1190
Wigmore, Lionel: *Struggle for the Snowy* ... 1086
Wilber, Donald N: *Annotated bibliography of Afghanistan* 51
Wildfowl Trust 1078
Wildfowlers' Association of Great Britain and Ireland 1078
The wildscape atlas of England and Wales 1236
Wiley, John and Sons, Limited 1237
Wilgus, A Curtis: *Historical atlas of Latin America* ... 615
Wilhelm, Fritz: *Allgemeine hydrogeographie* 777
Wilkins, P M, bibliography of Greater London 541
Wilkinson, H R: *Maps and diagrams* ... 823
Willem Janszoon Blaeu ... 218
Williams, Donovan: 'Clements Robert Markham ...' 836
Williams, F L: *Matthew Fontaine Maury* ... 838
Williams, H W: *Bibliography of printed Maori to 1900* 206
Williams *see* Baynton-Williams
Williamson, J A 984
The age of Drake 972
on Hakluyt 595
Wilson, H W, Company: *Biological and agricultural index* 217
Wilson, J T: 'Continental drift' 927
'The development of earth science' 811
Wilson, L G, ed: *Sir Charles Lyell's scientific journals* ... 801
Wind and current charts (Maury) 838
The Wineland voyages 1215
Winsor, J: *Bibliography of Ptolemy's Geography* 998
Wint, Guy: *Asia* ... 602
Winterbotham, H St J L: 'The national plans' 924
Wisconsin University, Geophysical and Polar Research Center 649
Wise, M J, on Sir Dudley Stamp 759
Witherall, J W: *French-speaking West Africa: a guide to official publications* 184

Witt, W: 'Deutscher planungsatlas' 378
Witwatersrand University, African Climatology Unit 321
Wöhlbier, Herbert, ed: *Worldwide directory of mineral industries* 647, 1274
Wolstenholme, M J: *An encyclopaedia of the iron and steel industry* 422
The wonder book of wool 708
Wood, D: *Economic aspects of pigmeat marketing* 655
Wood, H R B: *The British Isles* 1202
Europe 1202
Wood, Margaret: *Foreign maps and landscapes* 469
Wood, Peter A: *Character of a conurbation* ... 297
Wood 1268
Wood specimens 1268
Wood-Jones, F: *Corals and atolls* ... 921
Woods, B M: 'Map cataloging' 284
Woods, C S: *Native and introduced freshwater fishes* (New Zealand) 1016
Woods, K S: *The development of country towns in the South-west Midlands* 655
Woods, P D: *Essays in geography for Austin Miller* 460
Woods Hole Oceanographic Institution 370, 877, 994, 1076, 1238
Wool knowledge 708
Wool science review 708
The wool trade directory of the world 708
Wooldridge, S W 458, 1239
The geographer as scientist ... 488, 1239
Provinces of England, rev ed 997
and W G East: *The spirit and purpose of geography* 632, 1117, 1239
and Frederick Goldring: *The Weald* 331
and D L Linton: *Structure and surface drainage in South-East England* 786
Structure, surface and drainage in South-east England 657
Worcestershire, maps 505, 1008
Wordie, Sir James 1240

527

Wordie Collection of Polar Exploration 1240
Workington, L V: 'The water masses of the North Atlantic' 1076
World Aeronautical Chart 1:1M 1241
World atlas of agriculture 1243
World atlas of mountaineering 1244
World agricultural economics and rural sociology abstracts 336, 655, 1242
World bibliography of agricultural bibliographies 188
The world book of wool 708
World cartography 269, 695, 1192, 1245
The world cities 1246
World Conservation Committee 431
World Data Center 1247
World data survey A: Glaciology 45
World directory of geographers 1248
The world economic survey 1192, 1249
The world encompassed 596
World fisheries abstracts 468, 1250
World fishing 1251
World forestry atlas 1027, 1252
A world geography of irrigation 1253
World highways 698, 1254
World Inventory of Land Use Commission 1255
The world is round . . . 368
World Land Use Survey 113, 776, 867, 1118, 1255
World map, 1:2,500,000 1256
World map of climatology 129
The world map of Richard Haldingham . . . 607
World Meteorological Organization 31, 319, 1192, 1258, 1271
World mining 1259
The world of the soil 1260
The world of wool 708
World oil atlas (Westermann) 1232
World political geography 1261
World political patterns 1262
World population and food supply 1263
World Population Congress 1264
World Railways 1265
World reference maps series (Bartholomew) 164

World shipping . . . 1266
World survey of climatology 1267
World timber . . . 471
World timbers, ed: B J Rendle 1268
World trade annual 1269
World travel series (Bartholomew) 164
World vegetation 1270
World vegetation types 1212, 1270
World Weather Program 1271
World weather records 1272
World Weather Watch 1258, 1271
World wool digest 708
The world's landscapes 1273
Worldwide directory of mineral industries 647, 1274
World-wide geographies series 1128
Worrall, R D: 'Report on transportation and physical planning in the lower Swansea Valley' 799
Worth, R H: *Dartmoor* 363
Wright, F W: *The large Magellanic cloud* 1084
Wright, H E, jr and D G Frey: *The quaternary of the United States* 1002
Wright, J K 1275
The geographical lore at the time of the Crusades 508, 1275
Geography in the making . . . 45
Human nature in geography . . . 627, 648, 1275
and E T Platt: *Aids to geographical research* 36, 1275
Wright, W R and L V Workington: 'The water waves of the North Atlantic' 1076
Wrigley, G: *Tropical agriculture* 1182
The writing of geography 534
The writings and correspondence of the two Richard Hakluyts 1150
Writings in Irish history 184
Wu, Eugene: *Contemporary China* . . . 344
Wulf, Hanna: *The European bibliography* . . . 184, 437
Württemberg-Baden, maps 378
Wye College 1276
Wyllie, James 1276

Yale University: Africana collections 20, 665

South-east Asia collections 299
Studies in the history of science and medicine series 801
Yates, R A: *British weather in maps* 242
The year of the quiet sun 689
Yearbook of the United Nations 1192
Yeates, Maurice H: *The geography of economic activity* 539
　An introduction to quantitative analysis 711
Yoeli, Pinhas: 'Abraham and Yehuda Cresques and the Catalan atlas' 277
Yonge, Ena L: *A catalogue of early globes*... 567
York University, Toronto, Department of Sociology 727
Yorkshire, England: field studies 416
　maps 1008
Yorkshire and Humberside Planning Region: an atlas ... 133, 1277
Young, Anthony: *World vegetation* 1270
Young, Arthur: Board of Agriculture *Reports* 220
The Younghusband expedition 76
Yugoslavia 1107
　bibliography 182, 184
Yugoslavia: patterns of economic activity 167
　see also under Jugoslavia
Yukon bibliography 1278
Yukon Research Project series 1278

Yules, Henry: *Cathay and the way thither* 596

Zambia 286
Zanzibar 286, 788
Zaychikov, V T, ed: *The physical geography of China* 968
Zeiss, Carl, V E B, Jena 1279
Zeiss, H, ed: *Welt-seuchen atlas* 1230
Zeitschrift für geomorphologie 786
Zeitschrift für geopolitik 1280
Zelinsky Wilbur: *A bibliographic guide to population geography* 995
Zell, H M: *A bibliography of non-periodical literature on Sierra Leone*... 184
Zetland, land use survey 916
Ziegler, T: *Atlas of glaciers in South Norway* 123
Zimmerman, Irene: *Current national bibliographies of Latin America* 184
Zinc abstracts 3, 1281
Zinc Development Association 3, 1281
Zionism 723
Zon, Raphael and William N Sparhawk: *Forest resources of the world* 470
Zoogeography 164, 967, 1149
Zoogeography . . ., P J Darlington 1237
Zoology 1088
Zumstein katalog ... 269, 1282
Zumsteins landkartenhaus, Munich 1282